ADVANCED TECHNOLOGIES IN WASTEWATER TREATMENT

Advanced Technologies in Wastewater Treatment

MUNICIPAL WASTEWATER TREATMENT

Edited by

ANGELO BASILE

Institute of Membrane Technology (ITM) of the Italian National Research Council (CNR), Cosenza, Italy - Retired; General TAG, Ligornetto, Switzerland

ALFREDO CASSANO

Institute of Membrane Technology (ITM) of the Italian National Research Council (CNR), Rende, Italy

KAMRAN GHASEMZADEH

Chemical Engineering Department, Urmia University of Technology, Urmia, Iran

Institutes for Materials and Processes, University of Edinburgh, Scotland, United Kingdom

ELSEVIER

Elsevier
Radarweg 29, PO Box 211, 1000 AE Amsterdam, Netherlands
125 London Wall, London EC2Y 5AS, United Kingdom
50 Hampshire Street, 5th Floor, Cambridge, MA 02139, United States

ISBN: 978-0-443-24826-9

For information on all Elsevier publications
visit our website at https://www.elsevier.com/books-and-journals

Publisher: Candice Janco
Acquisitions Editor: Anita Koch
Editorial Project Manager: Ashi Jain
Production Project Manager: Sujatha Thirugnana Sambandam
Cover Designer: Miles Hitchen

Typeset by STRAIVE, India

Contents

Contributors

Ashique Ahmed CSIR-North East Institute of Science and Technology, Jorhat; Academy of Scientific and Innovative Research (AcSIR), Ghaziabad, India

Ahmad Aidan Department of Chemical and Biological Engineering, American University of Sharjah, Sharjah, United Arab Emirates

Sameer Al-Asheh Department of Chemical and Biological Engineering, American University of Sharjah, Sharjah, United Arab Emirates

Ahmed Alengebawy College of Engineering, Huazhong Agricultural University, Hongshan District, Wuhan, China

Fahad M. Al-Fadhli Department of Chemical Engineering, College of Engineering and Petroleum, Kuwait University, Safat, Kuwait

Abdallah Alhajar Department of Chemical and Biological Engineering, American University of Sharjah, Sharjah, United Arab Emirates

Nawaf S. Alhajeri Department of Environmental Sciences, College of Life Sciences, Kuwait University, Safat, Kuwait

Mehran Alizadeh Faculty of Chemical Engineering, Urmia University of Technology, Urmia, Iran

Niloofar Arabbaseri Department of Industrial Chemical & Environmental Engineering, Escuela Técnica Superior de Ingenieros Industriales, Universidad Politécnica de Madrid, Madrid, Spain

Angelo Basile Institute of Membrane Technology (ITM) of the Italian National Research Council (CNR), Cosenza, Italy - Retired; General TAG, Ligornetto, Switzerland

Jie-Yu Cao College of Architecture and Environment, Sichuan University, Chengdu, China

Alfredo Cassano Institute of Membrane Technology (ITM) of the Italian National Research Council (CNR), Rende, Italy

Ye Du College of Architecture and Environment, Sichuan University, Chengdu, China

Abdelbaky Hossam Elgarhy State Environmental Protection Key Laboratory of Soil Health and Green Remediation, College of Resources and Environment, Huazhong Agricultural University, Wuhan, China; Central Laboratory for Environmental Quality Monitoring (CLEQM), NWRC, Qalyobia, Egypt

Nourhan M. Elmahsoub Biotechnology Department, Faculty of Science, Mansoura University, Mansoura, Egypt

Mohamed Elsamadony Department of Mechanical Engineering; Interdisciplinary Research Center for Refining & Advanced Chemicals, King Fahd University of Petroleum & Minerals, Dhahran, Saudi Arabia

Mohamed Eraky College of Engineering, Huazhong Agricultural University, Hongshan District, Wuhan, China; Water Pollution Research Department, National Research Centre, Dokki, Giza, Egypt

Pablo Santiago Espiñeira Department of Industrial Chemical & Environmental Engineering, Escuela Técnica Superior de Ingenieros Industriales, Universidad Politécnica de Madrid, Madrid, Spain

Salar Fakhraddinfakhriazar Faculty of Chemical Engineering, Urmia University of Technology, Urmia, Iran

Patricia García-Muñoz Department of Industrial Chemical & Environmental Engineering, Escuela Técnica Superior de Ingenieros Industriales, Universidad Politécnica de Madrid, Madrid, Spain

Cheikh Gaye Department of Industrial Chemical & Environmental Engineering, Escuela Técnica Superior de Ingenieros Industriales, Universidad Politécnica de Madrid, Madrid, Spain

Sachin Rameshrao Geed CSIR-North East Institute of Science and Technology, Jorhat; Academy of Scientific and Innovative Research (AcSIR), Ghaziabad, India

Kamran Ghasemzadeh Chemical Engineering Department, Urmia University of Technology, Urmia, Iran; Institutes for Materials and Processes, University of Edinburgh, Scotland, United Kingdom

Ayşegül Yağmur Gören Department of Environmental Engineering, İzmir Institute of Technology, Izmir, Turkey

Nastaran Jafar Harasi Chemical Engineering Department, Urmia University of Technology, Urmai, Iran

Abdelrahman Ibrahim Agricultural and Biosystems Engineering Department, Faculty of Agriculture, Alexandria University, Alexandria, Egypt

Elham Jalilnejad Faculty of Chemical Engineering, Urmia University of Technology, Urmia, Iran

Nastaran Khademi Health, Safety and Environment, University of Salford, Manchester, United Kingdom

Sara Khademi Health, Safety and Environment, University of Salford, Manchester, United Kingdom

Tarek Mahrous College of Engineering, Huazhong Agricultural University, Hongshan District, Wuhan, China

Liane Yuri Kondo Nakada Department of Environmental Engineering, Institute of Geosciences and Exact Sciences, São Paulo State University (UNESP), Rio Claro, São Paulo, Brazil

Petteri Peltola VTT Technical Research Centre of Finland Ltd., Ruukinmestarintie 2, Espoo, Finland

Jesús Quintana-Terriza Department of Industrial Chemical & Environmental Engineering, Escuela Técnica Superior de Ingenieros Industriales, Universidad Politécnica de Madrid, Madrid, Spain

Yaşar Kemal Recepoğlu Department of Chemical Engineering, İzmir Institute of Technology, Izmir, Turkey

Jorge Rodríguez-Chueca Department of Industrial Chemical & Environmental Engineering, Escuela Técnica Superior de Ingenieros Industriales, Universidad Politécnica de Madrid, Madrid, Spain

Pooja Talukdar CSIR-North East Institute of Science and Technology, Jorhat; Academy of Scientific and Innovative Research (AcSIR), Ghaziabad, India

Ahmed Tawfik Department of Environmental Sciences, College of Life Sciences, Kuwait University, Safat, Kuwait

S.M. Sadati Tilebon Seawater Reverse Osmosis (SWRO) Desalination Plant Operator, Neka Power Generation Management Co., Shahid Salimi Neka Power Plant, Neka, Mazandaran, Iran

Tara Torabi Faculty of Chemical Engineering, Urmia University of Technology, Urmia, Iran

Khaled Touil State Environmental Protection Key Laboratory of Soil Health and Green Remediation, College of Resources and Environment, Huazhong Agricultural University, Wuhan, China

Rodrigo Custodio Urban Department of Environmental Engineering, Institute of Science and Technology, São Paulo State University (UNESP), São José dos Campos, São Paulo, Brazil

Bao-Jun Xu College of Architecture and Environment, Sichuan University, Chengdu, China

Chang-Jie Yuan College of Architecture and Environment, Sichuan University, Chengdu, China

Preface

Municipal wastewater treatment is a critical aspect of modern urban living, playing a vital role in ensuring the protection of public health, the environment, and the sustainability of our precious water resources. As cities expand and populations grow, the demand for efficient and effective wastewater treatment systems has never been greater. This necessity is driven by the need to manage the increasing volume and complexity of wastewater generated from households, industries, and commercial establishments. The discharge of untreated or inadequately treated wastewater can lead to severe environmental degradation, posing significant risks to aquatic ecosystems, drinking water sources, and overall public health.

With rapid urbanization and industrialization, the nature of municipal wastewater has become more complex, including a wide array of contaminants and emerging pollutants such as pharmaceuticals, personal care products, and microplastics. These contaminants present new challenges that traditional treatment methods may not effectively address. Consequently, there is a pressing need for advancements in treatment technologies that can efficiently remove these diverse pollutants and adapt to the changing characteristics of municipal wastewater.

This book, "Municipal Wastewater Treatment," aims to provide a comprehensive overview of the latest developments, technologies, and methodologies in the field of municipal wastewater treatment. It serves as a valuable resource for researchers, practitioners, and students who are engaged in or interested in the field of wastewater management. By compiling the most recent innovations and research findings, this book seeks to bridge the gap between current practices and emerging technologies, offering insights into the future directions of wastewater treatment.

In the chapters that follow, readers will find detailed discussions on a variety of cutting-edge technologies and processes designed to enhance the efficiency and effectiveness of municipal wastewater treatment. From advanced biological treatment systems to innovative membrane technologies, this book covers a broad spectrum of topics that are essential for addressing the multifaceted challenges of modern wastewater management. The aim is to equip readers with the knowledge and tools needed to implement sustainable and resilient wastewater treatment solutions in their communities.

Going a bit into detail, in the following, a brief summary of all the chapters is presented.

Chapter 1 (Nastaran Khademi, Sara Khademi, and Nastaran Jafar Harasi) introduces the various aspects of municipal wastewater treatment, illustrating the current state, treatment methodology principles, and the benefits of good wastewater management techniques. Several treatment technologies, including conventional methods to determine their limitations and shed light on wastewater cleanup, are analyzed. To resume, this chapter aims to give readers a holistic view of municipal wastewater treatment's many challenges and opportunities, enabling informed decision-making and sustainable water resource stewardship.

Chapter 2 (Liane Yuri Kondo Nakada and Rodrigo Custodio Urban) treats various aspects of the emerging contaminants (ECs) in municipal wastewater. ECs comprise chemicals or microorganisms that have not been subject to routine monitoring in the environment, have not been previously identified, or are being identified at varying levels compared to those anticipated. Pharmaceuticals and personal care products, polyfluoroalkyl substances, microplastics, and emerging viruses belong to ECs. If municipal wastewater treatment does not remove these ECs, treated effluent can become a source of environmental contamination. The objective of this chapter is focused on providing a concise overview of the two principal challenges associated with ECs in municipal wastewater: detection and removal.

Chapter 3 (Ashique Ahmed, Pooja Talukdar, and Sachin Rameshrao Geed) discusses in detail the application of Sequential Batch Bioreactors (SBRs) as an advanced technology in municipal wastewater treatment. It elucidates the operational principles and treatment phases of SBRs, focusing on their biological treatment mechanisms, which harness microorganisms for organic pollutant decomposition. Notably, aeration during reactions and nutrient removal underscore SBRs' adaptability to diverse water quality challenges. Real-world case studies and examples from municipal wastewater treatment plants illustrate the successful application of SBR technology.

Chapter 4 (Sameer Al-Asheh, Abdallah Alhajar, and Ahmad Aidan) focuses on the most important aspects related to membrane bioreactors (MBRs) for wastewater treatment, which combine conventional wastewater treatment technologies with membrane systems. As a result, these processes enhance the purity of the effluent and eliminate the need for advanced treatment methods. For example, by utilizing microfiltration or ultrafiltration, MBRs can achieve over 90% removal levels of biological oxygen demand (BOD) and chemical oxygen demand (COD). Factors that affect membrane performance, as well as the potential of MBRs in wastewater treatment, are also analyzed and discussed.

Chapter 5 (Yaşar Kemal Recepoğlu and Ayşegül Yağmur Gören) focuses on the aeration systems for wastewater treatment, including diffused, mechanical, natural, and jet aeration systems, each with distinct mechanisms and efficiencies. It also discusses the operational principles, advantages, and limitations of these systems, emphasizing their impact on energy consumption and treatment efficiency. By exploring the design considerations and future trends, this comprehensive chapter underscores the indispensable role of aeration systems in advancing sustainable wastewater management.

Chapter 6 (Nawaf S. Alhajeri, Ahmed Tawfik, Mohamed Elsamadony, and Fahad M. Al-Fadhli) stresses why discharging wastewater into the environment poses serious health risks, particularly in impoverished and low-income nations. Water scarcity in these regions incentivizes wastewater reuse to support livelihoods and economies. An overview of the design, application, performance, and limitations of moving bed biofilm reactors (MBBRs) is comprehensively presented. These systems can be employed to enhance existing wastewater treatment plants and are particularly effective for treating heavily polluted municipal and industrial wastewater.

Chapter 7 (Tara Torabi, Kamran Ghasemzadeh, Alfredo Cassano, and Angelo Basile) focuses on the application of membrane-based processes for municipal wastewater reuse, highlighting the need for effective treatment of wastewater to mitigate health and environmental risks posed by contaminants. Membrane technology, including microfiltration (MF), ultrafiltration (UF), nanofiltration (NF), and reverse osmosis (RO), is favored for its energy efficiency and scalability, despite challenges like membrane fouling. Advances in membrane bio-reactor (MBR) technology, especially anaerobic membrane bioreactors (AnMBRs), are also explored for their energy efficiency and treatment effectiveness.

Chapter 8 (Ahmed Tawfik, Nawaf S. Alhajeri, Fahad M. Al-Fadhli, and Mohamed Elsamadony) deals with the application of membrane aerated biofilm reactors (MABR) in wastewater treatment. MABR technology is a recently developed method for wastewater treatment, known for its efficiency in organic oxidation, simultaneous nitrification, and denitrification. Within the MABR system, the membrane acts as a biocarrier, immobilizing active bacteria and supplying oxygen to the attached microbes within the biofilm on its surface. Wastewater and aeration are primarily directed to the biofilm on the membrane surface in a counter-diffusion mode, creating a gradient of dissolved oxygen across the biofilm thickness, thus facilitating nitrification and denitrification processes in the reactor. The chapter explores various aspects of MABRs, including factors affecting performance, hybrid systems, membrane types and configurations, membrane modifications, membrane materials, and module design.

Chapter 9 (Ye Du, Bao-Jun Xu, Chang-Jie Yuan, and Jie-Yu Cao) concentrates on the progress of wastewater disinfection. Disinfection represents the ultimate phase of wastewater treatment, wherein achieving a delicate balance between microorganism inactivation and the formation of disinfection by-products (DBPs) is crucial. In addition to conventional methods such as chlorine, ozone, and ultraviolet (UV) disinfection, emerging technologies based on the use of ferrate (Fe(VI)) and persulfate are illustrated as promising approaches. Given the diverse nature of DBPs coupled with their low concentration levels and inherent identification challenges, it remains difficult to determine these compounds despite the continuous discovery of new ones. Therefore, adopting a systematic approach toward wastewater treatment development becomes essential to comprehensively address biological toxicity control rather than solely focusing on single DBP control. Achieving toxicity control during disinfection primarily involves (1) precursor removal; (2) masking reactive halogens; (3) activating disinfectants while generating free radicals to inhibit toxicity generation; and (4) eliminating residual toxicity postdisinfection.

Chapter 10 (Ahmed Alengebawy, Mohamed Eraky, Tarek Mahrous, Khaled Touil, Nourhan M. Elmahsoub, Abdelrahman Ibrahim, and Abdelbaky Hossam Elgarhy) provides an extensive review of the advancements achieved in anaerobic digestion for treating municipal wastewater. It examines the composition and difficulties related to municipal wastewater. And a thorough review is conducted on the potential of innovations, such as two-stage anaerobic digestion, high-rate anaerobic reactors, and the integration of anaerobic and aerobic processes, to enhance treatment efficiency. Furthermore, this work emphasizes the microbial ecology of anaerobic digestion systems, offering valuable information on microbial communities and their effect on treatment efficacy. It also explores methods for assessing and tracking performance, with a particular emphasis on key performance indicators and real-time monitoring technologies. Case studies implemented in different areas are also illustrated.

The various emerging technologies for sludge management are suggested in Chapter 11 (Rodrigo Custodio Urban and Liane Yuri Kondo Nakada). Sludge from wastewater treatment plants is a material with a highly variable composition, depending on the characteristics of the municipal wastewater and the type of treatment. Managing this waste is a complex, costly, and challenging task. Although landfilling sludge is considered a safe disposal method, it does not move toward a circular economy or sustainable development. Therefore, the focus here is on the beneficial uses of sewage sludge, including various land applications and reuse as a raw material for manufacturing/production purposes.

Chapter 12 (S.M. Sadati Tilebon, Kamran Ghasemzadeh, Alfredo Cassano, Angelo Basile) develops the aspects related to the use of forward osmosis (FO) for municipal wastewater treatment. FO is a process with minimal energy requirements and the possibility of producing by-products such as biomass and biogas, which in addition to producing cheap purified water can turn the municipal wastewater treatment process into a cost-effective process. Anyway, despite all the advantages, the low productivity of FO membranes leads to an increase in the process footprint as well as both investment and operating costs. As suggested by the Authors, further investigations concerning the efficiency and membrane fouling in long-term operation, the development of innovative membranes with higher productivity and selectivity, and the design of new hybrid processes are needed to make FO a mature technology.

Emerging wastewater disinfection methodologies are treated in Chapter 13 (Patricia García-Muñoz, Jesús Quintana-Terriza, Niloofar Arabbaseri, Pablo Santiago Espiñeira, Cheikh Gaye, and Jorge Rodríguez-Chueca). The increase in the world's population entails an increase in the consumption of water resources, the rise in the generation of wastewater, and the overall worsening of water quality. In this sense, the disinfection of wastewater allows, on one hand, to improve its quality for discharge into the hydraulic public domain, but also the possibility of its reuse for other human uses. However, conventional disinfection technologies such as chlorination or ozonation have certain limitations that make them less suitable for wastewater disinfection, such as the generation of disinfection by-products as a result of the reaction with organic matter. In this regard, technologies such as advanced oxidation processes, heterogeneous photocatalysis, carbonaceous catalysts, or plasma technology are presented as alternatives. Some of these technologies still require more years of development; others begin to be implemented on a real scale. This work addresses the fundamentals of these technologies and also reviews the state of the art and their level of development and implementation.

Interesting aspects of the modeling of municipal biological wastewater treatment are investigated in Chapter 14 (Elham Jalilnejad, Salar Fakhraddinfakhriazar, and Mehran Alizadeh). Municipal wastewater treatment companies are seeking accurate, enduring strategies to forecast how their treatment procedures might be improved in the context of increasing competition. In recent decades, mathematical models enhanced the financial and operational effectiveness of wastewater treatment plants. Predictive models that incorporate physical and theoretical understanding, such as the Monod, Contois, and Activated Sludge models, are primarily mechanistic models, and they have been created to forecast the efficiency of wastewater treatment plants. Nevertheless, these models are an oversimplification of reality, leading to ambiguity in the structure of the model and a continuous requirement for calibration. The progress in computational fluid dynamics (CFD), artificial intelligence (AI), and machine learning (ML) has facilitated the creation of advanced models capable of precisely forecasting system behavior, optimizing operational parameters, and improving the system's overall effectiveness. To summarize, in this work, both mechanistic and data-driven models utilized in municipal biological wastewater treatment systems are overviewed. And a complete discussion is also given about the practical applications of these models. Challenging issues associated with their implementation and future directions are also provided.

Chapter 15 (Petteri Peltola) illustrates the advancements in the technologies for municipal wastewater excess sludge. With an increasing emphasis on emission reduction, nutrient recycling, and clean energy generation, extensive efforts are currently underway to develop cost-effective, energy-efficient, and sustainable technologies for managing excess sewage sludge (SS) produced during wastewater treatment. Aligned with sustainable development goals and circular economy principles, these technologies aim to minimize SS and transform it from waste into a renewable resource for energy, nutrient, and material recovery. This chapter delves into recent research and explores promising pathways for the sustainable treatment and utilization of SS, with a particular emphasis on novel sludge-to-energy and sludge-to-resource recovery methods. It discusses key factors and trends shaping future developments and assesses the benefits, limitations, and potential of applicable technologies, paying specific attention to thermochemical conversion, energy and nutrient recovery, and sludge biorefinery. Despite promising development prospects, the technology readiness level (TRL) of these applications remains relatively low for widespread adoption. Hence, continued research, innovation, and validation are imperative to advance these technologies and pave the way for their successful industrial application.

Last but not least, the Editors are in debt for the hard work of the Authors and thank them for their fruitful cooperation in the preparation of this volume. Special thanks also to all the staff of Elsevier for their precious and irreplaceable help in preparing, step by step, this book.

Kamran Ghasemzadeh
Alfredo Cassano
Angelo Basile

1

Introduction to municipal wastewater treatments

Nastaran Khademi[a], **Sara Khademi**[a], **and Nastaran Jafar Harasi**[b]

[a]*Health, Safety and Environment, University of Salford, Manchester, United Kingdom.* [b]*Chemical Engineering Department, Urmia University of Technology, Urmai, Iran*

Abbreviations

BOD biological oxygen demand
COD chemical oxygen demand
DAF dissolved air flotation
MPN most probable number
O&M operation and maintenance
SDGs sustainable development goals
UV ultraviolet
WWTP wastewater treatment plant
WWTPs wastewater treatment plants

1.1 Introduction

The burgeoning human population and evolving lifestyles, among various other factors, have exerted profound impacts on both the quality and quantity of municipal wastewater generated. The widespread discharge of untreated wastewater, often laden with toxic organic and inorganic pollutants, into natural ecosystems poses a significant threat to human and aquatic life (Freitas et al., 2016; Vunain et al., 2019). The Sustainable Development Goal of providing access to clean water and sanitation has drawn more attention from researchers in recent years due to growing concerns about environmental degradation and related health risks, such as water pollution and a shortage of drinkable water (Abaliwano et al., 2008; Kansal and Kumari, 2014).

Municipal Wastewater Treatment. https://doi.org/10.1016/B978-0-443-24826-9.00001-2

It is disconcerting that approximately 1.1 billion individuals worldwide lack access to clean and safe drinking water, with a disproportionate number residing in less developed regions (Abaliwano et al., 2008; Kansal and Kumari, 2014). Consequently, there has been a concerted effort to develop and explore various techniques and processes for wastewater treatment, including precipitation, ion exchange, filtration, electrodialysis, electrochemical processes, coagulation, membrane technologies, and reverse osmosis (Crini and Lichtfouse, 2019). However, the widespread adoption of these technologies in developing countries is impeded by cost constraints and limited technical expertise required for their implementation (Kansal and Kumari, 2014; Crini and Lichtfouse, 2019). Moreover, many conventional treatment methods are marred by shortcomings such as excessive chemical usage, generation of hard-to-treat secondary wastes, and inefficacy in treating diverse pollutant compositions (Tripathi and Mishra, 2017; Valverde et al., 2018; Nwoba et al., 2020). A summary of the limitations associated with different treatment methods is presented in Table 1.1. While traditional methods like chlorination, electrochemical treatment, and membrane filtration have historically been employed in wastewater treatment, their inefficiency and high operational costs have become increasingly apparent (Palansooriya et al., 2020; Shah et al., 2020; Liu et al., 2020; Bhat et al., 2022). Consequently, advanced techniques such as adsorption, ultrafiltration, chemical precipitation, and coagulation have been developed to address these shortcomings (Bhat et al., 2022). However, it is worth noting that some of these advanced methods, particularly those involving coagulation and flocculation, give rise to sludge that necessitates appropriate disposal, thereby exacerbating environmental concerns and escalating treatment costs. Furthermore, these processes are energy-intensive, further contributing to the economic burden of wastewater treatment (Wu et al., 2020).

In general, significant strides have been made in the development of wastewater treatment technologies, there remain formidable challenges in ensuring their widespread adoption, particularly in resource-constrained settings. Addressing these challenges necessitates not only technological innovation but also a holistic approach encompassing environmental, economic, and social considerations (Basile and Ghasemzadeh, 2020).

The primary objective of municipal wastewater treatment plants (WWTPs) is to mitigate or eliminate the adverse environmental impacts of sewage discharge. Broadly defined, municipal wastewater or sewage comprises a mixture of domestic effluents, encompassing both dissolved and suspended matter

Table 1.1 Limitations and drawbacks of different technologies used in wastewater treatment (Alsawy et al., 2022).

Wastewater treatment technique	Limitations/disadvantages
Adsorption	– Some adsorbents are expensive – Regeneration process – Adsorbent disposal – Time-consuming
Coagulation and flocculation	– Operate at low pH so alkaline should be added – Not applicable to all soluble contaminants – Result secondary pollutants (e.g., sludge)
Reverse osmosis	– Energy-intensive – Not practical for pharmaceuticals and volatile organics – Treated water is acidic
Distillation	– Not efficient for contaminants having boiling point $>100\,°C$ – Most of pollutants remain behind
Microfiltration/ultrafiltration/nanofiltration	– Not suitable for dissolved inorganics – Rapid fouling and plugging – Energy-intensive process – Pretreatment process is needed – Regular cleaning is needed
Biological processes	– Limited to municipal and biodegradable wastewater – Time-consuming – Challenging to be controlled

(Rachid-Sally and Jayakody, 2008). Urban regions are the primary producers of wastewater, thus providing efficient sewage services to these places requires careful design and strong infrastructure (Abusoglu et al., 2012). Processes used for wastewater treatment frequently require significant amounts of energy and have direct effects on the environment and public health (Abusoglu et al., 2012). Forecasts indicate that the value of the worldwide water and wastewater market will reach USD 500 billion by 2028, from an expected USD 263.07 billion in 2020 (Statista, 2021). Despite notable advancements in wastewater treatment technologies, numerous urban centers grapple with formidable challenges in wastewater management, attributed to several key factors:

(i) Inadequacies in existing centralized WWTP infrastructure to accommodate the escalating volumes of wastewater from burgeoning urban settlements.

(ii) Inefficiencies in the removal of contaminants of emerging concern (CECs) by conventional municipal wastewater treatment plants (WWTPs).

(iii) Escalating complexities associated with the handling of sewage sludge (Shanmugam et al., 2022).

Municipal WWTPs have the potential to emerge as pivotal contributors to a circular economy through the integration of the 3R principles: reduce, reuse, and recycle. While the primary objective of WWTPs is to mitigate sewage pollution, these processes yield several potentially valuable byproducts, including treated effluent, biogas, and sludge (Shanmugam et al., 2022). Treated effluent holds promise for diverse end-use applications, while biogas can serve as a versatile fuel source for electricity generation, transportation, and cooking, or as a chemical feedstock. Moreover, sewage sludge can be repurposed either directly as a soil conditioner or via thermochemical/biochemical pathways to recover materials (e.g., hydrochar), energy (e.g., heat and syngas), and valuable resources such as phosphorus (Shanmugam et al., 2022).

1.2 What is municipal wastewater?

Municipal wastewater, often referred to as sewage or domestic wastewater, encompasses water that originates from household use. It can be defined as "the water supply of the community after it has been utilized in various applications and now contains constituents that make it unsuitable for most purposes without treatment" (Tchobanoglous et al., 2013). The primary component of municipal wastewater is water itself, accompanied by a mixture of feces, organic kitchen wastes, urine, and water used for cleaning, washing, and cooking activities. At the household level, municipal wastewater is typically categorized into three separate streams: black water, gray water, and yellow water, which are commonly collected together. Black water consists of a blend of feces, urine, and flush water, characterized by elevated concentrations of organic matter, nitrogen, and phosphorus. Gray water, originating from activities such as cooking, washing, and cleaning, is comparatively more diluted and constitutes the majority of the wastewater flow. Urine is sometimes collected separately and termed as yellow water (Pikaar et al., 2022).

Table 1.2 summarizes the relative contributions of waste streams generated at the household level to the total wastewater mass flow. Additionally, it presents the typical concentrations of various components found in each stream (Larsen et al., 2013).

Table 1.2 Relative contributions and average concentrations of major components in domestic wastewater (Larsen et al., 2013).

Parameter	Gray water	Urine	Feces
Typical daily flow (L/capita/day)	108	1.2[a]	0.2[a]
Contribution to overall flow rate (%)	99	0.9	0.1
Contribution to overall COD (%)	50	11	31
Contribution to overall nitrogen (%)	8	79	13
Contribution to overall phosphorus (%)	28	47	25
Contribution to overall potassium (%)	7	71	22
Water content (%)	99	95	77
COD[b] (mg/L)	620	10,236	155,000
BOD5[c] (mg/L)	279	4567	60,000
Total nitrogen (mg/L)	23	8661	7500
Total phosphorus (mg/L)	8.5	732	3000
Potassium (mg/L)	10	2047	4500

[a]Mass flow, expressed in g/capita/day.
[b]COD the entire amount of oxidizable materials in the wastewater is represented by the term COD, or chemical oxygen demand.
[c]BOD_5 biological oxygen demand, or BOD5, is the biodegradable portion of the oxidizable materials (measured after 5 days at 20°C).

Typical municipal wastewater is composed of approximately 99.9% water and 0.1% pollutants. Among these pollutants, around 60–80% exist in dissolved form, while the remainder is in suspended matter. The pollutants encompass a variety of substances, including mineral and organic matter, suspended solids, oil and grease, detergents, nitrogen (in forms such as ammonia, nitrate, and organic nitrogen), phosphorus in various forms, sulfur, phenols, and heavy metals. The organic matter is comprised of numerous compounds, and due to the impracticality of identifying and quantifying each one, it is often represented by indicators such as Biochemical Oxygen Demand (BOD) or Chemical Oxygen Demand (COD), which respectively measure the amount of oxygen required for biological and chemical decomposition of the organic matter.

Moreover, municipal wastewaters harbor significant quantities of bacteria and viruses, some of which are pathogenic. Raw wastewater typically contains a total bacteria count ranging from 10^7 to 10^8 Most Probable Number (MPN) per 100 mL. Similarly, it's challenging to identify and quantify all present organisms, thus indicator organisms are utilized. Fecal coliforms, although other indicators such as intestinal enterococci are also employed, are commonly used to represent pathogenic organisms. Fecal coliform counts in raw municipal wastewater typically fall within the range of 10^6 to 10^7 MPN/100 mL (Libhaber and Orozco-Jaramillo, 2012).

1.3 Principle of municipal wastewater treatment

Wastewater treatment stands as the cornerstone of any water reuse initiative, serving as the most economical means to mitigate health and environmental hazards linked with recycled water usage (Asano et al., 2007; Lazarova, 2005). The term "waste to resource" applies to the mechanisms by which wastewater is transformed into a suitable and useful form for both residential and commercial purposes. Principally, wastewater treatment encompasses three primary stages aimed at converting it into various products catering to diverse needs (Meena et al., 2019; Huang et al., 2012; Kivaisi, 2001; Lee et al., 2017). As depicted in Fig. 1.1, these stages include primary treatment, secondary treatment, and tertiary treatment. Additionally, preliminary treatment and sludge treatment are integrated at the initiation and culmination of the process, respectively (Chojnacka et al., 2020; Muga and Mihelcic, 2008; Surya Nair et al., 2021; Lu et al., 2020).

During primary treatment, wastewater is directed into holding tanks facilitating the settling of solid particles (sludge) at the bottom while chemical components such as oil rise to the surface (Muga and Mihelcic, 2008). This initial phase eliminates a significant proportion of impurities present in the wastewater. Secondary treatment involves the breakdown of solid waste using aerobic bacteria integrated into the treatment system (Hamed et al., 2004). Tertiary treatment entails filtering wastewater to eliminate nutrients and waste particles that could pose harm to the broader ecosystem. This stage may also encompass routing wastewater through additional lagoons to eliminate any residual impurities or chemicals before presenting the final product for the intended use (Silva, 2023).

The selection and design of appropriate wastewater treatment processes largely hinge upon the final effluent disposal option. The chosen treatment approach must align with the proposed disposal method, which could involve ocean or sea discharge, river or stream discharge, lake discharge, groundwater infiltration, irrigation reuse, or recycling for other applications. The scale and configuration of wastewater treatment plants are predicated on the

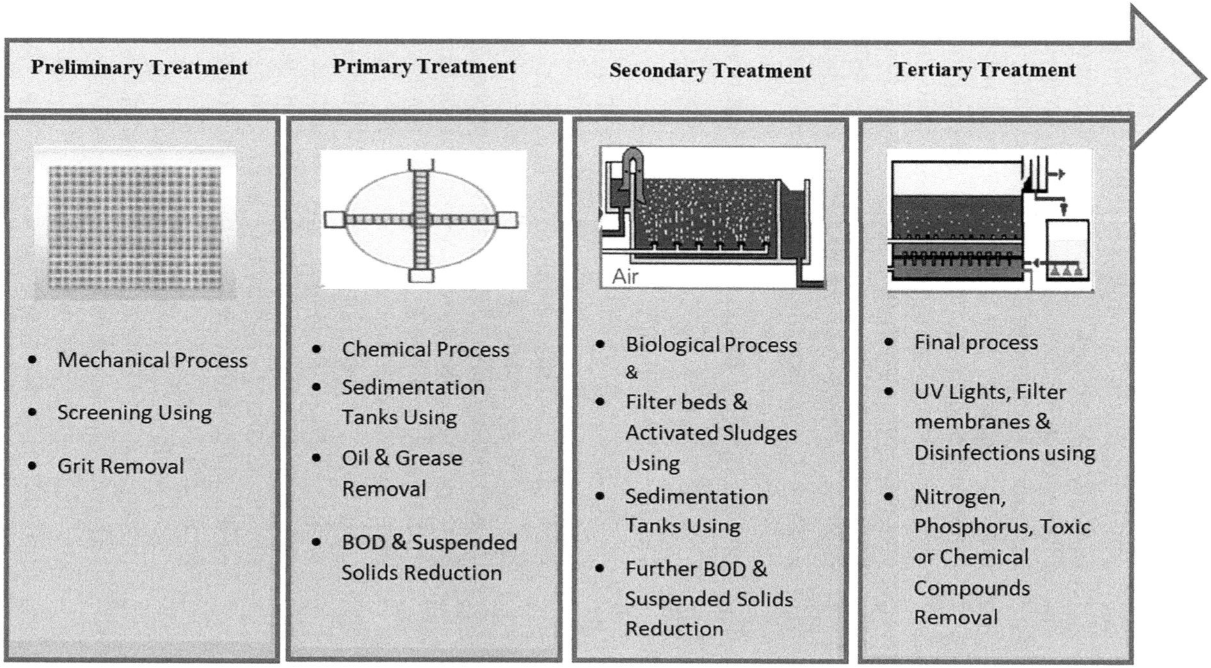

Fig. 1.1 Stages of wastewater treatment, from preliminary to tertiary treatment.

projected raw wastewater flow rate (in m^3/day) and the contaminant loads (in tons/day) present in the raw wastewater.

The foundational principles of wastewater treatment are illustrated in Fig. 1.1. Consequently, treatment is conventionally segmented into a sequence of preliminary, primary, secondary, and tertiary processes. Preliminary and primary treatments rely on physical processes, while secondary treatment is predicated on biological processes, and tertiary treatment typically incorporates physicochemical processes, although it may also integrate biological processes. Within each of these treatment units, several alternative processes exist to achieve similar outcomes, resulting in a multitude of potential overall treatment process schemes. Moreover, numerous treatment processes yield sludge as a by-product, and there are various options for sludge treatment and disposal. Consequently, the array of treatment method combinations is vast, with treatment plants featuring distinct unit processes achieving comparable results at substantially differing costs (Libhaber and Orozco-Jaramillo, 2012). The subsequent sections provide detailed insights into each treatment stage.

1.3.1 Preliminary treatment

The initial phase of the wastewater treatment process, preliminary treatment, marks the pivotal step in the process. This treatment phase entails the mechanical extraction of both coarse and fine solid materials (Silva, 2023). Except for septic tanks and household systems, all treatment processes necessitate some form of preliminary or screening procedure to eliminate large and floatable objects. Among the most commonly employed methods are screening and grit removal, which also serve to eliminate floatable material and rags (Libhaber and Orozco-Jaramillo, 2012). Wastewater is directed through a screening apparatus designed to trap various debris such as wood fragments, rags, plastic particles, and wires, among other materials (Suzenet et al., 2002; Kalair et al., 2021; Choudhary et al., 2011; Liang et al., 2017).

As preliminary treatment processes are illustrated in Fig. 1.2, this treatment facilitates the removal of over 60% of solid materials from the wastewater, with the percentage potentially exceeding this benchmark depending on the intended use of the final product. For instance, in the context of producing drinking water, the preliminary stage may eliminate upwards of 80% of solid materials, leaving only minimal residues for subsequent treatment stages (EPA, 2024; Choudhary et al., 2011; Demirbas et al., 2017; Łuczkiewicz et al., 2010; Mao et al., 2015). Following

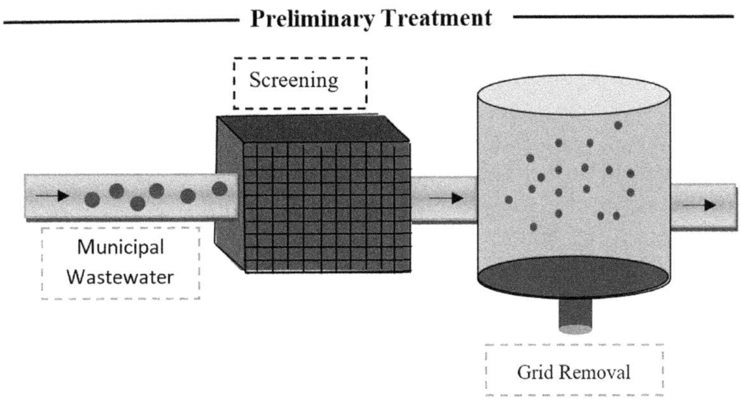

Fig. 1.2 Schematic of preliminary treatment.

extraction, the removed materials are typically disposed of either through burial or incineration. Environmental concerns associated with incineration have often led to burial being favored as the preferred method for solid material disposal (Villarín and Merel, 2020; Tortajada, 2020; Kesari et al., 2021).

Nonetheless, even burial presents environmental challenges, as it can lead to soil contamination by chemicals and nondegradable materials such as plastics. The quantity and nature of the solid materials extracted from the wastewater typically dictate the most appropriate disposal method. Various mechanisms exist to ensure the safe incineration of plastic materials, thereby mitigating potential environmental harm (Liao et al., 2021).

1.3.2 Primary treatment

The second stage of the wastewater treatment process called "Primary treatment" typically follows preliminary treatment and often precedes biological treatment in conventional secondary wastewater treatment facilities (Silva, 2023). Its primary objective is to diminish the proportion of suspended solids and BOD, achieved primarily through gravity settling. Sedimentation tanks, commonly known as primary clarifiers, are widely used for Primary treatment of domestic sewage. Within these tanks, suspended solids settle to the bottom, where they are then scraped to a central point and removed either by gravity or via a sludge pump. Concurrently, scum, primarily composed of oil and grease, rises to the surface and is collected by a mechanical arm at regular intervals. Alternatively, a Dissolved Air Flotation (DAF) process can serve as a primary treatment unit, particularly effective in removing oil and grease within a smaller footprint. In the DAF process, wastewater and air are pressurized and released into an open tank, resulting in the formation of small bubbles that ascend to the surface, carrying along with them light solids and oil particles, which are subsequently collected by a skimmer. The clarified liquid continues downstream for further treatment processes. While DAF is more commonly employed for industrial wastewater treatment, conventional sedimentation tanks typically remove 25–40% of BOD, 40–60% of total suspended solids, and around 50% of the bacterial load. The principal residuals collected include settled solids, scum, and oil (Libhaber and Orozco-Jaramillo, 2012).

According to Melo et al. (2019), Primary treatment involves the gradual passage of wastewater or effluent through grit tanks, allowing fine sand particles to settle. Nonetheless, finer sand particles may still remain suspended in the wastewater (Chojnacka et al., 2020; Kim et al., 2010; Baawain et al., 2020; Manning et al., 2016). To address these finer particles, wastewater undergoes further processing in large primary sedimentation tanks, where a significant portion of the solid material settles out to form sludge (Dutta et al., 2021). Al-Juaidi et al. (2014) reported that primary treatment typically removes about 60–70% of suspended solid materials. The liquid residue following primary sedimentation generally contains very fine solids in dissolved form (Fito and Van Hulle, 2021; Melin et al., 2006; Panepinto et al., 2016; Ratner and Rivera, 2004), necessitating secondary treatment to eliminate any dissolved particles from the water.

As Primary treatment processes are illustrated in Fig. 1.3, primary sedimentation is crucial for the removal of solid materials from water intended for various purposes, such as drinking and cooking. However, for activities like irrigation, wastewater may be deemed suitable for use if it lacks potentially harmful chemicals and solid materials (Kalair et al., 2021; Ratner and Rivera, 2004; Rosso et al., 2008). The cost of primary treatment typically amounts to about half that of full secondary treatment. Although primary treatment processes entail relatively simple mechanical procedures, they necessitate routine maintenance, primarily comprising the upkeep of pumps, sludge scrapers, scum collectors, and motors. The Operation and Maintenance (O&M) costs associated with primary treatment mainly stem from the treatment and disposal of sludge. The capital cost of the sludge treatment unit constitutes a significant portion of the overall expense of primary treatment (Libhaber and Orozco-Jaramillo, 2012).

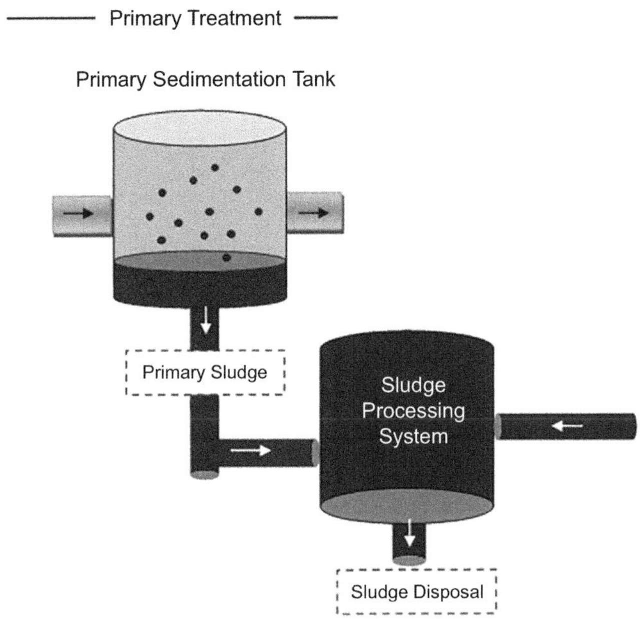

Fig. 1.3 Schematic of Primary treatment.

1.3.3 Secondary treatment

Between 70% and 90% of the suspended solids in the wastewater are to be removed during the secondary treatment stage (Silva, 2023). The secondary treatment procedures are shown in Fig. 1.4. Secondary treatment usually uses biological processes, utilizing microorganisms to break down organic materials (Baskar et al., 2022). Because they can supply the oxidative energy needed for the degradation of organic compounds, aerobic bacteria are frequently used (Alagha et al., 2020; Villarín and Merel, 2020; Dutta et al., 2021; Capodaglio, 2021; Liao et al., 2021; Sardana et al., 2019). Since aerobic bacteria need sufficient oxygen to metabolize organic material, the rate at which suspended materials are broken down by microbes is largely determined by the availability of oxygen (Silva, 2023). There are two main varieties of biological reactors used: connected growth reactors and suspended growth reactors. Diffused air systems, mechanical aerators, or other aeration techniques are utilized to supply oxygen to suspended-growth systems. Bacteria must then be extracted from the wastewater (Libhaber and Orozco-Jaramillo, 2012).

Two procedures can be used in secondary treatment to remove dissolved elements from wastewater (Baawain et al., 2020). Filter beds and activated sludge are two of these procedures. In filter

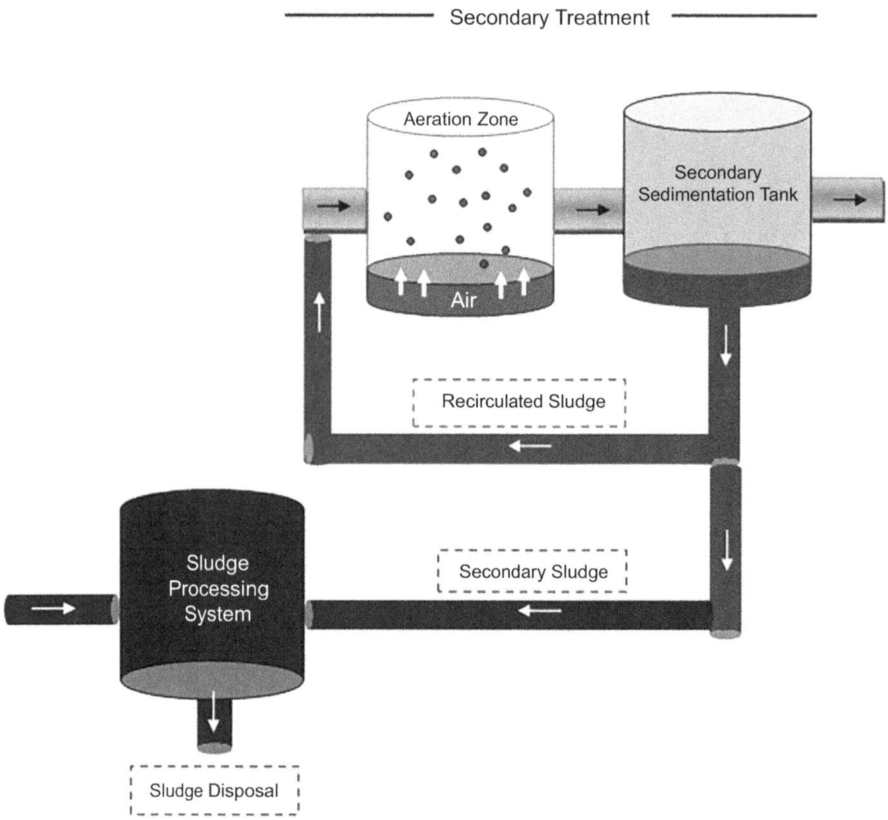

Fig. 1.4 Schematic of Secondary treatment.

beds, the wastewater is sprayed gradually across gravel or broken stone beds to increase the surface area needed for effective oxidation (Delanka-Pedige et al., 2021). The removal of suspended debris from wastewater collected at the base of the filter beds is accomplished by secondary sedimentation. In contrast, bacteria found in activated sludge are essential for the oxidation and digestion of suspended organic materials in wastewater (Silva, 2023). To settle the flocculated cell mass, a separate secondary sedimentation tank is incorporated into the majority of secondary treatment operations. The sedimentation tank overflow stream, or effluent, is either released into a receiving water body for ultimate disposal or goes on to downstream treatment facilities. In order to keep a viable bacterial concentration for metabolizing incoming organic material, sedimentation tank sludge is returned to the aeration tank; this procedure is known as return activated sludge. But in order to keep sludge concentrations from rising to unmanageable levels for sedimentation, part of the sludge needs to be removed and not put back into the aeration tank. The sludge that has been extracted is known as waste- or excess-activated sludge. Though they use permanent media—like plastic or stones—for the growth of bacteria, attached growth systems function similarly to suspended growth systems. This kind of reactor doesn't flood; instead, primary wastewater is sprayed over the media, where bacteria take up oxygen straight from the atmosphere. Aeration, or assisted oxygen supply, may be required on occasion. Aerobic treatment methods come in a variety of forms. Because they are naturally occurring systems that can provide secondary treatment, lagoons are often used in poor nations. Anaerobic lagoons function without oxygen and are powered by anaerobic bacteria that break down organic debris, whereas aerobic lagoons provide oxygen through photosynthesis produced by algae (Libhaber and Orozco-Jaramillo, 2012).

1.3.4 Tertiary treatment for nutrient removal

After primary and secondary treatment processes, tertiary treatment is an essential step in eliminating hazardous chemicals, such as phosphorus and nitrogen compounds, that may still be present (Silva, 2023). As shown in Fig. 1.5, tertiary treatment makes use of a range of instruments, including UV lamps, filter membranes, and additional disinfectants, to accomplish its goals (Tortajada, 2020; Dutta et al., 2021; Baskar et al., 2022; Capodaglio, 2021; Fito and Van Hulle, 2021). The main objective of tertiary treatment is to guarantee that the finished product is free of any dangerous chemicals that could endanger public health or the environment, especially in situations where wastewater is used for irrigation (Jamrah et al., 2008). Every instrument used

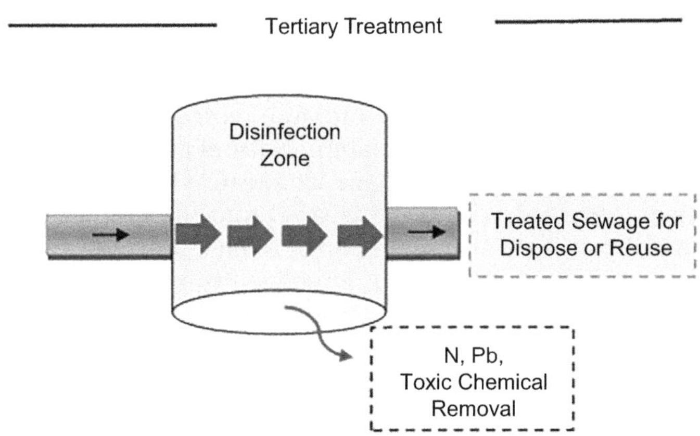

Fig. 1.5 Schematic of Tertiary treatment.

in tertiary treatment is made especially to target certain chemical substances found in wastewater. In addition, a thorough screening process is used to the finished product to ensure that it is free of any chemicals or possibly dangerous substances (Kamali et al., 2019). Chemical removal is a crucial factor to take into account before providing water to customers, as many cities across the world rely on treated wastewater to supplement clean and safe water sources for drinking (Silva, 2023).

While some nitrogen and phosphorus are removed by ordinary secondary treatment operations, larger removal rates require specialized techniques. These techniques include chemical precipitation for total phosphorus removal, partial phosphorus removal via biological processes, and nitrification and denitrification for nitrogen removal. Except in situations where discharges into enclosed water bodies, such as lakes, estuaries, or bays, may cause eutrophication, nutrient removal procedures are complex and expensive as compared to secondary treatment and are typically not needed in poor nations. Sometimes nitrification is required prior to river discharge in order to stop downstream nitrogenous oxygen consumption. The development of extra tanks, pipes, and recirculation pumps is one of the higher capital costs associated with nutrient removal. With nutrient removal, operating and maintenance expenses usually increase since more installations need to be maintained. The cost of chemicals used in phosphorus precipitation adds to operating and maintenance expenditures. Processes for removing nutrients require competent labor to run well.

The purpose of disinfection is to remove pathogens and other microorganisms from treated wastewater effluent. Chlorination, ultraviolet light, ozonation, and solar radiation are common techniques for naturally disinfecting lagoons. Both ozone and chlorine are strong oxidizers that rapidly destroy germs when they meet them. Ozone is added as a gas, whereas chlorine can be added as a liquid, tablet, or gas. Low-pressure mercury lamps are used to sterilize germs with UV radiation. Natural processes like sedimentation and solar radiation are what cause lagoons to get disinfected. When wastewater is released into the ground through infiltration basins, it also significantly reduces bacteria and pathogens as it seeps into the soil. Effective disinfection is possible with well-designed infiltration systems.

Feed rate modification, mechanical component maintenance, and chemical supply management are all part of disinfection installation maintenance. Systems that use chlorine need little maintenance, however UV radiation systems need to have their lamps changed and cleaned frequently. On-site ozone production necessitates sophisticated operation and significant electrical use. While UV radiation can be used to treat high turbidity effluents after sand filtration, chlorination is still the most often utilized disinfection technique in the world. Ozonation is very efficient and doesn't create any hazardous byproducts, but because of its possible risks, it needs to be handled carefully. Since chlorine gas poses a significant risk, liquid sodium hypochlorite is frequently used as a safer substitute. To protect the environment, dechlorination can be required to lower the amount of residual chlorine in wastewater.

Strong safety precautions need to be put in place at storage locations because chlorine gas poses a safety risk. These factors highlight how crucial it is to give environmental sustainability and safety a priority while treating wastewater (Libhaber and Orozco-Jaramillo, 2012).

1.4 Benefits of municipal wastewater treatment

Resource recovery from wastewater offers numerous environmental, economic, and social advantages. Environmental benefits include safeguarding and enhancing the quality of surface water bodies, curbing water consumption and related availability issues, mitigating greenhouse gas emissions, reducing the carbon footprint, minimizing landfill usage for disposing of treatment by-products, cutting waste generation, decreasing fossil fuel dependency, and facilitating nutrient recycling by returning them to the soil (Casiano Flores et al., 2018; Shore, 2017).

Furthermore, nutrient recovery in wastewater units reduces emissions associated with conventional fertilizer production processes (Tchobanoglous et al., 2015; Villarroel Walker et al., 2014). Particularly focusing on phosphorus recovery, it offers an alternative and renewable fertilizer source for food production, leading to significant cost savings in sludge handling and disposal while mitigating eutrophication in aquatic systems by maintaining low effluent phosphate concentrations. Eutrophication, often rampant in developing countries due to effluents with high nutrient concentrations, is a critical concern (Sarvajayakesavalu et al., 2018).

The escalating fertilizer prices, driven by phosphorus scarcity, adversely affect farmers and consumers globally, especially in developing nations unable to afford conventional chemical fertilizers (Sarvajayakesavalu et al., 2018). Therefore, utilizing phosphorus reclaimed from wastewater as fertilizers presents a promising alternative, bolstering food security, particularly in developing regions (Mayer et al., 2016). Data from the International Fertilizer Industry Association (Chrispim et al., 2019). underscores the significance of phosphorus recovery initiatives, notably in developing countries, where consumption surpasses production.

In a broader context, the sustainable management of resources, facilitated by resource recovery solutions, is pivotal for achieving Sustainable Development Goals (SDGs) (Avellan et al., 2017; Chrispim et al., 2019). Thus, alongside mitigating environmental and health concerns, integrating resource recovery into sewage treatment systems yields social benefits (Trimmer et al., 2017).

Wastewater collection, treatment, and disposal are all included in the global definition of "sanitation." Trimmer et al. (2017) attempted to analyze the multifaceted impacts of integrating resource recovery with sanitation goals. They found that substituting resource recovery technologies for current sewage treatment systems could result in a roughly 50–79% increase in global potential for nutrient and energy recovery. This suggests that resource recovery should be the focus of efforts in the development of new sewage treatment systems and the enhancement (retrofitting) of existing installations. Resource recovery not only offers environmental benefits but also economic gains through monetary returns and broader social benefits with indirect economic value (Andersson et al., 2016). Villarroel Walker et al. (2014) assessed the additional revenue generated by implementing various resource recovery technologies in London. Scenarios included sewage sludge pyrolysis for energy recovery and algae cultivation for fuel and fertilizer production. Results showed higher income from sludge pyrolysis compared to algae cultivation, although outcomes vary based on the market value of recovered products in different contexts.

Moreover, resource recovery reduces operational costs and may generate revenue by producing value-added products and potentially trading carbon credits (Breach and Simonovic, 2018). By-products like sludge, once costly to transport and dispose of, are now valued commodities. In summary, Table 1.3 reports the advantages of Municipal Wastewater Treatment.

1.5 Conclusion and future trends

The discourse surrounding municipal wastewater treatment underscores the critical need for sustainable and effective approaches to address the challenges stemming from the

Table 1.3 Advantages of municipal wastewater treatment.

Category	Advantages of municipal wastewater treatment
Environmental advantages	• Protects and improves surface water quality • Reduces water consumption and availability issues • Lowers greenhouse gas emissions • Decreases carbon footprint • Reduces landfill use for treatment by-products • Lowers fossil fuel use • Recycles nutrients, closing the cycle • Reduces eutrophication in aquatic systems
Economic advantages	• Provides alternative renewable fertilizer sources contributes • Saves costs on sludge handling and disposal • Reduces operational costs of treatment plants • Generates revenue from value-added products • Potential for commercializing carbon credits
Social advantages	• Increases food security, especially in developing countries • Contributes to achieving SDGs • Generates broad social benefits with indirect economic value

escalating volume of wastewater and the discharge of untreated contaminants into the environment. As the global water and wastewater sector continues to expand, innovative strategies are imperative to manage wastewater efficiently.

Despite advancements in wastewater treatment technologies, urban authorities encounter persistent challenges such as inadequate infrastructure for handling large wastewater volumes, ineffective removal of emerging contaminants, and mounting difficulties in sewage sludge management. However, municipal wastewater treatment plants possess the potential to transition toward a circular economy by embracing the principles of reduce, reuse, and recycle.

By implementing improved treatment methods, WWTPs can generate valuable byproducts like treated effluent, biogas, and sludge, which can be repurposed for various applications, thereby promoting resource recovery and environmental sustainability. Utilizing effluent for diverse purposes, harnessing biogas as a renewable energy source, and employing thermochemical or biochemical methods to recycle sludge illustrate the myriad benefits achievable through wastewater treatment processes.

This chapter underscores the profound impact of wastewater treatment on public health and the environment, stressing the necessity for a comprehensive approach to wastewater management. Municipal wastewater, comprising household waste products, underscores the significance of delivering high-quality sewer services to urban areas through meticulous planning and infrastructure development. Mitigating or eliminating the environmental ramifications of sewage underscores the pivotal role of wastewater treatment in safeguarding public health and environmental integrity.

Given the evolving landscape of wastewater management and the imperative for sustainable practices, stakeholders must collaborate, innovate, and invest in novel solutions to address prevailing challenges and forge a resilient and eco-friendly future. Municipal wastewater treatment stands poised to become a cornerstone of environmental sustainability and a catalyst for achieving global water security objectives through the adoption of advanced technologies, integration of circular economy principles, and emphasis on resource recovery.

References

Abaliwano, J.K., Ghebremichael, K.A., Amy, G.L., 2008. United Nations Educational, Scientific and Cultural Organization. vol. 5 Institute of Water Education.

Abusoglu, A., Demir, S., Kanoglu, M., 2012. Thermoeconomic assessment of a sustainable municipal wastewater treatment system. Renew. Energy 48, 424–435. https://doi.org/10.1016/j.renene.2012.06.005. 09601481.

Alagha, O., Allazem, A., Bukhari, A.A., Anil, I., Mu'azu, N.D., 2020. Suitability of SBR for wastewater treatment and reuse: pilot-scale reactor operated in different anoxic conditions. Int. J. Environ. Res. Public Health 17 (5), 1617. https://doi.org/10.3390/ijerph17051617. 16604601.

Al-Juaidi, A.E., Kaluarachchi, J.J., Mousa, A.I., 2014. Hydrologic-economic model for sustainable water resources management in a coastal aquifer. J. Hydrol. Eng. 19 (11). https://doi.org/10.1061/(ASCE)HE.1943-5584.0000960. 19435584. American Society of Civil Engineers (ASCE), United States. Available from: https://ascelibrary.org/journal/jhyeff.

Alsawy, T., Rashad, E., El-Qelish, M., Mohammed, R.H., 2022. A comprehensive review on the chemical regeneration of biochar adsorbent for sustainable wastewater treatment. npj Clean Water 5, 1–29. https://doi.org/10.1038/s41545-022-00172-3.

Andersson, K., Rosemarin, A., Lamizana, B., Kvarnstrom, E., McConville, J., Seidu, R., Dickin, S., 2016. Trimmer, Sanitation, Wastewater Management and Sustainability: From Waste Disposal to Resource Recovery. United Nations Environment Programme and Stockholm Environment Institute, Nairobi and Stockholm.

Asano, T., Burton, F.L., Leverenz, H., Tsuchihashi, R., Tchobanoglous, G., 2007. Water Reuse: Issues, Technologies, and Applications. McGraw-Hill.

Avellan, C.T., Ardakanian, R., Gremillion, P., 2017. The role of constructed wetlands for biomass production within the water-soil-waste nexus. Water Sci. Technol. 75 (10), 2237–2245. https://doi.org/10.2166/wst.2017.106. 02731223.

Baawain, M.S., Al-Mamun, A., Omidvarborna, H., Al-Sabti, A., Choudri, B.S., 2020. Public perceptions of reusing treated wastewater for urban and industrial applications: challenges and opportunities. Environ. Dev. Sustain. 22 (3), 1859–1871. 15732975. Springer, Oman. Available from: www.wkap.nl/journalhome.htm/1387-585X. https://doi.org/10.1007/s10668-018-0266-0.

Basile, A., Ghasemzadeh, K. (Eds.), 2020. Current Trends and Future Developments on (Bio-) Membranes. Recent Achievements in Wastewater and Water Treatments. Elsevier.

Baskar, A.V., Bolan, N., Hoang, S.A., Sooriyakumar, P., Kumar, M., Singh, L., Jasemizad, T., Padhye, L.P., Singh, G., Vinu, A., Sarkar, B., Kirkham, M.B., Rinklebe, J., Wang, S., Wang, H., Balasubramanian, R., Siddique, K.H., 2022. Recovery, regeneration and sustainable management of spent adsorbents from wastewater treatment streams: a review. Sci. Total Environ. 822, 1–24. https://doi.org/10.1016/j.scitotenv.2022.153555. 153555.

Bhat, S.A., Sher, F., Hameed, M., Bashir, O., Kumar, R., Vo, D.N., Ahmad, P., Lima, E.-C., 2022. Sustainable nanotechnology-based wastewater treatment strategies: achievements, challenges, and future perspectives. Chemosphere 288, 1–13. https://doi.org/10.1016/j.chemosphere.2021.132606. 132606.

Breach, P.A., Simonovic, S.P., 2018. Wastewater treatment energy recovery potential for adaptation to global change: an integrated assessment. Environ. Manag. 61, 624–636.

Capodaglio, A.G., 2021. Fit-for-purpose urban wastewater reuse: analysis of issues and available technologies for sustainable multiple barrier approaches. Crit. Rev. Environ. Sci. Technol. 51 (15), 1619–1666. https://doi.org/10.1080/10643389.2020.1763231. 10643389.

Casiano Flores, C., Bressers, H., Gutierrez, C., de Boer, C., 2018. Towards circular economy—a wastewater treatment perspective, the Presa Guadalupe case. Manag. Res. Rev. 41 (5), 554–571. 20408269. Emerald Group Holdings Ltd., Netherlands. Available from: http://www.emeraldinsight.com/Insight/viewContainer.do;jsessionid=746E76A1B81993BFFED6D07E37E888A6?containerType=Journal%26containerId=15002370. https://doi.org/10.1108/MRR-02-2018-0056.

Chojnacka, K., Witek-Krowiak, A., Moustakas, K., Skrzypczak, D., Mikula, K., Loizidou, M., 2020. A transition from conventional irrigation to fertigation with reclaimed wastewater: prospects and challenges. Renew. Sust. Energ. Rev. 130, 109959. https://doi.org/10.1016/j.rser.2020.109959. 13640321.

Choudhary, A.K., Kumar, S., Sharma, C., 2011. Constructed wetlands: an option for pulp and paper mill wastewater treatment. Elec. J. Env. Agricult. Food Chem. 10 (10), 3023–3037. 15794377. Available from: http://minerva.uvigo.es/index.php/ejeafche/article/view/120/118. India.

Chrispim, M.C., Scholz, M., Nolasco, M.A., 2019. Phosphorus recovery from municipal wastewater treatment: Critical review of challenges and opportunities for developing countries. Journal of environmental management 248, 1–18. https://doi.org/10.1016/j.jenvman.2019.109268. 109268.

Crini, G., Lichtfouse, E., 2019. Advantages and disadvantages of techniques used for wastewater treatment. Environ. Chem. Lett. 17 (1), 145–155. 16103661. Springer Verlag, France. Available from: https://www.springer.com/journal/10311. https://doi.org/10.1007/s10311-018-0785-9.

Delanka-Pedige, H.M.K., Munasinghe-Arachchige, S.P., Abeysiriwardana-Arachchige, I.S.A., Nirmalakhandan, N., 2021. Wastewater infrastructure for sustainable cities: assessment based on UN sustainable development goals (SDGs). Int J Sust Dev World 28 (3), 203–209. https://doi.org/10.1080/13504509.2020.1795006. 13504509.

Demirbas, A., Edris, G., Alalayah, W.M., 2017. Sludge production from municipal wastewater treatment in sewage treatment plant. Energy Sources A Recovery Util. Environ. Eff. 39 (10), 999–1006. 15567230. Taylor and Francis, Saudi Arabia. Available from: http://www.tandf.co.uk/journals/titles/15567036.asp. https://doi.org/10.1080/15567036.2017.1283551.

Dutta, D., Arya, S., Kumar, S., 2021. Industrial wastewater treatment: current trends, bottlenecks, and best practices. Chemosphere 285, 131245. https://doi.org/10.1016/j.chemosphere.2021.131245. 00456535.

EPA, 2024. Basic Information About Water Reuse. Environmental Protection Agency. Available from: https://www.epa.gov/waterreuse/basic-information-about-water-reuse#:~:text=Water%20reuse%20can%20provide%20alternatives,security%2C%20sustainability%2C%20and%20resilience. (Accessed 26 April 2024.

Fito, J., Van Hulle, S.W.H., 2021. Wastewater reclamation and reuse potentials in agriculture: towards environmental sustainability. Environ. Dev. Sustain. 23 (3), 2949–2972. 15732975. Springer Science and Business Media B.V., Belgium. Available from: www.wkap.nl/journalhome.htm/1387-585X. https://doi.org/10.1007/s10668-020-00732-y.

Freitas, J.H.E.S., de Santana, K.V., do Nascimento, A.C.C., de Paiva, S.C., de Moura, M.C., Coelho, L.C.B.B., de Oliveira, M.B.M., Paiva, P.M.G., do Nascimento, A.E., Napoleão, T.H., 2016. Evaluation of using aluminum sulfate and water-soluble *Moringa oleifera* seed lectin to reduce turbidity and toxicity of polluted stream water. Chemosphere 163, 133–141. 18791298. Elsevier, Brazil. Available from: www.elsevier.com/locate/chemosphere. https://doi.org/10.1016/j.chemosphere.2016.08.019.

Hamed, M.M., Khalafallah, M.G., Hassanien, E.A., 2004. Prediction of wastewater treatment plant performance using artificial neural networks. Environ. Model. Softw. 19 (10), 919–928. https://doi.org/10.1016/j.envsoft.2003.10.005.

Huang, J.J., Hu, H.Y., Lu, S.Q., Li, Y., Tang, F., Lu, Y., Wei, B., 2012. Monitoring and evaluation of antibiotic-resistant bacteria at a municipal wastewater treatment plant in China. Environ. Int. 42 (1), 31–36. 18736750. Elsevier, China. Available from: www.elsevier.com/locate/envint. https://doi.org/10.1016/j.envint.2011.03.001.

Jamrah, A., Al-Futaisi, A., Prathapar, S., Harrasi, A.A., 2008. Evaluating greywater reuse potential for sustainable water resources management in Oman. Environ. Monit. Assess. 137 (1–3), 315–327. https://doi.org/10.1007/s10661-007-9767-2. 15732959.

Kalair, A.R., Seyedmahmoudian, M., Stojcevski, A., Abas, N., Khan, N., 2021. Waste to energy conversion for a sustainable future. Heliyon 7 (10), e08155. https://doi.org/10.1016/j.heliyon.2021.e08155. 24058440.

Kamali, M., Persson, K.M., Costa, M.E., Capela, I., 2019. Sustainability criteria for assessing nanotechnology applicability in industrial wastewater treatment: current status and future outlook. Environ. Int. 125, 261–276. 18736750. Elsevier, Portugal. Available from: www.elsevier.com/locate/envint. https://doi.org/10.1016/j.envint.2019.01.055.

Kansal, S.K., Kumari, A., 2014. Potential of *M. oleifera* for the treatment of water and wastewater. Chem. Rev. 114 (9), 4993–5010. 15206890. American Chemical Society, India. Available from: http://pubs.acs.org/journal/chreay. https://doi.org/10.1021/cr400093w.

Kesari, K.K., Soni, R., Jamal, Q.M.S., Tripathi, P., Lal, J.A., Jha, N.K., Siddiqui, M.H., Kumar, P., Tripathi, V., Ruokolainen, J., 2021. Wastewater treatment and reuse: a review of its applications and health implications. Water Air Soil Pollut. 232 (5). 15732932. Springer Science and Business Media Deutschland GmbH, Finland.

Available from: http://www.kluweronline.com/issn/0049-6979/. https://doi.org/10.1007/s11270-021-05154-8.
Kim, J.H., Behera, S.K., Oh, S.Y., Park, H.S., 2010. Reuse potential of municipal wastewater treatment facility effluents for sustainable water resource management in Ulsan, Korea. Water Environ. J. 24 (4), 293–302. https://doi.org/10.1111/j.1747-6593.2009.00182.x. 17476593.
Kivaisi, A.K., 2001. The potential for constructed wetlands for wastewater treatment and reuse in developing countries: a review. Ecol. Eng. 16 (4), 545–560. 09258574. Elsevier, Tanzania. Available from: www.elsevier.com/inca/publications/store/5/2/2/7/5/1. https://doi.org/10.1016/S0925-8574(00)00113-0.
Larsen, T.A., Udert, K.M., Lienert, J., 2013. Source Separation and Decentralization for Wastewater Management. IWA Publishing.
Lazarova, V., 2005. Wastewater treatment for water recycling. In: Lazarova, V., Bahri, A. (Eds.), Irrigation With Recycled Water: Agriculture, Turfgrass and Landscape. CRC Press, Boca Raton, FL, USA, pp. 164–234.
Lee, J., Jeon, J.H., Shin, J., Jang, H.M., Kim, S., Song, M.S., Kim, Y.M., 2017. Quantitative and qualitative changes in antibiotic resistance genes after passing through treatment processes in municipal wastewater treatment plants. Sci. Total Environ. 605–606, 906–914. https://doi.org/10.1016/j.scitotenv.2017.06.250. 18791026. Elsevier B.V., South Korea. Available from: www.elsevier.com/locate/scitotenv.
Liang, Y., Zhu, H., Bañuelos, G., Yan, B., Zhou, Q., Yu, X., Cheng, X., 2017. Constructed wetlands for saline wastewater treatment: a review. Ecol. Eng. 98, 275–285. 09258574. Elsevier B.V., China. Available from: www.elsevier.com/inca/publications/store/5/2/2/7/5/1. https://doi.org/10.1016/j.ecoleng.2016.11.005.
Liao, Z., Chen, Z., Xu, A., Gao, Q., Song, K., Liu, J., Hu, H.Y., 2021. Wastewater treatment and reuse situations and influential factors in major Asian countries. J. Environ. Manag. 282. 10958630. Academic Press, China. Available from: https://www.sciencedirect.com/journal/journal-of-environmental-management. https://doi.org/10.1016/j.jenvman.2021.111976.
Libhaber, M., Orozco-Jaramillo, A., 2012. Sustainable Treatment and Reuse of Municipal Wastewater: For Decision Makers and Practising Engineers. IWA Publishing.
Liu, F., Zhu, S., Li, D., Chen, G., Ho, S.H., 2020. Detecting ferric iron by microalgal residue-derived fluorescent nanosensor with an advanced kinetic model. iScience 23 (6). https://doi.org/10.1016/j.isci.2020.101174. 25890042. Elsevier, China. Available from: www.cell.com/iscience.
Lu, Z., Loftus, S., Sha, J., Wang, W., Park, M.S., Zhang, X., Johnson, Z.I., Hu, Q., 2020. Water reuse for sustainable microalgae cultivation: current knowledge and future directions. Resour. Conserv. Recycl. 161. 18790658. Elsevier B.V., United States. Available from: www.elsevier.com/locate/resconrec. https://doi.org/10.1016/j.resconrec.2020.104975.
Łuczkiewicz, A., Jankowska, K., Fudala-Ksiazek, S., Olańczuk-Neyman, K., 2010. Antimicrobial resistance of fecal indicators in municipal wastewater treatment plant. Water Res. 44 (17), 5089–5097. 00431354. Elsevier, Poland. Available from: https://www.sciencedirect.com/science/journal/00431354. https://doi.org/10.1016/j.watres.2010.08.007.
Manning, L.J., Graham, D.W., Hall, J.W., 2016. Wastewater systems assessment. In: The Future of National Infrastructure: A System-of-Systems Approach. Cambridge University Press, United Kingdom, pp. 134–157. Available from: https://doi.org/10.1017/CBO9781107588745.

Mao, D., Yu, S., Rysz, M., Luo, Y., Yang, F., Li, F., Hou, J., Mu, Q., Alvarez, P.J.J., 2015. Prevalence and proliferation of antibiotic resistance genes in two municipal wastewater treatment plants. Water Res. 85, 458–466. 18792448. Elsevier, China. Available from: www.elsevier.com/locate/watres. https://doi.org/10.1016/j.watres.2015.09.010.

Mayer, B.K., Baker, L.A., Boyer, T.H., Drechsel, P., Gifford, M., Hanjra, M.A., Parameswaran, P., Stoltzfus, J., Westerhoff, P., Rittmann, B.E., 2016. Total value of phosphorus recovery. Environ. Sci. Technol. 50 (13), 6606–6620. 15205851. American Chemical Society, United States. Available from: http://pubs.acs.org/journal/esthag. https://doi.org/10.1021/acs.est.6b01239.

Meena, R.A.A., Kannah, R.Y., Sindhu, J., Ragavi, J., Kumar, G., Gunasekaran, M., Banu, J.R., 2019. Trends and resource recovery in biological wastewater treatment system. Bioresour. Technol. Rep. 7, 100235. https://doi.org/10.1016/j.biteb.2019.100235. 2589014X.

Melin, T., Jefferson, B., Bixio, D., Thoeye, C., De Wilde, W., De Koning, J., van der Graaf, J., Wintgens, T., 2006. Membrane bioreactor technology for wastewater treatment and reuse. Desalination 187 (1–3), 271–282. https://doi.org/10.1016/j.desal.2005.04.086. 00119164.

Melo, L.D.V., da Costa, E.P., Pinto, C.C., Barroso, G.R., Oliveira, S.C., 2019. Adequacy analysis of drinking water treatment technologies in regard to the parameter turbidity, considering the quality of natural waters treated by large-scale WTPs in Brazil. Environ. Monit. Assess. 191 (6). 15732959. Springer International Publishing, Brazil. Available from: https://link.springer.com/journal/10661. https://doi.org/10.1007/s10661-019-7526-9.

Muga, H.E., Mihelcic, J.R., 2008. Sustainability of wastewater treatment technologies. J. Environ. Manag. 88 (3), 437–447. 03014797. Academic Press, United States. Available from: https://www.sciencedirect.com/journal/journal-of-environmental-management. https://doi.org/10.1016/j.jenvman.2007.03.008.

Nwoba, E.G., Vadiveloo, A., Ogbonna, C.N., Ubi, B.E., Ogbonna, J.C., Moheimani, N.R., 2020. Algal cultivation for treating wastewater in African developing countries: a review. Clean Soil, Air, Water 48 (3). 18630669. Wiley-VCH Verlag, Nigeria. Available from: http://onlinelibrary.wiley.com/journal/10.1002/(ISSN)1863-0669. https://doi.org/10.1002/clen.202000052.

Palansooriya, K.N., Yang, Y., Tsang, Y.F., Sarkar, B., Hou, D., Cao, X., Meers, E., Rinklebe, J., Kim, K.H., Ok, Y.S., 2020. Occurrence of contaminants in drinking water sources and the potential of biochar for water quality improvement: a review. Crit. Rev. Environ. Sci. Technol. 50 (6), 549–611. 15476537. Taylor and Francis, South Korea. Available from: www.tandf.co.uk/journals/titles/10643389.asp. https://doi.org/10.1080/10643389.2019.1629803.

Panepinto, D., Fiore, S., Zappone, M., Genon, G., Meucci, L., 2016. Evaluation of the energy efficiency of a large wastewater treatment plant in Italy. Appl. Energy 161, 404–411. https://doi.org/10.1016/j.apenergy.2015.10.027. 03062619. Elsevier, Italy. Available from: http://www.elsevier.com/inca/publications/store/4/0/5/8/9/1/index.htt.

Pikaar, I., Guest, J., Ganigué, R., Jensen, P., Rabaey, K., Seviour, T., 2022. Verstraete, Resource Recovery From Water: Principles and Application. IWA Publishing.

Rachid-Sally, L., Jayakody, P., 2008. Drivers and characteristics of wastewater agriculture in developing countries: results from a global assessment. Research Report 127, IWMI.

Ratner, B.D., Rivera, A., 2004. Reasserting community: the social challenge of wastewater management in Panajachel, Guatemala. Hum. Organ. 63 (1), 47–56. https://doi.org/10.17730/humo.63.1.krkvu1q77km1hk8b. 00187259.

Rosso, D., Larson, L.E., Stenstrom, M.K., 2008. Aeration of large-scale municipal wastewater treatment plants: state of the art. Water Sci. Technol. 57 (7), 973–978. https://doi.org/10.2166/wst.2008.218. 02731223.

Sardana, A., Cottrell, B., Soulsby, D., Aziz, T.N., 2019. Dissolved organic matter processing and photoreactivity in a wastewater treatment constructed wetland. Sci. Total Environ. 648, 923–934. https://doi.org/10.1016/j.scitotenv.2018.08.138. 18791026. Elsevier B.V., United States. Available from: www.elsevier.com/locate/scitotenv.

Sarvajayakesavalu, S., Lu, Y., Withers, P.J.A., Pavinato, P.S., Pan, G., Chareonsudjai, P., 2018. Phosphorus recovery: a need for an integrated approach. Ecosyst. Health Sustain. 4 (2), 48–57. 23328878. Taylor and Francis, China. Available from: https://spj.science.org/journal/ehs. https://doi.org/10.1080/20964129.2018.1460122.

Shah, A.I., Din Dar, M.U., Bhat, R.A., Singh, J.P., Singh, K., Bhat, S.A., 2020. Prospectives and challenges of wastewater treatment technologies to combat contaminants of emerging concerns. Ecol. Eng. 152. https://doi.org/10.1016/j.ecoleng.2020.105882. 09258574. Elsevier B.V., India. Available from: www.elsevier.com/inca/publications/store/5/2/2/7/5/1.

Shanmugam, K., Gadhamshetty, V., Tysklind, M., Bhattacharyya, D., Upadhyayula, V.K.K., 2022. A sustainable performance assessment framework for circular management of municipal wastewater treatment plants. J. Clean. Prod. 339, 130657. https://doi.org/10.1016/j.jclepro.2022.130657. 09596526.

Shore, D., 2017. From waste treatment to resource recovery: a Chicago sustainability story. MRS Energy Sustain. 4 (1). https://doi.org/10.1557/mre.2017.8. 23292237. Springer Nature, United States. Available from: https://www.springer.com/journal/43581.

Silva, J.A., 2023. Wastewater treatment and reuse for sustainable water resources management: a systematic literature review. Sustainability 15 (14), 10940. https://doi.org/10.3390/su151410940. 20711050.

Statista, 2021. Water and Wastewater Treatment Market Size Worldwide in 2020. https://www-statista-com.salford.idm.oclc.org/statistics/1199744/market-size-water-and-wastewater-treatment-global/.

Surya Nair, K., Manu, B., Azhoni, A., 2021. Sustainable treatment of paint industry wastewater: current techniques and challenges. J. Environ. Manag. 296, 113105. https://doi.org/10.1016/j.jenvman.2021.113105. 03014797.

Suzenet, G., Tal, A., Boymanns, D., 2002. Sustainable water management for the city: technologies for improving domestic water supply. Built Environ. 28 (2), 138–151.

Tchobanoglous, G., Burton, F.L., Stensel, H.D., 2013. In: Metcalf & Eddy (Ed.), Wastewater Engineering Treatment and Reuse, fifth ed. McGraw-Hill Education, New York.

Tchobanoglous, G., Stensel, H.D., Tsuchihashi, R., Burton, F., Abu-Orf, M., Bowden, G., Pfrang, W., 2015. Metcalf & Eddy/AECOM: Tratamento de efluentes e recuperacao de recursos, fifth ed. AMGH Editora Ltda, Porto Alegre.

Tortajada, C., 2020. Contributions of recycled wastewater to clean water and sanitation Sustainable Development Goals. npj Clean Water 3 (1). 20597037. Nature Research, Singapore. Available from: https://www.nature.com/npjcleanwater/. https://doi.org/10.1038/s41545-020-0069-3.

Trimmer, J.T., Cusick, R.D., Guest, J.S., 2017. Amplifying progress toward multiple development goals through resource recovery from sanitation. Environ. Sci. Technol. 51 (18), 10765–10776. 15205851. American Chemical Society, United States. Available from: http://pubs.acs.org/journal/esthag. https://doi.org/10.1021/acs.est.7b02147.

Tripathi, A., Mishra, A.K., 2017. Knowledge and passive adaptation to climate change: an example from Indian farmers. Clim. Risk Manag. 16, 195–207. https://doi.org/10.1016/j.crm.2016.11.002. 22120963. Elsevier B.V., India. Available from: http://www.journals.elsevier.com/climate-risk-management/.

Valverde, K.C., Coldebella, P.F., Silva, M.F., Nishi, L., Bongiovani, M.C., Bergamasco, R., 2018. *Moringa oleifera* Lam. and its potential association with aluminium sulphate in the process of coagulation/flocculation and sedimentation of surface water. Int. J. Chem. Eng., 12–15.

Villarín, M.C., Merel, S., 2020. Paradigm shifts and current challenges in wastewater management. J. Hazard. Mater. 390, 122139. https://doi.org/10.1016/j.jhazmat.2020.122139. 03043894.

Villarroel Walker, R., Beck, M.B., Hall, J.W., Dawson, R.J., Heidrich, O., 2014. The energy-water-food nexus: strategic analysis of technologies for transforming the urban metabolism. J. Environ. Manag. 141, 104–115. https://doi.org/10.1016/j.jenvman.2014.01.054. 03014797.

Vunain, E., Masoamphambe, E.F., Mpeketula, P.M.G., Monjerezi, M., Etale, A., 2019. Evaluation of coagulating efficiency and water borne pathogens reduction capacity of *Moringa oleifera* seed powder for treatment of domestic wastewater from Zomba, Malawi. J. Environ. Chem. Eng. 7 (3). https://doi.org/10.1016/j.jece.2019.103118. 22133437. Elsevier, Malawi. Available from: http://www.journals.elsevier.com/journal-of-environmental-chemical-engineering/.

Wu, W., Shi, Y., Liu, G., Fan, X., Yang, Y., 2020. Recent development of graphene oxide based forward osmosis membrane for water treatment: a critical review. Desalination 491, 114452. https://doi.org/10.1016/j.desal.2020.114452. 00119164.

2

Emerging contaminants in municipal wastewater

Liane Yuri Kondo Nakada[a] and Rodrigo Custodio Urban[b]

[a]*Department of Environmental Engineering, Institute of Geosciences and Exact Sciences, São Paulo State University (UNESP), Rio Claro, São Paulo, Brazil.*

[b]*Department of Environmental Engineering, Institute of Science and Technology, São Paulo State University (UNESP), São José dos Campos, São Paulo, Brazil*

Abbreviations

AdVs	adenoviruses
AFFF	aqueous film-forming foam
AOP	advanced oxidative processes
BOD	biochemical oxygen demand
BPA	bisphenol A
CEC	contaminants of emerging concern
DEET	*N,N*-diethyl-*m*-toluamide
EC	emerging contaminants
EVs	enteroviruses
FD	filter discs
GAC	granular activated carbon
HAV	hepatitis A virus
LC-HRMS	liquid chromatography-high resolution mass spectrometry
LRV	$\log_{10}$ reduction values
MeP	methylparaben
MOF	metal organic framework
MP	microplastics
NoVs	noroviruses
PAC	powdered activated carbon
PCP	personal care products
PFAAS	perfluoroalkyl acids
PFAS	per- and polyfluoroalkyl substances
PFHxS	perfluorohexanesulfonate
PFOA	perfluorooctanoic acid
PFOS	perfluorooctanesulfonate
PPCP	pharmaceuticals and personal care products
PrP	propylparaben
ROS	reactive oxygen species

Municipal Wastewater Treatment. https://doi.org/10.1016/B978-0-443-24826-9.00002-4

STP	sewage treatment plant
TCS	triclosan
TRL	technology readiness level
UV	ultraviolet
WWTP	wastewater treatment plant

2.1 Introduction

Emerging contaminants (EC), also known as contaminants of emerging concern (CEC) (Sauvé and Desrosiers, 2014), are chemicals or microorganisms that are not commonly monitored in the environment, have not been identified previously or are being identified at different levels than expected, but have the potential to cause adverse effects on the environment or human health (Kumar et al., 2022; US EPA, 2024). The classification encompasses a range of potential pollutants, including pharmaceuticals, personal care products, per- and polyfluoroalkyl substances (PFAS), microplastics, surfactants, endocrine disruptors, and pathogenic microorganisms (US EPA, 2024).

The primary source of CEC pollution in the environment is wastewater effluent (Puri et al., 2023). Conventional wastewater treatment techniques have been demonstrated to be ineffective at removing many CECs, prompting the evaluation of several advanced hybrid technologies (Eniola et al., 2022; Kumar et al., 2022).

This chapter presents a brief literature review on emerging contaminants, particularly pharmaceuticals and personal care products (PPCPs), PFAS, microplastics and viruses, including fate and transport in the environment, occurrence and removal in wastewater treatment plants by conventional and advanced hybrid technologies.

2.2 Pharmaceuticals and personal care products (PPCPs)

PPCPs are used extensively in medicine, livestock and aquaculture worldwide, both for the treatment and prevention of diseases. As a result, they are frequently detected in the environment, with municipal sewage being one of the most important routes of introduction into the environment (Krishnan et al., 2021).

PPCPs can be classified into distinct categories based on their intrinsic characteristics. In general, pharmaceuticals are composed primarily of analgesics, antibiotics, anticancer drugs, anticonvulsants, antidepressants, antiseptics, anti-inflammatories, hormones, and lipid regulators, among other active ingredients. Personal hygiene products encompass a range of items, including disinfectants, fragrances, mosquito repellents, condoms, sunblocks, soaps, and shampoos (Krishnan et al., 2021).

The production of an active pharmaceutical substance and the formulation of the drug can result in the release of raw materials, intermediates and/or the active substance into the air, water (in wastewater) and soil (in the form of solid waste). It can be reasonably assumed that the primary means of introducing pharmaceutical substances into the environment is through the effluents produced during the cleaning process of the active pharmaceutical ingredient and the equipment utilized for coating, mixing, compressing and packaging the drugs. Nevertheless, it is believed that biological and chemical degradation processes, including biotransformation, mineralization, hydrolysis and photolysis, effectively eliminate the majority of drug residues prior to the discharge of effluents or solid waste from production into surface waters, sewage treatment plants or the disposal of waste in soil. It is considered that losses to the environment during the production or formulation of veterinary medicines are minimal, due to the implementation of practices by industries which serve to reduce the generation of waste and material losses (Crane et al., 2016).

The failure of sewage treatment processes to remove drugs, their metabolites (resulting from modifications to the chemical structure that occur in vivo, after administration of the drugs) and their transformation products (resulting from modifications to the chemical structure that occur after excretion into the environment) can result in their eventual reaching public water supplies. It is possible for active compounds to reach different environmental compartments from a variety of sources, including sewage treatment plant effluents, solid waste disposal and landfill leachate. The application of veterinary drugs to livestock, poultry and aquaculture can result in the introduction of these substances into the environment. This may occur when excreta are carried by rainwater in fields or through direct discharge, for example during meat processing or aquarium discharge. In the absence of degradation, these substances can migrate to the soil and groundwater, and subsequently reach surface water via surface runoff (Kümmerer, 2008a).

In the environment, interactions may occur between the drugs and the particles present, or even the formation of complexes, which may result in the drugs being undetected and even losing their antimicrobial activity. The following processes determine the fate of drugs in different environmental compartments: adsorption, photodegradation, biodegradation and hydrolysis. Therefore, the disappearance of a substance does not necessarily indicate biological or photochemical degradation. Indeed, adsorption represents an important route for eliminating drugs, with up to 50% adsorption in sewage treatment plant sludge. This has the potential to interfere with the distribution and bioavailability of these substances in the environment (transport of drugs bound to particles) (Fig. 2.1; Kümmerer, 2008a).

With regard to human consumption, any pharmaceutical component that is not metabolized or disintegrated by the body is excreted in urine and feces, and subsequently discharged into the sewer. In instances where municipal sewage is subjected to treatment, these substances may be transferred between phases and incorporated into the sludge, or alternatively, they may remain in the treated sewage effluent. The utilization of biosolids or treated sewage in agricultural practices, for instance, may result in the contamination of nearby water sources due to the potential for these substances to be carried to such locations (Krishnan et al., 2021).

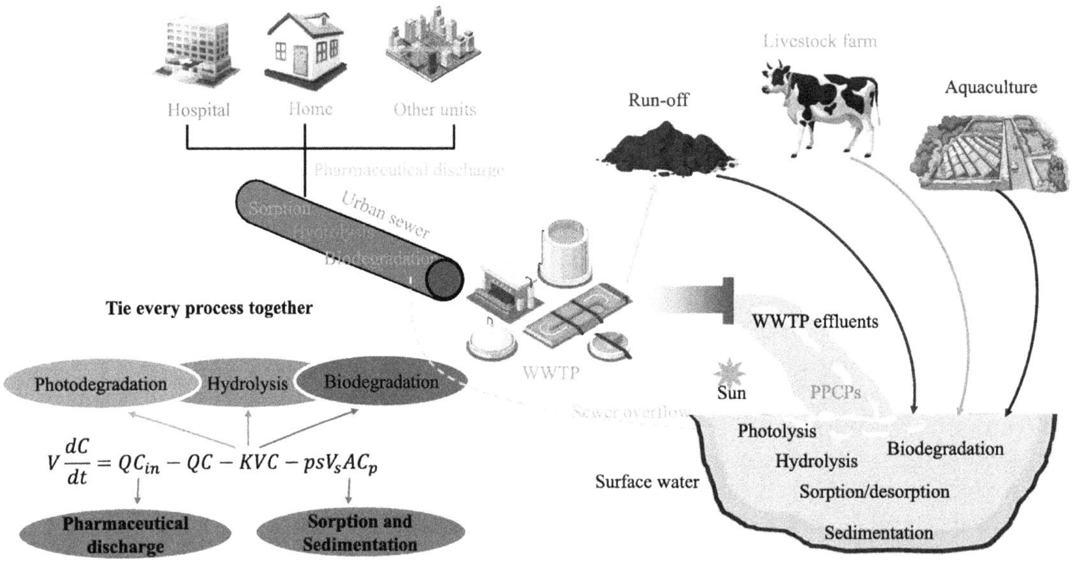

Fig. 2.1 Fate and transport of PPCPs in the environment (Bavumiragira et al., 2022). Reprinted with permission from Bavumiragira, J.P., Ge, J., Yin, H., 2022. Fate and transport of pharmaceuticals in water systems: a processes review. Sci. Total Environ. 823. https://doi.org/10.1016/j.scitotenv.2022.153635. ISSN:153-635, Copyright (2022), Elsevier.

The production process of personal care products (PCPs) bears resemblance to that of pharmaceuticals. A significant number of concerns pertaining to the release of drug residues into the environment also apply to PCPs, particularly given the extensive use of these products.

Despite the existence of numerous studies monitoring the presence of PCPs in sewage treatment plants, there is still a paucity of information regarding their effects on effluent treatment processes (Yang et al., 2017). Among the substances most commonly detected in sewage treatment plants (STP) tributaries and effluents worldwide, methylparaben (MeP) and propylparaben (PrP) stand out due to their use as preservatives, the bactericide triclosan (TCS), *N,N*-diethyl-*meta*-toluamide (DEET) used in insect repellents, various substances used as UV radiation filters and different types of fragrances (Yang et al., 2017).

It is also of interest to consider the impact of seasonal variation on the frequency of detection of PCPs in STPs. The occurrence of UV radiation filters is more frequently observed during periods of elevated temperatures, particularly during the summer season (Tsui et al., 2014). A comparable phenomenon can be observed with the utilization of alternative products, including insect repellents (Biel-Maeso et al., 2019). Such interactions may also influence the behavior of substances in the environment.

PPCPs are active biological components that have been designed to interact with a range of pathways and processes in humans and animals. Accordingly, the potential impact of active PPCPs on the environment and human health represents a persistent concern (Boxall et al., 2012). While there are no specific reports on the effect of PCPs on human health, it is possible that they occur in the long term (Sengar and Vijayanandan, 2022).

The presence of PPCPs has been documented in a multitude of locations across the globe, in a variety of matrices including drinking water, groundwater, tap water and wastewater (Cangola et al., 2024; Rehman et al., 2024; Sengar and Vijayanandan, 2022), sediments, sewage sludge and fish (Biel-Maeso et al., 2019).

The impact of PPCPs on aquatic ecosystems is more extensively researched. A number of studies have documented a range of adverse effects associated with the bioaccumulation of PPCPs, including the transmission of these chemicals to offspring, reduced growth, disruption to reproductive processes, behavioral anomalies, alterations in organ function, changes in enzyme activity and the feminisation of male fish (Rehman et al., 2024; Fig. 2.2). It is plausible that the effects on aquatic species are more pronounced due to the greater mobility of PPCPs within an ecosystem (Rehman et al., 2024).

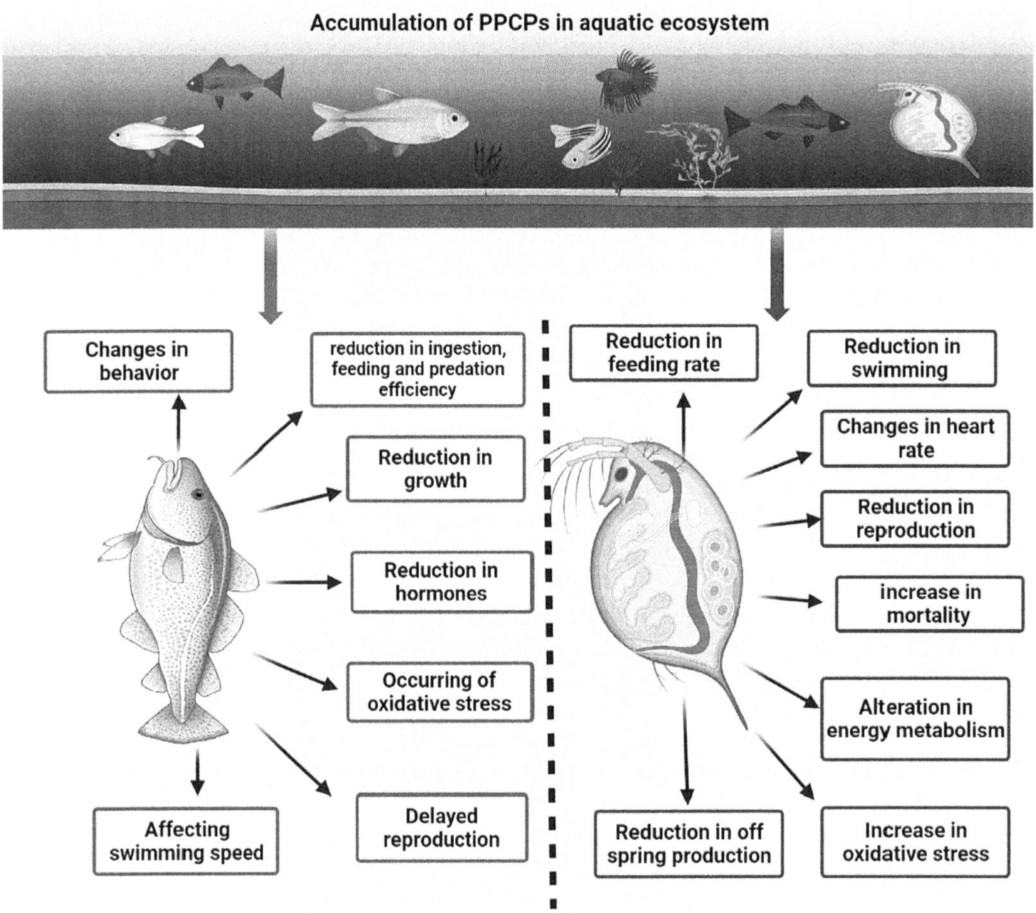

Fig. 2.2 Effects of PPCPs on Fish and Daphnia (Rehman et al., 2024). Reprinted with permission from Rehman, M.U., Nisar, B., Mohd Yatoo, A., Sehar, N., Tomar, R., Tariq, L., Ali, S., Ali, A., Mudasir Rashid, S., Bilal Ahmad, S., Aldossari, R.M., 2024. After effects of Pharmaceuticals and Personal Care Products (PPCPs) on the biosphere and their counteractive ways. Sep. Purif. Technol. 342, 126921. https://doi.org/10.1016/j.seppur.2024.126921, Copyright (2024), Elsevier.

2.3 Per- and polyfluoroalkyl substances (PFAS)

Perfluoroalkyl and polyfluoroalkyl substances (PFASs) represent a significant category of fluorinated chemical products that exhibit remarkable thermal stability, resistance to degradation, and hydrophobic and oleophobic properties (Chen et al., 2022; Zhou et al., 2024). These characteristics render these substances highly useful for industrial and commercial applications

(Zhou et al., 2024), as they enhance the durability and longevity of a diverse range of products, including textile coatings, fire-resistant foams, metal coatings, food packaging, and others (Chen et al., 2022).

However, the same characteristics that render PFAS useful for industrial purposes also render them a source of environmental concern. The environmental persistence, bioaccumulation and toxicity of these substances have resulted in their presence in a variety of environmental, aqueous and solid matrices (Chen et al., 2022; Zhang et al., 2024). A number of surveys have indicated that there may be in excess of 4800 variations of PFAS currently available on the market (Zhang et al., 2024). The challenging degradation of these substances makes their elimination from the human body a significant challenge. Consequently, they are likely to persist in the bloodstream indefinitely, which has led to the term "eternal chemicals" being used to describe PFAS (Fang et al., 2024).

These aspects give rise to significant concerns regarding the impact of PFAS on human health, including the weakening of immune systems and the increased risk of developing cancer (Zhang et al., 2024). Consequently, certain PFAS, including Perfluorooctanesulfonate (PFOS), Perfluorooctanoic acid (PFOA) and Perfluorohexanesulfonate (PFHxS), have been designated as Persistent Organic Pollutants in accordance with the Stockholm Convention, which imposes restrictions on their production, trade and utilization on a global scale (Zhang et al., 2024).

Among the potential avenues through which PFAS may enter the environment (Fig. 2.3), wastewater treatment plants represent a significant nexus between the commercial utilization of these substances and environmental matrices (Zhou et al., 2024). This is due to the fact that many conventional WWTPs are not efficacious at removing PFAS, resulting in the treated effluent becoming a secondary source, releasing the substances into the environment (Chen et al., 2022), or even the concentration of PFAS in biosolids, which may pose some risk in their agricultural application (Zhou et al., 2024).

In a study conducted in Shaanxi province, Chen et al. (2022) reached the conclusion that the ecological risk associated with PFAS residues in wastewater effluent is low. However, in sludge, this risk is classified as being between low and high. This underscores the significance of detecting and quantifying PFAS in sludge matrices. However, determining PFAS concentrations in sludge presents a significant challenge, given the high affinity these substances have for the extracellular polymeric substances present in the sludge (Zhou et al., 2024). It is therefore imperative

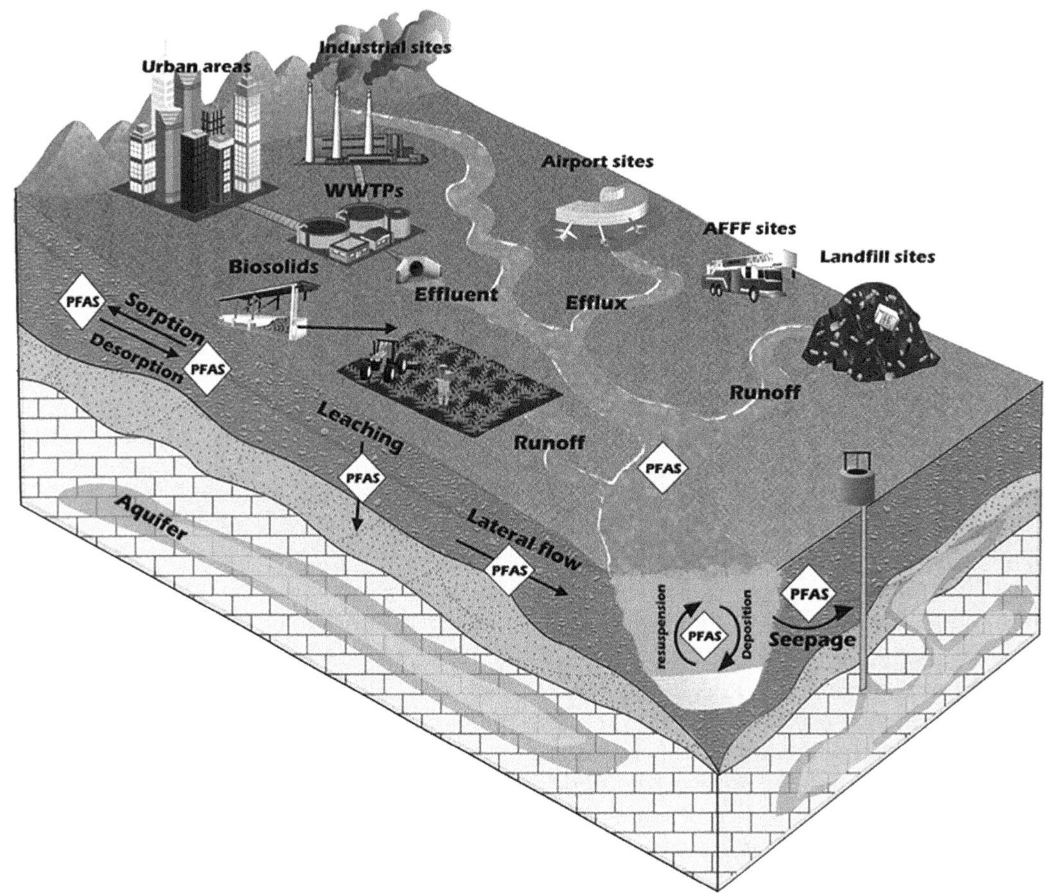

Fig. 2.3 Fate and transport of PFAS in the environment (Rafiei and Nejadhashemi, 2023). Reprinted with permission from Rafiei, V., Nejadhashemi, A.P., 2023. Watershed scale PFAS fate and transport model for source identification and management implications. Water Res. 240, 120073. https://doi.org/10.1016/j.watres.2023.120073, Copyright (2023), Elsevier.

that well-defined procedures are followed for the collection, preservation and preparation of samples, as well as for instrumental analysis. A summary of these steps can be found in Fig. 2.4. Further details of each are provided in Zhou et al. (2024).

2.4 Microplastics

Since the mid-20th century, plastics have been a crucial component of human civilisation, with applications spanning numerous domains. However, while their characteristics have enabled

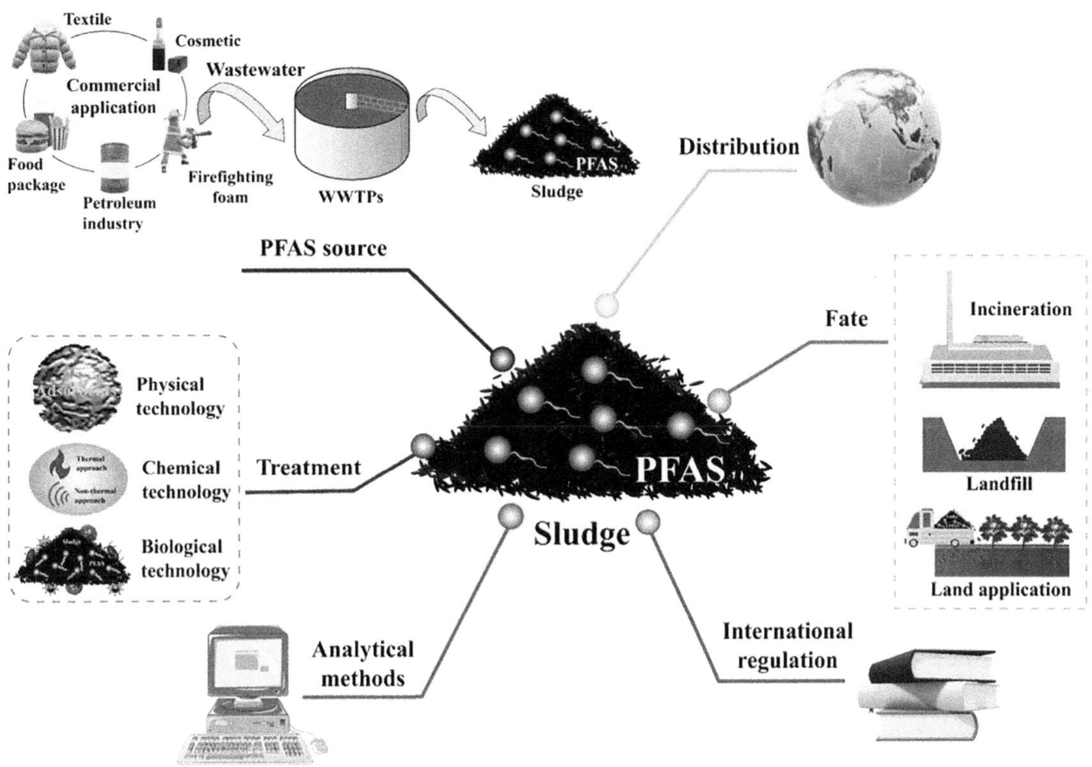

Fig. 2.4 Detection process for sludge-based PFAS (Zhou et al., 2024). Reprinted with permission from Zhou, T., Li, X., Liu, H., Dong, S., Zhang, Z., Wang, Z., Li, J., Nghiem, L.D., Khan, S.J., Wang, Q., 2024. Occurrence, fate, and remediation for per-and polyfluoroalkyl substances (PFAS) in sewage sludge: a comprehensive review. J. Hazard. Mater. 466, 133637. https://doi.org/10.1016/j.jhazmat.2024.133637, Copyright (2024), Elsevier.

unquestionable progress in food safety, project flexibility and other activities, their excessive use also leads to a significant presence in the environment, generating consequences that are not yet well defined.

Of the various forms of plastic debris in the environment, microplastics (MPs) represent a particularly concerning phenomenon. There is no universally accepted definition for classifying microplastics. However, the generally accepted understanding is that microplastics are plastic fragments smaller than 5 mm (Bui et al., 2020), with those smaller than 1 mm being referred to as small microplastics (Krishnan et al., 2023). Furthermore, there have been discussions about adopting the nomenclature of nanoplastics for fragments smaller than 1000 nm (Cai et al., 2021).

With regard to their provenance, microplastics can be classified as either primary or secondary. Primary microplastics are already produced in small sizes for a variety of domestic and industrial applications, including microspheres in personal care products, synthetic clothing fibers and abrasive products. Subsequently, these MPs are transferred to the environment. Secondary MPs, in contrast, are produced indirectly through processes occurring in the environment. These include fragmentation, mechanical abrasion or UV irradiation of large plastic items, as well as the decay of primary MPs (Bui et al., 2020; Carnevale Miino et al., 2024; Krishnan et al., 2023). In terms of shape, microplastics can be classified as fragments, pellets, fibers, films and foams (Bui et al., 2020).

In response to mounting concerns regarding the environmental and human health impacts of microplastics in water, several countries have enacted legislation banning the use of plastic microbeads in cosmetic products. In 2014, the Netherlands became the first country to prohibit the use of microplastics in personal care products. Subsequently, in December 2015, the United States incorporated a ban on the manufacture, packaging, and distribution of rinse-off cosmetics containing plastic microbeads into its legislative framework. Similarly, other countries, including Australia, Canada, Italy, South Korea, New Zealand, Sweden, and the United Kingdom, have adopted a comparable approach (OECD, 2020).

The most common routes by which microplastics reach freshwaters are surface runoff and treated or untreated sanitary sewage. Other potential pathways include combined sewer overflows, industrial effluents, degradation of plastic waste, and atmospheric deposition (Fig. 2.5). Nevertheless, insufficient data are available to enable the contribution of each source to be quantified.

The fragmentation of plastic debris from urban areas represents a significant source of microplastic formation. There is a paucity of data on the fragmentation and degradation rates of plastics in marine and freshwater environments. However, the considerable quantity of plastic observed in marine environments indicates that fragmentation represents a substantial source of microplastics, given that it is assumed that all this plastic will undergo degradation and fragmentation at some point.

In the environment, ultraviolet radiation and elevated temperatures can induce chemical alterations in plastics, rendering them brittle and thus more prone to fragmentation and degradation.

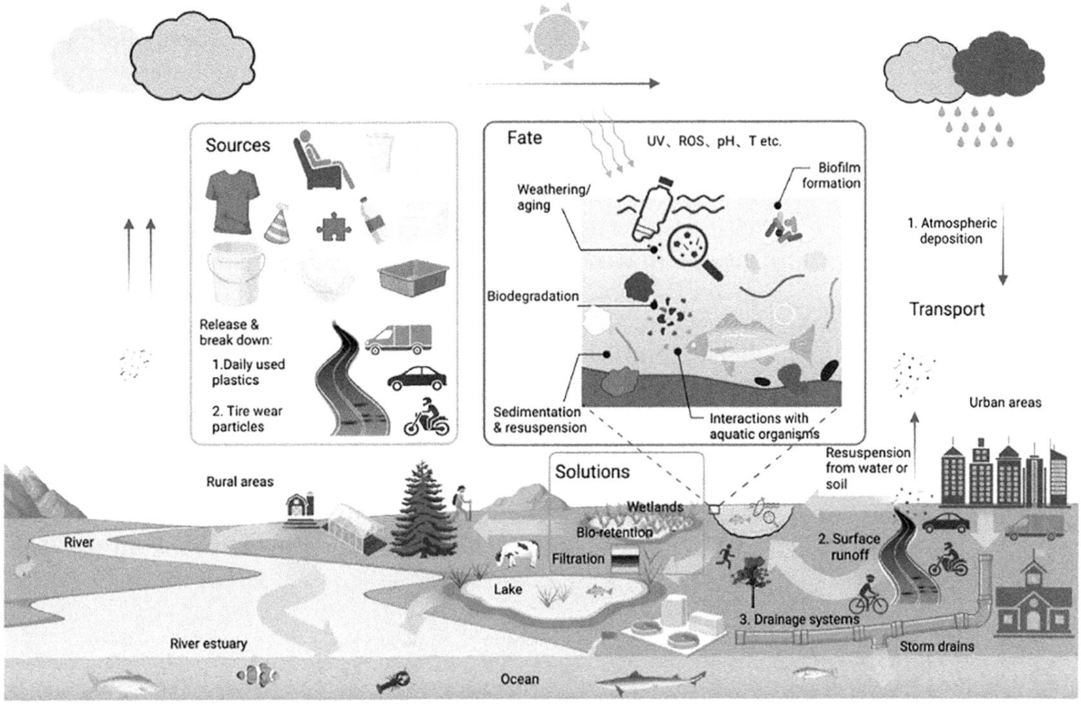

Fig. 2.5 Fate and transport of microplastics in the environment (Wei et al., 2023). Reprinted with permission from Wei, L., Yue, Q., Chen, G., Wang, J., 2023. Microplastics in rainwater/stormwater environments: influencing factors, sources, transport, fate, and removal techniques. TrAC Trends Anal. Chem. 165, 117147. https://doi.org/10.1016/j.trac.2023.117147, Copyright (2023), Elsevier.

2.5 Viruses

In the context of the COVID-19 pandemic, research into the presence of viruses in municipal sewage has assumed greater significance, particularly in relation to those associated with SARS-CoV-2 (Ahmed et al., 2020; Shaheen et al., 2022). Although the likelihood of waterborne transmission of the disease has been demonstrated to be low (Medema et al., 2020; Rimoldi et al., 2020), the monitoring of the virus in wastewater has been established as an important epidemiological tool (Zamhuri et al., 2022).

In a low-income country with limited capacity for SARS-CoV-2 testing, wastewater surveillance can be a valuable tool for monitoring the virus, as demonstrated by Barnes et al. (2023). The authors collected river water and wastewater samples over a period of two years, from May 2020 to May 2022. Their findings indicated that the peak detection of SARS-CoV-2 in these samples occurred prior to the peak of clinical cases. Furthermore, the research team was able to identify new variants of the virus before they were detected in patients. In light of these findings, the authors were able to conclude that wastewater surveillance can be employed to detect emerging waves, identify variants of concern and provide an alert system.

The findings of Barnes et al. (2023) can be extrapolated to other emerging viruses, emphasizing the necessity of monitoring viruses in wastewater. Furthermore, the presence of pathogenic viruses in wastewater represents a significant cause for concern, given the potential risks to human health and the environment (Corpuz et al., 2020). A number of human pathogenic viruses have been identified in wastewater samples from locations around the world (Fig. 2.6). These include enteric viruses, Caliciviridae (Noroviruses (NoVs)), Picornaviridae (Enteroviruses (EVs)), Hepatitis A virus (HAV) and Adenoviridae (Adenoviruses (AdVs)).

One of the primary sources of pathogenic viruses in wastewater is human fecal matter from infected individuals (Hellmér et al., 2014; Symonds et al., 2009). However, waterborne viruses from

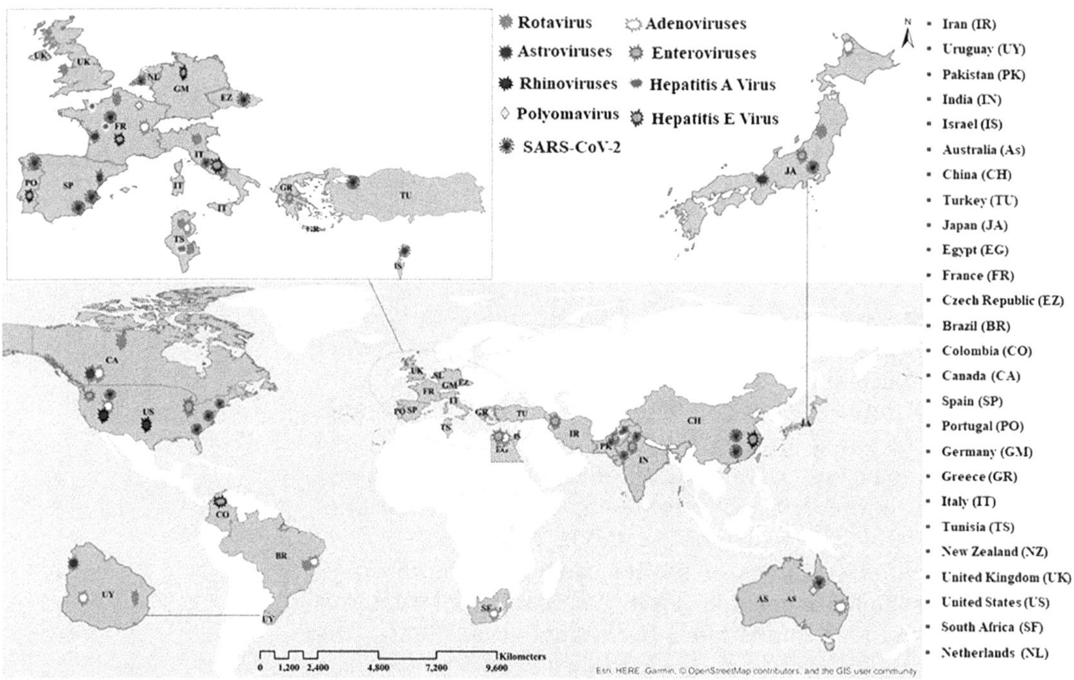

Fig. 2.6 Diversity of human viruses identified in wastewater systems in various locations around the world (Ali et al., 2021). Reprinted with permission from Ali, W., Zhang, H., Wang, Z., Chang, C., Javed, A., Ali, K., Du, W., Niazi, N. K., Mao, K., Yang, Z., 2021. Occurrence of various viruses and recent evidence of SARS-CoV-2 in wastewater systems. J. Hazard. Mater. 414, 125439. https://doi.org/10.1016/j.jhazmat.2021.125439, ISSN: 0304-3894, Copyright (2021), Elsevier.

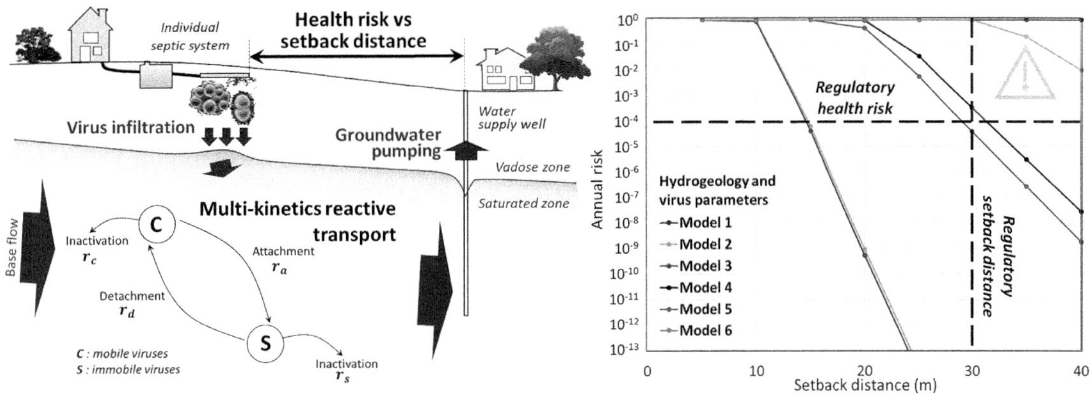

Fig. 2.7 Fate of viruses in aquifers (Rafini et al., 2023). Reprinted with permission from Rafini, S., Chesnaux, R., Lompe, K.M., Barbeau, B., Claveau-Mallet, D., Richard, D., 2023. Modeling the fate of viruses in aquifers: multi-kinetics reactive transport, risk assessment, and governing parameters. Sci. Total Environ. 903, 166276. https://doi.org/10.1016/j.scitotenv.2023.166276, Copyright (2023), Elsevier.

animals or other sources can also be identified (Corsi et al., 2014). The issue of viral contamination in wastewater is a significant concern, particularly with regard to the potential for viruses to be introduced into water bodies that are used for recreation and as a source of raw water for water treatment plants. Additionally, there is a risk of contamination of water resources used for irrigation and aquifer recharge (Fig. 2.7). Furthermore, there is a possibility of viral contamination of sludge used in agricultural practices and exposure of workers at wastewater treatment plants (WWTPs) (Corpuz et al., 2020).

The process of detecting and quantifying viruses in wastewater is a complex and challenging undertaking. A variety of methods may be employed, including epifluorescence microscopy, transmission electron microscopy, pulsed field gel electrophoresis, immunofluorescence assay, flow cytometry, traditional cell culture, and molecular methods. Furthermore, some methods yield quantitative data, whereas others provide qualitative data. The responses of each method may vary depending on the matrix under consideration, which could include raw sewage, treated sewage, or sludge, for instance. Molecular methods for quantifying viruses are based on the determination of selected segments of the virus's genetic material, which enables the identification of viruses even in the event of inactivation. Methods based on cell culture or immunological techniques are employed to ascertain the viability of viruses (Corpuz et al., 2020). Consequently, a significant challenge in the control and monitoring of viruses in wastewater systems is the identification of these pathogens.

2.6 Removal in WWTP

In addition to the challenges of detecting emerging contaminants, the sanitation sector is also concerned about the removal of ECs due to the difficulties of establishing accurate and low-cost analytical methods for diverse substances and organisms. It is of paramount importance to guarantee that treated effluents do not constitute a source of contamination to the environment and a risk to human health. This objective can only be achieved through the effective operation of wastewater treatment systems and the development of innovative techniques that ensure the production of an effluent of superior quality and safety. Consequently, the removal efficiency of ECs in wastewater treatment plants, across their various operational units, is the subject of numerous studies.

Wastewater treatment can be divided into a number of stages, including preliminary treatment, primary treatment, secondary treatment, and tertiary treatment. Additionally, advanced techniques may be employed, and the sludge generated in the processes can be treated. Secondary treatment is fundamentally a biological process, whereas preliminary and primary treatments are primarily physical and chemical. The specific characteristics of tertiary treatment and advanced treatments may vary depending on the pollutants that require removal. It can thus be observed that a diverse range of processes and operations can be employed for the removal of ECs, spanning from conventional to advanced techniques (see Figs. 2.8 and 2.9). Fig. 2.10 provides an overview of

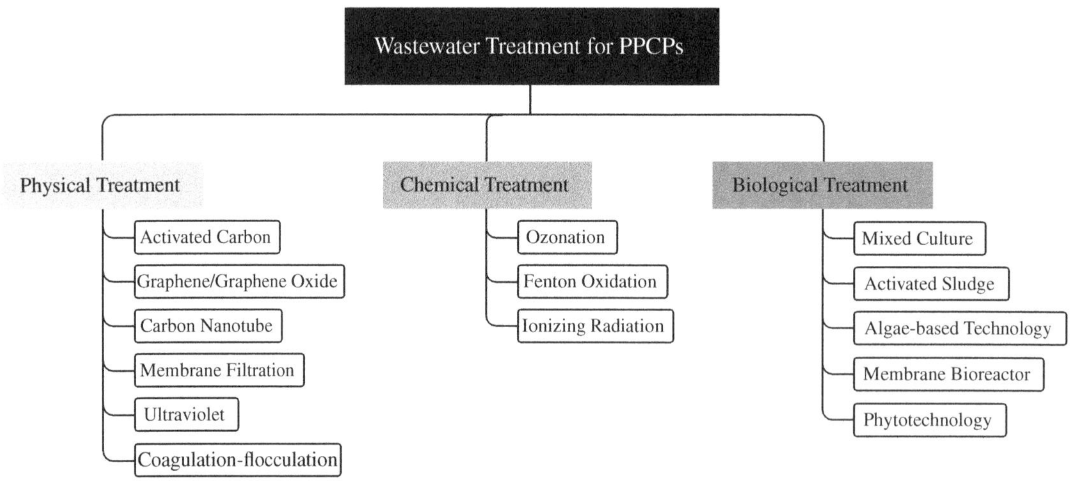

Fig. 2.8 Physical, chemical and biological treatment processes for PPCPs in wastewater (Falahi et al., 2022). Reprinted with permission from Falahi O.A.A.L., Abdullah, S.R.S., Hasan, H.A., Othman, A.R., Ewadh, H.M., Kurniawan, S.B., Imron, M.F., 2022. Occurrence of pharmaceuticals and personal care products in domestic wastewater, available treatment technologies, and potential treatment using constructed wetland: a review. Process Saf. Environ. Prot. 168, 1067–1088. https://doi.org/10.1016/j.psep.2022.10.082, Copyright (2022), Elsevier.

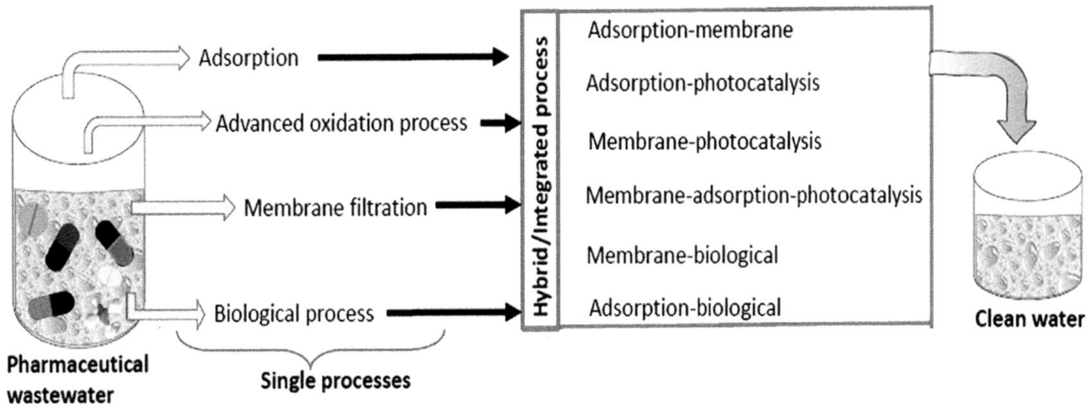

Fig. 2.9 Conventional and advanced hybrid technologies for pharmaceutical wastewater treatment (Eniola et al., 2022). Reprinted with permission from Eniola, J.O., Kumar, R., Barakat, M.A., Rashid, J., 2022. A review on conventional and advanced hybrid technologies for pharmaceutical wastewater treatment. J. Clean. Prod. 356, 131826. https://doi.org/10.1016/j.jclepro.2022.131826, Copyright (2022), Elsevier.

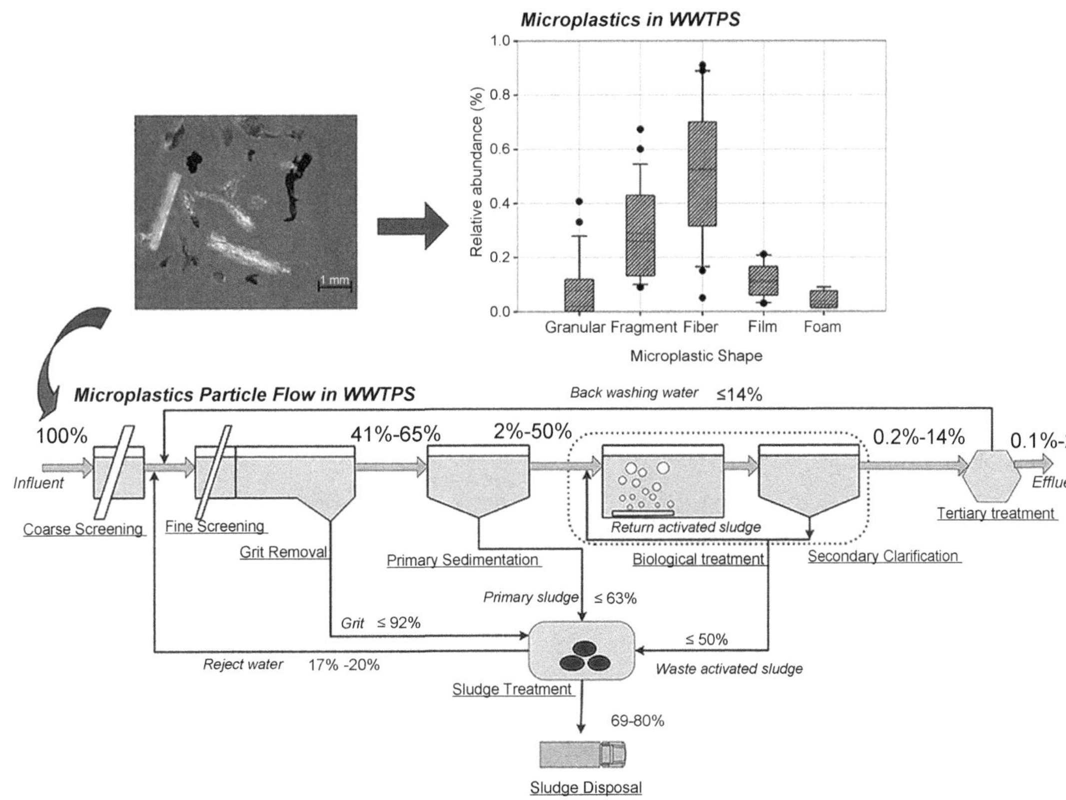

Fig. 2.10 Microplastics particle flow in wastewater treatment plants (Sun et al., 2019). Reprinted with permission from Sun, J., Dai, X., Wang, Q., van Loosdrecht, M.C.M., Ni, B.J., 2019. Microplastics in wastewater treatment plants: detection, occurrence and removal. Water Res. 152, 21–37. https://doi.org/10.1016/j.watres.2018.12.050, Copyright (2019), Elsevier.

a wastewater treatment plant (WWTP), with the estimated removal of microplastics at each stage.

The most common method of assessing the removal efficacy of ECs in sewage treatment stages is through a mass balance, which involves determining the concentration of ECs in the effluent and in the influent. This approach can be applied to a range of contaminants, including PFAS (Zhang et al., 2024), microplastics (Talukdar et al., 2024) and other substances.

The efficacy of conventional sewage treatment for the elimination of ECs is contingent upon the specific characteristics of each individual contaminant. While physical-chemical processes are highly effective for the removal of microplastics (Fig. 2.10), the overall treatment process exhibits low efficiency for PPCPs and negligible efficiency for PFAS (Fig. 2.11). Furthermore, the technology employed at the wastewater treatment plant (WWTP) has a significant impact on the removal of viruses. In a study

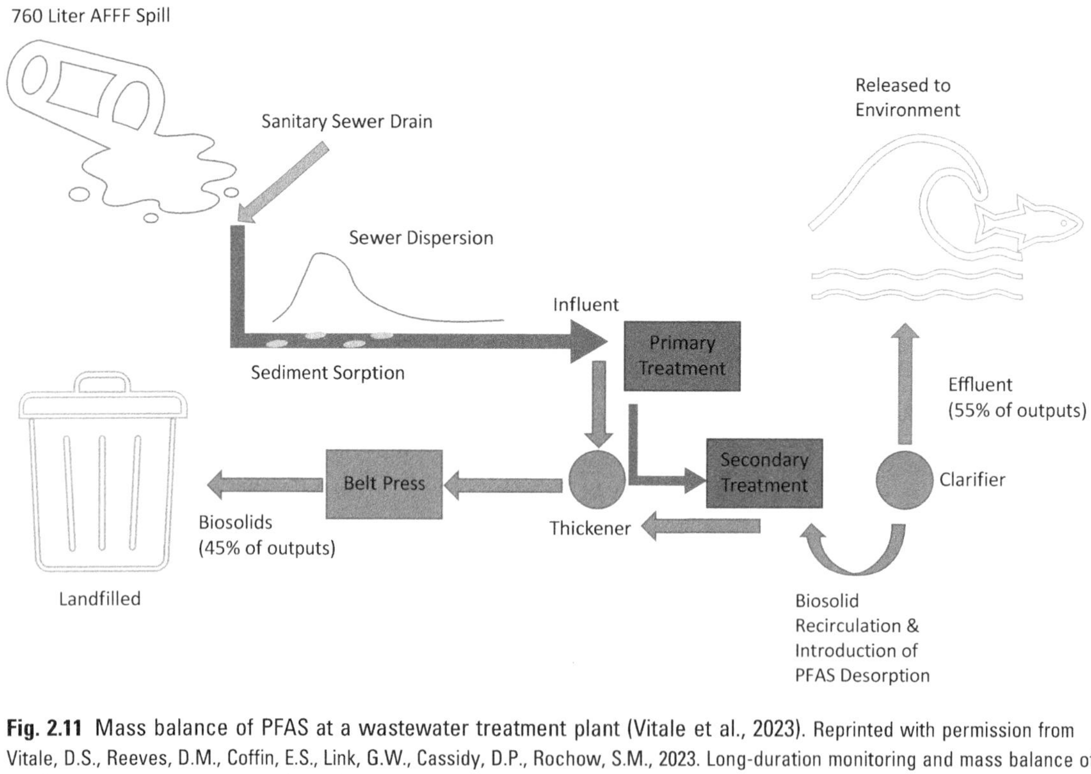

Fig. 2.11 Mass balance of PFAS at a wastewater treatment plant (Vitale et al., 2023). Reprinted with permission from Vitale, D.S., Reeves, D.M., Coffin, E.S., Link, G.W., Cassidy, D.P., Rochow, S.M., 2023. Long-duration monitoring and mass balance of PFAS at a wastewater treatment plant following the release of aqueous film-forming foam concentrate. Water Res. 242, 120268. https://doi.org/10.1016/j.watres.2023.120268, Copyright (2023), Elsevier.

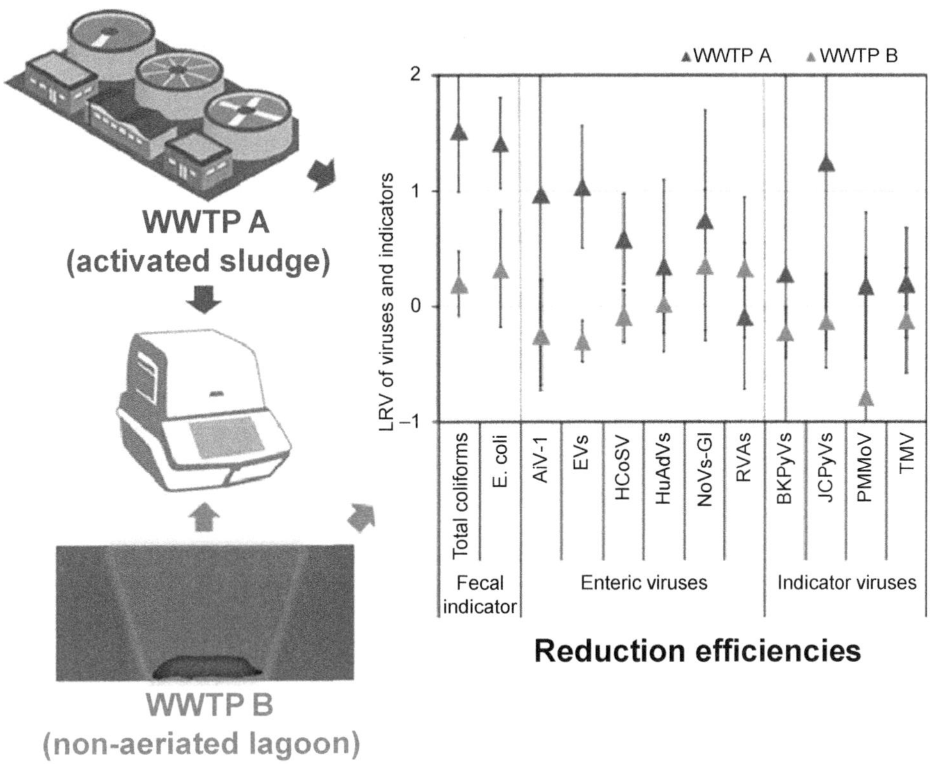

Fig. 2.12 Log_{10} reduction values (LRV) of viruses and indicators (Tandukar et al., 2021). Reprinted from Tandukar, S., Ghaju Shrestha, R., Malla, B., Sthapit, N., Sherchand, J.B., Sherchan, S.P., Haramoto, E., 2021. Virus reduction at wastewater treatment plants in Nepal. Environ. Challenges, 5, 100281. https://doi.org/10.1016/j.envc.2021.100281, Under a Creative Commons License (2021).

conducted in Nepal, Tandukar et al. (2021) observed that an activated sludge wastewater treatment plant (WWTP) demonstrated superior virus and indicator removal rates compared to a non-aerated lagoon WWTP (Fig. 2.12).

2.6.1 Preliminary and primary treatment

In this section, the preliminary and primary treatment stages will be discussed together, as they are closely interrelated. The preliminary and primary treatment of municipal wastewater is designed to separate floating materials, heavy inorganic settleable solids, oils and grease (Talukdar et al., 2024), in order to prevent overloading of subsequent treatment stages. In order to achieve this, a variety of operational techniques, including sorting, grading, primary sedimentation and flotation, can be employed in different configurations, according to the specific requirements of each WWTP (Talukdar et al., 2024).

Primary treatment has been demonstrated to be effective in the removal of microplastics, with a range of 50–98% removal efficiency (Sun et al., 2019). In general, microplastics (MPs) can be removed in the fine screen stages (mesh size 2.5–10 mm), in the skimming of oils and greases, and in the surface skimming of primary clarifiers. This is particularly the case when the MPs are light. They can also be indirectly removed through MPs trapped in sand particles and in the flocs formed in the sedimentation stage (Sun et al., 2019; Talukdar et al., 2024). The direct removal rates in grit chambers are low due to the low densities of the MPs (Talukdar et al., 2024). Nevertheless, the elevated MP removal efficiencies observed at this stage are more pronounced for fibrous MPs than for microspheres, fragments, and leaf-shaped MPs. This underscores the significance of subsequent treatment stages for comprehensive MP removal (Bui et al., 2020).

In general, the preliminary/primary stages of wastewater treatment have been observed to exhibit relatively poor removal efficiencies for the other ECs. This is attributable to the chemical or biological characteristics of these substances and microorganisms. For instance, the removal capacity of PPCPs in these stages is diminished due to the hydrophilic nature of the majority of PPCPs. However, there are exceptions, such as the removal of specific types of fragrances and other compounds that can be adsorbed in primary treatment processes (Yang et al., 2017).

2.6.2 Secondary treatment

The biological treatment stage is typically the principal component of a municipal effluent treatment plant, as it is during this phase that the biological decomposition of organic matter occurs. The biological treatment process may be based on either aerobic or anaerobic conditions, depending on the configuration of the individual wastewater treatment plant (WWTP). Furthermore, a considerable reduction in biochemical oxygen demand (BOD) is anticipated, which will consequently lead to a reduction in the polluting potential of the wastewater. Nevertheless, it should be noted that not all of the ECs are eliminated at this stage.

The activated sludge process is the most prevalent of the various aerobic treatment technologies (Wang and Wang, 2016). However, other processes, including membrane bioreactors and sequential batch reactors, have also demonstrated notable efficiency (Kumar et al., 2022). Anaerobic processes include anaerobic sludge reactors and anaerobic film reactors (Kumar et al., 2022).

The removal of PPCPs in biological processes occurs through partitioning, adsorption, biotransformation and biodegradation

(Yang et al., 2017). However, the efficiency of these processes is highly variable and dependent on a number of factors, including the physicochemical properties of the compounds, the environmental conditions, the configuration of the biological reactor and the operating parameters (Wang and Wang, 2016).

The utilization of activated sludge has the potential to achieve removal efficiencies of up to 85% for a range of PPCPs (Kumar et al., 2022). However, compounds resistant to biodegradation, such as carbamazepine, are minimally affected by biological treatment (Wang and Wang, 2016). It is also pertinent to consider the potential for variation in the removal efficiency of the same compounds, even when the treatment operating conditions are maintained. This phenomenon can be attributed to a number of different factors, underscoring the necessity for long-term monitoring programs to enhance the capacity to anticipate PPCP loads in effluent from WWTPs (Wang and Wang, 2016).

In contrast, PFAS have a complex interaction with conventional biological treatment plants. The operating conditions of a WWTP, including hydraulic and biological retention time, feed concentration, temperature, flow rate and suspended solids, can influence the fate of PFAS. Furthermore, extended retention periods and elevated temperatures have the potential to result in the formation of perfluoroalkyl acids. Conversely, elevated mixed liquor suspended solids (5,000 mg/L) and extended sludge retention periods (45 days) may confer the capacity to markedly enhance PFAS removal in a WWTP utilizing the activated sludge process (Lenka et al., 2021).

The type of treatment employed at the WWTP is a further crucial consideration. In a study conducted on six municipal WWTPs situated in the upper reaches of the Guanting Reservoir along the Yanghe River in China, Chen et al. (2018) observed significant discrepancies in the annual average PFAS removal efficiencies among WWTPs that employed the Cyclic Activated Sludge System (removal efficiency of 32.2%), in comparison to those utilizing the Orbal Oxidation Ditch (removal efficiency of 17.5%) and Anaeroxic-Anoxic-Oxic (removal efficiency of −1.49%).

In addition to external factors and operating parameters, the type of substance is another relevant factor affecting the efficiency of PFAS removal in biological processes. To illustrate, perfluoroalkyl acids (PFAAS) represent the most prevalent class of PFAS, are typically present in anionic forms, are non-volatile, and are resistant to biodegradation. It should be noted that neutral PFAS, such as perfluoroalkyl sulfonamidoethanols, are volatile or semi-volatile and may even be released into the atmosphere as a result of the processes occurring within the WWTP. Accordingly, further study is required to ascertain the efficiency of the various processes employed in municipal wastewater treatment plants for the removal of PFAS (Lenka et al., 2021).

Conversely, the removal efficiency of microplastics in activated sludge systems has been observed to range between 42% and 98% (Bui et al., 2020). In contrast to PPCPs and PFAS, the removal of microplastics is primarily physical. Microplastics are removed by adhering to the suspended solids of the process, which are subsequently sedimented or can even serve as biomedia, enabling biofilm growth on their surface (Bui et al., 2020). This latter mechanism is worthy of further investigation, as biofilm growth on the surface of MPs has the potential to modify their density and surface properties, thereby rendering them more susceptible to subsequent sedimentation (Sun et al., 2019).

The leaching of bisphenol A (BPA) from polyvinyl chloride (PVC) microplastics (MPs) has been demonstrated to cause toxicity in the environment, as well as limiting the activity of heterotrophic and nitrifying bacteria in activated sludge systems (Bui et al., 2020; Li et al., 2015). This, in conjunction with the necessity for a more profound understanding of the decomposition capabilities of MPs within biological treatment, underscores the imperative for further research in this domain.

Additionally, viruses can be eliminated through aerobic biological processes in a manner analogous to that of PMs. A diverse range of virus genera can be adsorbed onto the biological flocs formed in activated sludge processes, with a Log Removal Virus range of 0.6 to 4.24. Conversely, virus removal in anaerobic processes is achieved through thermal inactivation at temperatures exceeding 45°C, with Log Virus Removal values ranging from 0.45 to 6.0. This data is significant because numerous WWTPs lack tertiary treatments, particularly disinfection (Plaza-Garrido et al., 2022).

In general, it can be considered that secondary treatment can help to remove ECs, but as its main objective is not this, it is not possible to guarantee its complete efficiency. This demonstrates the need to implement specific tertiary and/or advanced treatment stages in order to achieve greater control of ECs in the treated effluent.

2.6.3 Tertiary treatment

Tertiary sewage treatment is employed in wastewater treatment plants with the objective of enhancing the quality of the treated effluent. This is achieved by controlling specific characteristics or pollutants that have not yet reached the desired levels of quality. Such processes may include the removal of additional nutrients, such as phosphorus and nitrogen, the disinfection of the effluent, or the physical and chemical removal of particles. It is also possible to consider tertiary treatment as a form of post-treatment. It should be noted that the implementation of tertiary treatment is not a universal requirement for all WWTPs, as the specific design of each plant may dictate otherwise.

The following tertiary treatment processes have the potential to remove ECs: sand filtration, advanced oxidative processes (AOPs), membrane separation, activated carbon adsorption, ion exchange, chemical oxidation, dissolved air flotation, ultrafiltration, reverse osmosis, and disinfection processes (Kumar et al., 2022; Sun et al., 2019; Yang et al., 2017). It is pertinent to highlight that among the aforementioned processes, only those involving photochemical and oxidative mechanisms have the capacity to degrade compounds. The remaining processes are solely capable of compound removal and phase transfer (Kümmerer, 2008b).

The rate of removal of MPs by sand filtration is not significant (Bui et al., 2020) nor is the removal of PPCPs (Yang et al., 2017). Conversely, rapid sand filters comprising a minimum of three layers have been demonstrated to effectively remove low-concentration range MPs (Bui et al., 2020).

Advanced oxidation processes (AOPs) represent a pivotal technique for the mineralization of recalcitrant organic contaminants (Talukdar et al., 2024). In this context, AOPs will be discussed in conjunction with other chemical oxidation processes.

Chemical oxidation has the capacity to remove a range of PPCPs and PFAS through the activation of processes that result in the generation of highly reactive radical species. Some processes have already been successfully employed for the degradation of PPCPs and PFAS, including: Fenton oxidation, whether or not aided by UV radiation and H_2O_2; persulfate oxidation via direct electron transfer or activation into reactive radicals; and ozone oxidation (Fang et al., 2024; Wang and Wang, 2016). Furthermore, photocatalytic degradation schemes have demonstrated efficacy in the removal of PFAS (Venkatesh-Reddy et al., 2024).

Ozone is the most prevalent oxidation method employed for the degradation of PPCPs. The process is dependent on the non-selective oxidizing activity of hydroxyl radicals, which are responsible for the elimination of the target components. The results of several studies demonstrate that ozone is an effective method for removing the majority of PPCPs, with a removal efficiency exceeding 90% (Wang and Wang, 2016).

Additionally, ozonation can be employed to decompose the polymers that constitute MPs into oxygen-containing functional groups, with decomposition values exceeding 90% in exposure times between 30 and 90 min. However, other studies have indicated that the polymer chains were merely broken, rather than undergoing complete decomposition of the compounds. It is therefore imperative to undertake a rigorous assessment of the viability of ozone as a treatment option, taking into account its high operational costs and the potential by-products generated during the process (Bui et al., 2020).

Additionally, some AOPs have been demonstrated to be effective for the degradation of MPs. In light of the above, several techniques have been proposed for this purpose, including photocatalysis in an aqueous matrix, carbocatalytic oxidation and hydrothermal hydrolysis (Talukdar et al., 2024).

Adsorption by activated carbon is a process that is commonly employed in the treatment of water, but it can also be deployed as a tertiary treatment for the removal of ECs in municipal wastewater. In general, adsorption is a physical process that is frequently employed for the removal of trace organic pollutants in water. It is also one of the principal processes utilized for the elimination of PPCPs in the environment (Wang and Wang, 2016).

Activated carbon can be utilized in two distinct forms: powdered activated carbon (PAC) and granular activated carbon (GAC). The adsorption capacity of PPCPs by activated carbon is contingent upon the hydrophobicity and charge of the PPCPs (Wang and Wang, 2016). Four mechanisms have been identified for the removal of PFAS: (i) dispersion in the liquid phase, (ii) mass transfer through the solid phase, (iii) pore diffusion within an adsorbent and (iv) electrostatic interface between the pollutant and the exchange site (Barisci and Suri, 2021). Conversely, given the characteristics of MPs, it is unlikely that they will be removed by adsorption by activated carbon (Talukdar et al., 2024).

Two operational issues are associated with the utilization of activated carbon: (i) a reduction in adsorption capacity, (ii) the deterioration of activated carbon in complex sewage systems (Wang and Wang, 2016). Nevertheless, it can be employed in post-treatment systems and for the recovery of effluents for reuse (Barisci and Suri, 2021). In light of these considerations, the efficacy of utilizing activated carbon for the sole purpose of removing ECs in municipal wastewater treatment plants remains to be fully evaluated.

Ion exchange can be employed as a post-treatment for effluents of satisfactory quality that require the removal of PFAS, for instance. In this instance, ion exchange resins (in particular, anion exchange resins) can be employed due to their molecular structure and their simultaneous anion exchange and adsorption removal mechanisms (Barisci and Suri, 2021).

While ultrafiltration, nanofiltration and reverse osmosis can be employed for the removal of microplastics (MPs), they have significant drawbacks, including high costs and removal efficiency flux (Talukdar et al., 2024). Reverse osmosis and ultrafiltration represent promising alternatives for the removal of PPCPs from wastewater. However, ultrafiltration and microfiltration have been observed to exhibit low removal rates, which can be attributed to their pore size (Yang et al., 2017).

The operation of filter discs (FD) has been demonstrated to present a significant challenge in the removal of PPCPs. This is due to difficulties encountered during backwashing and the formation of biofilm on their surface. Nevertheless, the use of 20 μm pore size filters has been shown to yield favorable outcomes (Bui et al., 2020).

The removal of viruses from the treated effluent can be achieved through the application of chemical disinfection. The use of sodium hypochlorite has been demonstrated to be an effective method for inactivating viruses, provided that it is employed under the optimal operating conditions (Plaza-Garrido et al., 2022). Furthermore, the application of ozone and ultraviolet radiation can also exert a disinfectant effect.

It is also worth noting that a number of other advanced techniques are currently being studied around the world, although their application at the municipal wastewater treatment plant scale remains limited. Barisci and Suri (2021) present a number of technologies that are at different Technology Readiness Levels (TRLs), which could potentially be applied in the future to remove PFAS. These include electro-oxidation, sonochemical degradation (ultrasound), plasma-based technologies, heterogeneous photocatalysis using nanomaterials, and carbon nanotubes (new generation adsorbents).

Talukdar et al. (2024) present a number of emerging technologies that are specific to MPs, including: algal mass colonization; bioinspired molecules; metal-organic framework (MOF)-based foams; and photocatalytic micromotors.

2.7 Conclusions and future trends

The detection and removal of emerging contaminants from municipal wastewater represents a significant contemporary challenge. While the concentrations of these substances in municipal wastewater are typically very low, concerns about their potential impacts on human health and the environment have led to significant scrutiny of their presence in treated effluents. In the event that municipal wastewater treatment is not efficacious in the removal of emerging contaminants, it can become a significant conduit for their dissemination in the environment.

The initial significant challenge is the monitoring of these substances. The analytical methods employed for the determination of chemical substances are typically complex and expensive. Identifying microplastics presents a significant challenge, particularly given the labor-intensive nature of the required methods. The identification of viruses typically employs molecular methods. Furthermore, the aforementioned methods are specific and therefore unsuitable for the general identification of ECs. The necessity for more straightforward, more rapid, or less costly methodologies is paramount to more effectively regulating the prevalence of ECs in municipal wastewater and formulating strategies to enhance their removal and ascertain the sources of contamination. It is therefore imperative that future research in this area be conducted.

The second significant challenge is the effective removal of these substances from municipal wastewater. Once more, the considerable diversity of substances classified as ECs renders this an exceptionally intricate undertaking. While microplastics can be removed in primary treatment, the same cannot be said for PPCPs and PFAS, which are so diverse and complex that they can not only fail to be removed, but also undergo transformation into substances with greater toxicity or environmental risk during the course of sewage treatment. The removal of viruses can be achieved through disinfection processes; however, failures in these processes can pose a significant risk to public and environmental health.

The biological (secondary) treatment that forms the basis of sewage treatment plants is typically ineffective at removing and degrading ECs, with a few exceptions for biodegradable substances. It is therefore imperative that tertiary and advanced treatments are implemented in order to guarantee the removal and degradation of ECs in general. However, the considerable variability of their characteristics, as previously stated, renders this challenge particularly complex. Notwithstanding the sophistication of the techniques employed, the removal of different classes of ECs remains an elusive goal. The use of advanced oxidation processes (AOP) is a promising avenue of research, as it has the potential to be used for a broader range of treatments. However, further studies are needed to gain a deeper understanding of the by-products of oxidation and the synergistic effects of interactions between different substances.

Despite the inherent difficulties, it is crucial to propose and validate hybrid technologies for the removal and decomposition of ECs, as well as new technologies for the identification and joint monitoring of ECs. Additionally, it is essential to develop lower-cost alternatives that facilitate the expansion of their use in WWTPs in diverse locations.

References

Ahmed, W., Angel, N., Edson, J., Bibby, K., Bivins, A., O'Brien, J.W., Choi, P.M., Kitajima, M., Simpson, S.L., Li, J., Tscharke, B., Verhagen, R., Smith, W.J.M., Zaugg, J., Dierens, L., Hugenholtz, P., Thomas, K.V., Mueller, J.F., 2020. First confirmed detection of SARS-CoV-2 in untreated wastewater in Australia: a proof of concept for the wastewater surveillance of COVID-19 in the community. Sci. Total Environ. 1879-1026. 728, 138764. https://doi.org/10.1016/j.scitotenv.2020.138764. Elsevier B.V., Australia. Available from: www.elsevier.com/locate/scitotenv.

Ali, W., Zhang, H., Wang, Z., Chang, C., Javed, A., Ali, K., Du, W., Niazi, N.K., Mao, K., Yang, Z., 2021. Occurrence of various viruses and recent evidence of SARS-CoV-2 in wastewater systems. J. Hazard. Mater. 0304-3894. 414, 125439. https://doi.org/10.1016/j.jhazmat.2021.125439.

Barisci, S., Suri, R., 2021. Occurrence and removal of poly/perfluoroalkyl substances (PFAS) in municipal and industrial wastewater treatment plants. Water Sci. Technol. 84 (12), 3442–3468. https://doi.org/10.2166/WST.2021.484.

Barnes, K.G., Levy, J.I., Gauld, J., Rigby, J., Kanjerwa, O., Uzzell, C.B., Chilupsya, C., Anscombe, C., Tomkins-Tinch, C., Mbeti, O., Cairns, E., Thole, H., McSweeney, S., Chibwana, M.G., Ashton, P.M., Jere, K.C., Meschke, J.S., Diggle, P., Cornick, J., et al., 2023. Utilizing river and wastewater as a SARS-CoV-2 surveillance tool in settings with limited formal sewage systems. Nat. Commun. 2041-1723. 14 (1), 1–11. https://doi.org/10.1038/s41467-023-43047-y. Nature Research, Malawi. Available from: https://www.nature.com/ncomms/.

Bavumiragira, J.P., Ge, J., Yin, H., 2022. Fate and transport of pharmaceuticals in water systems: a processes review. Sci. Total Environ. 0048-9697. 823, 153635. https://doi.org/10.1016/j.scitotenv.2022.153635.

Biel-Maeso, M., Corada-Fernández, C., Lara-Martín, P.A., 2019. Removal of personal care products (PCPs) in wastewater and sludge treatment and their occurrence in receiving soils. Water Res. 1879-2448. 150, 129–139. https://doi.org/10.1016/j.watres.2018.11.045. Elsevier Ltd, Spain. Available from: www.elsevier.com/locate/watres.

Boxall, A.B.A., Rudd, M.A., Brooks, B.W., Caldwell, D.J., Choi, K., Hickmann, S., Innes, E., Ostapyk, K., Staveley, J.P., Verslycke, T., Ankley, G.T., Beazley, K.F., Belanger, S.E., Berninger, J.P., Carriquiriborde, P., Coors, A., DeLeo, P.C., Dyer, S.D., Ericson, J.F., et al., 2012. Pharmaceuticals and personal care products in the environment: what are the big questions? Environ. Health Perspect. 120 (9), 1221–1229. https://doi.org/10.1289/EHP.1104477).

Bui, X.T., Vo, T.D.H., Nguyen, P.T., Nguyen, V.T., Dao, T.S., Nguyen, P.D., 2020. Microplastics pollution in wastewater: characteristics, occurrence and removal technologies. Environ. Technol. Innov. 19, 101013. https://doi.org/10.1016/J.ETI.2020.101013.

Cai, H., Xu, E.G., Du, F., Li, R., Liu, J., Shi, H., 2021. Analysis of environmental nanoplastics: progress and challenges. Chem. Eng. J. 410, 128208. https://doi.org/10.1016/J.CEJ.2020.128208.

Cangola, J., Abagale, F.K., Cobbina, S.J., 2024. A systematic review of pharmaceutical and personal care products as emerging contaminants in waters: the panorama of West Africa. Sci. Total Environ. 911, 168633. https://doi.org/10.1016/J.SCITOTENV.2023.168633.

Carnevale Miino, M., Galafassi, S., Zullo, R., Torretta, V., Rada, E.C., 2024. Microplastics removal in wastewater treatment plants: a review of the different approaches to limit their release in the environment. Sci. Total Environ. 930, 172675. https://doi.org/10.1016/J.SCITOTENV.2024.172675.

Chen, S., Zhou, Y., Meng, J., Wang, T., 2018. Seasonal and annual variations in removal efficiency of perfluoroalkyl substances by different wastewater treatment processes. Environ. Pollut. 242, 2059–2067. https://doi.org/10.1016/J.ENVPOL.2018.06.078.

Chen, W., Yang, F., Hu, E., Yang, C., Sun, C., Li, M., 2022. Occurrence, fate and risk assessment of per- and polyfluoroalkyl substances in wastewater treatment plants in Shaanxi, China. Environ. Pollut. 314, 120226. https://doi.org/10.1016/J.ENVPOL.2022.120226.

Corpuz, M.V.A., Buonerba, A., Vigliotta, G., Zarra, T., Ballesteros, F., Campiglia, P., Belgiorno, V., Korshin, G., Naddeo, V., 2020. Viruses in wastewater: occurrence, abundance and detection methods. Sci. Total Environ. 745, 140910. https://doi.org/10.1016/J.SCITOTENV.2020.140910.

Corsi, S.R., Borchardt, M.A., Spencer, S.K., Hughes, P.E., Baldwin, A.K., 2014. Human and bovine viruses in the Milwaukee River watershed: hydrologically relevant representation and relations with environmental variables. Sci. Total Environ. 490, 849–860. https://doi.org/10.1016/J.SCITOTENV.2014.05.072.

Crane, M., Boxall, A.B.A., Barrett, K., 2016. Veterinary Medicines in the Environment. pp. 1–199, https://doi.org/10.1201/9781420084771/VETERINARY-MEDICINES-ENVIRONMENT-KATIE-BARRETT-ALISTAIR-BOXALL-MARK-CRANE.

Eniola, J.O., Kumar, R., Barakat, M.A., Rashid, J., 2022. A review on conventional and advanced hybrid technologies for pharmaceutical wastewater treatment. J. Clean. Prod. 356, 131826. https://doi.org/10.1016/j.jclepro.2022.131826.

Falahi, O.A.A.L., Abdullah, S.R.S., Hasan, H.A., Othman, A.R., Ewadh, H.M., Kurniawan, S.B., Imron, M.F., 2022. Occurrence of pharmaceuticals and personal care products in domestic wastewater, available treatment technologies, and potential treatment using constructed wetland: a review. Process Saf. Environ. Prot. 168, 1067–1088. https://doi.org/10.1016/j.psep.2022.10.082.

Fang, J., Li, S., Gu, T., Liu, A., Qiu, R., Zhang, W., xian., 2024. Treatment of per- and polyfluoroalkyl substances (PFAS): a review of transformation technologies and mechanisms. J. Environ. Chem. Eng. 12 (1), 111833. https://doi.org/10.1016/J.JECE.2023.111833.

Hellmér, M., Paxéus, N., Magnius, L., Enache, L., Arnholm, B., Johansson, A., Bergström, T., Norder, H., 2014. Detection of pathogenic viruses in sewage provided early warnings of hepatitis A virus and norovirus outbreaks. Appl. Environ. Microbiol. 80 (21), 6771–6781. https://doi.org/10.1128/AEM.01981-14/ASSET/4846F34D-CE94-4C48-8885-90C1E4043682/ASSETS/GRAPHIC/ZAM9991057710004.JPEG.

Krishnan, R.Y., Manikandan, S., Subbaiya, R., Biruntha, M., Govarthanan, M., Karmegam, N., 2021. Removal of emerging micropollutants originating from pharmaceuticals and personal care products (PPCPs) in water and wastewater by advanced oxidation processes: a review. Environ. Technol. Innov. 23, 101757. https://doi.org/10.1016/J.ETI.2021.101757.

Krishnan, R.Y., Manikandan, S., Subbaiya, R., Karmegam, N., Kim, W., Govarthanan, M., 2023. Recent approaches and advanced wastewater treatment technologies for mitigating emerging microplastics contamination—a critical review. Sci. Total Environ. 858, 159681. https://doi.org/10.1016/J.SCITOTENV.2022.159681.

Kumar, R., Qureshi, M., Vishwakarma, D.K., Al-Ansari, N., Kuriqi, A., Elbeltagi, A., Saraswat, A., 2022. A review on emerging water contaminants and the application of sustainable removal technologies. Case Stud. Chem. Environ. Eng. 6, 100219. https://doi.org/10.1016/J.CSCEE.2022.100219.

Kümmerer, K., 2008a. Pharmaceuticals in the environment—a brief summary. In: Pharmaceuticals in the Environment, pp. 3–21, https://doi.org/10.1007/978-3-540-74664-5_1.

Kümmerer, K., 2008b. Strategies for reducing the input of pharmaceuticals into the environment. In: Pharmaceuticals in the Environment, pp. 411–418, https://doi.org/10.1007/978-3-540-74664-5_25.

Lenka, S.P., Kah, M., Padhye, L.P., 2021. A review of the occurrence, transformation, and removal of poly- and perfluoroalkyl substances (PFAS) in wastewater treatment plants. Water Res. 199, 117187. https://doi.org/10.1016/J.WATRES.2021.117187.

Li, K., Wei, D., Zhang, G., Shi, L., Wang, Y., Wang, B., Wang, X., Du, B., Wei, Q., 2015. Toxicity of bisphenol A to aerobic granular sludge in sequencing batch reactors. J. Mol. Liq. 209, 284–288. https://doi.org/10.1016/J.MOLLIQ.2015.05.046.

Medema, G., Heijnen, L., Elsinga, G., Italiaander, R., Brouwer, A., 2020. Presence of SARS-coronavirus-2 RNA in sewage and correlation with reported COVID-19 prevalence in the early stage of the epidemic in the Netherlands. Environ. Sci. Technol. Lett. 7 (7), 511–516. https://doi.org/10.1021/ACS.ESTLETT.0C00357/ASSET/IMAGES/LARGE/EZ0C00357_0002.JPEG.

OECD, 2020. Microbeads in Cosmetics. https://www.oecd.org/en/publications/ocean-policies_17318a16-en/microbeads-in-cosmetics_609ea0bf-en.html.

Plaza-Garrido, A., Limaico, M., Villamar-Ayala, C.A., 2022. Influence of wastewater treatment technologies on virus removal under a bibliometric-statistical analysis. J. Water Process Eng. 47, 102642. https://doi.org/10.1016/J.JWPE.2022.102642.

Puri, M., Gandhi, K., Kumar, M.S., 2023. Emerging environmental contaminants: a global perspective on policies and regulations. J. Environ. Manag. 332, 117344. https://doi.org/10.1016/J.JENVMAN.2023.117344.

Rafiei, V., Nejadhashemi, A.P., 2023. Watershed scale PFAS fate and transport model for source identification and management implications. Water Res. 240, 120073. https://doi.org/10.1016/J.WATRES.2023.120073.

Rafini, S., Chesnaux, R., Lompe, K.M., Barbeau, B., Claveau-Mallet, D., Richard, D., 2023. Modeling the fate of viruses in aquifers: multi-kinetics reactive transport, risk assessment, and governing parameters. Sci. Total Environ. 903, 166276. https://doi.org/10.1016/j.scitotenv.2023.166276.

Rehman, M.U., Nisar, B., Mohd Yatoo, A., Sehar, N., Tomar, R., Tariq, L., Ali, S., Ali, A., Mudasir Rashid, S., Bilal Ahmad, S., Aldossari, R.M., 2024. After effects of Pharmaceuticals and Personal Care Products (PPCPs) on the biosphere and their counteractive ways. Sep. Purif. Technol. 342, 126921. https://doi.org/10.1016/j.seppur.2024.126921.

Rimoldi, S.G., Stefani, F., Gigantiello, A., Polesello, S., Comandatore, F., Mileto, D., Maresca, M., Longobardi, C., Mancon, A., Romeri, F., Pagani, C., Cappelli, F., Roscioli, C., Moja, L., Gismondo, M.R., Salerno, F., 2020. Presence and infectivity of SARS-CoV-2 virus in wastewaters and rivers. Sci. Total Environ. 744, 140911. https://doi.org/10.1016/J.SCITOTENV.2020.140911.

Sauvé, S., Desrosiers, M., 2014. A review of what is an emerging contaminant. Chem. Cent. J. 8 (1), 1–7. https://doi.org/10.1186/1752-153X-8-15/FIGURES/2.

Sengar, A., Vijayanandan, A., 2022. Human health and ecological risk assessment of 98 pharmaceuticals and personal care products (PPCPs) detected in Indian surface and wastewaters. Sci. Total Environ. 807, 150677. https://doi.org/10.1016/J.SCITOTENV.2021.150677.

Shaheen, M.N.F., Elmahdy, E.M., Shahein, Y.E., 2022. The first detection of SARS-CoV-2 RNA in urban wastewater in Giza, Egypt. J. Water Health 20 (8), 1212–1222. https://doi.org/10.2166/WH.2022.098.

Sun, J., Dai, X., Wang, Q., van Loosdrecht, M.C.M., Ni, B.J., 2019. Microplastics in wastewater treatment plants: detection, occurrence and removal. Water Res. 152, 21–37. https://doi.org/10.1016/j.watres.2018.12.050.

Symonds, E.M., Griffin, D.W., Breitbart, M., 2009. Eukaryotic viruses in wastewater samples from the United States. Appl. Environ. Microbiol. 75 (5), 1402–1409. https://doi.org/10.1128/AEM.01899-08/ASSET/8FFE9656-E93A-4B49-951B-8CD62BCC5131/ASSETS/GRAPHIC/ZAM0050996870001.JPEG.

Talukdar, A., Kundu, P., Bhattacharya, S., Dutta, N., 2024. Microplastic contamination in wastewater: sources, distribution, detection and remediation through physical and chemical-biological methods. Sci. Total Environ. 916, 170254. https://doi.org/10.1016/J.SCITOTENV.2024.170254.

Tandukar, S., Ghaju Shrestha, R., Malla, B., Sthapit, N., Sherchand, J.B., Sherchan, S.-P., Haramoto, E., 2021. Virus reduction at wastewater treatment plants in Nepal. Environ. Challenges 5, 100281. https://doi.org/10.1016/j.envc.2021.100281.

Tsui, M.M.P., Leung, H.W., Lam, P.K.S., Murphy, M.B., 2014. Seasonal occurrence, removal efficiencies and preliminary risk assessment of multiple classes of organic UV filters in wastewater treatment plants. Water Res. 53, 58–67. https://doi.org/10.1016/J.WATRES.2014.01.014.

US EPA, 2024. Treatment and Control of Contaminants Research. Retrieved September 4, 2024, from https://www.epa.gov/water-research/treatment-and-control-contaminants-research.

Venkatesh-Reddy, C., Kumar, R., Chakrabortty, P., Karmakar, B., Pottipati, S., Kundu, A., Jeon, B.H., 2024. A critical science mapping approach on removal mechanism and pathways of per- and poly-fluoroalkyl substances (PFAS) in water and wastewater: a comprehensive review. Chem. Eng. J. 492, 152272. https://doi.org/10.1016/J.CEJ.2024.152272.

Vitale, D.S., Reeves, D.M., Coffin, E.S., Link, G.W., Cassidy, D.P., Rochow, S.M., 2023. Long-duration monitoring and mass balance of PFAS at a wastewater treatment plant following the release of aqueous film-forming foam concentrate. Water Res. 242, 120268. https://doi.org/10.1016/j.watres.2023.120268.

Wang, J., Wang, S., 2016. Removal of pharmaceuticals and personal care products (PPCPs) from wastewater: a review. J. Environ. Manag. 182, 620–640. https://doi.org/10.1016/J.JENVMAN.2016.07.049.

Wei, L., Yue, Q., Chen, G., Wang, J., 2023. Microplastics in rainwater/stormwater environments: influencing factors, sources, transport, fate, and removal techniques. TrAC Trends Anal. Chem. 165, 117147. https://doi.org/10.1016/j.trac.2023.117147.

Yang, Y., Ok, Y.S., Kim, K.H., Kwon, E.E., Tsang, Y.F., 2017. Occurrences and removal of pharmaceuticals and personal care products (PPCPs) in drinking water and water/sewage treatment plants: a review. Sci. Total Environ. 596–597, 303–320. https://doi.org/10.1016/J.SCITOTENV.2017.04.102.

Zamhuri, S.A., Soon, C.F., Nordin, A.N., Ab Rahim, R., Sultana, N., Khan, M.A., Lim, G.P., Tee, K.S., 2022. A review on the contamination of SARS-CoV-2 in water bodies: transmission route, virus recovery and recent biosensor detection techniques. Sens. Bio-Sens. Res. 36, 100482. https://doi.org/10.1016/J.SBSR.2022.100482.

Zhang, Y., Zhou, Y., Dong, R., Song, N., Hong, M., Li, J., Yu, J., Kong, D., 2024. Emerging and legacy per- and polyfluoroalkyl substances (PFAS) in fluorochemical wastewater along full-scale treatment processes: source, fate, and ecological risk. J. Hazard. Mater. 465, 133270. https://doi.org/10.1016/J.JHAZMAT.2023.133270.

Zhou, T., Li, X., Liu, H., Dong, S., Zhang, Z., Wang, Z., Li, J., Nghiem, L.D., Khan, S.J., Wang, Q., 2024. Occurrence, fate, and remediation for per-and polyfluoroalkyl substances (PFAS) in sewage sludge: a comprehensive review. J. Hazard. Mater. 466, 133637. https://doi.org/10.1016/j.jhazmat.2024.133637.

3

Sequential batch bioreactors

Ashique Ahmed[a,b], Pooja Talukdar[a,b], and Sachin Rameshrao Geed[a,b]

[a]CSIR-North East Institute of Science and Technology, Jorhat, India. [b]Academy of Scientific and Innovative Research (AcSIR), Ghaziabad, India

Abbreviations

AGS	aerobic granular sludge
AMBR	algal-based membrane bioreactors
AMR	antimicrobial resistance
AnMBR	anaerobic membrane bioreactors
ANN	artificial neural network
AOA-SBR	anaerobic-aerobic-anoxic sequencing batch reactor
AOB	ammonia oxidizing bacteria
ARWW	ammonium rich wastewater
ASP	activated sludge process
BFT	bio-floc technology
BNR	biological nutrient removal
BOD	biochemical oxygen demand
BPC	biopolymer cluster content
CDM	clean development mechanism
COD	chemical oxygen demand
DEHP	diethylhexyl phthalate
DMSO	dimethyl sulfoxide
DO	dissolved oxygen
DWTP	decentralized wastewater treatment plants
EBPR	enhanced biological phosphorus removal
EDC	endocrine disrupting chemicals
FLX	fluoxetine
FNA	free nitrous acid
GACSBR	granular-activated carbon-SBR
HRT	hydraulic retention time
MaB	microalgae and bacteria
mAb	monoclonal antibody
MB	methylene blue
MBBR	moving bed biofilm reactor
MBR	membrane bioreactor
MCDA	multicriteria decision analysis
MEA	monoethanolamine
MUD	ethanol-utilizing denitrifiers
ORP	oxidation-reduction potential
PAO	phosphate accumulating organisms
PBCSBR	porous biomass carrier sequential batch bioreactor

Municipal Wastewater Treatment. https://doi.org/10.1016/B978-0-443-24826-9.00003-6

PCP	pentachlorophenol
PSW	plastic solid waste
RAS	return activated sludge
RO	reverse osmosis
RPMW	recycled paper mill wastewater
SBBGR	sequencing batch biofilter granular reactor
SBR	sequential batch bioreactors
SNAD	simultaneous partial nitrification, and denitrification system
SRT	solid retention time
SW	swine wastewater
TF	trickling filters
TFT-LCD	thin-film transistor liquid crystal display
TMAH	tetra-methyl ammonium hydroxide
TSS	total suspended solids
UASB	up-flow anaerobic sludge blanket

Symbols

Jth,c	threshold flux for compressibility
pH	potential of hydrogen
Q	daily influent flow rate
S_0	initial substrate concentration
t	total cycle time
V	volume of the reactor
V_f	feed volume per cycle
V_{full}	entire reactor volume

3.1 Introduction

Due to population expansion and industrialization, it is projected that the world's water demand would increase by 55% by the year 2050 (Colella et al., 2021). The pressing issues of water scarcity, resource depletion, and global warming have spurred significant global attention toward technologies aimed at producing clean water and clean energy (Chung et al., 2012). Among these technologies, the treatment of municipal wastewater has emerged as a promising avenue for addressing multiple challenges simultaneously. In addition to cleaning water, municipal wastewater treatment facilities have the capacity to generate renewable energy on the basis of net production. These treatment facilities are able to produce recyclable and reusable byproducts in addition to producing useful energy carriers by utilizing the energy content found in organic contaminants in raw municipal wastewater (Shizas and Bagley, 2004). One of the most prevalent types of wastewater is municipal wastewater, which is distinguished by a high concentration of organic matter (Van Lier, 2008). As environmental concerns escalate, regulatory bodies worldwide are imposing stricter regulations on pollution discharge, particularly with a focus on waste reduction (Chan et al., 2009).

Municipal wastewater treatment can utilize different reactor systems like aerobic, anaerobic, or a mix of both. These systems effectively treat wastewater while also allowing for energy generation and resource recovery. As industries aim to meet regulations and adopt sustainable practices, municipal wastewater treatment emerges as a promising solution, offering environmental benefits and economic advantages through energy production and adherence to international standards. Aerobic systems are particularly suitable for treating wastewater with low levels of biodegradable pollutants, while anaerobic systems are more suitable for high-strength wastewaters (Chan et al., 2009). The merits of anaerobic treatment become more pronounced when dealing with influents surpassing these thresholds, as anaerobic systems typically require less energy and offer potential for bio-energy and nutrient recovery (Cakir and Stenstrom, 2005). Aerobic systems are proficient in removing soluble biodegradable organic matter and generating well-flocculated biomass, resulting in reduced effluent suspended solids concentration. However, anaerobic systems typically yield lower effluent quality compared to aerobic systems (Wang et al., 2005). Thus, the choice between aerobic and anaerobic treatment technologies depends on the desired end product (such as effluent quality or biogas production) and the characteristics of the incoming wastewater (influent, i.e., municipal wastewater).

Many biological wastewater treatment techniques, such as the built wetlands (CWs), moving bed biofilm reactor (MBBR), trickling filters (TFs), and activated sludge process (ASP), have been used over the years (Rashid et al., 2021; Saidulu et al., 2022). A recent study by Saidulu et al. (2022) examined how 3D printing technology can be applied in wastewater treatment. They specifically looked at the benefits of 3D printed bio-carriers and were found to have demonstrated successful biofilm growth and increased degrading efficiency (Saidulu et al., 2022). Despite the promising performance of these techniques in addressing various targeted contaminants, their widespread adoption is hindered by significant energy and space requirements. Notably, ASP-based processes are particularly energy-intensive where 47–70% of the energy used is accounted for by aeration alone, resulting in elevated operational expenses and challenges in managing the disposal of large volumes of sludge (Sánchez et al., 2021). Hence, there is a pressing need to optimize wastewater treatment techniques by minimizing energy and space requirements while simultaneously augmenting their efficiency.

Recent advancements in technology, operational practices, and regulatory measures have led to a shift from traditional suspended-growth activated sludge systems to more efficient alternatives like sequencing batch reactors (SBRs). This transition is prominent in areas with low or fluctuating flow conditions. SBRs offer a sophisticated approach to biological wastewater treatment, combining metabolic reactions and solid-liquid separation within a single tank in a carefully defined sequence over time (Aziz et al., 2011; Jaramillo et al., 2018; Morgenroth and Wilderer, 1998). This innovative design allows for precise control over exposure conditions for microorganisms, akin to a fed-batch system, where parameters such as exposure time, concentration amplitude and frequency can be finely tuned independently of inflow patterns (Singh and Srivastava, 2011; Wilderer et al., 2001). Compared to conventional activated sludge processes, SBRs offer substantial cost savings, slashing expenses by over 60% in operational costs while delivering exceptional effluent quality within a remarkably short aeration period. Unlike traditional systems, which typically require 3 to 8 h of aeration, SBRs streamline the process, achieving comparable or superior results in a fraction of the time (Ng et al., 1993). Initially developed for small communities and addressing high-strength industrial waste, SBR technology has witnessed widespread adoption across various waste sources, including dilute ones. Notably, SBRs with suspended biomass configurations demonstrate superior performance in carbon removal in contrast to conventional suspended growth systems (Mohan et al., 2005).

In recent years, SBR technology has evolved with numerous modifications to tackle the increasing variety of pollutants in wastewater. This chapter aims to provide insights into SBR technology, exploring recent advancements in its design, merits, and its application for municipal wastewater treatment. The objective is to foster a deeper comprehension, which is crucial for developing optimized municipal wastewater treatment systems tailored to individual plant scenarios and future design considerations. For wastewater management techniques to be sustainable and for wastewater treatment methods to be effective, a thorough understanding is essential. Fig. 3.1 shows the schematic diagram of an SBR setup.

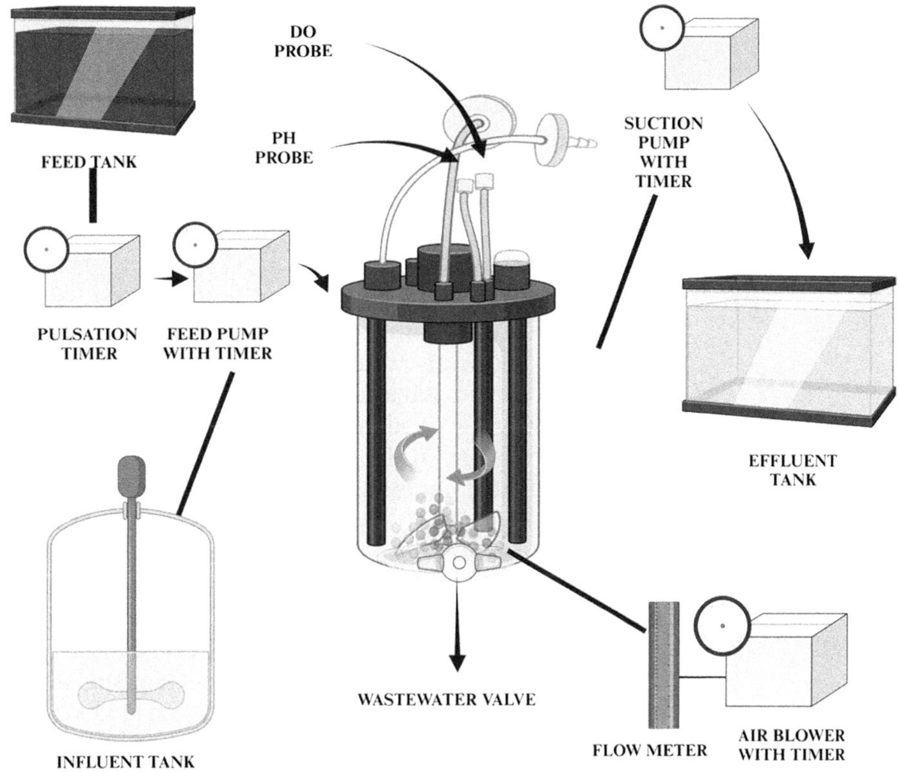

Fig. 3.1 Schematic diagram of an SBR setup. Created with BioRender.com.

3.2 Membrane bioreactor technology

Membrane bioreactor (MBR) technology stands at the forefront of wastewater treatment. It harmoniously integrates biological processes with membrane filtration to generate top-tier effluent. In MBR systems, microorganisms target organic pollutants, while ultrafiltration or microfiltration membranes trap suspended solids, microorganisms, and pathogens. Acknowledged for its effectiveness in industrial water treatment and reuse, MBR technology amalgamates the activated sludge process with micro- and ultrafiltration (Hoinkis et al., 2012). Its ability to deliver top-quality product water while occupying minimal space makes it an appealing option. Submerged MBR systems, notable for their resilience and versatility, are gaining prominence (Hoinkis et al., 2012). Recent achievements in MBR technology mark significant progress in wastewater treatment, with notable advancements across critical aspects. Efforts have been focused on enhancing treatment performance, addressing membrane fouling, and overcoming technical limitations. Innovative strategies, including membrane surface modifications and hybrid MBR systems, have emerged to tackle challenges and pave the way for more robust installations. Moreover, a focus on optimizing energy balance and conducting techno-economic assessments underscores a commitment to sustainability and cost-effectiveness.

Recent studies have showcased the potential of optimizing membrane bioreactor (MBR) technology to address challenges associated with membrane fouling. Shen et al. (2023) introduced innovative solutions to tackle membrane fouling issues in MBRs through structural optimization and in situ aeration techniques. Their utilization of computational fluid dynamics modeling enhances shear stress distribution and antifouling potential by optimizing membrane module configuration and aeration methods, marking significant strides in advancing MBR technology for more efficient and sustainable wastewater treatment. In a similar way, Jankowska et al. (2024) emphasized the crucial role of backwash in combating biofouling in submerged ceramic membrane bioreactors, focusing on *Saccharomyces cerevisiae* separation. Their comprehensive optimization efforts result in impressive flux recovery rates, reinforcing the efficacy of backwash procedures. Furthermore, their insights into membrane fouling mechanisms aid in refining biofouling control strategies. Meanwhile, Jiao et al. (2024) explored the combined effects of biochar addition and filtration mode optimization on mitigating membrane fouling in high-solid anaerobic membrane bioreactors treating food waste. Their findings suggest that optimizing the filtration-relaxation ratio (F/R) significantly improves membrane fouling control, with biochar integration altering sludge properties and enhancing fluid shear force on the membrane surface, facilitating pollutant separation during relaxation stages. Lastly, Bounaga et al. (2024) investigated elemental sulfur biorecovery from phosphogypsum using an O_2-based membrane biofilm reactor. Their study, optimizing reactor parameters and analyzing metagenomic data, identifies *Thauera* as the dominant genus in both synthetic and biogenic sulfide media biofilms. Results demonstrate high sulfide-oxidation efficiency and conversion rates, highlighting the potential of O_2-based membrane biofilm reactors for sulfur biorecovery from phosphogypsum leachate and similar sulfidogenic effluents.

Recent studies have emphasized the potential for significant energy savings through the utilization of membrane bioreactor (MBR) technology. Research on direct-flow membrane bioreactor optimization underscores the critical importance of assessing the threshold flux for compressibility (*Jth,c*) to achieve efficient filtration. Their findings highlight the role of cake compressibility in fouling dynamics, advocating for shorter filtration times (5–7 min) to enhance process productivity (Díaz et al., 2023). Additionally, their identification of suspension properties, such as biopolymer cluster content (BPC), offers valuable insights for long-term optimization strategies, emphasizing the potential of *Jth,cas* a guiding metric for optimizing filtration performance. In a separate study, Asrami et al. (2024) conducted a comprehensive energy balance survey to evaluate the design and auto-thermal thermophilic aerobic digestion of algal-based membrane bioreactors (AMBRs) for landfill leachate treatment across various hydraulic retention times (HRTs). Their aim was to maximize AMBR's energy efficiency, microalgal biomass production, and removal efficiency using artificial neural network (ANN) models. Experimental and simulation results underscored the dependence of biomass production on HRT, with decreased HRT leading to increased *Chlorella vulgaris* biomass productivity. Moreover, efficient removal of chemical oxygen demand (COD) from algal biomass was demonstrated. This study investigated energy-efficient landfill leachate treatment using algal-based membrane bioreactors, optimizing organic loading rates and demonstrating potential auto-thermality at high COD removal levels.

Numerous studies have extensively explored the techno-economic analysis of MBR technology, shedding light on its cost-effectiveness and applicability across diverse sectors. These investigations yield valuable insights into optimizing MBR systems for efficient wastewater treatment and industrial processes. For instance, Jones and Gerogiorgis (2023) conducted a meticulous examination focused on enhancing bioreactor-filtration systems for monoclonal antibody (mAb) production, emphasizing the critical phase of primary post-cultivation biomass recovery. Their findings, supported by a thorough technoeconomic analysis, underscore the advantages of specific configurations, such as the integration of fed-batch reactors and stacked membrane microfilters, in achieving streamlined mAb production. Meanwhile, Zhang et al. (2023) advocated for the adoption of anaerobic membrane bioreactors (AnMBRs) as the primary treatment technology in decentralized wastewater treatment plants (DWTPs), emphasizing their comparable treatment efficiencies and technoeconomic costs to other biological technologies, albeit acknowledging the associated challenges that require mitigation for widespread adoption. Additionally, Ghangrekar et al. (2023) highlighted the important role of moving bed biofilm reactors (MBBRs) in modern wastewater treatment methodologies, showcasing their reliability and operational simplicity. Moreover, their study provides significant insights into the technological advancements and

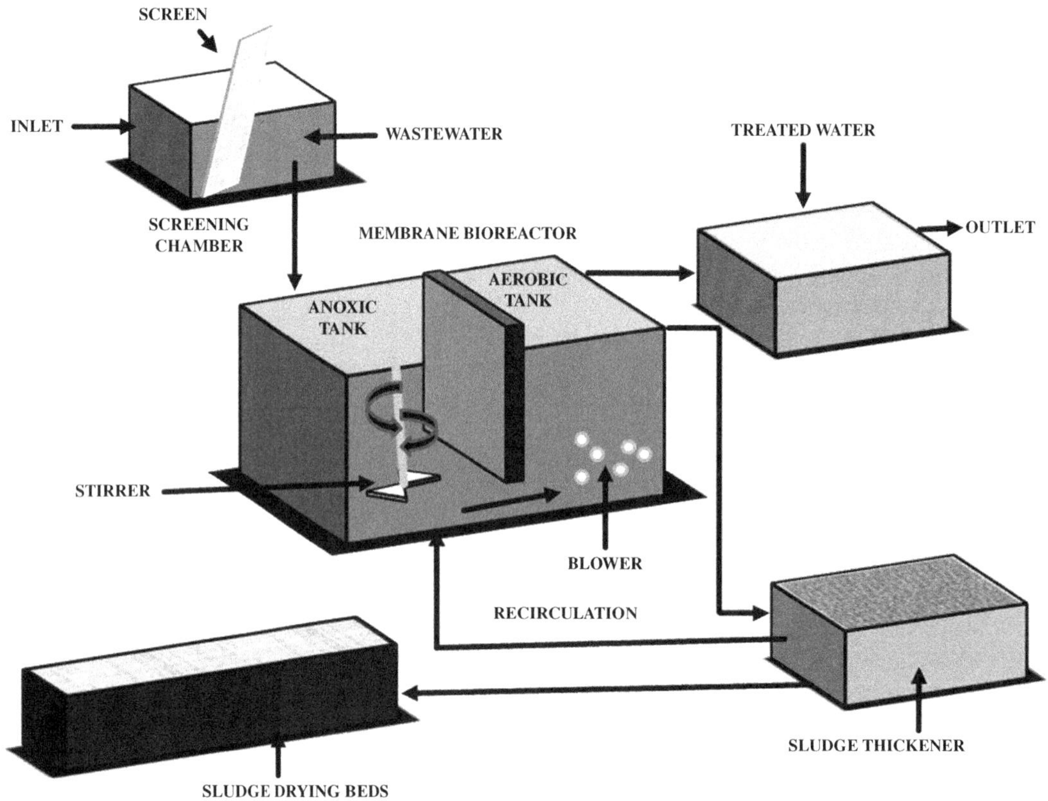

Fig. 3.2 Overview of an MBR system used for treatment of wastewater.

adaptations within MBBR systems, coupled with a detailed techno-economic analysis, emphasizing the financial implications of implementing MBBR technology. Fig. 3.2 illustrates the overview of an MBR system used for treatment of wastewater.

3.3 Sequential batch bioreactors (SBRs): An advanced technology

Sequencing batch reactor (SBR) systems lack stability and are designed for treating large volume of wastewater utilizing suspended-growth biological methods. An activated sludge (AS) system that is combined with a clarifier is incorporated in these systems. They function in a form of intermittent aeration, enabling many metabolic events to take place in a single tank.

The process involves a timed control sequence aimed at effectively separating solid and liquid phases, ensuring efficient wastewater treatment (Xiang et al., 2016). SBRs function as biological wastewater treatment units under which metabolic processes and the separation of solids and liquids occur within a single tank, following a distinct and repetitive time sequence (Singh and Srivastava, 2011). SBR processes are known for their ability to reduce operating costs by over 60% compared to conventional activated sludge processes, while also achieving high-quality effluent within a significantly shorter aeration period (Ng et al., 1993). The SBR system's design takes into account the input load, the amount of biomass and its settling properties, reactor size, aeration setup, and the distribution of treatment phase time. They are fitted with diffusers, valves for inlet and outlet, apparatus for oxygenation, and devices mechanically engineered for removal of sludge to ensure accuracy (Mahvi, 2008). The conventional SBR is an essential part of the nitrification-denitrification process, where ammonia is oxidized to nitrite, subsequently converting it to nitrate, leading to the formation of nitrogen gas (Duan et al., 2020). A SBR system's performance may be affected by a multitude of factors, including influent properties, organic loading rate, availability of carbon sources, pH levels, oxidation-reduction potential (ORP), solid retention time (SRT), dissolved oxygen (DO), hydraulic retention time (HRT), feeding patterns, cycle duration, settling characteristics, anoxic and oxic conditions, and temperature (de Sousa Rollemberg et al., 2019). SBR technology is used in a single tank to remove suspended solids (SS), nutrients, and organic and inorganic contaminants (Ong et al., 2020). The SBR system necessitates a compact footprint, operates within a single basin and functions automatically with advanced process control capabilities (Abubakar et al., 2016). Characterized by its simplicity and flexibility in both configuration and operation, it provides cost-effective solutions with low installation and operational costs. Furthermore, it demonstrates robustness, high biomass retention, decreased energy consumption, and resilience to diverse loading shocks and toxicity. It allows for the simultaneous removal of organics, nitrogen, and phosphorus (Jagaba et al., 2021; Mojiri et al., 2018). Treatment efficiency in SBR systems is significantly influenced by environmental and operational parameters, such as aeration, agitation, superficial gas velocity, shock loads, HRT, SRT, cycle duration. Advancements are required to improve the overall efficiency of traditional SBRs. The advantages of both the connected and suspended growth phases are combined in the engineering of the hybrid SBR systems. Biofilm-based biological systems offer significant benefits in treating high-strength wastewater due to their ability to achieve higher biomass concentrations per reactor unit. Comparable to aerobic SBR in terms of pace and anaerobic nature, an ASBR lacks aeration during the reaction phase. Instead, as substrates are broken down anaerobically, methane and carbon dioxide are produced. The feeding, reaction, settling, and discharge processes are the successive steps in the cyclic process that runs the ASBR. The primary benefits of this type of operation are its high biogas yield, ease of usage, effective effluent quality control and flexibility. Granular Activated Carbon-SBR is a product that uses a physical adsorption mechanism to remove organic pollutants that are volatile, semi-volatile, and non-volatile from wastewater (Singh and Srivastava, 2011). Owing to its effectiveness in municipal wastewater treatment, the Sequencing Batch Biofilter Granular Reactor (SBBGR) has been the subject of substantial research (Di Iaconi et al., 2008). It also addresses the challenges associated with conventional biological methods, such as excessive sludge generation, large space demands, and inadequate characteristics for sludge settling (De Sanctis et al., 2013).

Industrial SBR variants that are frequently used are the Intermittent Cycle Extended Aeration System, which is used to remove BOD, nitrogen (N), and phosphorus (P); Aqua SBR, which is also used in removal of BOD, N, and P; and C-Tech (Cyclic Activated Sludge Treatment), which is used to remove ammonia and biochemical oxygen demand (BOD). Laboratory-scale tests have shown that SBRs are effective in removing nitrate, ammonia, and COD from landfill leachate (Ranjan et al., 2016). Granular biomass systems are used in the recently created SBBGR, a technique for treating wastewater. It is a time-based technique that integrates all of the biological treatment stages, including nitrogen and carbon removal, into a single tank. This technique does not require a secondary tank, in contrast to traditional approaches such as activated sludge systems (Lotito et al., 2012).

3.4 Operational principles of SBRs

Whereas solid/liquid separation and metabolic reactions take place successively inside the same reactor, the SBR functions with an emphasis on time rather than space, in contrast to continuous-flow systems. Each of the five separate operating stages (fill, react, settle, draw/decant and idle) of the SBR technique has a specified duration and are discussed below. The SBR may go into an idle phase in situations when there is no wastewater input, like in industrial settings. The general design of an SBR process flow is depicted in Fig. 3.3.

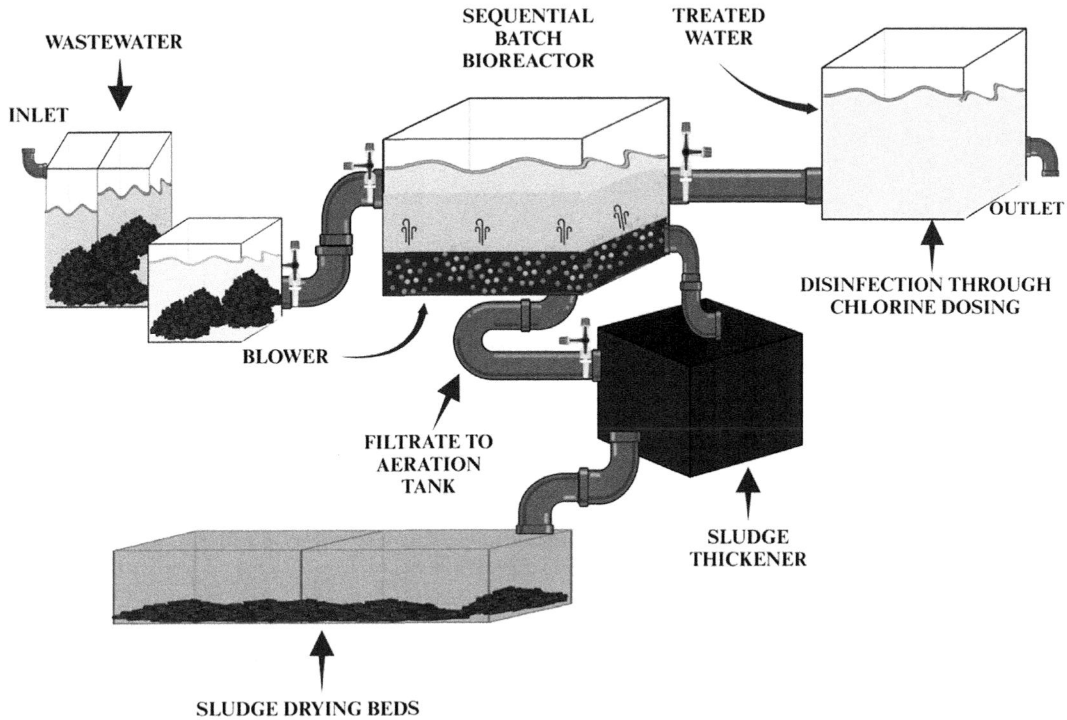

Fig. 3.3 General design of an SBR process flow. Created with BioRender.com.

3.4.1 Fill

During the Fill stage, the residual active biomass from the previous cycle is combined with untreated raw wastewater that has been introduced to the tank. This stage provides three distinct methods that are aerated fill, mixed fill, and static fill, based on the wastewater's properties and the necessary goals for removing organic matter and nutrients (Al-Rekabi et al., 2007). These can be used separately or in combination. The static fill approach produces a situation similar to plug flow by introducing influent wastewater into the SBR without agitation. This produces a high F/M ratio (Food to Microorganisms), which is comparable to an ASP's selection compartment. This environment promotes the growth of bacteria that form flocs, while suppressing filamentous ones, leading to improved settling properties for the sludge. Moreover, static fill conditions foster a favorable environment for phosphate accumulating organisms (PAO), resembling a "feast" situation, which plays a crucial role in biological phosphorus removal, as discussed subsequently. The fill period, which usually lasts for 25% of the entire cycle time, varies depending on variables including the number of tanks, SBR capacity, and effluent flow rate (Singh and Srivastava, 2011). During optimization, changing the fill duration may have an effect on SBR productivity (Coelho et al., 2000).

3.4.2 React

The primary biological reactions that break down organic matter are completed during the React phase. Additionally, it is often engineered to achieve significant nutrient removal. The presence or lack of air controls the phase's activity by determining whether the environment is anaerobic, anoxic, or aerobic. One may use variations like aerated react modes and mixed react modes. The finalization of aerobic reactions that were started during aerated fill occurs during the aerated react mode. This usually entails nitrification, the process of changing ammonia into nitrite and finally nitrate (Dutta and Sarkar, 2015). A combination of anoxic and anaerobic conditions may exist in the reactor during the mixed react mode in addition to aerobic conditions. The process of denitrification, which turns nitrate-nitrogen into nitrogen gas, is facilitated by anoxic environments. The "famine"-like phase produced by anaerobic circumstances encourages the elimination of phosphorus. This stage can constitute over 50% of the total cycle time (Singh and Srivastava, 2011).

3.4.3 Settle

In this case, there is no inflow or outflow and the whole aeration tank acts as a clarifier. The quiescent conditions promote better solid separation compared to conventional clarifiers. This phase has no flow entering or leaving the tank (Al-Rekabi et al., 2007). As a result, the sludge settles to form a flocculent mass, creating a distinct interface with the clear supernatant. The settling period usually ranges from 0.5 to 1.5 h, stopping the accumulation of gas from causing the solid blanket to float (Singh and Srivastava, 2011).

3.4.4 Draw (decant)

The reactor releases the cleared supernatant as effluent once it has settled. In addition to removing extra waste activated sludge, the withdrawal mechanism must stop floating material from being released. The draw period, which lasts 45 min on average, can be anything between 5% and 30% of the whole cycle time (Irvine et al., 1989; Singh and Srivastava, 2011).

3.4.5 Idle

An idle period refers to the duration between a draw and a fill (Al-Rekabi et al., 2007; Singh and Srivastava, 2011). To finish the fill cycle before moving on to another unit, this step is essential, especially in multi-tank systems. The SBR's functionality is determined by the particular objectives established for the treatment location. For instance, in order to manage filamentous bulking, there must be a protracted starving period after the respond phase with high concentration gradients. Short fill phases and large volumetric exchange ratios can be used to achieve high initial substrate concentrations. On the other hand, when working with strong incoming sewage, a long fill phase and modest volumetric exchange ratios produce low variation in substrate concentration during the cycle period, which can be helpful in minimizing substrate inhibition. For nutrient removal systems, using several fill phases can be advantageous. For instance, the COD added during the second fill phase can be utilized for denitrifying the nitrogen added during the initial fill phase. According to achieving satisfactory effluent quality in activated sludge processes like SBR hinges on two critical factors (Schwarzenbeck et al., 2005). Firstly, it depends on the oxidation of organic matter by active microorganisms, resulting in the production of carbon dioxide, water, and inert materials as final byproducts. Secondly, it relies on the flocculation of biomass and other suspended materials into sufficiently large, dense units that can coagulate and settle at the bottom of the reactor, facilitating the production of high-quality effluent. Hence, the key to meeting treatment objectives for a given wastewater lies in operating at optimal design and operational parameters, such as figuring out the right amount of cycles every day and the duration of treatment phases. Fig. 3.4 depicts the five operational stages of an SBR.

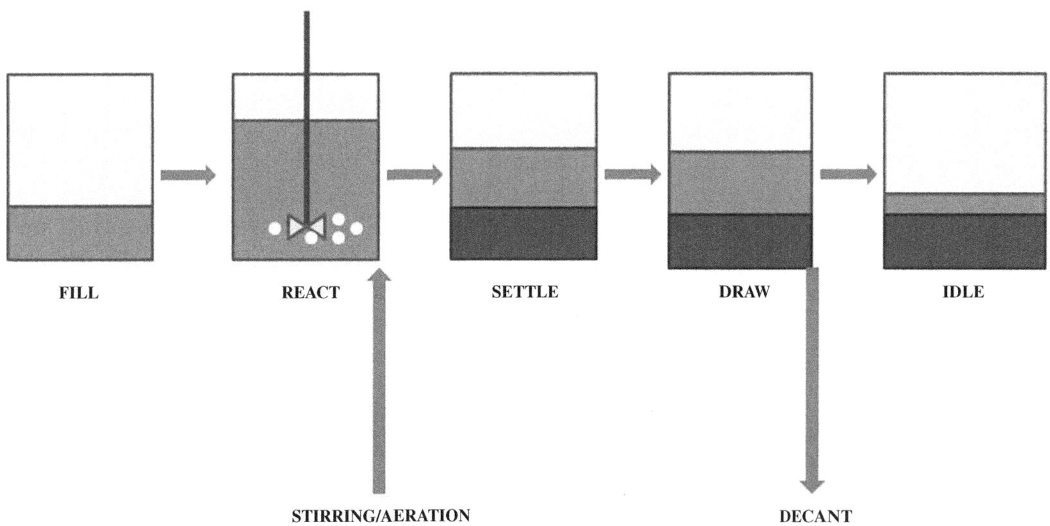

Fig. 3.4 Five operational stages of an SBR.

3.5 Biological treatment of wastewater using SBR

Of late, the release of nutrients like phosphorous (P) and nitrogen (N) has received a lot of attention because of their role in eutrophication and other adverse changes to aquatic ecosystems. To better understand, develop, and improve the biological nutrient removal process, numerous studies have been conducted. Because SBR combines anoxic, anaerobic, and aerobic processes and all of these take place inside a single tank in course of a treatment cycle, it has proven to be a viable substitute for continuous-flow systems in biological nutrient removal (BNR) (Singh and Srivastava, 2011). The petroleum industry has increased its capacity to exploit oil and gas in deep waters increasing the risk of environmental pollution due to the production of effluents that has high salinity levels with organic compounds (Hansen and Davies, 1994). The biological processes have been considered the best remedy due to their cost/benefit ratio for the treatment of such complex effluents. Freire et al. (2001) have reported the treatment of an effluent containing a mixture of oil field wastewater and sewage using a SBR that was quite effective in removing ammonium to about 95% and COD removal efficiencies in the range of 30–50%. Gray water is defined as wastewater without any input from toilets and it accounts for about 75% of all wastewater of the combined residential sewage (Al-Jayyousi, 2003; Christova-Boal et al., 1996; Nolde, 2000). Lamine et al. (2007) investigated the application of SBR technology for treating domestic gray water that has effectively removed nutrients and promoted biodegradation of organic matter for the same showing 90% COD removal. Sarvajith and Nancharaiah (2022) have demonstrated that when activated carbon is added, it could enhance the growth of microorganisms and establishment of enhanced biological phosphorus removal (EBPR) pathway in aerobic granular SBRs. Solid municipal garbage degrades and becomes leachate, a complex liquid, when precipitation seeps through it. Heavy metals and organic molecules, both of which can persist in the ecosystem for extended periods, are present in this fluid. In landfill design, suitable liners are incorporated to reduce the amount of toxins released into the environment from leachate (Bhambore and Kumar, 2022). In study by Ranjan et al. (2016), municipal wastewater and ancient landfill leachate were directly co-treated and was found that the removal effectiveness of nitrate and ammonia was over 83% and 93%, respectively, while the reduction of COD was in the range of 60–70% (Ranjan et al., 2016). In a study published in 2021, Altieri et al. (2021) assessed the effluent produced by washing plastic solid waste (PSW) that are non-hazardous using a SBBGR. The researchers concentrated on the wastewater's general characteristics, the production of sludge, and the effectiveness of pollutant removal. Notably, significant removal efficiencies were noted for parameters like 93.9% of COD, 99.6% of TSS, and 99.6% of NH_3, which showed a notable improvement over traditional treatment techniques with reductions roughly five times greater. Different types of SBRs used for biological treatment of wastewater have shown in Table 3.1.

Table 3.1 Different types of SBRs used for biological treatment.

System used	Type of industrial wastewater	Findings	References
Aerobic SBR utilizing indigenous bacterial consortium	Reactive Orange M2R (ROM2R)	COD, BOD and color reduction observed at 57%, 43%, and 70%, respectively	Jagwani et al. (2018)
Trametes versicolor attached to polyurethane foam in SBR	Vinyl sulfone azo dye's decolorization performance (RB5)	Dye decolorization is seen as a result of a shift in the chemical structure of the dye and is ascribed to enzymatic activity	Lemus-Gómez et al. (2018)
Utilizing SBR technology for biological biodegradation (activated sludge process running in batch mode)	Textile wastewater comprising of suspended solids, organic and inorganic matter, and heavy metals	80–95% COD; 91% BOD removal	Abu-Ghunmi and Jamrah (2006)
SBR employing activated sludge as inoculum	Wastewater containing Methylene Blue (MB)	(COD) and MB removal efficiency were 56% and 93%, respectively	Ma et al. (2011)
SBR	Dairy wastewater	90% COD removal	Mahvi (2008)
SBR	Landfill leachate treatment	98.2% ammonia removal, 45.9% COD reduction, and 33.5% phenol removal	Mojiri et al. (2014)
SBBR using cyanide degrading microbes	Cyanide waste streams	20 mg/l of cyanide removal with a 48-h cycle time	White et al. (2000)
Hybrid sequencing batch reactor (HSBR)	Sewage biological treatment	Ammonium removal efficiency ranges from 50% to 99%, whereas TCOD removal efficiency ranges from 73% to 96%	da Costa et al. (2008)

Continued

Table 3.1 Different types of SBRs used for biological treatment—Cont'd

System used	Type of industrial wastewater	Findings	References
Non-powdered activated carbon and powdered activated carbon SBR	Landfill leachate treatment	Achievement of 64.1%, 71.2%, 81.4%, and 1.33% removal of COD, color, NH_3-N, and TDS, respectively	Aziz et al. (2011)
Photo-sequencing batch reactor (PSBR) enriched with microalgae, ammonia oxidizing bacteria (AOB) and methanol-utilizing denitrifiers (MUD)	Treating ammonium rich wastewater (ARWW)	NH_4 + oxidation and NO_2 − reduction	Arun et al. (2019)
Lagoon based Sequencing Batch Reactor (SBR) Technology	Petrochemical wastewater treatment	SBR technology allowed the low cost retro-fit of an existing lagoon system to a fully functional nitrifying denitrifying activated sludge plant without the need for any major concrete structures	Hudson et al. (2001)
SBR	Tannery wastewater	80–82% COD removal	Ganesh et al. (2006)
SBR	Shrimp wastewater	In just 8 days, the initial chemical oxygen demand (COD) concentration of 1201 mg/L dropped to 32 mg/L	Boopathy et al. (2007)
Aerobic fixed-film sequencing batch reactor	Brewery wastewater	Removed between 83 and 92% of the brewery wastewater TBOD	Nguyen et al. (2000)
Batch reactors using activated carbon particles for the sequencing of aerobic granular sludge	Synthetic wastewater	>95% removal of ammonium, >95% removal of TN, and 95% removal of P	Sarvajith and Nancharaiah (2022)
Sequencing batch reactor	Fluoxetine (FLX)	$71.3 \pm 2.1\%$ biological phosphorous removal	Zhao et al. (2021)
SBR combined with coagulation	Wastewater from oil mills and landfill leachate	91.82% TSS, 94.25% NH_3-N, 84.89% COD, and 95% heavy metal removal	Abuabdou et al. (2021)
Landfill leachate and municipal wastewater	SBR	A removal efficiency of 60–70% for COD, 83% and 93% for nitrate, and ammonia	Ranjan et al. (2016)
Sewage wastewater	SBBGR	91.0% COD removal	Piergrossi et al. (2018)

3.6 Adaptability of SBRs: Role of aeration

Aerobic methods are preferred for treating dye wastewater due to their feasibility and cost-effectiveness. Among the aerobic biological approaches utilized for this purpose are aerated lagoons, activated sludge systems, and biofilm processes (Al-Dhawi et al., 2022). An activated sludge system represents an aerobic treatment method wherein oxygen is introduced in a reactor via aeration tank. This setup is aided by a settling tank, facilitating the settling of biological flocs to segregate biological sludge from the clarified treated water (Al-Dhawi et al., 2023). With the potential to address the primary shortcomings of the activated flocculent sludge process, more attention has been attracted by the aerobic granulation technology in the industry for wastewater treatment. Since they provide dissolved oxygen (DO) and hydraulic shear force, the aeration intensity has a major impact on the formation of aerobic granules (Dulekgurgen et al., 2008; Tay et al., 2004). Prior to ring fission, oxygen from O_2 is integrated into the aromatic ring of organic molecules by enzymes that are secreted by wastewater-dwelling microorganisms (mono- and dioxygenase) (Khouni et al., 2012). Heterotrophic bacteria denitrify nitrate without the requirement for oxygen (anoxic react phase), whereas autotrophic bacteria nitrify ammonium by requiring molecular oxygen dissolved in the mixed liquid by aeration (oxic react phase) (Khin and Annachhatre, 2004). Additionally, for heterotrophic bacteria to biodegrade organic waste (COD), the reactor must get enough oxygen, typically 2 mg/L DO. Therefore, it seems that there is a best aeration technique to remove organic matter and nitrogen from wastewater at the same time within a certain limit (Cho et al., 2010). When DO is high during aeration times in an intermittently aerated SBR, ammonium-nitrogen (NH_4-N) is oxidized to oxidized nitrogen (NO_2-N and NO_3-N) by aerobic nitrifiers. When DO falls to a low enough level during subsequent mixing periods, anoxic denitrifiers convert oxidized N to N_2 gas (Li et al., 2008). Periods of feast and famine characterize this kind of batch operation mode, which has been demonstrated to be advantageous for the development of stable aerobic granules. A feast period indicates the presence of an external carbon source in the bulk liquid, whereas a famine period indicates the absence of an external carbon source. The profile of dissolved oxygen (DO) in a single SBR cycle revealed that DO was near saturation during the famine period and low during the feast period (Arrojo et al., 2004; Beun et al., 2002). It appears that the microorganisms' physiology and kinetics should vary during the feast and famine eras based on the discernible variations in external carbon concentrations and DO. As a result, there should be a difference in the aeration requirements during these two times (Liu et al., 2006). To tackle the issues such as poor sludge settling and excessive sludge production in traditional aerobic systems, a novel aerobic process utilizing biofilm, known as attached-growth biomass systems, has emerged. These systems offer advantages in nitrogen removal and reduced biomass loss even under low temperature conditions (Payandeh et al., 2017).

3.7 Merits of SBRs

The primary advantages of Sequencing Batch Reactors (SBR) are as follows:

i. They can be run autonomously with superior process control capabilities, have a small footprint, and can function in a single basin without the requirement for a secondary clarifier (Abd Nasir et al., 2019; Alattabi et al., 2017).

ii. SBR systems have minimal installation and running expenses since they are easy to configure and operate. Additionally, they have excellent bulking control and a great tolerance to different loading shocks (Li et al., 2019; Mojiri et al., 2018).

iii. SBRs are resilient, retaining more biomass, using less energy, and withstanding toxicity. They make it easier to remove organics simultaneously such as N, and P and allow for the SRT separation from the HRT (Chao et al., 2020; Dutta and Sarkar, 2015).

Moreover, the SBR process offers notable advantages, particularly in terms of space efficiency and operational flexibility. One key benefit is the compact nature of the system, as it integrates treatment functions into single tanks rather than requiring separate units for aeration, clarification, and return activated sludge (RAS) pumping. These tanks typically have square or rectangular shapes, facilitating common-wall construction and minimizing site requirements. Moreover, this configuration allows for easy facility expansion in modular increments, enhancing scalability. Another significant advantage lies in the precise control and adaptability afforded by the SBR process. Unlike traditional systems where reaction time is tied to flow rates, the SBR process enables adjustment to meet specific process objectives. Through manipulation of oxygen supply and mixing patterns, operators can create alternating aerobic and anoxic conditions within the reactor, facilitating targeted removal of nitrogen and phosphorus. This level of control enhances treatment efficiency and supports the optimization of nutrient removal processes.

3.8 Real-world applications: Case studies

Real-world applications of Sequencing Batch Reactors (SBR) technology showcase its effectiveness across diverse industries and environmental challenges. Through case studies, tangible examples have been explored, where SBR systems have revolutionized wastewater treatment, offering sustainable solutions for wastewater.

Case study I (treatment capacity expansion using SBR)

A conventional Activated Sludge Process (ASP) was designed for the treatment of 104,000 m^3/day for population of 130,000 in 1982. In 1982, total capacity of ASP plant was consumed and in 1989 capacity expansion was needed. In 1991, ASP was received 138,000 m^3/day, which more than 32% excess than the design flow of 104,000 m^3/day. Then authority decided to expand the treatment capacity to 160,000 m^3/day which will take at least 350 day. In 1991, two effluent storage flat bottom rectangular tanks (110 × 30 × 4.75) were converted to SBR (Ellis et al., 1992).

Case study II (conversion of ASP to SBR)

Two full scale SBR plants at Culver, Indiana, United States have been operated for treatment domestic wastewater flow of 1340 m^3/day. The important Fill period subdivided into Fill (1) which is a mixed but an-aerated period; the influent BOD was 400 mg/L, total P 12 mg/L and total N 60 mg/L. The BOD loading was 0.60 kg/m^3/day. The effluent total P and total N was 1.2 and 8.0 mg/L, respectively. They noted denitrification in stage 1 and 3, while nitrate concentration was found to increase in stage 2. Release of phosphorous observed in stage-1, while in stage-2, about 2–4% extra P was found to contain in sludge. The release and uptake of P in stage 3 and 4 were not found to be significant (Irvine et al., 1983).

Case study III

The owner of the mixed-use building in the Andheri, Mumbai, Camlin complex has decided to develop a central Sewage Water Treatment Plant (STP) to handle wastewater from the building's residential and commercial portions. This program attempts to make highly treated sewage suitable for several purposes, such as irrigation for landscaping, toilet flushing, and a partial replacement of the makeup water required for the building's cooling towers, which powers the central air conditioning system. SBR was chosen due to its compact design and improved capacity for removing nutrients. Furthermore, a number of uses for the treated water were necessary, such as cooling towers, landscaping, and toilet flushing. In addition to the cooling tower make-up water feed pumps and pressure sand filter feed pumps, air blowers were fitted for aeration. In addition to cooling tower make-up water feed pumps and pressure sand filter feed pumps, air blowers were fitted for aeration. A softener and a sodium hypochlorite dosing system were installed to address the hardness of the water. The performance parameter results is found to be BOD <5.0 mg/L, COD <34.0 mg/L, TSS < 10.0 mg/L, pH 7.85, Detergent <1.0 mg/L, Oil & Grease <1.0 mg/L (Kalpataru Construction Overseas).

Case study IV

The problem of handling vegetable oil waste from product spills, tank washings, and dock runoff was encountered by International Matex Tank Terminal (IMTT). The terminal employed an offsite treatment process to deal with this issue; however, because vegetable oil contained both soluble and insoluble components, the most practical and cost-effective treatment strategy involved pre-treatment using an oil/water separator, followed by an induced air flotation system to remove insoluble components. The OMNIFLO® Sequencing Batch Reactor (SBR) technology was chosen by IMTT to handle the leftover vegetable oil from spills, tank washings, and dock runoff. A fill-and-draw method, also known as a non-steady state ASP, is used in the biological treatment process. It comprises batch treatment activities after wastewater is pumped into one or more reactor basins within a predetermined amount of time. Inside a single reactor basin, the OMNIFLO SBR system accomplishes equalization, aeration, and clarifying sequentially. This technique allows bacteria to settle while the water is being decanted from the top by combining mixing and aeration into a single basin. After then, as the bacteria eat the organic substance, they multiply. To keep the concentration at its ideal level for the removal of organic materials, bacteria must be removed. After that, this bacterium is put in a tank and given air. The bacteria will devour one another and reduce their mass if there are no other organic molecules present. This process termed as digesting. The remaining bacteria, which we refer to as sludge, needs to be disposed of. Still, this sludge is a useful fertilizer and will be applied to a neighboring nursery for fertilization. With its automated control package, the OMNIFLO SBR system reliably generates good effluent and requires little operator attention (Evoqua, Louisiana).

Case study V

A well-known corporation Coffee House, Services Station, Nottingham in the field of hospitality and the end user for this project, had a requirement to treat their commercial wastewater which was rich in food waste (dairy produce). Their premises were previously served by a traditional media based sewage treatment plant that could not cope with unexpected peaks in organics, as this was affecting the floating media and limiting the efficiency of the treatment. As with any commercial site, the incoming wastewater flow was expected to be rich in organics and nitrogenous compounds, with the former not easily degradable due to its particulate nature such as oils, grease and fats. The periodic operation of the site also mandated a solution that does not necessarily follow a continuous aeration pattern, as this may contribute to higher electricity bills, as well as an increased carbon footprint. A hybrid Rewatec SBR wastewater treatment plant, combined with an integrated grease separator was installed. Partial nutrients removal for improved effluent quality was observed (BOD: 20 mg/L, SS: 30 mg/L, Ammonia: 20 mg/L) (Premier Tech Group).

3.9 SBR contributions to wastewater treatment strategies

The goal of wastewater treatment is to enhance water quality to meet safety standards post-treatment. This process involves reducing pollutants level in the water. This pivotal process entails the reduction of pollutant levels, particularly targeting suspended solids which not only contaminate water bodies but also impede the flow in channels and pipes once settled. Moreover, a significant focus lies on diminishing the quantities of biodegradable organic matter, as quantified by parameters such as Biological Oxygen Demand (BOD), to ensure ecological balance (Guillossou et al., 2019).

Selecting the optimal treatment process involves navigating a multitude of potential options and interrelated factors. This complexity yields a vast array of feasible design and operational combinations. Compounding this challenge is the necessity to fulfill various objectives, encompassing effluent standards, local conditions, investment and operational costs, and environmental considerations. Consequently, determining the most suitable plant design poses a formidable undertaking, as noted by Flores-Alsina et al. (2010).

3.9.1 Contributions to the ongoing development of wastewater treatment strategies

The ongoing development of wastewater treatment strategies has been significantly influenced by Sequencing Batch Reactors (SBR). In response to the challenge posed by dense microbial communities, including both pathogenic and non-pathogenic organisms, within wastewater treatment plants, SBR technology has emerged as a pivotal contributor.

SBR technology's operational flexibility has led to its widespread use in many treatment processes. This design's adaptability allows for a single tank to perform biological treatment, secondary clarification, and even flow equalization by adjusting the operational phase duration and aeration period. It is a flexible therapeutic choice because of its versatility. To increase the application of SBR technology even more, researchers have been experimenting with mixing several treatment technologies at the laboratory scale in recent years. A variant of SBR known as sequencing batch biofilm reactor (SBBR) combines suspended and attached growth, creating a hybrid system. In SBBR, biofilm develops at the interface of solid and fluid by adhering to a supporting material. In this configuration, slower-growing microorganisms can flourish irrespective of the bio-aggregates' sedimentation properties and HRT. The size and type of support material selected will depend on the treatment objectives and the properties of the wastewater. The reactor fluid either suspends the support material or places it inside the reactor. The only phases in an SBBR cycle are Fill, React, and Draw in order to keep the reactor's plug flow conditions intact. The settling time of an SBR can be compared to the amount of time required for cleaning the support medium. Sadly, because there's a chance of severe head loss along with sloughing off, SBBR systems might not be appropriate for influents with high Total Suspended Solids (TSS) or when excessive microbial growth is anticipated. Following the successful leachate treatment from the Georgswerder landfill, Germany using the first pilot-scale SBBR, numerous installations have been implemented, showcasing the technology's potential (Smith and Wilderer, 1987).

In aquaculture, where the expenses of protein feed for fish and wastewater treatment pose financial issues, modified SBR systems based on bio-floc technology (BFT) have found interesting applications (De Schryver and Verstraete, 2009; Emerenciano et al., 2013). Biofloc refers to a unique aggregation of microorganisms capable of assimilating nitrogen containing compounds from wastewater and converting them into microbial protein, which can serve as food for fish. Studies have demonstrated that when configured as an external growth reactor for bio-floc, an SBR can attain nitrogen removal efficiencies of up to 98% in laboratory conditions, provided that the C/N ratio is maintained within the range of 10 to 15 (Luo et al., 2013). Moreover, BFT in the SBR reactor made it easier for nitrogen in aquaculture suspended particles to be converted into bacterial biomass (De Schryver and

Verstraete, 2009). It is possible to use this biomass as fish feed, which would increase nitrogen efficiency and remove over 100% of the nitrate in just 6h. Using SBR and SBBR, industrial effluent containing phenolic compounds was treated. One such chemical is p-nitrophenol (PNP), that is hazardous and is frequently used as a synthesis intermediate in the dye, pharmaceutical, and agricultural industries. With 0.368 kg/m^3/day of loading rate, both systems demonstrated complete removal of PNP when using polyethylene rings in the SBBR, up to an influent concentration of 350 mg/L (Goh et al., 2009). But there was only a slight impact on the average removal efficiency for NH_3-N in the SBR and SBBR, which dropped to 86% and 96%, respectively.

Wastewater with a high nitrogen concentration and low COD can be successfully treated by using SBR in the anaerobic ammonium oxidation, simultaneous partial nitrification, and denitrification (SNAD) system. This method was used to clean industrial effluent from optoelectronics that had a C/N ratio of almost 0.2 (Daverey et al., 2012). In a different study, wastewater from the production of thin-film transistor liquid crystal displays (TFT-LCD) was treated to remove pollutants such as tetra-methyl ammonium hydroxide (TMAH), monoethanolamine (MEA), and dimethyl sulfoxide (DMSO) (Lei et al., 2010). Two distinct SBR systems were used in this study: aerobic and anoxic/oxic (A/O). While DMSO and TMAH could only be effectively degraded in anaerobic and aerobic environments, respectively, MEA could be easily degraded under all three of the conditions tested. In hybrid systems, to improve the effectiveness of nutrient removal and encourage high biomass, the use of anoxic/oxic phases that are time-sequenced such as the Porous Biomass Carrier SBR (PBCSBR), is currently being studied (Shin and Park, 1991). In an additional research, dairy manure was treated using a novel biomass retention approach that used natural fibers as biofilm carriers. This strategy was evaluated in a psychrophilic ASBR treating flushed dairy manure, and despite a short HRT of 6 days and low temperature, a higher methane yield was obtained despite the low temperature and short HRT of 6 days. For the purpose of retaining biomass, ASBRs are renowned for their ability to separate HRT from SRT. Furthermore, a particular order of ASBR operations was employed to apply selective pressures for the purpose of immobilization of microorganisms (Liu et al., 2005; Ma et al., 2013; Wang et al., 2011).

Another application used reverse osmosis (RO) in conjunction with the photo-Fenton process and aerobic SBR to recover wastewater from the textile sector allowed for total internal water reuse (Blanco et al., 2014). The following results were obtained from a study on wastewater with azo dyes that used a combination of anoxic-aerobic React phases in an SBR to remove COD and biodecolorize simultaneously: a shorter anoxic phase improved COD removal, while a longer anoxic React phase encouraged decolorization (Al-Amrani et al., 2014). Granular-activated carbon-SBR (GACSBR), another intriguing method, has shown promise in the treatment of dye containing textile wastewater. GAC's physical adsorption process facilitates the treatment of dye-containing wastewater. The efficacy of direct dye removal was enhanced by the incorporation of exogenous organic carbon (Sirianuntapiboon and Sansak, 2008). The public's concern over endocrine disruptor chemicals (EDCs) in wastewater, public water supplies, and surface water has grown significantly in recent years.

Endocrine Disrupting Chemicals (EDCs) comprise a group of compounds capable of disrupting the regular functioning of the endocrine system in both humans and animals, even at extremely low concentrations, such as sub-μg/L levels in water (Liu et al., 2009). As an approach to removing EDCs from wastewater, the SBR process offers promise because it can create anaerobic, aerobic and anoxic conditions all within the same basin. Within the SBR process tank, it offers an abundance of opportunities for microorganisms capable of decomposing EDCs to thrive and sustain such dynamic environmental conditions. It has been found that longer hydraulic and solid retention times (SRT and HRT) remove more EDC. Mainly, this is because those kinds of environments let slow-growing microbes that can use the EDCs grow properly.

3.9.2 Design, operation, and optimization of treatment facilities in urban environments

Ensuring the seamless function of wastewater treatment facilities in urban settings demands a comprehensive approach encompassing design, operation, and optimization. Effluent standards compliance, cost-efficiency, and adaptability to evolving conditions are central to the design process, which must incorporate considerations ranging from seasonal fluctuations to anticipated regulatory changes. Integrating theory, practical design elements, operational strategies, and maintenance protocols is essential for the holistic management of treatment processes, ensuring optimal performance and sustainability.

3.9.2.1 Design

The SBR process epitomizes a variant of activated sludge treatment characterized by a reactor of variable volume, operating cyclically with predefined phases. Embedded within this system are fundamental parameters crucial to its operational efficacy, briefly outlined below.

In a study by Jasim (2020), a wastewater treatment facility tailored to the population of Al-Hay city was meticulously designed. Employing the GPS X program, the design and simulation of the Al-Hay station's processes were executed, with key parameters meticulously considered. Design elements encompassing the aeration tank, equalization basin, grit chamber, oil and grease removal unit, and secondary settling tank were meticulously crafted. Subsequently, calculations were conducted for vital operational parameters including mean cell residence time, aeration tank volume, hydraulic retention time, food to microorganism (F/M) ratio, return sludge flow rate, sludge production, and oxygen demand, across various operating conditions (Jasim, 2020).

In another study by Gherghel et al. (2020), a novel approach rooted in multicriteria decision analysis (MCDA) was developed. Utilizing two simplified schemes and four comprehensive schemes, the study explored various configurations ranging from extended aeration activated sludge processes to aerobic sludge stabilization. Through the resolution of the MCDA problem, a comprehensive scheme featuring primary sedimentation, biological activated sludge under low organic load, and anaerobic sludge digestion emerged as the optimal solution for treating large municipal wastewater streams.

3.9.2.2 Reactor volume

The significance of reactor volume in wastewater treatment processes is underscored by recent studies exploring various reactor configurations. Putra et al. (2020) highlighted the potential of anaerobic baffled reactor (ABR) with a volume of 60 l for effectively treating fishmeal wastewater. Similarly, Bakraoui et al. (2020) focused into the application of an Up-flow Anaerobic Sludge Blanket (UASB) digester, designed with a capacity of 70 L, specifically tailored for the treatment of Recycled paper mill wastewater (RPMW). These studies exemplify the critical role of reactor volume in optimizing treatment efficiency for diverse wastewater streams.

The SBR system features a reactor with variable volume capabilities. The entire reactor volume, denoted as V_{full}, is comprised of two distinct and controllable fractions. Firstly, there's the initial volume (V_o), which represents the volume occupied by settled sludge at the cycle's conclusion. Secondly, there's the final volume (V_f), which corresponds to the wastewater volume introduced and decanted during each cycle (Orhon et al., 2005). Consequently, during operation, the reactor volume attains its peak level by the conclusion of the feeding phase.

3.9.2.3 Number of basins

The SBR process can be implemented using either a single basin or multiple basins arranged in parallel, depending on the influent flow rate. Typically, wastewater treatment facilities employing SBR technology utilize two reactors operating in parallel. Each SBR follows identical phases in each cycle, and the cycles per day as well as each phase duration can be adjusted to meet hydraulic, organic, and nitrogenous loading requirements. An equalization tank may be used as a temporary holding tank for wastewater prior to getting discharged into an SBR during the fill phase. The scheduling of cycles for the two SBRs needs to be coordinated such that while one SBR is undergoing its react, settle, and decant phases, the second SBR is in its fill phase (Gerardi, 2011). In practice, expanding the number of reactors will inevitably escalate the overall capital expenditure. Nonetheless, deploying multiple tanks becomes practically feasible, especially in larger plants, as it offers enhanced flexibility to manage fluctuating flow conditions.

3.9.2.4 Number of cycles

The frequency of cycles per day is a crucial parameter in the design and operation of SBRs. It directly determines both the total cycle duration and the feed volume per cycle, V_f.

The following formula can be used to get the daily cycle count:

$$\text{No . of cycle} = 1/t$$

where, t represents the total cycle time and

$$V_f = Q/\text{No . of cycle}$$

where, Q stands for the daily influent flow rate.

The amount and strength of the influent wastewater determine how many cycles are performed in an SBR each day. Consequently, an increase in wastewater flow necessitates fewer cycles, whereas a reduction in wastewater strength decreases the cycle numbers (Gerardi, 2011). Typically, most SBR operations run between four and six cycles every day.

3.9.2.5 Hydraulic retention time (HRT)

The hydraulic retention time (HRT) in Sequencing Batch Reactors (SBRs) is often quantified as the ratio of volume to flow rate (Q), mirroring the approach used in continuous flow activated sludge systems. This similarity extends to the description of Solids Retention Time (SRT) and Organic Loading Rate (OLR), establishing a cohesive framework for operational analysis.

In a study by Soroosh et al. (2022), a consortium of native microalgae and bacteria (MaB) was deployed in SBRs to treat real wastewater. The investigation involved adjusting HRT from 3 days for the initial five batches to 2 days for the subsequent six batches. Notably, the SBR operated with a 2-day HRT exhibited improved pH balance, rendering this retention time advantageous for the single-step MaB treatment of dilute wastewaters.

Concurrently, Wang et al. (2021) examined the influence of HRT on aerobic granular sludge (AGS) performance in treating swine wastewater (SW) within SBRs. Assessing HRTs ranging from 4.8 to 16h across multiple AGSBRs, the study found enhanced sedimentation performance and biomass concentration when HRT was reduced from 16 to 8h (Wang et al., 2021).

Additionally, Yakamercan and Aygün (2021) conducted research on the removal of diethylhexyl phthalate (DEHP) and pentachlorophenol (PCP) from real textile wastewater, investigating the effects of HRT and Solid Retention Time (SRT). Optimal conditions for DEHP removal were identified at 48h for HRT and 20days for SRT, achieving an impressive removal efficiency of 83.27%. These studies collectively underscore the pivotal role of HRT in shaping the performance and efficiency of wastewater treatment processes.

3.9.2.6 Length of phases in a cycle

The SBR process may function in a variety of environmental conditions, including aerobic periods for the removal of COD and nutrients, as well as mixed phases that, depending on the amount of nitrate, can be either anoxic or anaerobic. These conditions can be changed by energy input. SBR systems are versatile and can be used for both carbon and nutrient removal through simple manipulation of operating parameters. The process options for SBR systems employed for carbon removal alone and combined with nitrification are quite similar, as both rely on aerobic metabolisms. Furthermore, organic carbon removal and nitrification can occur simultaneously in the same reactor during a single aerobic process phase within the cyclic operation. This adaptability and flexibility of operation have always been key advantages of the SBR technology. Achieving the desired system performance largely depends on appropriately selecting the sequence and duration of different phases and periods within the cycle, as well as making correct choices regarding design parameters such as Solids Retention Time (SRT). The SRT can be calculated by dividing the amount of solids that are kept in the reactor by the amount of solids that are removed from it every day (Grady Jr. et al., 2011). The primary difference lies in the energy demands of such plants.

In a study by Liu et al. (2020), the efficacy of nitrogen (N), phosphorus (P), and carbon (C) removal was scrutinized by adjusting cycle times in an anaerobic-aerobic-anoxic sequencing batch reactor (AOA-SBR) system. Results indicated promising removal efficiencies for C, N, and P with a 6-h cycle period, showcasing average removal efficiencies of 96.32% for COD, 94.33% for TN, and 96.81% for TP. These findings underscore the importance of cycle duration in optimizing nutrient removal performance within SBR systems (Liu et al., 2020).

3.9.2.7 Optimization

Conventional suspended-growth activated sludge systems, such as SBRs, are engineered and run with a substantial Solid Retention Time (SRT) ranging from 10 to 30 days (Metcalf, 2003). These extended SRT periods necessitate long Hydraulic Retention Times (HRT), resulting in large reactor volumes and footprints, leading to significantly investments in capitals and increase in the demands for energy (Ge et al., 2013). Moreover, the Organic Loading Rates (OLR) achieved in such treatments tend to be low, typically falling within the range of 0.5 to 1.5 g of Chemical Oxygen Demand (COD) per liter per day (Grady Jr. et al., 2011). Although a shorter SRT can increase the production of waste sludge, thereby elevating sludge treatment costs, recent research suggests that the sludge generated from aerobic processes operating at lower SRTs can be effectively utilized as a substrate for anaerobic digestion, leading to value recovery through methane production (Ge et al., 2013). According to the fundamental principles of activated sludge processes, incase of a given wastewater with specific flow rates and compositions, reduction in the HRT minimizes SBR volume and enhances OLR, The relationship between SBR volume, HRT, and OLR is expressed by the following equation:

$$\mathrm{HRT} = V/Q = S_0/OLR$$

where, V is the volume of the reactor, Q is the daily influent flow rate and S_0 is the initial substrate concentration.

The concentration of biomass within the reactor increases when the HRT is decreased while the SRT remains constant. This outcome is anticipated because the overall biomass quantity is regulated by the SRT, thus reducing the reactor volume (via shorter HRT) results in higher biomass concentration. However, there's a limit to how much the HRT can be reduced while maintaining a fixed SRT due to the adverse effects it can have. For instance, higher biomass concentration diminishes solids settling rate and escalates aeration demands per reactor volume unit. Consequently, reducing the HRT with a fixed SRT may potentially lead to process failure. To mitigate these challenges posed by short HRT, adjustments in SRT can be made as it influences the reactor's biomass amount. Nonetheless, decreasing the SRT is constrained by the desired effluent quality. Beyond a certain threshold, reducing SRT compromises substrate removal efficiency, impacting effluent quality negatively. SRT is, at its core, the most important design parameter that impacts the way activated sludge systems, such as SBRs, operate. SRT controls the effluent's substrate concentration, the oxygen consumed, and the generation of sludge, given a particular flow rate of the wastewater and composition. The SBR volume is determined by HRT, on the other hand, for a fixed SRT value, which also sets the reactor's biomass content.

3.10 Conclusions and future trends

The utilization of SBRs presents a promising solution to the pressing challenges associated with municipal wastewater treatment. Through a comprehensive examination of SBR technology, this chapter has underscored its capacity to enhance efficiency, adaptability, and sustainability in wastewater treatment processes. The operational principles of SBRs have been elucidated, emphasizing the important role of microorganisms in decomposing organic pollutants. The inclusion of aeration during the reaction phase and the potential for nutrient removal further highlight the adaptability of SBRs in addressing specific water quality challenges. Real-world case studies and examples have demonstrated the successful application of SBR technology in municipal wastewater treatment plants. As municipalities worldwide strive for sustainable wastewater management solutions, this chapter emerges as a valuable resource. By providing insights into the advancements and practical considerations of SBRs, it significantly contributes to the ongoing development of wastewater treatment strategies.

The future of SBRs in municipal wastewater treatment presents promising advancements across multiple fronts, including automation, sensor integration, energy efficiency, and resource recovery. With a focus on the modular and scalable designs of SBRs, along with improvements in treatment performance and integration with smart grid systems, SBR technology is poised to maintain its crucial role in meeting the changing demands of wastewater treatment. Furthermore, the adoption of decentralized wastewater treatment solutions will expand the reach of SBR technology, providing cost-effective and sustainable solutions across diverse environments.

References

Abd Nasir, M.A., Jahim, J.M., Abdul, P.M., Silvamany, H., Maaroff, R.M., Yunus, M.-F.M., 2019. The use of acidified palm oil mill effluent for thermophilic biomethane production by changing the hydraulic retention time in anaerobic sequencing batch reactor. Int. J. Hydrogen Energy, 3373–3381. 03603199. Elsevier Ltd, Malaysia International. Available from: http://www.journals.elsevier.com/international-journal-of-hydrogen-energy/. https://doi.org/10.1016/j.ijhydene.2018.06.149.

Abuabdou, S.M.A., Yew, Y.H., Ahmad, W., Aun, N.C., Bashir, M.J.K., 2021. Treatment of tropical stabilized landfill leachate by adsorption using powdered activated carbon: isothermal and kinetic studies. IOP Conf. Ser.: Earth Environ. Sci. 799 (1), 012032. https://doi.org/10.1088/1755-1315/799/1/012032. 1755-1307.

Abubakar, S., Latiff, A., Lawal, I., Jagaba, A., 2016. Aerobic treatment of kitchen wastewater using sequence batch reactor (SBR) and reuse for irrigation landscape purposes. Am. J. Eng. Res. 5, 23–31.

Abu-Ghunmi, L.N., Jamrah, A.I., 2006. Biological treatment of textile wastewater using sequencing batch reactor technology. Environ. Model. Assess. 11 (4), 333–343. 14202026. Springer Netherlands, Jordan. Available from: http://www.kluweronline.com/issn/1420-2026. https://doi.org/10.1007/s10666-005-9025-3.

Al-Amrani, W.A., Lim, P.E., Seng, C.E., Wan Ngah, W.S., 2014. Factors affecting biodecolorization of azo dyes and COD removal in anoxic-aerobic REACT operated sequencing batch reactor. J. Taiwan Inst. Chem. Eng. 45 (2), 609–616. https://doi.org/10.1016/j.jtice.2013.06.032. 18761070.

Alattabi, A.W., Harris, C.B., Alkhaddar, R.M., Hashim, K.S., Ortoneda-Pedrola, M., Phipps, D., 2017. Improving sludge settleability by introducing an innovative, two-stage settling sequencing batch reactor. J. Water Process. Eng. 20, 207–216. 22147144. Elsevier Ltd, United Kingdom. Available from: http://www.journals.elsevier.com/journal-of-water-process-engineering/. https://doi.org/10.1016/j.jwpe.2017.11.004.

Al-Dhawi, B.N.S., Kutty, S.R.M., Baloo, L., Almahbashi, N.M.Y., Ghaleb, A.A.S., Jagaba, A.H., Kumar, V., Saeed, A.A.H., 2022. Treatment of synthetic wastewater by using submerged attached growth media in continuous activated sludge reactor system. Int. J. Sustain. Build. Technol. Urban Dev. 13 (1), 2–10. 20937628. Sustainable Building Research Center, Malaysia. Available from: https://www.sbt-durabi.org/articles/article/8Kew/. 10.22712/susb.20220002.

Al-Dhawi, B.N.S., Kutty, S.R.M., Baloo, L., Almahbashi, N.M.Y., Ghaleb, A.A.S., Jagaba, A.H., Kumar, V., Saeed, A.A.H., 2023. Phosphorus removal from synthetic wastewater by using palm oil clinker as media in continuous activated sludge. Adv. Sci. Technol. Innov., 41–44. https://doi.org/10.1007/978-3-031-26580-8_8. 25228722. 2023/01. Springer Nature, Malaysia. Available from: https://www.springer.com/series/15883.

Al-Jayyousi, O.R., 2003. Greywater reuse: towards sustainable water management. Desalination 156 (1–3), 181–192. https://doi.org/10.1016/S0011-9164(03)00340-0.

Al-Rekabi, W.S., Qiang, H., Qiang, W.W., 2007. Review on sequencing batch reactors. Pak. J. Nutr. 6 (1), 11–19. 16805194. Asian Network for Scientific Information,

China. Available from: http://www.pjbs.org/pjnonline/fin585.pdf. https://doi.org/10.3923/pjn.2007.11.19.

Altieri, V.G., De Sanctis, M., Sgherza, D., Pentassuglia, S., Barca, E., Di Iaconi, C., 2021. Treating and reusing wastewater generated by the washing operations in the non-hazardous plastic solid waste recycling process: advanced method vs. conventional method. J. Environ. Manag. 284, 112011. https://doi.org/10.1016/j.jenvman.2021.112011. 03014797.

Arrojo, B., Mosquera-Corral, A., Garrido, J.M., Méndez, R., 2004. Aerobic granulation with industrial wastewater in sequencing batch reactors. Water Res. 38 (14–15), 3389–3399. 00431354. Elsevier Ltd, Spain. Available from: https://www.sciencedirect.com/science/journal/00431354. https://doi.org/10.1016/j.watres.2004.05.002.

Arun, S., Arul Manikandan, N., Pakshirajan, K., Pugazhenthi, G., 2019. Novel short-cut biological nitrogen removal method using an algae-bacterial consortium in a photo-sequencing batch reactor: process optimization and kinetic modelling. J. Environ. Manag. 250, 109401. https://doi.org/10.1016/j.jenvman.2019.109401. 03014797.

Asrami, M.R., Pirouzi, A., Nosrati, M., Hajipour, A., Zahmatkesh, S., 2024. Energy balance survey for the design and auto-thermal thermophilic aerobic digestion of algal-based membrane bioreactor for landfill leachate treatment (under organic loading rates): experimental and simulation-based ANN and NSGA-II. Chemosphere 347, 140652. https://doi.org/10.1016/j.chemosphere.2023.140652. 00456535.

Aziz, S.Q., Aziz, H.A., Yusoff, M.S., Bashir, M.J.K., 2011. Landfill leachate treatment using powdered activated carbon augmented sequencing batch reactor (SBR) process: optimization by response surface methodology. J. Hazard. Mater. 189 (1–2), 404–413. https://doi.org/10.1016/j.jhazmat.2011.02.052. 03043894.

Bakraoui, M., Karouach, F., Ouhammou, B., Aggour, M., Essamri, A., El Bari, H., 2020. Biogas production from recycled paper mill wastewater by UASB digester: optimal and mesophilic conditions. Biotechnol. Rep. 25, e00402. https://doi.org/10.1016/j.btre.2019.e00402. 2215017X.

Beun, J.J., van Loosdrecht, M.C.M., Heijnen, J.J., 2002. Aerobic granulation in a sequencing batch airlift reactor. Water Res. 36 (3), 702–712. https://doi.org/10.1016/s0043-1354(01)00250-0. 00431354.

Bhambore, N., Kumar, M.S., 2022. Municipal solid waste generation, management scenarios, and leachate treatment using sequencing batch biofilter granular reactor. Process Saf. Environ. Prot. 167, 454–468. https://doi.org/10.1016/j.psep.2022.09.027.

Blanco, J., Torrades, F., Morón, M., Brouta-Agnésa, M., García-Montaño, J., 2014. Photo-Fenton and sequencing batch reactor coupled to photo-Fenton processes for textile wastewater reclamation: feasibility of reuse in dyeing processes. Chem. Eng. J. 240, 469–475. https://doi.org/10.1016/j.cej.2013.10.101. 13858947.

Boopathy, R., Bonvillain, C., Fontenot, Q., Kilgen, M., 2007. Biological treatment of low-salinity shrimp aquaculture wastewater using sequencing batch reactor. Int. Biodeterior. Biodegradation 59 (1), 16–19. https://doi.org/10.1016/j.ibiod.2006.05.003. 09648305.

Bounaga, A., Alsanea, A., Danouche, M., Rittmann, B.E., Zhou, C., Boulif, R., Zeroual, Y., Benhida, R., Lyamlouli, K., 2024. Elemental sulfur biorecovery from phosphogypsum using oxygen-membrane biofilm reactor: bioreactor parameters optimization, metagenomic analysis and metabolic prediction of the biofilm activity. Bioresour. Technol. 400, 130680. https://doi.org/10.1016/j.biortech.2024.130680. 09608524.

Cakir, F.Y., Stenstrom, M.K., 2005. Greenhouse gas production: a comparison between aerobic and anaerobic wastewater treatment technology. Water Res. 39 (17), 4197–4203. https://doi.org/10.1016/j.watres.2005.07.042. 00431354.

Chan, Y.J., Chong, M.F., Law, C.L., Hassell, D.G., 2009. A review on anaerobic-aerobic treatment of industrial and municipal wastewater. Chem. Eng. J. 155 (1–2), 1–18. https://doi.org/10.1016/j.cej.2009.06.041. 13858947.

Chao, C., Zhao, Y., Keskar, J., Ji, M., Wang, Z., Li, X., 2020. Simultaneous removal of COD, nitrogen and phosphorus and the tridimensional microbial response in a sequencing batch biofilm reactor: with varying C/N/P ratios. Biochem. Eng. J. 154, 107215. https://doi.org/10.1016/j.bej.2019.04.017. 1369703X.

Cho, M.H., Lee, J., Kim, J.H., Lim, H.C., 2010. Optimal strategies of fill and aeration in a sequencing batch reactor for biological nitrogen and carbon removal. Korean J. Chem. Eng. 27 (3), 925–929. https://doi.org/10.1007/s11814-010-0122-x. 02561115.

Christova-Boal, D., Eden, R.E., McFarlane, S., 1996. An investigation into greywater reuse for urban residential properties. Desalination 106 (1–3), 391–397. https://doi.org/10.1016/S0011-9164(96)00134-8. 00119164.

Chung, T.S., Li, X., Ong, R.C., Ge, Q., Wang, H., Han, G., 2012. Emerging forward osmosis (FO) technologies and challenges ahead for clean water and clean energy applications. Curr. Opin. Chem. Eng. 1 (3), 246–257. 22113398. Elsevier Ltd., Singapore. Available from: http://www.elsevier.com/wps/find/journaldescription.cws_home/725837/description#description. https://doi.org/10.1016/j.coche.2012.07.004.

Coelho, M.A.Z., Russo, C., Araújo, O.Q.F., 2000. Optimization of a sequencing batch reactor for biological nitrogen removal. Water Res. 34 (10), 2809–2817. https://doi.org/10.1016/s0043-1354(00)00010-5. 00431354.

Colella, M., Ripa, M., Cocozza, A., Panfilo, C., Ulgiati, S., 2021. Challenges and opportunities for more efficient water use and circular wastewater management. The Case of Campania Region, Italy. J. Environ. Manag. 297, 113171. https://doi.org/10.1016/j.jenvman.2021.113171. 03014797.

da Costa, R.H.R., Souto, V.S., Prelhaz, A.T.S., Neto, L.G.L., Wolff, D.B., 2008. Utilization of a hybrid sequencing batch reactor (HSBR) as a decentralized system of domestic wastewater treatment. Water Sci. Technol. 57 (12), 1951–1956. https://doi.org/10.2166/wst.2008.318. 0273-1223.

Daverey, A., Su, S.H., Huang, Y.T., Lin, J.G., 2012. Nitrogen removal from opto-electronic wastewater using the simultaneous partial nitrification, anaerobic ammonium oxidation and denitrification (SNAD) process in sequencing batch reactor. Bioresour. Technol. 113, 225–231. https://doi.org/10.1016/j.biortech.2011.12.004. 18732976.

De Sanctis, M., Beccari, M., Di Iaconi, C., Majone, M., Rossetti, S., Tandoi, V., 2013. Study of performances, stability and microbial characterization of a sequencing batch biofilter granular reactor working at low recirculation flow. Bioresour. Technol. 129, 624–628. https://doi.org/10.1016/j.biortech.2012.12.110. 09608524.

De Schryver, P., Verstraete, W., 2009. Nitrogen removal from aquaculture pond water by heterotrophic nitrogen assimilation in lab-scale sequencing batch reactors. Bioresour. Technol. 100 (3), 1162–1167. https://doi.org/10.1016/j.biortech.2008.08.043. 09608524.

de Sousa Rollemberg, S.L., Barros, A.R.M., de Lima, J.P.M., Santos, A.F., Firmino, P.I.M., dos Santos, A.B., 2019. Influence of sequencing batch reactor configuration on aerobic granules growth: engineering and microbiological aspects. J. Clean. Prod. 238. 09596526. Elsevier Ltd, Brazil. Available from: https://www.journals.elsevier.com/journal-of-cleaner-production. https://doi.org/10.1016/j.jclepro.2019.117906.

Díaz, O., Vera, L., González, E., 2023. Optimization of the operating parameters in a membrane bioreactor operated in direct-flow mode by assessing threshold flux for compressibility. Chem. Eng. Sci. 274, p.118675. https://doi.org/10.1016/j.ces.2023.118675.

Di Iaconi, C., De Sanctis, M., Rossetti, S., Ramadori, R., 2008. Technological transfer to demonstrative scale of sequencing batch biofilter granular reactor (SBBGR) technology for municipal and industrial wastewater treatment. Water Sci. Technol. 58 (2), 367–372. https://doi.org/10.2166/wst.2008.395. 0273-1223.

Duan, Y., Liu, Y., Zhang, M., Li, Y., Zhu, W., Hao, M., Ma, S., 2020. Start-up and operational performance of the partial nitrification process in a sequencing batch reactor (SBR) coupled with a micro-aeration system. Bioresour. Technol. 296, 122311. https://doi.org/10.1016/j.biortech.2019.122311. 09608524.

Dulekgurgen, E., Yilmas, M., Wilderer, P.A., 2008. Shape and surface topology of anaerobic/aerobic granules influenced by shearing conditions. In: 4th IWA Specialized Conference on Sequencing Batch Reactor Technology, pp. 311–320.

Dutta, A., Sarkar, S., 2015. Sequencing batch reactor for wastewater treatment: recent advances. Curr. Pollut. Rep. 1 (3), 177–190. https://doi.org/10.1007/s40726-015-0016-y. 2198-6592.

Ellis, T.G., Soulsby, M.R., Harper, P.C., 1992. The selection and design of sequencing batch reactors for treatment capacity expansion in Abu Dhabi. Water Sci. Technol. 26 (5–6), 1105–1114. https://doi.org/10.2166/wst.1992.0552. 0273-1223.

Emerenciano, M., Gaxiola, G., Cuzo, G., 2013. Biofloc Technology (BFT): A Review for Aquaculture Application and Animal Food Industry. InTech, https://doi.org/10.5772/53902.

Flores-Alsina, X., Gallego, A., Feijoo, G., Rodriguez-Roda, I., 2010. Multiple-objective evaluation of wastewater treatment plant control alternatives. J. Environ. Manag. 91 (5), 1193–1201. https://doi.org/10.1016/j.jenvman.2010.01.009. 03014797.

Freire, D.D.C., Cammarota, M.C., Sant'Anna, G.L., 2001. Biological treatment of oil field wastewater in a sequencing batch reactor. Environ. Technol. 22 (10), 1125–1135. https://doi.org/10.1080/09593332208618203. 0959-3330.

Ganesh, R., Balaji, G., Ramanujam, R.A., 2006. Biodegradation of tannery wastewater using sequencing batch reactor—respirometric assessment. Bioresour. Technol. 97 (15), 1815–1821. https://doi.org/10.1016/j.biortech.2005.09.003. 09608524.

Ge, H., Batstone, D.J., Keller, J., 2013. Operating aerobic wastewater treatment at very short sludge ages enables treatment and energy recovery through anaerobic sludge digestion. Water Res. 47 (17), 6546–6557. 18792448. Elsevier Ltd., Australia. Available from: https://www.sciencedirect.com/science/journal/00431354. https://doi.org/10.1016/j.watres.2013.08.017.

Gerardi, M.H., 2011. Troubleshooting the Sequencing Batch Reactor. John Wiley & Sons.

Ghangrekar, M.M., Sathe, S.M., Chakraborty, I., 2023. Moving Bed Biofilm Bioreactors for Wastewater Treatment: Trends and Technoeconomic/Sustainability Aspects Material-Microbes Interactions: Environmental Biotechnological Perspective. Elsevier, India, pp. 425–437. Available from: https://shop.elsevier.com/books/material-microbes-interactions/aryal/978-0-323-95124-1. https://doi.org/10.1016/B978-0-323-95124-1.00001-2.

Gherghel, A., Teodosiu, C., Notarnicola, M., De Gisi, S., 2020. Sustainable design of large wastewater treatment plants considering multi-criteria decision analysis and stakeholders' involvement. J. Environ. Manag. 261, 110158. https://doi.org/10.1016/j.jenvman.2020.110158. 03014797.

Goh, C.P., Seng, C.E., Sujari, A.N.A., Lim, P.E., 2009. Performance of sequencing batch biofilm and sequencing batch reactors in simultaneous p-nitrophenol and nitrogen removal. Environ. Technol. 30 (7), 725–736. 1479487X. Taylor and Francis Ltd., Malaysia. Available from: http://www.tandf.co.uk/journals/titles/09593330.asp. https://doi.org/10.1080/09593330902911689.

Grady Jr., C.L., Daigger, G.T., Love, N.G., Filipe, C.D., 2011. Biological Wastewater Treatment. CRC Press.

Guillossou, R., Le Roux, J., Mailler, R., Vulliet, E., Morlay, C., Nauleau, F., Gasperi, J., Rocher, V., 2019. Organic micropollutants in a large wastewater treatment plant: what are the benefits of an advanced treatment by activated carbon adsorption in comparison to conventional treatment? Chemosphere 218, 1050–1060. https://doi.org/10.1016/j.chemosphere.2018.11.182. 00456535.

Hansen, B., Davies, S., 1994. Review of potential technologies for the removal of dissolved components from produced water. Chem. Eng. Res. Des. 72 (2), 176–188.

Hoinkis, J., Deowan, S.A., Panten, V., Figoli, A., Huang, R.R., Drioli, E., 2012. Germany membrane bioreactor (MBR) technology—a promising approach for industrial water reuse. Procedia Eng. 33, 234–241. https://doi.org/10.1016/j.proeng.2012.01.1199. 8777058.

Hudson, N., Doyle, J., Lant, P., Roach, N., de Bruyn, B., Staib, C., 2001. Sequencing batch reactor technology: the key to a BP refinery (Bulwer Island) upgraded environmental protection system—a low cost lagoon based retro-fit. Water Sci. Technol. 43 (3), 339–346. https://doi.org/10.2166/wst.2001.0155. 0273-1223.

Irvine, R.L., Ketchum, L.H., Breyfogle, R., Barth, E.F., 1983. Municipal application of sequencing batch treatment. J. Water Pollut. Control Fed. 55 (5), 484–488. 00431303.

Irvine, R.L., Ketchum Jr., L.H., Asano, T., 1989. Sequencing batch reactors for biological wastewater treatment. Crit. Rev. Environ. Sci. Technol. 18 (4), 255–294. https://doi.org/10.1080/10643388909388350.

Jagaba, A.H., Kutty, S.R.M., Lawal, I.M., Abubakar, S., Hassan, I., Zubairu, I., Umaru, I., Abdurrasheed, A.S., Adam, A.A., Ghaleb, A.A.S., Almahbashi, N.M.Y., Al-dhawi, B.N.S., Noor, A., 2021. Sequencing batch reactor technology for landfill leachate treatment: a state-of-the-art review. J. Environ. Manag. 282, 111946. https://doi.org/10.1016/j.jenvman.2021.111946. 03014797.

Jagwani, J., Johnson, J., Datta, M., Lakshmi, B.J., 2018. Bacterial community dynamics involved in Reactive Orange M2R dye degradation using a real time quantitative PCR and scale up studies using sequence batch reactor. Biorem. J. 22 (1–2), 63–71. https://doi.org/10.1080/10889868.2018.1476452. 1088-9868.

Jankowska, K., Domingo-Félez, C., Prado-Rubio, O.A., Skiadas, I.V., Woodley, J.M., Pinelo, M., 2024. Effect of backwash as a strategy for biofouling control in the submerged ceramic membrane bioreactor for high-density cultivations: process optimization and fouling mechanism at pilot scale. Sep. Purif. Technol. 338. 18733794. Elsevier B.V., Denmark. Available from: https://www.sciencedirect.com/science/journal/13835866. https://doi.org/10.1016/j.seppur.2024.126428.

Jaramillo, F., Orchard, M., Muñoz, C., Zamorano, M., Antileo, C., 2018. Advanced strategies to improve nitrification process in sequencing batch reactors—a review. J. Environ. Manag. 218, 154–164. 10958630. Academic Press, Chile. Available from: https://www.sciencedirect.com/science/journal/03014797. https://doi.org/10.1016/j.jenvman.2018.04.019.

Jasim, N.A., 2020. The design for wastewater treatment plant (WWTP) with GPS X modelling. Cogent Eng. 7 (1), 1723782. https://doi.org/10.1080/23311916.2020.1723782.

Jiao, C., Chen, H., Liu, Y., Zhao, H., Li, Q., Wang, G., Chen, R., Li, Y.-Y., 2024. Synergistic effects of biochar addition and filtration mode optimization on mitigating membrane fouling in high-solid anaerobic membrane bioreactors. Sci. Total Environ. 926, 171648. https://doi.org/10.1016/j.scitotenv.2024.171648. 00489697.

Jones, W., Gerogiorgis, D.I., 2023. Technoeconomic comparison of optimised bioreactor-filtration systems for mAb production. Comput. Chem. Eng. 179, 108438. https://doi.org/10.1016/j.compchemeng.2023.108438. 00981354.

Khin, T., Annachhatre, A.P., 2004. Novel microbial nitrogen removal processes. Biotechnol. Adv. 22 (7), 519–532. 07349750. Elsevier Inc., Thailand. Available from: www.elsevier.com/inca/publications/store/5/2/5/4/5/5/index.htt. https://doi.org/10.1016/j.biotechadv.2004.04.003.

Khouni, I., Marrot, B., Amar, R.B., 2012. Treatment of reconstituted textile wastewater containing a reactive dye in an aerobic sequencing batch reactor using a novel bacterial consortium. Sep. Purif. Technol. 87, 110–119. https://doi.org/10.1016/j.seppur.2011.11.030. 13835866.

Lamine, M., Bousselmi, L., Ghrabi, A., 2007. Biological treatment of grey water using sequencing batch reactor. Desalination 215 (1–3), 127–132. https://doi.org/10.1016/j.desal.2006.11.017. 00119164.

Lei, C.N., Whang, L.M., Chen, P.C., 2010. Biological treatment of thin-film transistor liquid crystal display (TFT-LCD) wastewater using aerobic and anoxic/oxic sequencing batch reactors. Chemosphere 81 (1), 57–64. 00456535. Elsevier Ltd, Taiwan. Available from: www.elsevier.com/locate/chemosphere. https://doi.org/10.1016/j.chemosphere.2010.07.001.

Lemus-Gómez, L.E., Martínez-Trujillo, M.A., Membrillo-Venegas, I., García-Rivero, M., 2018. Performance analysis of azo dye decolorization by immobilized trametes versicolor in a sequencing batch reactor. Environ. Eng. Sci. 35 (12), 1322–1328. 15579018. Mary Ann Liebert Inc., Mexico. Available from: www.liebertonline.com/ees. https://doi.org/10.1089/ees.2018.0033.

Li, J., Healy, M.G., Zhan, X., Norton, D., Rodgers, M., 2008. Effect of aeration rate on nutrient removal from slaughterhouse wastewater in intermittently aerated sequencing batch reactors. Water Air Soil Pollut. 192 (1–4), 251–261. https://doi.org/10.1007/s11270-008-9652-9. 15732932.

Li, S., Fei, X., Cao, L., Chi, Y., 2019. Insights into the effects of carbon source on sequencing batch reactors: performance, quorum sensing and microbial community. Sci. Total Environ. 691, 799–809. https://doi.org/10.1016/j.scitotenv.2019.07.191.

Liu, Y., Wang, Z.W., Qin, L., Liu, Y.Q., Tay, J.H., 2005. Selection pressure-driven aerobic granulation in a sequencing batch reactor. Appl. Microbiol. Biotechnol. 67 (1), 26–32. https://doi.org/10.1007/s00253-004-1820-2.

Liu, Y.Q., Tay, J.H., Moy, B.Y.P., 2006. Characteristics of aerobic granular sludge in a sequencing batch reactor with variable aeration. Appl. Microbiol. Biotechnol. 71 (5), 761–766. https://doi.org/10.1007/s00253-005-0209-1. 01757598.

Liu, Z.-h., Kanjo, Y., Mizutani, S., 2009. Removal mechanisms for endocrine disrupting compounds (EDCs) in wastewater treatment—physical means, biodegradation, and chemical advanced oxidation: a review. Sci. Total Environ. 407 (2), 731–748. https://doi.org/10.1016/j.scitotenv.2008.08.039.

Liu, S., Daigger, G.T., Liu, B., Zhao, W., Liu, J., 2020. Enhanced performance of simultaneous carbon, nitrogen and phosphorus removal from municipal wastewater in an anaerobic-aerobic-anoxic sequencing batch reactor (AOA-SBR) system by alternating the cycle times. Bioresour. Technol. 301 (122750). https://doi.org/10.1016/j.biortech.2020.122750.

Lotito, A.M., Fratino, U., Mancini, A., Bergna, G., Di Iaconi, C., 2012. Effective aerobic granular sludge treatment of a real dyeing textile wastewater. Int. Biodeterior. Biodegrad. 69, 62–68. https://doi.org/10.1016/j.ibiod.2012.01.004. 09648305.

Luo, G.Z., Avnimelech, Y., Pan, Y.F., Tan, H.X., 2013. Inorganic nitrogen dynamics in sequencing batch reactors using biofloc technology to treat aquaculture sludge. Aquac. Eng. 52, 73–79. https://doi.org/10.1016/j.aquaeng.2012.09.003. 01448609.

Ma, D.Y., Wang, X.H., Song, C., Wang, S.G., Fan, M.H., Li, X.M., 2011. Aerobic granulation for methylene blue biodegradation in a sequencing batch reactor. Desalination 276 (1–3), 233–238. https://doi.org/10.1016/j.desal.2011.03.055. 00119164.

Ma, J., Yu, L., Frear, C., Zhao, Q., Li, X., Chen, S., 2013. Kinetics of psychrophilic anaerobic sequencing batch reactor treating flushed dairy manure. Bioresour. Technol. 131, 6–12. https://doi.org/10.1016/j.biortech.2012.11.147.

Mahvi, A.H., 2008. Sequencing batch reactor: a promising technology in wastewater treatment. Iran. J. Environ. Health Sci. Eng. 5 (2), 79–90. Iran. Available from: 17352746 http://www.bioline.org.br/pdf?se08017.

Metcalf, W., Metcalf and Eddy, 2003. Wastewater Engineering: Treatment and Reuse. McGraw Hill.

Mohan, S.V., Rao, N.C., Prasad, K.K., Madhavi, B.T.V., Sharma, P.N., 2005. Treatment of complex chemical wastewater in a sequencing batch reactor (SBR) with an aerobic suspended growth configuration. Process Biochem. 40 (5), 1501–1508. https://doi.org/10.1016/j.procbio.2003.02.001. 13595113.

Mojiri, A., Aziz, H.A., Zaman, N.Q., Aziz, S.Q., Zahed, M.A., 2014. Powdered ZELIAC augmented sequencing batch reactors (SBR) process for co-treatment of landfill leachate and domestic wastewater. J. Environ. Manag. 139, 1–14. 10958630. Academic Press, Malaysia. Available from: https://www.sciencedirect.com/journal/journal-of-environmental-management. https://doi.org/10.1016/j.jenvman.2014.02.017.

Mojiri, A., Ohashi, A., Ozaki, N., Kindaichi, T., 2018. Pollutants removal from synthetic wastewater by the combined electrochemical, adsorption and sequencing batch reactor (SBR). Ecotoxicol. Environ. Saf. 161, 137–144. https://doi.org/10.1016/j.ecoenv.2018.05.053.

Morgenroth, E., Wilderer, P.A., 1998. Sequencing batch reactor technology: concepts, design and experiences (abridged). Water Environ. J. 12 (5), 314–320. https://doi.org/10.1111/j.1747-6593.1998.tb00192.x. 1747-6585.

Ng, W.J., Sim, T.S., Ong, S.L., Ng, K.Y., Ramasamy, M., Tan, K.N., 1993. Efficiency of sequencing batch reactor (SBR) in the removal of selected microorganisms from domestic sewage. Water Res. 27 (10), 1591–1600. https://doi.org/10.1016/0043-1354(93)90105-Q. 00431354.

Nguyen, A.L., Duff, S.J.B., Sheppard, J.D., 2000. Application of feedback control based on dissolved oxygen to a fixed-film sequencing batch reactor for treatment of brewery wastewater. Water Environ. Res. 72 (1), 75–83. 10614303. Water Environment Federation, Canada. Available from: http://www.ingentaconnect.com/content/wef/wer. https://doi.org/10.2175/106143000X137130.

Nolde, E., 2000. Greywater reuse systems for toilet flushing in multi-storey buildings—over ten years experience in Berlin. Urban Water 1 (4), 275–284. https://doi.org/10.1016/s1462-0758(00)00023-6. 14620758.

Ong, C., Lee, K., Chang, Y., 2020. Biodegradation of mono azo dye-reactive Orange 16 by acclimatizing biomass systems under an integrated anoxic-aerobic REACT sequencing batch moving bed biofilm reactor. J. Water Process Eng. 36, 101268. https://doi.org/10.1016/j.jwpe.2020.101268. 22147144.

Orhon, D., Karahan, O., Zengin, G., Olsson, O., Bauer, M., 2005. Mechanism and Design of Sequencing Batch Reactors for Nutrient Removal. IWA Publishing.

Payandeh, P.E., Mehrdadi, N., Dadgar, P., 2017. Study of biological methods in landfill leachate treatment. Open J. Ecol. 07 (09), 568–580. https://doi.org/10.4236/oje.2017.79038. 2162-1985.

Piergrossi, V., De Sanctis, M., Chimienti, S., Di Iaconi, C., 2018. Energy recovery capacity evaluation within innovative biological wastewater treatment process. Energy Convers. Manag. 172, 529–539. https://doi.org/10.1016/j.enconman.2018.07.013. 01968904.

Putra, A.A., Watari, T., Maki, S., Hatamoto, M., Yamaguchi, T., 2020. Anaerobic baffled reactor to treat fishmeal wastewater with high organic content. Environ. Technol. Innov. 17, 100586. https://doi.org/10.1016/j.eti.2019.100586. 23521864.

K. Ranjan, S. Chakraborty, M. Verma, J. Iqbal, R. Naresh Kumar, Co-treatment of old landfill leachate and municipal wastewater in sequencing batch reactor (SBR): effect of landfill leachate concentration Water Qual. Res. J. Can. 51 (4) (2016), 377–387, 12013080. IWA Publishing, India. Available from: http://wqrjc.iwaponline.com/content/ppiwawqrjc/51/4/377.full.pdf. https://doi.org/10.2166/wqrjc.2016.020.

Rashid, R., Shafiq, I., Akhter, P., Iqbal, M.J., Hussain, M., 2021. A state-of-the-art review on wastewater treatment techniques: the effectiveness of adsorption method. Environ. Sci. Pollut. Res. 28 (8), 9050–9066. 16147499. Springer Science and Business Media Deutschland GmbH, Pakistan. Available from: https://link.springer.com/journal/11356. https://doi.org/10.1007/s11356-021-12395-x.

Saidulu, D., Srivastava, A., Gupta, A.K., 2022. Enhancement of wastewater treatment performance using 3D printed structures: a major focus on material composition, performance, challenges, and sustainable assessment. J. Environ. Manag. 306, 114461. https://doi.org/10.1016/j.jenvman.2022.114461. 03014797.

Sánchez, J.B., Vuono, M., Dionisi, D., 2021. Model-based comparison of sequencing batch reactors and continuous-flow activated sludge processes for biological wastewater treatment. Comput. Chem. Eng. 144, 107127. https://doi.org/10.1016/j.compchemeng.2020.107127. 00981354.

Sarvajith, M., Nancharaiah, Y.V., 2022. Enhancing biological nitrogen and phosphorus removal performance in aerobic granular sludge sequencing batch reactors by activated carbon particles. J. Environ. Manag. 303, 114134. https://doi.org/10.1016/j.jenvman.2021.114134. 03014797.

Schwarzenbeck, N., Borges, J.M., Wilderer, P.A., 2005. Treatment of dairy effluents in an aerobic granular sludge sequencing batch reactor. Appl. Microbiol. Biotechnol. 66 (6), 711–718. https://doi.org/10.1007/s00253-004-1748-6. 0175-7598.

Shen, L., Qihang, W., Ye, Q., Lin, H., Zhang, J., Chen, C., Yue, R., Teng, J., Hong, H., Liao, B.-Q., 2023. Superior performance of a membrane bioreactor through innovative in-situ aeration and structural optimization using computational fluid dynamics. Water Res. 243, 120353. https://doi.org/10.1016/j.watres.2023.120353. 00431354.

Shin, H.S., Park, H.S., 1991. Enhanced nutrient removal in porous biomass carrier sequencing batch reactor (PBCSBR). Water Sci. Technol. 23 (4–6), 719–728. https://doi.org/10.2166/wst.1991.0522. 02731223. IWA Publishing South, Korea https://iwaponline.com/wst.

Shizas, I., Bagley, D.M., 2004. Experimental determination of energy content of unknown organics in municipal wastewater streams. J. Energy Eng. 130 (2), 45–53. https://doi.org/10.1061/(ASCE)0733-9402(2004)130:2(45. 07339402.

Singh, M., Srivastava, R.K., 2011. Sequencing batch reactor technology for biological wastewater treatment: a review. Asia Pac. J. Chem. Eng. 6 (1), 3–13. https://doi.org/10.1002/apj.490. 1932-2135.

Sirianuntapiboon, S., Sansak, J., 2008. Treatability studies with granular activated carbon (GAC) and sequencing batch reactor (SBR) system for textile wastewater containing direct dyes. J. Hazard. Mater. 159 (2–3), 404–411. https://doi.org/10.1016/j.jhazmat.2008.02.031. 03043894.

Smith, R., Wilderer, P.A., 1987. Treatability of Hazardous Landfill Leachate Using Sequencing Batch Reactors With Silicone Membrane Oxygenation. 41st Purdue Indust. Lewis Publ.

Soroosh, H., Otterpohl, R., Hanelt, D., 2022. Influence of hydraulic retention time on municipal wastewater treatment using microalgae-bacteria flocs in sequencing batch reactors. Bioresour. Technol. Rep. 17, 100884. https://doi.org/10.1016/j.biteb.2021.100884. 2589014X.

Tay, J.H., Liu, Q.S., Liu, Y., 2004. The effect of upflow air velocity on the structure of aerobic granules cultivated in a sequencing batch reactor. Water Sci. Technol. 49 (11–12), 35–40. https://doi.org/10.2166/wst.2004.0798. 0273-1223.

Van Lier, J.B., 2008. High-rate anaerobic wastewater treatment: diversifying from end-of-the-pipe treatment to resource-oriented conversion techniques. Water Sci. Technol. 57 (8), 1137–1148. https://doi.org/10.2166/wst.2008.040. 02731223.

Wang, L.K., Hung, Y.-T., Lo, H.H., Yapijakis, C., 2005. Waste Treatment in the Food Processing Industry. CRC Press.

Wang, Z.W., Ma, J., Chen, S., 2011. Bipolar effects of settling time on active biomass retention in anaerobic sequencing batch reactors digesting flushed dairy manure. Bioresour. Technol. 102 (2), 697–702. https://doi.org/10.1016/j.biortech.2010.08.045. 09608524.

Wang, X., Li, J., Zhang, X., Chen, Z., Shen, J., Kang, J., 2021. Impact of hydraulic retention time on swine wastewater treatment by aerobic granular sludge sequencing batch reactor. Environ. Sci. Pollut. Res. 28, 5927–5937. https://doi.org/10.1007/s11356-020-10922-w.

White, D.M., Pilon, T.A., Woolard, C., 2000. Biological treatment of cyanide containing wastewater. Water Res. 34 (7), 2105–2109. https://doi.org/10.1016/S0043-1354(99)00362-0. 00431354.

Wilderer, P.A., Irvine, R.L., Goronszy, M.C., 2001. Sequencing Batch Reactor Technology.

Xiang, Y., Shao, Z., Kang, W., Zou, B., Chai, H., 2016. Effect of biofilm density on nitrous oxide emissions and treatment efficiency on sequencing batch biofilm reactor. Water Air Soil Pollut. 227 (9). https://doi.org/10.1007/s11270-016-3009-6. 0049-6979.

Yakamercan, E., Aygün, A., 2021. Fate and removal of pentachlorophenol and diethylhexyl phthalate from textile industry wastewater by sequencing batch biofilm reactor: effects of hydraulic and solid retention times. J. Environ. Chem. Eng. 9 (4), 105436. https://doi.org/10.1016/j.jece.2021.105436. 22133437.

Zhang, S., Tchalala, M.R., Cheng, H., Medina, J.S., Xiong, Y., Hong, P.Y., 2023. Anaerobic membrane bioreactor as the core technology for future decentralized wastewater treatment plants. Curr. Opin. Chem. Eng. 42. 22113398. Elsevier Ltd., Saudi Arabia. Available from: http://www.elsevier.com/wps/find/journaldescription.cws_home/725837/description#description. https://doi.org/10.1016/j.coche.2023.100975.

Zhao, J., Yuan, Q., Sun, Y., Zhang, J., Zhang, D., Bian, R., 2021. Effect of fluoxetine on enhanced biological phosphorus removal using a sequencing batch reactor. Bioresour. Technol. 320, 124396. https://doi.org/10.1016/j.biortech.2020.124396. 09608524.

4

Membrane bioreactors for wastewater treatment

Sameer Al-Asheh, Abdallah Alhajar, and Ahmad Aidan
Department of Chemical and Biological Engineering, American University of Sharjah, Sharjah, United Arab Emirates

Abbreviations

AEM	anion exchange membrane
ADMBR	aerobic dynamic membrane bioreactor
AFMBR	aerobic fluidized membrane bioreactor
AMBR	aerobic membrane bioreactor
AnMBRs	anaerobic membrane bioreactors
AOP	advanced oxidation process
BOD	biological oxygen demand
CEM	cation exchange membrane
COD	chemical oxygen demand
DS	draw solution
F/M	food to microorganism ratio
FO	forward osmosis
HRT	hydraulic retention time
MBR	membrane bioreactor
MBPR	membrane photobioreactors
MC	membrane crystallization
MCRT	mean cell residence time
MD	membrane distillation
MF	microfiltration
MFC	microbial fuel cell
MLSS	mixed liquor suspended solids
NF	nanofiltration
ODMPs	osmotically driven membrane processes
OMBR	osmotic membrane biological reactor
PEM	proton exchange membrane
PRO	pressure retarded osmosis
PV	pervaporation
RO	reverse osmosis
SRT	solid retention time
TOC	total organic carbon
TPV	thermopervaporation
TMP	transmembrane pressure
UF	ultrafiltration
VSS	volatile suspended solid

Municipal Wastewater Treatment. https://doi.org/10.1016/B978-0-443-24826-9.00004-8

Symbols

A	water permeability constant of the membrane
J_w	water flux, mole/m^2 s
K_s	half saturation constant for biodegradable substrate, $\frac{\text{gCOD}}{\text{m}^3}$
k_d	decay coefficient, $\frac{\text{gVSS}}{\text{gVSS day}}$
$R_{g,net}$	net growth rate, $\frac{\text{gVSS}}{\text{m}^3\text{ day}}$
S	biodegradable substrate concentration, $\frac{\text{gCOD}}{\text{m}^3}$
X	biomass concentration, $\frac{\text{gVSS}}{\text{m}^3}$
ΔP	applied pressure, kPa
$\Delta\pi$	osmotic pressure difference, kPa
μ_m	maximum specific growth rate, day^{-1}
σ	reflection coefficient

4.1 Introduction

A membrane bioreactor (MBR) is an advanced technology used for wastewater treatment that combines biological treatment processes with membrane filtration. It represents a significant advancement in wastewater treatment technology and is widely employed in various industrial, municipal, and decentralized settings.

The MBR system consists of two primary components: a biological reactor and a membrane filtration unit. The biological reactor employs microorganisms to break down organic pollutants present in the wastewater. This biological process converts organic matter into microbial biomass and gases like carbon dioxide and methane. The membrane filtration unit, on the other hand, acts as a physical barrier that separates treated water from the biomass.

The membranes used in MBRs are typically made of materials like polymeric, ceramic, or composite substances. They come in various configurations, including flat sheet, hollow fiber, and tubular types. These membranes have fine pores that effectively filter out suspended solids, pathogens, and other contaminants, producing high-quality effluent.

One of the key advantages of MBR technology is its compactness and efficiency. It eliminates the need for secondary clarifiers and allows for a significantly smaller footprint compared to conventional wastewater treatment plants. Additionally, MBRs produce a consistently high-quality effluent that meets strict regulatory standards, making it suitable for reuse in various applications.

Moreover, MBR systems are highly adaptable and can be tailored to handle a wide range of influent characteristics and flow rates. They are effective in treating both municipal and industrial wastewater, including challenging streams with high levels of contaminants.

Despite these advantages, MBR technology also presents some challenges, such as the potential for membrane fouling, which requires regular maintenance and cleaning. However, ongoing research and development efforts are focused on addressing these challenges and further improving the efficiency and cost-effectiveness of MBR systems.

Overall, MBRs play a crucial role in modern wastewater treatment strategies, offering a sustainable and efficient solution for the purification of water resources in a variety of settings. This review intends to present the different aspects related to MBR, including membrane technology, membrane fouling and characteristics, and the different technologies for BMR and their applications in wastewater treatment.

4.2 Membrane technology

A membrane is a material that is permeable or semi-permeable, enabling certain species to pass through (Scott, 1995). The permeate contains the species that successfully crossed the membrane, while the retentate consists of the species that the membrane prevented from passing through. The performance of a membrane is governed by two key factors: flux and retention. Flux refers to the rate at which fluid flows through the membrane per unit area. Retention is the amount of solute in the feed that has been separated by the membrane. This can be evaluated using a parameter called selectivity, which compares the flow rates of different species across the membrane. Therefore, an ideal membrane is one that is optimal to maximize both flux and retention (Scott, 1995).

4.2.1 Types of membrane separation processes

Membranes owe their distinctiveness to their role in establishing a separation phase. This phase, which can exist in solid, liquid, or gaseous states, is an intermediary between the permeate and the retentate. It may possess an electrical charge or remain neutral, and its porosity, whether porous or non-porous, depends on the specific characteristics of the species intended for separation

(Scott, 1995). To achieve the separation, a driving force is applied across the membrane, typically using a pressure, concentration, or temperature gradient. In certain membrane applications such as electrodialysis, an electrical potential is necessary to propel the separation process forward.

4.2.1.1 Pressure driven membrane processes

Pressure-driven membrane processes are recognized as the most prominent separation methods in membrane technology. Their importance lies in the use of hydraulic pressure to expedite the separation process (Nath, 2017). These membranes are essentially microporous and are categorized based on the size of their pores. The various types of pressure-driven membranes and their applications by their pore size distributions are as follows:

- Reverse Osmosis (RO): These membranes have pore sizes ranging from 0.0001 to 0.001 μm (Nath, 2017). They find considerable use in water desalination due to their pores being smaller than sodium chloride but larger than water.
- Nanofiltration (NF): Their pore sizes range from 0.0002 to 0.002 μm (Chuntanalerg et al., 2019). These pores reject divalent molecules such as calcium carbonate, making them indispensable for water softening. Additionally, they are a crucial component in removing heavy toxic metals such as mercury, arsenic, and lead.
- Ultrafiltration (UF): With pore sizes ranging from 0.005 to 0.1 μm (Arvanitoyannis et al., 2008), these membranes prevent the passage of microorganisms, including bacteria, viruses, and parasites.
- Microfiltration (MF): Their pore sizes range from 0.1 to 10 μm (Pal, 2020). They are particularly employed in food processing industries, ensuring the removal of waterborne parasites such as Protozoa and Giardia.
- This method employs a non-porous membrane, allowing the passage of species that are soluble in the membrane. It is widely utilized in the extraction of water and small volatile compounds.

4.2.1.2 Concentration driven membrane processes

Concentration-driven membrane processes, such as osmotically driven membrane processes (ODMPs), utilize the concentration difference across a membrane to create a flow of substances. These processes are commonly employed in water treatment and power generation, such as forward osmosis (FO) and pressure

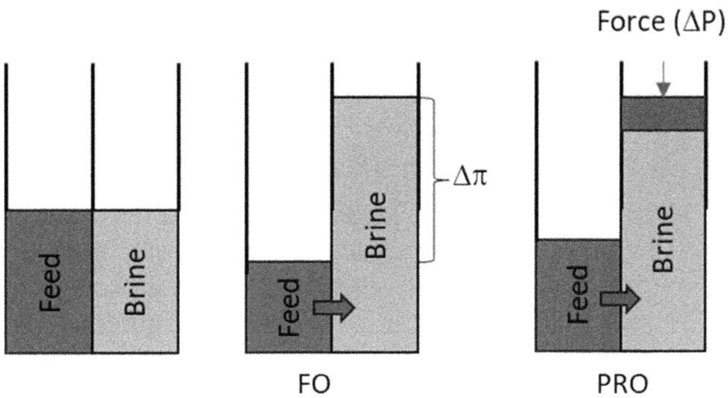

Fig. 4.1 Solvent flows in FO and PRO.

retarded osmosis (PRO), respectively. In both FO and PRO, mass flow is caused by the osmotic pressure difference between a diluted feed solution and a concentrated draw solution, or brine, as indicated in Fig. 4.1.

The flux for the ODMPs can be determined using the following equation:

$$J_w = A(\sigma\Delta\pi - \Delta P)$$

where, J_w is the water flux, A is the water permeability constant of the membrane, σ is the reflection coefficient, $\Delta\pi$ is the osmotic pressure difference, and ΔP is the applied pressure. In FO, a non-porous semipermeable membrane is an obstacle between the feed solution and the draw solution. As water passes through it, the membrane selectively filters out contaminants. The draw solution is subsequently re-concentrated and reused to sustain the separation process. Similarly, Pressure Retarded Osmosis (PRO) operates on this fundamental principle, but it incorporates a turbine positioned in the high-pressure flow to generate hydroelectric power.

One obstacle encountered with ODMPs is the accumulation of salt at the membrane's surface, resulting in the phenomenon known as concentration polarization. This phenomenon creates a new concentration gradient, resulting in a reverse flux that hinders the separation process by reducing water flow and obstructing the membrane. Nevertheless, research indicates that the effectiveness of ODMPs is significantly influenced by the choice of membrane (Cath et al., 2013). It has been determined that densely packed, thin, minimally porous, and robustly supported membranes are optimal for enhancing the performance of

ODMPs, as they achieve high solute rejection, minimize crystallization polarization, facilitate high solvent flux, and withstand hydraulic pressure effectively.

4.2.1.3 Temperature driven membrane processes

Membrane processes driven by temperature differentials (thermal-driven) achieve the separation of volatile substances from the initial solution through a process involving evaporation across a membrane followed by condensation on the opposite side of the membrane (Volkov et al., 2016). This separation mechanism is based on maintaining the feed solution at a high temperature, while the permeate is kept at a lower temperature. This principle involves membrane distillation (MD), membrane crystallization (MC), pervaporation (PV), and thermopervaporation (TPV). MD utilizes primary application in water treatment and desalination, providing nearly complete removal of non-volatile substances, independence from salinity levels for water transportation, and eliminating the need for disinfection stages. MC provides a controlled pathway for nucleation, where a porous membrane allows the desired species to pass through, while the rate of supersaturation is regulated on the opposite side of the membrane (Drioli et al., 2012). Furthermore, PV and TPV utilize dense, non-porous membranes that facilitate the absorption of selectively targeted species (Volkov et al., 2016). These processes are particularly utilized for the extraction of volatile organic compounds.

4.2.1.4 Electrically driven membrane processes

Electrically powered membrane processes utilize electric potential to facilitate the separation process (Rype, 2003). Consequently, these processes are primarily applicable to the separation of charged substances, where the segregation of positively and negatively charged ions is essential. One notable application is the removal of salts from seawater. Additionally, there is ongoing research into their potential use in hydraulic fracturing. The membrane selectively allows the passage of cations or anions, with each being attracted to the oppositely charged terminal of the electric potential source, as illustrated in Fig. 4.2.

4.2.2 Types of synthetic membranes

Membranes and filters function similarly, however, filters are typically referred to as media that separate suspended particulates larger than 1 to 10 μm. Moreover, membranes have a wider range of applications and can be classified into five types, which are briefly described below.

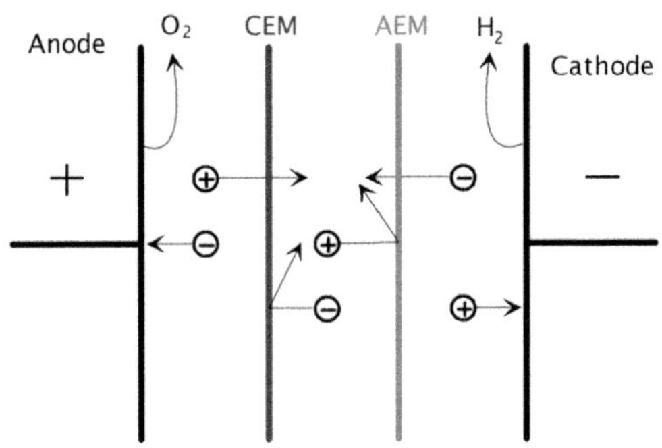

Fig. 4.2 Schematic of an electrically driven membrane process. *CEM*, cation exchange membrane; *AEM*, anion exchange membrane.

4.2.2.1 Microporous membranes

Microporous membranes share a similar structure with filters. They possess rigid structures with randomly distributed pores (Nath, 2017). However, these pores are very small, with diameters of up to 10 μm. The principle of their operation is straightforward: particles with a diameter greater than the membrane pores are rejected, thus achieving the desired separation. The two kinds of microporous membranes are isotropic and anisotropic. The pores of anisotropic membranes are the same size, while anisotropic membranes have different pore sizes throughout their structure.

4.2.2.2 Asymmetric membranes

Asymmetric membranes, such as skinned membranes, are characterized by a thin skin layer on the surface of the membrane, typically with a thickness of 0.1 to 1 μm (Nath, 2017; Manickam and McCutcheon, 2017). This skin layer is supported by a layer that lacks voids. The separation process in asymmetric membranes is not solely based on the size of the membrane's pores, but also based on the properties of the membrane material itself. The two types of asymmetric membranes are integrally skinned and non-integrally skinned. The skin layer is porous and formed through a phase inversion process. In addition, non-integrally skinned membranes have homogeneous skin layers that are deposited from a solution.

4.2.2.3 Thin film composite

Thin film composites were primarily utilized for reverse osmosis (RO) and nanofiltration (NF) applications. These membranes are constructed by combining a thin-colored polymer skin with a microporous film for support (Lau et al., 2012). The microporous film has fewer pores than the thickness of the skin layer, which enhances adhesion. Thin film composites made a significant improvement in RO applications by providing improved performance in terms of pH stability, pressure, temperature, and salt rejection compared to the previously used cellulose acetate membranes. However, it is worth noting that thin film composites possess a greater susceptibility to biofouling (Nath, 2017).

4.2.2.4 Electrically charged membranes

Electrically charged membranes are created using gels that carry either a positive or negative charge. Membranes with a positive charge are known as anion exchange membranes (AEMs), while those with a negative charge are referred to as cation exchange membranes (CEMs) (Nath, 2017; Barbhuiya et al., 2021). These charged membranes attract ions that are either oppositely charged or negatively charged. Charged membranes have a porous or non-porous structure, although they typically consist of very fine micropores.

4.2.2.5 Inorganic membranes

Inorganic membranes, also known as ceramic membranes, are specifically designed to function in high temperatures and harsh conditions. They can operate at temperatures exceeding 1000 °C and are resistant to corrosive liquids and gases (Nath, 2017; Goh and Ismail, 2018). These membranes have an asymmetrical structure with two or three distinct porosity levels. A mesoporous layer is often applied beneath the active microporous layer to reduce surface roughness.

4.2.3 Membrane fouling

Membranes provide effective solutions to various industrial challenges. However, membrane fouling is a significant obstacle that limits their applications and hinders their effectiveness in certain situations. The extent of fouling is influenced by several factors, such as the nature of the membrane, properties of the influent, operating conditions, and the methods employed for membrane cleaning. These factors collectively contribute to the challenge of membrane fouling and need to be carefully considered to mitigate its impact.

4.2.3.1 Fouling phenomena

Fouling in membrane processes refers to the obstruction of the membrane's pores by solute molecules, which hinders the diffusion of the solvent. This obstruction leads to a decrease in

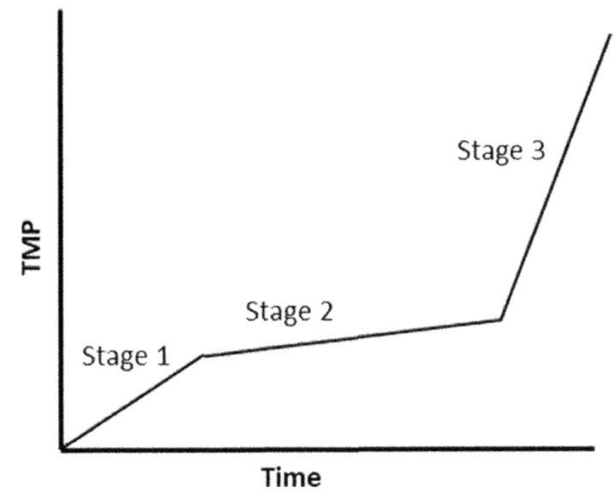

Fig. 4.3 TMP's three-stage jump pattern.

permeation flux or an increase in transmembrane pressure (TMP), both of which serve as indicators of fouling (Al-Asheh et al., 2021). Generally, TMP is considered a more reliable indicator since most processes operate at a constant flux.

TMP exhibits two distinct patterns: a two-stage jump and a three-stage jump. The two-stage jump initially consists of a slight TMP increase due to the adsorption of particles into the membrane's pores, while still maintaining the local flux below a critical threshold. Once the local flux surpasses this critical point, there is a significant increase in TMP, indicating that the membrane is now blocked and requires cleaning. The three-stage jump, on the other hand, displays a rapid and minor TMP increase caused by the swift pore obstruction in the initial stage. The subsequent two jumps follow a similar pattern as the two-stage TMP jump. In the second stage, TMP rises at a more gradual and steady pace until it eventually exceeds the critical flux, leading to the third stage characterized by a sharp TMP increase. Fig. 4.3 provides a visual representation of these TMP jump configurations.

4.2.3.2 Fouling rate

Fouling in membranes progresses through various stages. The smallest pores become obstructed. Similarly, molecules adhere to the inner surfaces of these pores. As time progresses, flowing molecules accumulate over those already inside the pores. Finally, a cake layer appears. These stages are portrayed in Fig. 4.4 (Wang et al., 2014). Characterizing each stage is challenging; therefore, the overall fouling tendency is determined through the fouling rate. This rate is defined as the TMP per unit of time. The

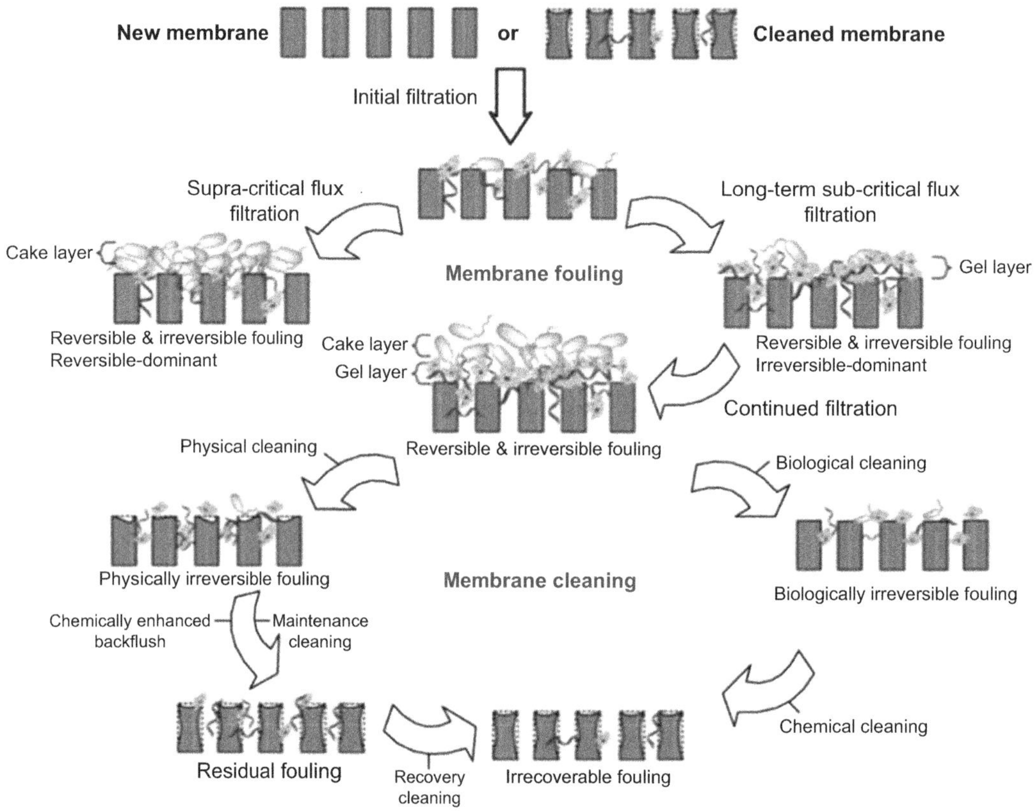

Fig. 4.4 Membrane fouling and cleaning stages (Wang et al., 2014).

operating flux and fouling rate have a proportional relationship. As the operating flux increases, the fouling rate decreases until it reaches the critical flux, at which point the fouling rate experiences a sudden and significant increase.

4.2.3.3 Classification of fouling

Fouling in membrane processes can be categorized by three factors: flux recovery after cleaning, fouling location, and solid deposition pattern (Maaz et al., 2019).

The process of restoring the flux following a cleaning procedure (Wang et al., 2014). It can be further categorized into three categories: reversible, irreversible, and irrecoverable fouling. Reversible fouling indicates that the flux can be easily restored through a simple physical cleaning technique, such as backwashing, pressure relaxation, or air scouring. When the fouling layer becomes strongly established, the reversible fouling becomes irreversible. Irreversible fouling necessitates biological cleaning, as physical methods are not sufficient. In cases where the membrane has been fouled over a prolonged period, it cannot be recovered, as it cannot be cured.

The spatial distribution of fouling on the membrane is a concern. It can be classified into three forms: clogging, cake layer, or internal pore fouling. Insufficiently designed membrane modules may lead to clogging, where particles accumulate within the pores, causing flow to decrease and increasing TMP. A cake layer is formed when particles adhere to the membrane's surface, resulting in a secondary filter that enhances permeate purity but also reduces flux and hinders the separation. The third form, internal pore fouling, involves particles adhering to the walls of the internal pores. In addition, cake layer formation is the most prevalent in describing fouling (Al-Asheh et al., 2021).

The solid deposition pattern refers to the way solids adhere to the membrane. This classification is closely related to fouling location, encompassing cake layer formation, pore narrowing, and pore plugging. Pore narrowing involves particles adhering to the inner pore walls, since their diameters are smaller than the pore itself. Similarly, pore plugging refers to the blocking of the pores due to particles with a larger diameter (Goh and Ismail, 2018).

4.2.3.4 Factors affecting membrane fouling

Fouling in MBRs is primarily influenced by three key factors: membrane characteristics, operating conditions, and microbial characteristics.

Membrane characteristics

Membrane pores are intended to be slightly larger than the desired size of the molecules intended to pass through. In applications such as wastewater treatment, hydrophilic membranes are usually preferred over hydrophobic ones. This preference is based on the fact that hydrophilic membranes exhibit strong interactions with the components of wastewater, resulting in increased separation rates. However, it is important to note that hydrophilic membranes are also associated with higher fouling rates.

Moreover, the choice of membrane type significantly influences fouling reduction. The majority of commercially available membranes are polymeric, making them less capable of achieving the harsh conditions commonly encountered in wastewater, such as high temperatures or extreme pH levels. As previously mentioned, inorganic membranes are more suitable for such anticipated conditions; however, they have a higher cost and limited configuration choices. In the context of wastewater treatment, there is a preference for higher packing density, a better criterion met by polymeric membranes as opposed to inorganic ones, which provide modules with lower packing density (Yue et al., 2015; Fan et al., 2018).

Operating conditions

The HRT stands as a pivotal parameter in wastewater treatment in which microorganisms are used to degrade organic matter. This parameter is defined as the ratio of the size of the wastewater treatment reactor to the inlet volumetric flow rate, essentially indicating the duration of the wastewater remains inside the treatment reactor. A higher HRT results in a more prolonged interaction between the resident microorganisms and the substances in the wastewater. The food-to-microorganism ratio (F/M) is closely linked to it. With an increase in HRT, F/M decreases, implying a reduced flow rate and subsequently less nourishment available to microorganisms. Therefore, a lower level of fouling is expected.

Furthermore, the mean cell residence time (MCRT) also plays a role in fouling. The duration of microorganisms, or activated sludge, in the wastewater treatment clarifier is measured by MCRT. It is directly related to the mixed liquor suspended solids (MLSS), which is the concentration of suspended solids within the reactor containing the activated sludge. Elevated MLSS concentration could pose a risk to the membrane, as it is characterized by a viscous fluid that can easily obstruct it. It is noteworthy that enhancing MCRT reduces the presence of substances in the reactor, thereby reducing fouling tendencies. Therefore, it is essential to optimize the reactor design to enhance MCRT while avoiding the adverse effects of excessively high MLSS concentration (Hamedi et al., 2019).

Microbial characteristics

Microorganisms, biomass, or dissolved biopolymers can be characterized by microorganisms, biomass, or dissolved biopolymers. Their quantification can be achieved by assessing the Mixed Liquor Suspended Solids (MLSS) concentration in the bioreactor. As noted earlier, a high MLSS concentration heightens the likelihood of fouling. The biomass gradually accumulates on the membrane's surface, enhancing the thickness of the cake layer. However, this layer is constrained to a specific thickness due to the shear forces exerted on the membrane by the aeration source (Al-Asheh et al., 2021).

4.3 Membrane biological reactors (MBRs)

MBRs combine the primary components of traditional secondary wastewater treatment into a single process utilizing membrane separation (Melin et al., 2006). Within the MBR, the activated sludge reactor functions similar to a conventional secondary treatment reactor but eliminates the need for a secondary clarifier. Two main MBR configurations exist, as depicted in Fig. 4.5. The first, pioneered by Smith et al. in 1969 (Radjenović et al., 2008), is the side-stream configuration where an ultrafiltration membrane sits outside the reactor to separate treated water from activated sludge. Although this setup initially provided exceptionally pure water, it came with high energy consumption and challenges in addressing membrane fouling. To overcome these limitations, Yamamoto et al. introduced the second configuration in 1989, employing a submerged membrane (Yamamoto et al., 1989). Here, a hollow fiber membrane resides inside the bioreactor, eliminating the requirement for a pressurizing pump to transfer effluent from the aeration tank to the membrane. Instead, a suction pump at the membrane intermittently removes water. Subsequent studies have aimed to improve MBR performance by reducing both capital and operational expenses while addressing membrane fouling (Al-Asheh et al., 2021).

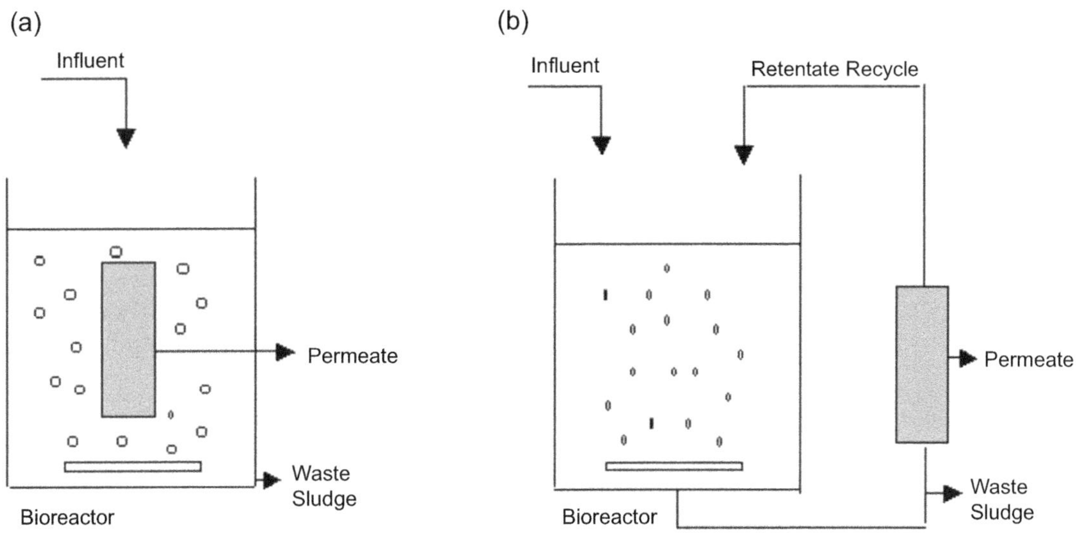

Fig. 4.5 Configuration of MBRs: (A) submerged, (B) side-stream (Melin et al., 2006). Copyright 2006 Elsevier. License Number 5710580046154.

4.3.1 Types of MBRs

MBRs are advanced wastewater treatment systems that combine biological processes with membrane filtration technology. Various types of MBR configurations exist, each offering distinct advantages and applications in wastewater treatment processes. These are discussed below.

4.3.1.1 Aerobic membrane bioreactors (AMBRs)

An AMBR combines an aerated activated sludge tank with a membrane, effectively removing suspended solids through membrane filtration and utilizing microbial activity to eliminate dissolved chemical substances. For instance, carbon is metabolized by microbes, and Nitrosomonas bacteria aid in ammonia removal through nitrification and denitrification, while other substances like phosphate are consumed by microbial action. Oxygen is crucial for these processes, hence the aeration of the reactor (Achilli and Holloway, 2015). AMBRs provide consistently high-quality output, occupy less space, and nearly completely remove suspended solids, resulting in water suitable for irrigation or further treatment for potable use. However, membrane fouling poses a significant challenge.

Dvořák et al. (2013) investigated the application of AMBRs for pharmaceutical and chemical wastewater treatment. With an initial COD concentration of 1898 ppm, the wastewater underwent treatment with a HRT of 14 days and a SRT ranging from 30 to 51 days, achieving an impressive 80% COD removal. Blending industrial wastewater with municipal wastewater helped lower initial COD levels, aiding the treatment process. The system effectively handled a maximum of 30% industrial wastewater content in the influent.

Ahmadi et al. (2019) employed AMBRs for oil refinery wastewater treatment, which had a COD range of 195 to 590 ppm in the influent. They varied the HRT between 12 and 24 h and the Mixed Liquor Suspended Solids (MLSS) concentration between 6.5 and 8.5 g/L to evaluate their impact on system performance. Overall, the system demonstrated COD removal rates between 74% and 97%, with HRT being the most influential factor.

4.3.1.2 Anaerobic MBRs (AnMBRs)

AnMBRs are esteemed for their ability to achieve treatment results similar to AMBRs while substantially saving energy (Smith et al., 2012). An AnMBR combines an anaerobic bioreactor with a membrane, which can be either submerged within the bioreactor or set up in a side-stream configuration. However, most designs advocate for submerging the membrane in a separate tank external to the anaerobic reactor, from which water can be pumped to the membrane chamber. This setup ensures that the membrane cleaning process doesn't disrupt the anaerobic conditions of the wastewater.

Despite demonstrating notable energy efficiency, AnMBRs encounter limitations due to the necessity of maintaining anaerobic conditions (Smith et al., 2012). Sustaining anaerobic environments requires temperatures ranging from 30 to 60 °C (Huang et al., 2018). Consequently, AnMBR applications have been confined to regions with naturally occurring mesophilic temperatures, avoiding the energy-intensive process of heating wastewater. Additionally, a SRT is crucial, potentially leading to membrane fouling from extended membrane usage. Increasing the SRT results in heightened biomass concentration, reducing permeate flux. Furthermore, it increases concentrations of soluble microbial products and extracellular polymeric substances, both significant contributors to membrane fouling. Nevertheless, AnMBRs have demonstrated the ability to achieve over 90% removal of (COD) (Hu and Stuckey, 2006).

Wijekoon et al. (2011) investigated AnMBRs for treating high-strength molasses-based wastewater. The study subjected the wastewater to varying organic loading rates ranging from 2 to 22 kg COD/m^3d, operating at a temperature of 55 °C and a HRT of 16 h. Results indicated that COD removal increased as the organic loading rate rose from 5 to 8 kg COD/m^3d, showing a 3% improvement. However, further increases in loading rate led to a decline in removal efficiency. Ultimately, COD removal peaked at 81% and then decreased to 61% at the highest loading rate. Similarly, Biological Oxygen Deman (BOD_5) removal initially reached 96% but later decreased to 89%.

Lin et al. (2011) conducted a study on AnMBRs for municipal wastewater treatment. The influent had a COD concentration ranging from 342 to 527 ppm, with a pH of 7 and a temperature of 30 °C. The research demonstrated an impressive COD removal of 90%. However, it was noted that the system incurred a high life-cycle cost primarily due to expenses associated with the membrane.

In another study, An et al. (2009) investigated the application of AnMBRs for municipal wastewater treatment. The influent wastewater had a COD concentration of 259.5 ppm, ammoniacal nitrogen at 27.5 ppm, and total phosphorus at 4.2 ppm. The AnMBR achieved a COD removal of 70%, no removal of ammoniacal nitrogen, and a 24% removal of total phosphorus.

4.3.1.3 Factors affecting MBRs performance

In MBRs, two key factors significantly influence their performance. Firstly, the chemical properties of micropollutants in the wastewater are crucial. These micropollutants encompass a range of contaminants, including pharmaceuticals, industrial chemicals, and various organic or inorganic substances. The MBR's effectiveness in removing these micropollutants depends on factors such as its design, the specific membranes utilized, and the characteristics of the contaminants themselves.

Secondly, the HRT is a critical operational parameter. The HRT denotes the duration that wastewater remains within the MBR system, directly impacting treatment capacity and efficiency. A longer HRT typically enables more comprehensive biological and physical treatment processes, leading to higher contaminant removal rates.

Overall, comprehending and optimizing these factors are imperative for attaining the desired treatment levels in MBR systems.

4.3.2 Biokinetics

Biokinetics encompasses the rate at which microbial organisms utilize substrates present in wastewater to foster growth and treat the influent. In this context, microorganisms serve as catalysts, generating biomass while reproducing themselves. Their significance lies in their ability to forecast and estimate biological performance and treatments. Additionally, the reaction rate is pivotal for determining the bioreactor volume required to achieve the desired biomass concentration.

Biokinetics mainly revolve around microbial growth and substrate utilization rates. Microbial growth occurs through the metabolism of biodegradable substrates in the wastewater. However, the influent also contains non-biodegradable substrates, necessitating removal through techniques like flocculation or filtration. Furthermore, microorganisms naturally undergo decay during growth, affecting the net growth rate, which can be assessed using the following equations:

$$R_{g,net} = R_{growth} - R_{decay} \tag{1}$$

$$R_{g,net} = \frac{dX}{dt} = \frac{\mu_m SX}{K_s + S} - k_d X \tag{2}$$

where $R_{g,net}$ is the net growth rate $\left(\frac{gVSS}{m^3\ day}\right)$, X is the biomass concentration $\left(\frac{gVSS}{m^3}\right)$, S is the biodegradable substrate concentration $\left(\frac{gCOD}{m^3}\right)$, μ_m is the maximum specific growth rate (day^{-1}), K_s is the half saturation constant for biodegradable substrate $\left(\frac{gCOD}{m^3}\right)$, and k_d is the decay coefficient $\left(\frac{gVSS}{gVSS\ day}\right)$. Table 4.1 summarizes the biokinetic coefficients of different MBR plants for different bacteria.

4.3.3 Hydraulic retention time (HRT)

HRT refers to the duration that wastewater remains within the bioreactor. As mentioned earlier, a shorter HRT can decrease the likelihood of membrane fouling, thus enhancing the performance of MBR systems. In a study conducted by Isma et al. (2014), synthetic wastewater underwent treatment using an AMBR. The process maintained a solid retention time of 30 days, while the HRT was varied at intervals of 4, 8, and 12 h.

The study's findings revealed that a 12-h HRT resulted in the highest removal efficiencies for various parameters, including COD (95%), BOD5 (93%), total suspended solids (TSS) (99%), NH3-N (97%), and PO_4^{3-} (81%). Across all HRT durations, removal efficiencies above 90% were observed for most parameters, except for PO_4^{3-}. In this case, 4 and 8-h HRTs exhibited removal percentages of 62% and 68%, respectively (Isma et al., 2014).

4.3.4 Chemical properties of micropollutants

The chemical properties of micropollutants are pivotal in their removal process. These contaminants, found in wastewater, often exhibit hydrophobic traits, leading them to repel water. As a result, these hydrophobic pollutants tend to adhere to sludge, suspended solids, and particulate matter during treatment. This characteristic aids in the treatment process, enabling these substances to settle in the clarifier or be removed by the membrane (Cirja et al., 2008).

Table 4.1 Biokinetic coefficient of different bacteria in different MBRs.

MBR characteristics	Biomass type	Yield coefficient	Maximum specific growth rate	Half-saturation coefficient	Decay coefficient	Refs.
Submerged hollow fiber ultrafiltration MBR worked at HRT = 9.5 h, T = 14.7 °C, MLSS concentration = 6.6 g/L	Heterotrophic bacteria	0.4887 $\left(\frac{mgVSS}{mgCOD}\right)$	0.0141 (h^{-1})	7.467 $\left(\frac{mgO_2}{L}\right)$	0.0521 (day^{-1})	Leyva-Diaz et al. (2020)
	Ammonium-oxidizing bacteria	1.191 $\left(\frac{mgVSS}{mgN}\right)$	0.1612 (h^{-1})	0.2204 $\left(\frac{mgN}{L}\right)$		
	Nitrite-oxidizing bacteria	0.6473 $\left(\frac{mgVSS}{mgN}\right)$	0.0786 (h^{-1})	0.324 $\left(\frac{mgN}{L}\right)$		
Submerged hollow fiber ultrafiltration MBR HRT = 9.5 h, MLSS concentration = 3.3 g/L	Heterotrophic bacteria	0.4609 $\left(\frac{mgVSS}{mgCOD}\right)$	0.01917 (h^{-1})	16.47 $\left(\frac{mgO_2}{L}\right)$	Total bacteria decay = 0.03043 (day^{-1})	Leyva-Díaz et al. (2015)
	Nitrifying bacteria	1.0389 $\left(\frac{mgO_2}{mgN}\right)$	0.27193 (h^{-1})	0.9329 $\left(\frac{mgN}{L}\right)$		
	Nitrite-oxidizing bacteria	0.77913 $\left(\frac{mgO_2}{mgN}\right)$	0.11244 (h^{-1})	0.4364 $\left(\frac{mgN}{L}\right)$		
Immersed MBR operated at HRT = 13 h, MLSS concentration = 5 g/L under steady state condition	*Serratia liquefaciens* and *Aeromonas hydrophila* (predominant bacteria)	0.567 $\left(\frac{mg}{mg}\right)$	0.0233 (h^{-1})	326.14 $\left(\frac{mgCOD}{L}\right)$	0.062 (day^{-1})	Al-Malack (2006)

Microfiltration MBR operated at HRT = 33 h, $T = 25\,°C$	Heterotrophic biomass	$0.756\ \left(\frac{mgCOD}{mgCOD}\right)$	$3.687\ \left(\frac{mg}{g.h}\right)$		0.353 (day^{-1})	Ruiz et al. (2011)
Moving bed biofilm reactor combined with MBR operated at HRT = 26.47 h, $T = 15\,°C$, and MLSS concentration = 2.9 g/L	Heterotrophic biomass	$0.5041\ \left(\frac{mgVSS}{mgCOD}\right)$	0.00484 (h^{-1})	$0.96\ \left(\frac{mgO_2}{L}\right)$	Total bacteria decay = 0.04844 (day^{-1})	Leyva-Díaz et al. (2013)
	Autotrophic biomass	$0.77718\ \left(\frac{mgO_2}{mgN}\right)$	0.02632 (h^{-1})	$0.76\ \left(\frac{mgN}{L}\right)$		
MBR plant with hollow fiber membrane operated at HRT = 8 h, $T = 27\,°C$, and MLSS concentration = 1.3 g/L	Heterotrophic biomass	$0.703\ \left(\frac{gVSS}{gCOD}\right)$			0.02 (day^{-1})	Huang et al. (2020)
Bioreactor plant by using glucose as substrate at $T = 20\,°C$	*Shewanella baltica* KB30	$0.6681\ \left(\frac{mgVSS}{mgCOD}\right)$	0.0840 (h^{-1})	$1.608\ \left(\frac{mgO_2}{L}\right)$		Leyva-Díaz et al. (2017)

4.4 Recent developments in MBRs

Recent developments in MBRs have focused on enhancing efficiency, improving membrane technology, and optimizing system design for better performance in wastewater treatment. These are discussed below.

4.4.1 Integrated microbial fuel cell—Membrane biological reactor (MFC-MBR) system

Microbial fuel cells (MFCs) are devices that utilize bacteria to break down various organic compounds, generating an electric current (Logan et al., 2006). These cells contain an anode where organic substrates undergo oxidation, producing protons and electrons. While electrons flow through an external circuit to the cathode, generating current, protons pass through a proton exchange membrane (PEM) to the cathode. At the cathode, protons combine with oxygen, forming water.

MFCs hold promise for converting biomass into energy sources during wastewater treatment. However, their development faces challenges in terms of cost and technical aspects. Integrating MFCs into existing wastewater treatment processes enhances their performance. This led to the development of the MFC-MBR system, where the bioreactor serves as the cathode chamber. In a study by Wang et al. (2014), an MFC without a PEM was placed in the aeration tank of an MBR. The results showed that the MFC did not hinder the MBR's performance. The aeration tank promoted biofilm formation on the cathode, improving oxygen's role as an oxidizer and significantly contributing to generating sufficient current.

The primary factor affecting the adoption of MFCs is the cost associated with the PEM. The model proposed by Wang et al. (2014), illustrated in Fig. 4.6, eliminates the need for a PEM, making integration cost-effective. Therefore, MFC-MBR systems demonstrate high potential for sustainable energy generation and wastewater treatment simultaneously.

4.4.2 Osmotic membrane biological reactor (OMBR)

Osmotic membrane biological reactors (OMBRs) set themselves apart from MBRs by utilizing forward osmosis (FO) for separation. They employ a non-porous membrane and a draw solution (DS) to establish the necessary concentration gradient for wastewater movement, as depicted in Fig. 4.7. FO membranes offer improved rejection rates, reduced fouling, and efficient

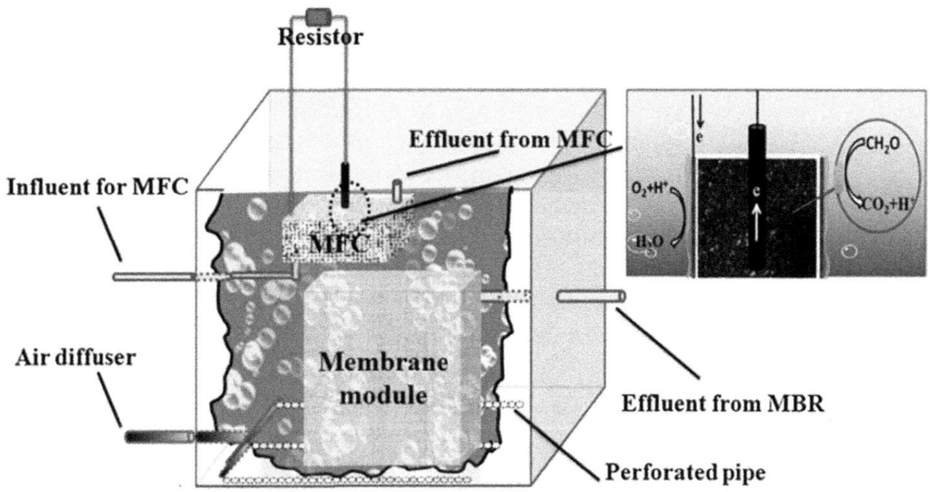

Fig. 4.6 Schematic of the MFC-MBR system developed by Wang et al. (2014). Copyright 2014 Elsevier. License Number 5710570711218.

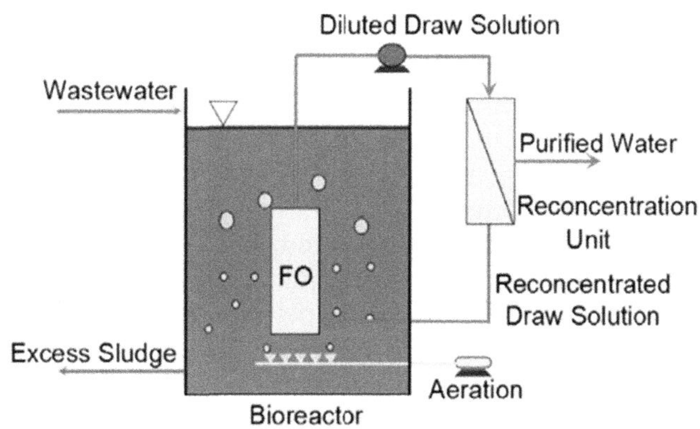

Fig. 4.7 Schematic diagram of OMBR (Wang et al., 2016). Copyright 2016 Elsevier. License Number 5710570896165.

energy utilization. However, they introduce an obstacle known as salinity build-up, which refers to the accumulation of salts on the membrane's surface. This leads to concentration polarization, triggering a reverse flux due to the emerging concentration gradient on the membrane (Wang et al., 2016).

OMBRs demonstrate impressive performance in the literature. They achieve a remarkable 98% removal of total organic carbon (TOC) at concentrations below 5 mg/L (Wang et al., 2016). Nitrogen removal surpasses 80% due to the inherent properties of FO membranes and the activity of ammonia and nitrite oxidizing bacteria. Moreover, OMBRs nearly completely eliminate phosphate due to FO membrane efficiency in rejecting calcium, which aids in phosphate precipitation. In contrast, conventional MBRs can only contribute to around 60% phosphate removal. Consequently, the performance of OMBRs underscores their ability to produce higher-purity water compared to typical MBRs. However, their drawbacks arise from low water flux associated with FO and the accumulation of high salt levels. Salt accumulation not only obstructs the separation process by creating a reverse concentration gradient but also significantly diminishes microbial activity within the activated sludge (Wang et al., 2016).

4.4.3 Other MBR systems

Other investigations were conducted on MBRs and their hybrid systems including aerobic dynamic MBRs (ADMBR), aerobic fluidized MBRs (AFMBR), and membrane photobioreactors (MBPR). Table 4.2 summarizes these investigations involving the source of wastewater used.

4.5 Conclusions and future trends

Membrane bioreactors have been an increasingly popular technology in wastewater treatment due to their efficiency in producing high-quality treated water. The MBR can be employed for the treatment of wastewater generated by different industries with an emphasis on the removal of COD, BOD_5, and TOC. The MBR performance is dependent on the pH of the wastewater, sludge concentration, and operating temperature. Originality of wastewater and membrane type and characteristics also play significant roles in MBR performance.

Looking ahead, several trends are expected to shape the future of MBRs:

1. **Advancements in Membrane Technology:** Continued developments in membrane materials, such as nanomaterials or enhanced polymers, can improve membrane performance, durability, and fouling resistance. These advancements may lead to longer membrane lifespans and lower operational costs.
2. **Energy Efficiency and Sustainability:** There's a growing emphasis on making wastewater treatment processes more energy-efficient and environmentally friendly. Future MBRs may incorporate energy recovery systems, such as anaerobic digestion or membrane aerated biofilm reactors, to reduce energy consumption and enhance sustainability.

Table 4.2 Operational and performance of different MBR configurations with different feedstock.

Configuration	Feedstock	Operating condition	Performance	Refs.
AnMBR (external)	Synthetic molasses wastewater	• $T = 35–37$ °C, HRT = 30 h in pilot scale • Silver (as nanoparticles) coated PVDF • Backwashing the membrane with the air and produced gas from an anaerobic baffled reactor	• COD removal = 82% • 91% reduction in EPS formation on the nanocomposite PVDF membrane surface	Amouamouha and Badalians Gholikandi (2018)
AnMBR (submerged)	Synthetic wastewater	• $T = 35.7$ °C, HRT = 10 days in lab scale • Flat sheet membrane with 0.014 m^2 filtration area • 2 kg COD/m^3/d as organic loading rate (OLR)	• COD removal = 99.5% • Provide shorter start-up period, higher biogas production, and better COD removal in comparison to external one	Ersahin et al. (2017)
AnMBR (external)	Pharmaceutical wastewater	• $T = 14–38$ °C, HRT = 36 h in pilot scale • Hollow-fiber membrane with 1 m^2 surface area • 3.48 kg COD/m^3/d as OLR	• TCOD removal = 88–92.5% • 77–171 L/d for biogas production	Huang et al. (2018)
Aerobic MBR	Pharmaceutical wastewater	• Reactor was operated until clogging (around 1 month) • Hollow-fiber membrane with 0.125 m^2 surface area • Using natural quorum quencher (i.e., endophytic *Penicillium restrictum*) to control biofouling and increase removal efficiency of antibiotics	• COD removal = 82% • Longer operation before clogging happened in comparison to the reactor without the addition of *P. restrictum* (35 days and 24 days respectively) • 4.5% increase in removal efficiency of antibiotics	Fakhri et al. (2021)
Aerobic MBR	Municipal wastewater	• SRT = 50 days, HRT = 10 h in pilot scale • PVDF flat sheet membrane with 0.1 m^2 surface area • Average TSS of 9.6 and 10.1 g/L without and with the addition of powdered activated carbon (PAC) to evaluate the effect of fouling	• COD removal = 89% • 19% increase in the critical flux by using PAC in comparison to sludge without PAC	Remy et al. (2010)
Aerobic dynamic MBR (ADMBR)	Municipal wastewater	• SRT = 106 day, HRT = 0.87 day, and MLSS = 5 g/L	• COD removal = 92.8%	Isik et al. (2021)

Continued

Table 4.2 Operational and performance of different MBR configurations with different feedstock—Cont'd

Configuration	Feedstock	Operating condition	Performance	Refs.
Aerobic fluidized MBR bioreactor (AFMBR)	Synthetic wastewater	• Hollow-fiber membrane with 0.015 m^2 surface area • 0.66 kg COD/m^3/d as OLR • $T = 37.1$ °C, HRT = 12 h in a lab scale • PVDF tubular membrane • 0.6 kg COD/m^3/d as OLR • Using polymeric materials as fluidized agents	• Faster clogging and decreasing COD removal efficiency were experienced by increasing sludge concentration • COD removal = 87.6% • Major foulant material on the membrane surface corresponds to EPS which fluidization wasn't sufficient enough to eliminate pore blockage • The low energy consumption and stability of PVDF provide an excellent surface for biofilm formation	Kim et al. (2020)
Membrane photobioreactor (MPBR)	High ammonia nitrogen wastewater	• $T = 25$–30 °C, HRT = 24 h, and SRT = 30 d • Using novel hollow fiber membrane with enhanced antifouling properties and hydrophilicity (i.e., polyvinylpyrrolidone (PVP)-graphene oxide (GO)/PVDF)	• COD removal = 93% • Increase permeability and flux rate in filtration in comparison to normal PVDF • The great potential of PVP-GO/PVDF for high-density chlorella cultivation and effective treatment of high nitrogen ammonia wastewater	Wu et al. (2020)
MFC-MBR	Synthetic wastewater	• Submerging anode and cathode in the submerged aerobic MBR at both sides of the membrane module • Using PVDF hollow fiber membrane with 0.2 m^2 surface area	• 4.4% increase in COD removal efficiency when compared to control MBR • Maximum power density and average voltage were 2.18 w/m^3 and 0.15 v, respectively • Improvement in filterability and dewaterability of the sludge • Alleviate the membrane fouling	Tian et al. (2015)
MFC-MBR	Synthetic medium	• $T = 30$ °C • Integration of anode as microfiltration membrane in side stream crossflow AnMBR configuration • Using stainless steel filtration membrane	• Achieving fourfold higher current density in comparison to experiment without filtration (6 A/m^2 for 0.5 μm filter grade) • Improving the current density was mainly due to permeate flow	Madjarov et al. (2018)

3. **Resource Recovery:** There's a growing focus on recovering valuable resources from wastewater. MBRs could be part of integrated systems designed to recover energy (via biogas production), nutrients (like phosphorus and nitrogen), or even clean water for reuse in non-potable applications.
4. **Modularity and Scalability:** Modular MBR systems that can be easily scaled up or down based on demand might gain popularity. This scalability allows for flexibility in adapting to varying wastewater treatment needs, whether for municipalities or industrial applications.
5. **Improved Fouling Mitigation:** Fouling remains a challenge in MBRs, impacting efficiency and maintenance. Future innovations may include better fouling-resistant membranes, enhanced cleaning techniques, or novel pretreatment methods to mitigate fouling and extend membrane lifespan.
6. **Hybrid Systems:** Integration of MBR technology with other treatment processes, such as advanced oxidation processes (AOPs) or membrane distillation, could lead to hybrid systems with enhanced performance and tailored treatment capabilities.

Overall, the future of MBRs in wastewater treatment appears promising, driven by advancements in technology, a focus on sustainability, and the continual quest for more efficient and effective wastewater treatment solutions.

References

Achilli, A., Holloway, R.W., 2015. Aerobic Membrane Bioreactor. Springer Science and Business Media LLC, pp. 1–3, https://doi.org/10.1007/978-3-642-40872-4_7-1.

Ahmadi, M., Zoroufchi Benis, K., Faraji, M., Shakerkhatibi, M., Aliashrafi, A., 2019. Process performance and multi-kinetic modeling of a membrane bioreactor treating actual oil refinery wastewater. J. Water Process Eng. 28, 115–122. 22147144. Elsevier Ltd, Iran. Available from: http://www.journals.elsevier.com/journal-of-water-process-engineering/. https://doi.org/10.1016/j.jwpe.2019.01.010.

Al-Asheh, S., Bagheri, M., Aidan, A., 2021. Membrane bioreactor for wastewater treatment: a review. Case Stud. Chem. Environ. Eng. 4. Elsevier Ltd, United Arab Emirates. Available from: www.journals.elsevier.com/case-studies-in-chemical-and-environmental-engineering/. https://doi.org/10.1016/j.cscee.2021.100109. 26660164.

Al-Malack, M.H., 2006. Determination of biokinetic coefficients of an immersed membrane bioreactor. J. Membr. Sci. 271 (1–2), 47–58. https://doi.org/10.1016/j.memsci.2005.07.008. 03767388.

Amouamouha, M., Badalians Gholikandi, G., 2018. Assessment of anaerobic nanocomposite membrane bioreactor efficiency intensified by biogas backwash. Chem. Eng. Process. Process Intensif. 131, 51–58. 02552701. Elsevier B.V., Iran,

Available from: www.elsevier.com/inca/publications/store/5/0/4/0/8/1/index.htt. https://doi.org/10.1016/j.cep.2018.07.004.

An, Y., Wang, Z., Wu, Z., Yang, D., Zhou, Q., 2009. Characterization of membrane foulants in an anaerobic non-woven fabric membrane bioreactor for municipal wastewater treatment. Chem. Eng. J. 155 (3), 709–715. https://doi.org/10.1016/j.cej.2009.09.003. 13858947.

Arvanitoyannis, I.S., Kassaveti, A., Ladas, D., 2008. Food waste treatment methodologies. In: Waste Management for the Food Industries. Elsevier, Greece, pp. 345–410, https://doi.org/10.1016/B978-012373654-3.50009-2. Available from: http://www.sciencedirect.com/science/book/9780123736543.

Barbhuiya, N.H., Misra, U., Singh, S.P., 2021. Synthesis, fabrication, and mechanism of action of electrically conductive membranes: a review. Environ. Sci.: Water Res. Technol. 7 (4), 671–705. 20531419. Royal Society of Chemistry, India. Available from: http://pubs.rsc.org/en/journals/journal/ew. https://doi.org/10.1039/d0ew01070g.

Cath, T.Y., Elimelech, M., McCutcheon, J.R., McGinnis, R.L., Achilli, A., Anastasio, D., Brady, A.R., Childress, A.E., Farr, I.V., Hancock, N.T., Lampi, J., Nghiem, L.D., Xie, M., Yip, N.Y., 2013. Standard methodology for evaluating membrane performance in osmotically driven membrane processes. Desalination 312, 31–38. https://doi.org/10.1016/j.desal.2012.07.005. 00119164.

Chuntanalerg, P., Bureekaew, S., Klaysom, C., Lau, W.-J., Faungnawakij, K., 2019. Nanomaterial-Incorporated Nanofiltration Membranes for Organic Solvent Recovery. Elsevier BV, pp. 159–181, https://doi.org/10.1016/b978-0-12-814503-6.00007-0.

Cirja, M., Ivashechkin, P., Schäffer, A., Corvini, P.F.X., 2008. Factors affecting the removal of organic micropollutants from wastewater in conventional treatment plants (CTP) and membrane bioreactors (MBR). Rev. Environ. Sci. Biotechnol. 7 (1), 61–78. https://doi.org/10.1007/s11157-007-9121-8. 15691705.

Drioli, E., Di Profio, G., Curcio, E., 2012. Progress in membrane crystallization. Curr. Opin. Chem. Eng. 1 (2), 178–182. 22113398. Elsevier, Italy. Available from: http://www.elsevier.com/wps/find/journaldescription.cws_home/725837/description#description. https://doi.org/10.1016/j.coche.2012.03.005.

Dvořák, L., Svojitka, J., Wanner, J., Wintgens, T., 2013. Nitrification performance in a membrane bioreactor treating industrial wastewater. Water Res. 47 (13), 4412–4421. 18792448. Elsevier, Czech Republic. Available from: https://www.sciencedirect.com/science/journal/00431354. https://doi.org/10.1016/j.watres.2013.03.053.

Ersahin, M.E., Tao, Y., Ozgun, H., Gimenez, J.B., Spanjers, H., van Lier, J.B., 2017. Impact of anaerobic dynamic membrane bioreactor configuration on treatment and filterability performance. J. Membr. Sci. 526, 387–394. 18733123. Elsevier B.V., Netherlands. Available from: www.elsevier.com/locate/memsci. https://doi.org/10.1016/j.memsci.2016.12.057.

Fakhri, H., Shahi, A., Ovez, S., Aydin, S., 2021. Bioaugmentation with immobilized endophytic *Penicillium restrictum* to improve quorum quenching activity for biofouling control in an aerobic hollow-fiber membrane bioreactor treating antibiotic-containing wastewater. Ecotoxicol. Environ. Saf. 210. 10902414. Academic Press, Turkey. Available from: http://www.elsevier.com/inca/publications/store/6/2/2/8/1/9/index.htt. https://doi.org/10.1016/j.ecoenv.2020.111831.

Fan, H., Xiao, K., Mu, S., Zhou, Y., Ma, J., Wang, X., Huang, X., 2018. Impact of membrane pore morphology on multi-cycle fouling and cleaning of hydrophobic and hydrophilic membranes during MBR operation. J. Membr. Sci. 556, 312–320. 18733123. Elsevier B.V., China. Available from: www.elsevier.com/locate/memsci. https://doi.org/10.1016/j.memsci.2018.04.014.

Goh, P.S., Ismail, A.F., 2018. A review on inorganic membranes for desalination and wastewater treatment. Desalination 434, 60–80. https://doi.org/10.1016/j.desal.2017.07.023. 00119164.

Hamedi, H., Ehteshami, M., Mirbagheri, S.A., Rasouli, S.A., Zendehboudi, S., 2019. Current status and future prospects of membrane bioreactors (MBRs) and fouling phenomena: a systematic review. Can. J. Chem. Eng. 97 (1), 32–58. 1939019X. Wiley-Liss Inc., Iran. Available from: http://onlinelibrary.wiley.com/journal/10.1002/(ISSN)1939-019X. https://doi.org/10.1002/cjce.23345.

Hu, A.Y., Stuckey, D.C., 2006. Treatment of dilute wastewaters using a novel submerged anaerobic membrane bioreactor. J. Environ. Eng. 132 (2), 190–198. https://doi.org/10.1061/(asce)0733-9372(2006)132:2(190. 0733-9372.

Huang, B., Wang, H.C., Cui, D., Zhang, B., Chen, Z.B., Wang, A.J., 2018. Treatment of pharmaceutical wastewater containing B-lactams antibiotics by a pilot-scale anaerobic membrane bioreactor (AnMBR). Chem. Eng. J. 341, 238–247. 13858947. Elsevier B.V., China. Available from: www.elsevier.com/inca/publications/store/6/0/1/2/7/3/index.htt. https://doi.org/10.1016/j.cej.2018.01.149.

Huang, S., Pooi, C.K., Shi, X., Varjani, S., Ng, H.Y., 2020. Performance and process simulation of membrane bioreactor (MBR) treating petrochemical wastewater. Sci. Total Environ. 747. 18791026. Elsevier B.V., Singapore. Available from: www.elsevier.com/locate/scitotenv. https://doi.org/10.1016/j.scitotenv.2020.141311.

Isik, O., Batyrow, M., Abdelrahman, A.M., Orman, I., Ozgun, H., Ersahin, M.E., Pasaoglu, M.E., Demir, I., Koyuncu, I., 2021. Dynamic membrane bioreactor performance for treatment of municipal wastewaters at different sludge concentrations. Environ. Technol. Innov. 22. 23521864. Elsevier B.V., Turkey. Available from: http://www.journals.elsevier.com/environmental-technology-and-innovation/. https://doi.org/10.1016/j.eti.2021.101452.

Isma, A., Idris, A., Omar, R., Razreena, A.R.P., 2014. Effects of SRT and HRT on treatment performance. Int. J. Chem. Mol. Nucl. Mater. Metall. Eng. 8, 488–492.

Kim, M., Lam, T.Y.C., Tan, G.Y.A., Lee, P.H., Kim, J., 2020. Use of polymeric scouring agent as fluidized media in anaerobic fluidized bed membrane bioreactor for wastewater treatment: system performance and microbial community. J. Membr. Sci. 606. 18733123. Elsevier B.V., South Korea. Available from: www.elsevier.com/locate/memsci. https://doi.org/10.1016/j.memsci.2020.118121.

Lau, W.J., Ismail, A.F., Misdan, N., Kassim, M.A., 2012. A recent progress in thin film composite membrane: a review. Desalination 287, 190–199. https://doi.org/10.1016/j.desal.2011.04.004. 00119164.

Leyva-Díaz, J.C., Calderón, K., Rodríguez, F.A., González-López, J., Hontoria, E., Poyatos, J.M., 2013. Comparative kinetic study between moving bed biofilm reactor-membrane bioreactor and membrane bioreactor systems and their influence on organic matter and nutrients removal. Biochem. Eng. J. 77, 28–40. https://doi.org/10.1016/j.bej.2013.04.023. 1369703X.

Leyva-Díaz, J.C., González-Martínez, A., González-López, J., Muñío, M.M., Poyatos, J.M., 2015. Kinetic modeling and microbiological study of two-step nitrification in a membrane bioreactor and hybrid moving bed biofilm reactor–membrane bioreactor for wastewater treatment. Chem. Eng. J. 259, 692–702. https://doi.org/10.1016/j.cej.2014.07.136. 13858947.

Leyva-Díaz, J.C., Poyatos, J.M., Barghini, P., Gorrasi, S., Fenice, M., 2017. Kinetic modeling of *Shewanella baltica* KB30 growth on different substrates through respirometry. Microb. Cell Factories 16 (1). 14752859. BioMed Central Ltd., Spain. Available from: http://www.microbialcellfactories.com/home/. https://doi.org/10.1186/s12934-017-0805-7.

Leyva-Diaz, J.C., Fenice, M., Poyatos, J.M., 2020. Respirometric method for kinetic modeling of ammonium-oxidizing and nitrite-oxidizing bacteria in a membrane bioreactor. AICHE J. 66, 189.

Lin, H., Chen, J., Wang, F., Ding, L., Hong, H., 2011. Feasibility evaluation of submerged anaerobic membrane bioreactor for municipal secondary wastewater treatment. Desalination 280 (1–3), 120–126. https://doi.org/10.1016/j.desal.2011.06.058. 00119164.

Logan, B.E., Hamelers, B., Rozendal, R., Schröder, U., Keller, J., Freguia, S., Aelterman, P., Verstraete, W., Rabaey, K., 2006. Microbial fuel cells: methodology and technology. Environ. Sci. Technol. 40 (17), 5181–5192. https://doi.org/10.1021/es0605016. 0013936X.

Maaz, M., Yasin, M., Aslam, M., Kumar, G., Atabani, A.E., Idrees, M., Anjum, F., Jamil, F., Ahmad, R., Khan, A.L., Lesage, G., Heran, M., Kim, J., 2019. Anaerobic membrane bioreactors for wastewater treatment: novel configurations, fouling control and energy considerations. Bioresour. Technol. 283, 358–372. 18732976. Elsevier, Pakistan. Available from: www.elsevier.com/locate/biortech. https://doi.org/10.1016/j.biortech.2019.03.061.

Madjarov, J., Götze, A., Zengerle, R., Kerzenmacher, S., 2018. Simultaneous use of a crossflow filtration membrane as microbial fuel cell anode—permeate flow leads to 4-fold increased current densities. Bioresour. Technol. 257, 274–280. 18732976. Elsevier, Germany. Available from: www.elsevier.com/locate/biortech. https://doi.org/10.1016/j.biortech.2018.02.032.

Manickam, S.S., McCutcheon, J.R., 2017. Understanding mass transfer through asymmetric membranes during forward osmosis: a historical perspective and critical review on measuring structural parameter with semi-empirical models and characterization approaches. Desalination 421, 110–126. 00119164. Elsevier B.V., United States. Available from: https://www.journals.elsevier.com/desalination. https://doi.org/10.1016/j.desal.2016.12.016.

Melin, T., Jefferson, B., Bixio, D., Thoeye, C., De Wilde, W., De Koning, J., van der Graaf, J., Wintgens, T., 2006. Membrane bioreactor technology for wastewater treatment and reuse. Desalination 187 (1–3), 271–282. https://doi.org/10.1016/j.desal.2005.04.086. 00119164.

Nath, K., 2017. Membrane Separation Processes. PHI Learning Private Limited, Delhi.

Pal, P., 2020. Introduction to Membrane-Based Technology Applications. Elsevier BV, pp. 71–100, https://doi.org/10.1016/b978-0-12-819455-3.00002-9.

Radjenović, J., Matošić, M., Mijatović, I., Petrović, M., Barceló, D., 2008. Membrane bioreactor (MBR) as an advanced wastewater treatment technology. In: Handbook of Environmental Chemistry, Volume 5: Water Pollution. 5. Springer-Verlag, Berlin, Heidelberg, pp. 37–101 (14336863; the print ISBN: 978-3-540-46684-0; and the online ISBN: 978-3-540-46685-7) https://doi.org/10.1007/698_5_093.

Remy, M., Potier, V., Temmink, H., Rulkens, W., 2010. Why low powdered activated carbon addition reduces membrane fouling in MBRs. Water Res. 44 (3), 861–867. 00431354. Elsevier, Netherlands. Available from: https://www.sciencedirect.com/science/journal/00431354. https://doi.org/10.1016/j.watres.2009.09.046.

Ruiz, L.M., Arévalo, J., Parada, J., González, D., Moreno, B., Pérez, J., Gómez, M.A., 2011. Respirometric assays of two different MBR (microfiltration and ultrafiltration) to obtain kinetic and stoichiometric parameters. Water Sci. Technol. 63 (11), 2478–2485. https://doi.org/10.2166/wst.2011.578. 0273-1223.

Rype, J.-U., 2003. Modelling of Electrically Driven Membrane Processes (Ph.D. thesis). Technical University of Denmark, Kgs. Lyngby, Denmark (223 pp; print ISBN: 87-90142-93-4).

Scott, K., 1995. Handbook of Industrial Membranes. Elsevier Advanced Technology (ISBN: 978-1856172332).

Smith, A.L., Stadler, L.B., Love, N.G., Skerlos, S.J., Raskin, L., 2012. Perspectives on anaerobic membrane bioreactor treatment of domestic wastewater: a critical review. Bioresour. Technol. 122, 149–159. https://doi.org/10.1016/j.biortech.2012.04.055. 18732976.

Tian, Y., Li, H., Li, L., Su, X., Lu, Y., Zuo, W., Zhang, J., 2015. In-situ integration of microbial fuel cell with hollow-fiber membrane bioreactor for wastewater treatment and membrane fouling mitigation. Biosens. Bioelectron. 64, 189–195. 18734235. Elsevier Ltd, China. Available from: www.elsevier.com/locate/bios. https://doi.org/10.1016/j.bios.2014.08.070.

Volkov, A.V., Novitsky, E.G., Borisov, I.L., Vasilevsky, V.P., Volkov, V.V., 2016. Porous condenser for thermally driven membrane processes: gravity-independent operation. Sep. Purif. Technol. 171, 191–196. https://doi.org/10.1016/j.seppur.2016.07.038. 13835866.

Wang, Z., Ma, J., Tang, C.Y., Kimura, K., Wang, Q., Han, X., 2014. Membrane cleaning in membrane bioreactors: a review. J. Membr. Sci. 468, 276–307. 18733123. Elsevier, China. Available from: www.elsevier.com/locate/memsci. https://doi.org/10.1016/j.memsci.2014.05.060.

Wang, X., Chang, V.W.C., Tang, C.Y., 2016. Osmotic membrane bioreactor (OMBR) technology for wastewater treatment and reclamation: advances, challenges, and prospects for the future. J. Membr. Sci. 504, 113–132. 18733123. Elsevier, China. Available from: www.elsevier.com/locate/memsci. https://doi.org/10.1016/j.memsci.2016.01.010.

Wijekoon, K.C., Visvanathan, C., Abeynayaka, A., 2011. Effect of organic loading rate on VFA production, organic matter removal and microbial activity of a two-stage thermophilic anaerobic membrane bioreactor. Bioresour. Technol. 102 (9), 5353–5360. https://doi.org/10.1016/j.biortech.2010.12.081. 09608524.

Wu, W., Zhang, X., Qin, L., Li, X., Meng, Q., Shen, C., Zhang, G., 2020. Enhanced MPBR with polyvinylpyrrolidone-graphene oxide/PVDF hollow fiber membrane for efficient ammonia nitrogen wastewater treatment and high-density Chlorella cultivation. Chem. Eng. J. 379. 13858947. Elsevier B.V China. Available from: www.elsevier.com/inca/publications/store/6/0/1/2/7/3/index.htt. https://doi.org/10.1016/j.cej.2019.122368.

Yamamoto, K., Hiasa, M., Mahmood, T., Matsuo, T., 1989. Direct solid-liquid separation using hollow fiber membrane in an activated sludge aeration tank. Water Sci. Technol. 21 (4–5), 43–54. https://doi.org/10.2166/wst.1989.0209. 02731223.

Yue, X., Koh, Y.K.K., Ng, H.Y., 2015. Effects of dissolved organic matters (DOMs) on membrane fouling in anaerobic ceramic membrane bioreactors (AnCMBRs) treating domestic wastewater. Water Res. 86, 96–107. 18792448. Elsevier, Singapore. Available from: www.elsevier.com/locate/watres. https://doi.org/10.1016/j.watres.2015.07.038.

5

Aeration systems for wastewater treatments

Yaşar Kemal Recepoğlu[a] and Ayşegül Yağmur Gören[b]

[a]Department of Chemical Engineering, İzmir Institute of Technology, Izmir, Turkey. [b]Department of Environmental Engineering, İzmir Institute of Technology, Izmir, Turkey

Abbreviations

AI	artificial intelligence
BOD	biological oxygen demand
COD	chemical oxygen demand
CSC	circular stepped cascade
DO	dissolved oxygen
DWP	dynamic wet pressure
EAAS	extended aeration activated sludge
EDI	Environmental Dynamics International
FBDA	fine-bubble diffused air
HLR	hydraulic loading rate $=Q/A$ ($m^3/m^2/s$)
HRT	hydraulic retention time
IoT	implement internet of things
LCA	lifecycle assessment
MCRT	mean cell retention time
MFC	microbial fuel cell
MWWT	municipal wastewater treatment
NDSAE	non-dimensional standard aeration efficiency
PCSC	pooled circular stepped cascade
PPCSC	perforated pooled circular stepped cascade
PTA	perforated tray aerator
PV	photovoltaic
SA	surface aeration
SAE	standard aeration efficiency
SAS	showering aeration system
SOTR	standard O_2 transfer rate
SP	spacing between trays
TN	total nitrogen
TSS	total suspended solid
UFCW	up-flow constructed wetland
UV	ultraviolet
VOC	volatile organic compound

Municipal Wastewater Treatment. https://doi.org/10.1016/B978-0-443-24826-9.00005-X

VSSFCW	vertical subsurface flow constructed wetland
WWT	wastewater treatment

Symbols

$\mathbf{A}$	interface area of the thin film (m^2)
a	specific area (the ratio between the interface area and the liquid volume) (m^2/m^3)
a_f	specific area of the thin film (m^2/m^3)
$\mathbf{A_j}$	cross-sectional area of the water jet (m^2)
BO	bond number relative to the inside diameter of the nozzle (dimensionless)
$\mathbf{BO_o}$	bond number relative to the outside diameter (wetting diameter) of the nozzle (dimensionless)
C	real O_2 concentration in the liquid media (mg/L)
$\mathbf{C_O}$	initial O_2 concentration (mg/L)
$\mathbf{C_1}$	DO concentration of water at the exit from tray 1 (mg/L)
$\mathbf{C_2}$	DO concentration of water at the exit from tray 2 (mg/L)
$\mathbf{C_{2i}}$	influent DO to the ith tray (mg/L)
$\mathbf{C_D}$	coefficient of discharge through nozzles (0.66) (dimensionless)
$\mathbf{C_j}$	concentration of DO in the jth tank (mg/L)
$\mathbf{C_{j-1}}$	concentration of DO in the (j−1)th tank (mg/L)
$\mathbf{C_S}$	saturation concentration of the gas in water (mg/L)
$\mathbf{C_V}$	coefficient of velocity through nozzles (0.99) (dimensionless)
$C^*_\infty O_2$	concentration in liquid media at equilibrium (mg/L)
D	dispersion coefficient
d	hole diameter (m)
$\mathbf{D_1}$	initial diameter of the water stream (m)
$\mathbf{D_2}$	jet diameter reaching the subsequent tray (m)
$\mathbf{D_j}$	mean jet diameter (m)
$\mathbf{D_L}$	diffusivity of air in water (m^2/s)
$\mathbf{dC/dt\ O_2}$	concentration change rate in the liquid media
δ	thickness of a thin laminar film (m)
Fr	Froud number (dimensionless)
g	gravitational acceleration (9.81 m/s^2) (m/s^2)
H	cascade height (overall system height, aeration tank depth) (m)
h	height of each step (m)
h′	corrected film height (m)
$\mathbf{h_s}$	tray side height (m)
$\mathbf{h_W}$	height of water film over tray (m)
i	number of trays ranging from 1 to N
k	a constant for water injected into the air (0.37)
$\mathbf{K_L}$	coefficient for liquid mass transfer (m/s)
$\mathbf{K_La_T}$	coefficient for the total volumetric O_2 transfer at a specific temperature
$\mathbf{K_La_{20}}$	coefficient for the total volumetric O_2 transfer at 20 °C
$\mathbf{(k_La)_j}$	overall mass transfer coefficient for O_2 in the jth tank
L	total aeration tank length (m)
$\mathbf{L_j}$	jet length (m)
N	number of cascade steps (trays, tanks)
n	number of holes
n′	corrected number of holes per tray

N_e	number of enclosures within each step
υ_w	kinematic viscosity of water (m^2/s)
P	energy consumption of the aerator (kW)
P_e	percentage coverage of each step's circumference by enclosure (%)
Q	pump discharge (total flow rate) (m^3/s)
q_s	average sewage flow rate (m^3/s)
Q_{hole}	flow rate per hole (m^3/s)
r	recycle ratio (dimensionless)
R_b	radius of the bottommost circular step (m)
R_j	DO consumption rate of microorganisms
R_t	radius of the topmost circular step (m)
ρ	liquid density (kg/m^3)
ρ_a	mass density of air (kg/m^3)
ρ_w	mass density of water (997.2 kg/m^3 at 24 °C) (kg/m^3)
σ	surface tension of the liquid (0.073 N/m for water at 24 °C) (N/m)
σ_w	surface tension of water (0.073 N/m at 24 °C) (N/m)
S	liquid surface renewal rate
T	temperature (°C)
t	time (s)
t_e	exposure time of laminar film to the gas (s)
t_f	thin film aeration time (s)
θ	temperature correction factor (1.024 for pure water)
u	average fluid velocity (m/s)
V	volume of water (m^3)
V_f	liquid volume over a tray (m^3)
v_j	mean jet velocity (m/s)
υ_1	initial velocity upon exit from the nozzle (m/s)
υ_2	jet velocity reaching the subsequent tray (m/s)
υ_1'	corrected initial exit velocity from the nozzle (m/s)
W	aeration width (m)
We	Weber number (dimensionless)
We_c	critical Weber number (dimensionless)

5.1 Introduction

Aeration systems play a critical role in wastewater treatment, where transforming contaminated water into a reusable resource is a fundamental environmental challenge. In most wastewater treatment facilities, aeration constitutes the predominant share of energy expenses within the plant, with costs between 45% and 75% of the overall energy budget (Rosso and Stenstrom, 2006). Aeration systems facilitate the infusion of oxygen into a liquid medium through either the diffusion of gas across a gas-liquid interface or the dissolution of gas into the liquid solution via a semi-permeable membrane to remove pollutants and enhance biological processes (Rosso et al., 2006). This crucial step is driven by the pressing need to address the escalating concerns surrounding water quality, environmental sustainability, and public health.

It is impossible to overestimate the importance of wastewater treatment, considering the burgeoning global population and the increasing strain on freshwater resources. As urbanization and industrialization continue unabated, the discharge of wastewater containing many contaminants poses a grave threat to ecosystems and human well-being. Proper wastewater treatment is necessary for mitigating the ecological impact and is a cornerstone for safeguarding public health (Freeman et al., 2018). In this landscape, aeration emerges as a key protagonist in the narrative of wastewater treatment. Its role is multifaceted, encompassing the essential task of oxygenating water, fostering an environment conducive to microbial activity, and accelerating the breakdown of organic pollutants. Beyond mere oxygenation, aeration systems contribute significantly to the mixing and circulation of water within treatment units, ensuring uniform exposure to treatment processes and enhancing the overall efficiency of pollutant removal.

Fundamentally, aeration includes the feed of air or O_2 into wastewater to provide promising environments for the metabolic activity of aerobic microorganisms to grow. These microorganisms utilize O_2 to break down complex organic compounds into more straightforward, less harmful substances through biological oxidation. This process enhances the overall efficiency of wastewater treatment by accelerating the decomposition of pollutants and facilitating the removal of organic matter. The introduction of O_2 also helps prevent the occurrence of anaerobic conditions, which can lead to the production of malodorous gases and undesirable by-products. A well-aerated wastewater treatment system ensures the maintenance of aerobic conditions, promoting the growth of beneficial bacteria that contribute to water purification (Tchobanoglus et al., 2003; Mara and Pearson, 1998). Overall, the objectives of aeration in wastewater treatment are multifaceted, encompassing critical processes that contribute to the overall effectiveness of wastewater purification. One primary objective is the promotion of aerobic microbial activity. Aeration involves the introduction of air or O_2 into wastewater, creating an O_2-rich environment that supports the growth and metabolic functions of aerobic microorganisms. These microorganisms are crucial in biologically oxidizing organic pollutants, breaking them into less harmful forms. Another crucial objective is the oxygenation of wastewater, ensuring the maintenance of aerobic conditions and preventing the occurrence of anaerobic pockets. This supports the prevention of malodorous gas production and undesirable by-products. Namely, a well-aerated system ensures the sustained activity of these microorganisms, enhancing the efficiency of treatment and ensuring pollutant removal from the water (Eddy et al., 2014). Additionally, aeration aims to facilitate the removal of volatile compounds, such as volatile organic compounds (VOCs), by promoting their transport from the liquid to the gas phase. By facilitating the volatilization of these compounds, aeration contributes to reducing environmental impact and ensures that the discharged water meets regulatory standards (Chen et al., 2020). Furthermore, the aeration process contributes to adequate mixing and circulation, which helps achieve uniform treatment throughout the wastewater. Uniform mixing prevents the formation of stagnant zones, ensuring that all parts of the wastewater receive adequate O_2 for microbial activity. These objectives collectively enhance the efficiency of the wastewater treatment process.

As we delve into the intricacies of aeration systems for wastewater treatment in this chapter, we aim to unravel the mechanisms, types, and operational aspects that make aeration an indispensable tool in pursuing clean and safe water. By exploring design considerations, applications, and future trends, we endeavor to underscore the vital role that aeration systems play in transforming wastewater into a valuable resource, aligning with the global imperative for sustainable water management.

5.2 Aeration systems

In this section, diffused, mechanical, natural, jet, and packaged aeration systems are considered and comprehensively discussed, considering their working principles. Moreover, the benefits and limitations of these systems are evaluated with their efficiencies to understand the applicability of the systems in real-scale applications.

5.2.1 Diffused aeration systems

In the diffused aeration system, the bubbles are dispersed from diffusers at the reactor base, and O_2 is transferred from the rising bubbles to the mixed liquid to meet the O_2 needs of the water body. A matrix of membranes or porous plates placed close to the bottom of the tank to maximize gas-to-water contact makes up the most often utilized diffuser system. Moreover, the diffused aeration system introduces gas/air into the water by contacting gas bubbles using the propeller aspirators (Fig. 5.1A) and submersible aerators (Fig. 5.1B).

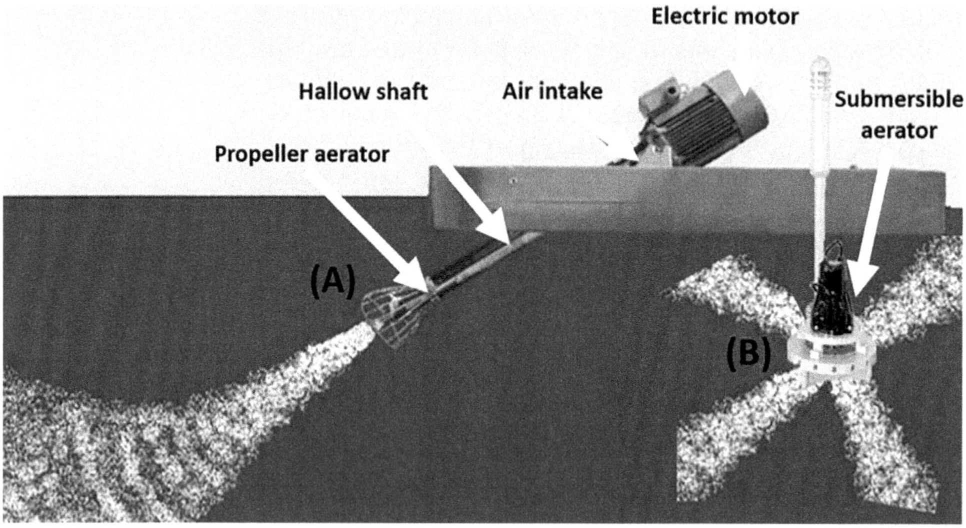

Fig. 5.1 Aerator types in diffused aeration systems: (A) propeller aspirators and (B) submersible aerators.

Moreover, the fine and coarse bubble diffuser types are widely utilized in diffused aeration systems. The fine bubble diffusers produce small bubbles that provide a high surface area for gas exchange, making them very efficient. They are often used in wastewater treatment plants. Basically, the fine bubble diffusers release air through small pores, producing tiny bubbles. These fine bubbles have a high surface area-to-volume ratio, enhancing O_2 transfer efficiency. As bubbles rise through the water column, O_2 diffuses into the wastewater, driven by the concentration gradient between the gas phase (bubble) and the liquid phase (water). Fine bubble diffusers generate many small bubbles that provide greater surface area for O_2 transfer than larger bubbles. The small bubble size increases the retention time of the bubbles in the water, allowing more O_2 to dissolve. Although this type of diffuser provides higher O_2 transfer efficiency, better O_2 distribution, and reduced energy consumption per unit of O_2 transferred, they are susceptible to clogging and fouling. Therefore, they require regular maintenance to sustain performance. On the other hand, coarse bubble diffusers produce larger bubbles and are less efficient than fine bubble diffusers but are more resistant to clogging and fouling. Moreover, they are used where higher agitation is needed. Namely, the coarse bubble diffusers produce larger bubbles with a lower surface area-to-volume ratio. These diffusers are less prone to clogging but provide lower O_2 transfer efficiency. Coarse bubbles rise more rapidly to the surface, reducing the contact time between the bubbles and the wastewater. While the O_2 transfer per bubble is less efficient, the simpler design often results in lower maintenance requirements, and hence, they are less susceptible to fouling, but more energy is required to achieve the same oxygenation levels as fine bubble systems.

Overall, systems that employ diffused air and release fine bubbles are more efficient than those that release coarse bubbles. This is because smaller bubbles have a bigger surface area than larger bubbles in the surrounding water. A high surface area facilitates greater O_2 absorption because O_2 diffuses into water at the surface. Also, diffused-air systems perform better in deep ponds than shallow ones (Boyd and Moore, 1993). An analysis of the operation of the wastewater treatment plant in Frankenmuth, Michigan, for instance, which has six reservoirs with a volume of 413 m^3 each and 0.97 fine bubble diffuser/m^3 reservoir, revealed that the average power used for aeration by fine bubbles is only 48% of the power used by coarse bubbles; the average air flow rate used by fine bubbles diffuser is 60% of the flow rate used by coarse bubbles (Oprina et al., 2010).

5.2.2 Mechanical aeration systems

Mechanical aeration involves using devices to physically agitate the water, promoting O_2 transfer through increased surface area and turbulence. Mechanical aeration systems can be classified into two main categories based on the type of aerator: surface aerators and subsurface aerators. The surface aerators agitate the surface of the water to increase the air-water interface, promoting O_2 transfer, and they are commonly used in ponds and lagoons. On the other hand, subsurface aerators include mixers and agitators that operate below the water surface, creating turbulence and mixing and enhancing aeration. This section comprehensively discusses the working principles and types of surface and subsurface aerators.

5.2.2.1 Subsurface aeration

Several subsurface and/or diffused aeration systems have been built and developed to dissolve O_2 into wastewater since the early 1900s when the activated sludge process was introduced (Al-Ahmady, 2006). These have varied from straightforward individual holes and slots drilled in a pipe segment to intricate devices composed of fused small-diameter particles (Fig. 5.2). The subsurface aeration systems release bubbles at the bottom of the water body, allowing them to rise by gravity. Fundamentally, lakes, reservoirs,

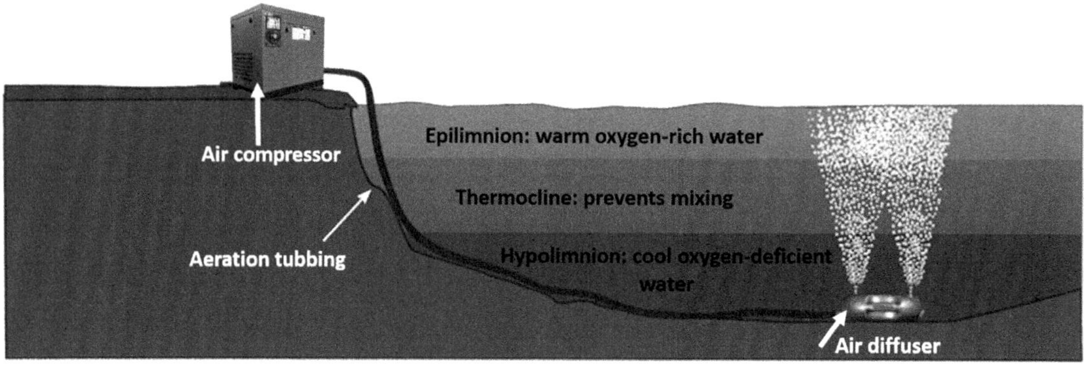

Fig. 5.2 Schematic diagram of the subsurface aeration system in a natural pond.

and wastewater treatment plants utilize subsurface aeration systems to raise the dissolved O_2 (DO) content and drive water circulation. At almost an 11 degrees angle of spreading, a free, turbulent bubble-plume is created by subsurface diffusers releasing air or pure O_2 bubbles, and this bubble-plume rises to the surface of the water due to buoyant forces (Fischer et al., 1979). In this system, water is drawn in by the rising bubble plume, resulting in lateral surface spreading and vertical circulation. As the bubbles rise from the diffuser to the surface of the water, O_2 is transferred from the diffuser to the water through the bubble interfaces. The free surface of the air-water interface also experiences O_2 transfer because of turbulence brought on by water circulation and bubble-plume motion.

O_2 transfer efficiency in aeration systems, a critical factor in the overall performance of wastewater treatment plants, is influenced by several factors, including bubble size, contact time, water temperature, turbulence, and pressure. Smaller bubbles, which provide a higher surface area-to-volume ratio, enhance O_2 transfer efficiency; hence, fine bubble diffusers, designed to produce smaller bubbles, produce more efficient oxygenation than coarse bubble diffusers. Contact time, the duration that air bubbles remain in contact with wastewater, is crucial for O_2 transfer, with longer contact times allowing more O_2 to dissolve into the water. Factors such as the depth of the aeration tank, bubble size, and aeration system design influence contact time. Water temperature affects both the solubility of O_2 and the metabolic activity of microorganisms: higher temperatures reduce O_2 solubility but increase microbial activity rates, leading to higher O_2 demand. Thus, optimal aeration system design must account for seasonal temperature variations to maintain efficient O_2 transfer. Moreover, turbulence enhances O_2 dispersion throughout the wastewater, ensures uniform O_2 distribution, prevents the formation of anaerobic zones, and promotes efficient biological treatment. Increased atmospheric pressure or deeper water depths can improve O_2 dissolution by enhancing the partial pressure of O_2, thereby driving more O_2 into the water. However, maintaining higher pressures requires more energy input, necessitating a balance between efficiency and energy consumption.

In addition, compared to other aeration systems, the subsurface aeration systems provide better aeration related to substantial air bubble entrainment, great turbulent mixing, and long exposure time (Roy et al., 2021). By pushing water from the bottom to the top and eliminating VOCs from the pond bottom, the cascade aerators enhance the quality of the water. Furthermore, minimal maintenance is needed since it has no moving parts (Roy et al., 2022a). Subsurface aeration systems are frequently employed in water treatment plants to remove manganese, iron, dissolved gases, and VOCs (Rozainy et al., 2015). Consequently, compared to other types of aerators, subsurface aeration systems have one of the best energy consumption rates (El-Zahaby and El-Gendy, 2016). The subsurface aeration systems can be classified as cascade, tray, and weir-type aerators (Fig. 5.3).

Namely, mechanical aeration has a slight aeration performance since its massive energy consumption of the electrically powered blades (Alkhalidi et al., 2016). In this regard, cascade aeration is an additional technique that uses a similar mass transfer process. This kind of aeration occurs spontaneously in streams and waterfalls when water is discharged at a height that causes it to break into small droplets. Therefore, many wastewater treatment facilities use cascade aeration since it requires no power input. Particularly, a cascade aerator that uses the altitude difference of a contiguous region is an efficient method for a passive treatment system since it is easy to design and doesn't need to be maintained (Geroni et al., 2012). Previous research used large-scale studies to illustrate the applicability of the cascade aerator and its impact on the oxidation pond's efficiency (Cravotta, 2007). Extensive research has been conducted on the hydrodynamic behavior of water throughout cascade aeration (Chanson, 2009; Moulick et al., 2010). The studies assumed that maintaining a consistent and effective aeration process required an external power source or a height differential. To effectively integrate a cascade aerator into a passive treatment system, it is crucial to investigate methods to optimize performance given the existing height difference or to evaluate how drop height

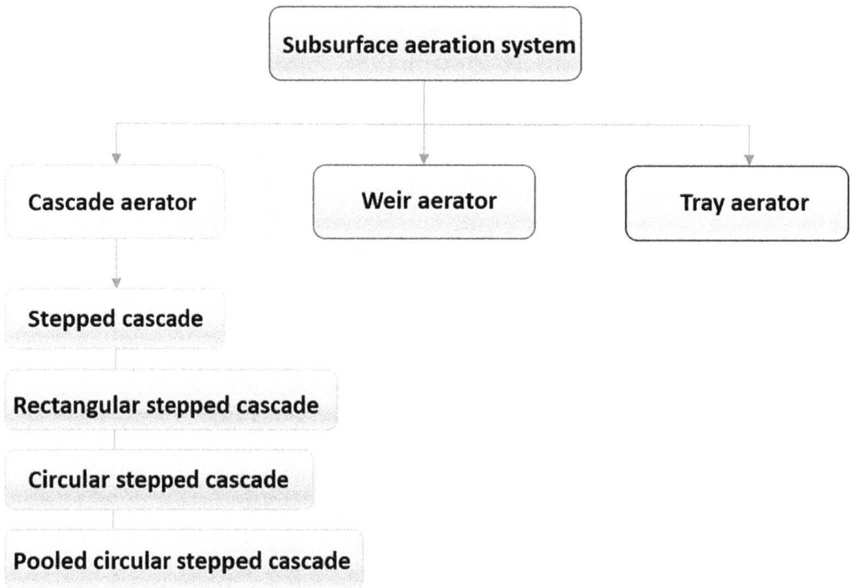

Fig. 5.3 Classification of various subsurface aeration systems.

affects the remediation efficiency of the pond, particularly as achieving the necessary topographical conditions may not be possible at all locations (Yim et al., 2014).

Moreover, the weir aerators have been utilized in the subsurface aeration systems. For instance, Gameson (1957) suggested using weirs to accelerate the aeration process. A free water jet from a weir plummet falls into running water downstream, trapping air bubbles in the process and causing aeration, which is the mechanism that raises the DO in the water. In a separate study, Singh and Kumar (2022) investigated a piano key weir's function in enhancing water's O_2 content. Their findings showed that piano key weir aeration efficiency improves as drop height increases. According to different research by Baylar and Bağatur (2000), the triangle notch weir performs better than other weirs in terms of raising the O_2 content of water. Furthermore, tray aerators are an integral component of subsurface aeration systems, which enhance O_2 transfer efficiency in wastewater treatment processes and natural water bodies. These devices consist of multiple perforated trays stacked vertically, through which water cascades, breaking into smaller droplets. As the water descends through the trays, it is repeatedly exposed to the air, significantly increasing the surface area for O_2 transfer. The design of tray aerators facilitates a high degree of turbulence and mixing, which enhances the oxygenation process and promotes the removal of volatile substances. Additionally, tray aerators can help strip unwanted gases such as carbon dioxide and hydrogen sulfide, thereby improving water quality. Their effectiveness in O_2 transfer makes them suitable for use in various applications, including municipal and industrial wastewater treatment, fish farming, and lake and pond aeration. The modular design also allows for easy scaling and maintenance, making tray aerators a versatile and efficient solution in subsurface aeration systems.

5.2.2.2 Surface aeration

Surface aerators primarily function by disrupting the water surface to aid in O_2 transfer through various techniques, such as using a pump to spray water into the air, forcing it through nozzles, or employing a hydraulic leap. Generally, bubbles smaller than 5 mm in diameter are classified as fine bubbles, whereas coarse bubbles are larger, with diameters reaching up to 50 mm (Padrón-Páez et al., 2016). Moreover, the most common surface aeration devices are the spiral, paddle-wheel-type, pump-sprayers, and vertical pump aerators, which are commercially available (Fig. 5.4; Roy et al., 2022b). Paddle-type aeration systems aerate the water either by continuously splashing air into the

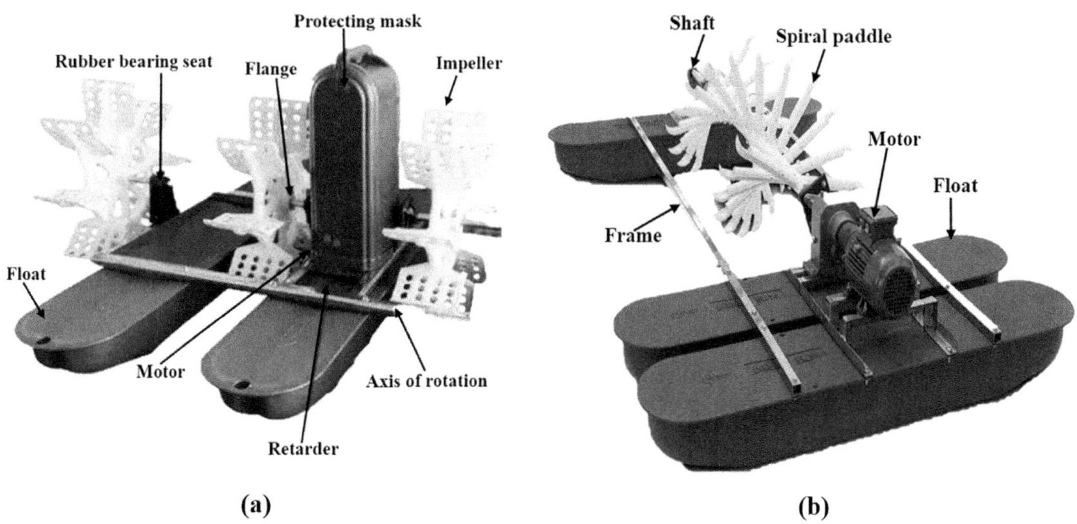

Fig. 5.4 Surface aeration devices: (A) paddle wheel-type aerator and (B) spiral wheel-type aerator.

atmosphere or by agitating the pond directly. As water particles are exposed to the air and return to the water's surface, the rotating wheel of the system further enhances the oxygenation of the water body.

Furthermore, thermal stratification is eradicated due to the constant circulation that builds up at the bottom and water splashing at the surface to manage temperature. During aeration, the overabundance of O_2 is also released by splashing, and supersaturation is sustained; thus, the DO concentration in the water body can be controlled. A spiral wheel aerator is another surface aerator, an enhanced/modified version of paddle-type aerators. In this system, the water body is splashed using the spiral paddles (spines) at constant spinning with the help of the shaft. A similar mechanism occurs with the paddle-type aeration system. However, the length and speed of the spiral paddles determine how much O_2 dissolves into the water, and as a result, longer spines will have more DO. Furthermore, it prevents feed from sinking to the bottom of the water body.

Consequently, these aerators use rotating impellers or blades to churn the water, creating droplets and increasing the water surface area exposed to air. The mechanical action disrupts the water surface, allowing more O_2 to enter the water. They are effective at mixing and oxygenating large volumes of water and have robust and relatively simple designs. However, higher energy consumption and the potential for aerosol formation and spray, which can cause odor issues, are drawbacks of this type of aerator.

5.2.3 Natural aeration systems

Natural aeration systems use the principles of natural environmental processes to enhance the O_2 levels in water bodies, promoting the health of aquatic ecosystems and improving water quality. Unlike mechanical or subsurface aeration methods that rely on human-made equipment to facilitate gas exchange, natural aeration leverages wind action, water flow, and biological activity. Wind-driven surface agitation, such as waves and ripples, increases the interface between air and water, allowing for efficient O_2 transfer, but natural aeration performance can significantly change related to the seasons (Fig. 5.5). Similarly, waterfalls, rapids, and flowing streams naturally incorporate air into the water column through physical turbulence. Biological activity, particularly photosynthesis by aquatic plants and algae, also plays a crucial role; during daylight hours, these organisms produce O_2 as a byproduct, contributing significantly to the DO content. Natural aeration systems are sustainable and energy-

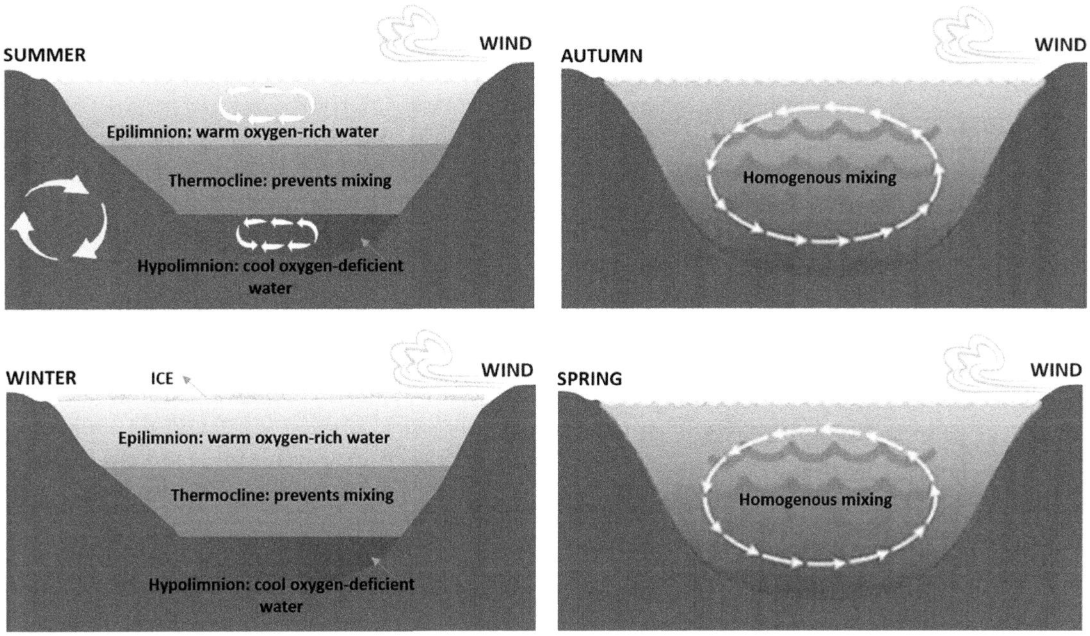

Fig. 5.5 Natural aeration in ponds with wind at different seasons.

efficient, relying on inherent ecological processes without the need for external energy inputs or complex machinery. However, their efficiency can be influenced by various factors such as climatic conditions, water temperature, and pollutants in the water bodies. While natural aeration is often sufficient for maintaining beneficial O_2 levels in many natural and semi-natural water bodies, it may need to be supplemented with artificial aeration methods in heavily polluted or densely populated aquatic environments to ensure optimal water quality (Dawen and Nabi, 2024).

5.2.4 Jet aeration systems

Jet aeration systems are advanced mechanical aeration technologies used primarily in wastewater treatment to enhance O_2 transfer and promote efficient mixing. These systems operate by injecting high-velocity jets of air or O_2-enriched air into the water, creating fine bubbles that maximize the surface area for gas exchange. The process begins with a high-pressure pump that forces water through specially designed nozzles, where it is mixed with air before being discharged into the aeration tank. The resulting high-speed jet stream produces significant turbulence and mixing, ensuring a thorough O_2 distribution throughout the water column. This efficient mixing not only improves O_2 transfer rates but also helps break down organic matter and remove nutrients by maintaining optimal conditions for microbial activity. Jet aeration systems are known for their high energy efficiency, flexibility, and ability to handle varying loads, making them suitable for a several purposes, such as in aquaculture, municipal, and industrial wastewater treatment. Additionally, the modular design of these systems allows for easy installation and maintenance, and their operational parameters can be adjusted to meet specific treatment requirements. The high O_2 transfer efficiency and robust mixing capabilities of jet aeration systems make them a preferred choice for facilities seeking to enhance aeration performance and achieve stringent effluent quality standards.

In jet aeration systems, a submerged two-phase zone forms beneath the water surface when a water jet enters a still water pool after traversing a gas layer and incorporating substantial amounts of air. This effect results from the interaction between the hydrodynamic forces of the jet and the surrounding air. Downward water jets are employed in various industrial and environmental applications, including air bubble flotation processes. In wastewater treatment facilities, jet aeration systems use activated sludge to process household and animal waste. Additionally, overfall jets from weirs, spillways, and other hydraulic structures aerate and purify water by absorbing O_2 from the air (Radkevich et al., 2021; Shukla et al., 2020).

5.2.5 Packaged aeration systems

Packaged aeration systems are pre-engineered, self-contained units designed to deliver efficient aeration for wastewater treatment and various water quality applications. These systems integrate all necessary components, including blowers, diffusers, and control panels, into a single compact unit, facilitating straightforward installation, operation, and maintenance. Packaged aeration systems typically employ diffused aeration techniques, where air is forced through fine-pore diffusers submerged in the treatment tank, creating fine bubbles that enhance O_2 transfer rates due to the increased surface area and prolonged contact time with the water. The pre-engineered design ensures that all components are optimally matched and configured, resulting in consistent performance and reliability. These systems are particularly advantageous for small to medium-sized facilities, such as small municipal wastewater plants, industrial wastewater treatment, and decentralized or remote locations where custom-built systems may be impractical or cost-prohibitive. Packaged aeration systems often have advanced control features, including automated monitoring and adjustment capabilities, which help maintain optimal DO levels and energy efficiency. The modular nature allows for scalability, enabling facilities to expand or upgrade their aeration capacity. Overall, packaged aeration systems provide a significant solution that combines ease of use, effectiveness, and adaptability, making them a valuable option for enhancing water treatment processes (John et al., 2024; Rosso et al. 2023).

5.3 Aeration system design

Aeration is a critical procedure in most wastewater treatment facilities, constituting the most significant portion of energy expenses, typically ranging from 45% to 75% of the total energy outlay for the plant (Cardoso et al., 2021; Musvoto and Ikumi, 2016). Aeration systems enable O_2 transfer into a liquid medium using different methods, such as agitation with mixers or turbines or by releasing air through large orifices or porous materials. The interaction between the gas and liquid results in falling droplets and rising coarse bubbles, each featuring significant velocity gradients at the gas-liquid interface. These interfaces are classified into high-flow regime interfaces, like those with falling droplets and rising coarse bubbles, and low-flow regime interfaces, characterized by fine bubbles with minimal velocity gradients at the interface (Rosso et al., 2018). Because of their excellent energy efficiency, fine pore diffusers, which create tiny bubbles, have emerged as the industry standard for aeration in wastewater treatment. Submerged turbines and jet diffusers, two further types of aeration devices, likewise produce fine bubbles without tiny holes. Instead, they apply mechanical energy to break down larger bubbles into smaller ones, placing them in the high-flow regime interfaces category (Marcelino et al., 2023; John et al., 2022; Gu et al., 2023).

The mean cell retention time (MCRT) is the paramount process parameter influencing aeration efficiency. MCRT correlates directly with biomass concentration, thereby regulating O_2 needs. Alpha factors and aeration efficiency, denoting the ratio of mass transfer between process water and clean water, exhibit higher values at elevated MCRTs. When operated at increased MCRTs, biological nutrient removal processes demonstrate enhanced aeration efficiency (Rosso and Stenstrom, 2005). Incorporating anoxic and anaerobic selectors in nutrient removal plants provides advantages beyond just nutrient removal and enhanced settling characteristics. These selectors utilize the carbon in wastewater to remove rapidly acting surfactants, significantly reducing the need for O_2 transfer. Research has shown that O_2 transfer efficiency is positively related to the MCRT, negatively related to the airflow rate per diffuser, and positively related to geometric parameters such as diffuser submergence, number, and surface area (Rosso et al., 2005, 2018). MCRT is crucial for determining the net O_2 requirement in wastewater treatment, as it influences the removal of contaminants that reduce O_2 transfer, such as surfactants. Systems with higher MCRTs tend to eliminate or sorb surfactants earlier in the process, thereby improving overall O_2 transfer efficiency. As a result, increasing the MCRT leads to higher O_2 demands, better removal of biodegradable organics, and improved O_2 transfer efficiency. The increased O_2 requirement is often balanced or surpassed by the savings gained from higher transfer efficiency (Bencsik et al., 2022; Rosso et al., 2023).

Almost all modern activated sludge process designs now incorporate anoxic or anaerobic selectors. These selectors not only improve the removal of influent substrates and reduce filamentous organisms but also enhance O_2 transfer efficiency. Recent studies have shown that systems with a higher MCRT are more effective in removing anthropogenic compounds, including pharmaceuticals (Van Loosdrecht et al., 2015; Stensel and Makinia, 2014; Wanner et al., 2014).

Moreover, the airflow rate significantly impacts the fluid dynamics of bubbles: a higher airflow rate per diffuser or orifice leads to the formation of larger bubbles, resulting in a reduced surface-to-volume ratio and increased bubble rise velocity. Consequently, this yields a smaller gas-to-liquid area and shorter bubble residence time, diminishing mass transfer. Geometry also plays a crucial role in efficiency, as greater submergence and tank coverage (the ratio between diffusing area and total tank area) lead to extended mass transfer time and increased surface area (Bencsik et al., 2022; Rosso et al., 2023; Liew et al., 2020).

Various techniques have been employed to clean fine-pore diffusers, each differing in complexity and associated costs. The simplest approach involves dewatering the aeration tank and conducting diffuser cleaning from the tank's upper surface. Known as tank top hosing, this method effectively eliminates accumulated biological slime, typically resulting in either full or partial restoration of efficiency. Acid cleaning becomes necessary when inorganic precipitates (such as silica, calcium carbonate, or gypsum) have led to scaling. Manual washing with diluted hydrochloric acid (typically 10% to 15% by weight) is commonly used. In contrast, acid gas cleaning, achieved by injecting hydrochloric acid gas or acetic acid into air distribution lines, presents an alternative option. Over time, the O_2 transfer efficiency of fine-pore diffusers naturally declines. Concurrently, the back pressure, often called dynamic wet pressure (DWP), typically experiences an increase, occasionally to a significant extent. The increase in DWP can be attributed to the clogging of pores in ceramic diffusers or to permanent alterations in orifice characteristics for polymeric membranes. Consequently, cleaning fine-pore diffusers becomes almost invariably necessary to restore process efficiency and mitigate power costs. Analysis of 94 field tests reveals that efficiency diminishes over time, with the most significant rate of decline typically occurring within the initial 24 months of operation. This decline in efficiency has been quantified and factored into cost analyses, with the net present worth compared against cleaning expenses. Cleaning frequency is consistently higher for systems with higher fouling rates, and the optimal cleaning interval ranges from as short as 9 months to a maximum of 24 months (Rosso et al., 2008).

5.3.1 Artificial (passive) aeration systems

Passive aeration units operate without electrical energy. Artificial aeration is typically achieved through three main methods: gravity aeration, using cascade aerators; diffused aeration, with devices like propeller aspirators and submersible aerators; and surface aeration with paddle wheel and spiral aerators, vertical pumps, and sprayers. Various aeration systems have been developed to improve the energy efficiency of the O_2 mass transfer process. These systems generally follow two principles: (i) aeration by splashing water into the air, which includes devices such as paddle wheel aerators, vertical pumps, pump sprayers, and gravity aerators, and (ii) aeration by bubbling air into water, which involves systems like propeller aspirators and diffused aeration systems.

5.3.1.1 Gravity aeration system

Gravity aeration systems harness gravitational forces to facilitate air infusion into water, thereby optimizing gas exchange processes. Among these systems, cascade aerators, weir aerators, and tray aerators are prominent, each employing unique mechanisms to improve O_2 transfer efficiency. The subsequent sections will provide a comprehensive analysis of these aerators, examining their operational principles, design parameters, and performance characteristics.

Cascade aerators

Cascade aerators stand out for their superior aeration capabilities. This is attributed to their high turbulent mixing, extended exposure time, and significant entrainment of air bubbles. Cascade aerators, a specific type of gravity aerator, are typically employed as pre- or post-aeration systems. When site constraints and hydraulic conditions allow for gravity flow, the most cost-effective method to increase DO levels is often a cascade aeration system or a propeller-aspirator pump. Cascade aerators enhance water quality by promoting the upward displacement of water from the bottom to the top, effectively removing VOCs accumulated at the pond bottom. Their design, characterized by the absence of moving parts, translates to reduced maintenance requirements. Cascade aerators, a subtype of gravity aerators, enjoy widespread popularity in water treatment plants due to their effectiveness in removing iron, manganese, dissolved gases, and VOCs from water. Overall, the main advantage of cascade aerators is their low energy consumption (Oh et al., 2016).

For instance, the stepped cascades dissipate high stream velocities and enhance air circulation. They are primarily employed for pre- or post-aeration purposes. Water flows through a series of steps, and the resulting turbulence from water falling onto these steps disrupts the air-water interface, promoting O_2 transfer and removing VOCs such as methane, chlorine, dissolved iron, manganese, carbon dioxide, and hydrogen sulfide. Stepped cascade aerators can remove 60–70% of carbon dioxide and are particularly effective at increasing the DO content of water. As the height of the steps increases, the flow tends to adopt the nappe flow regime. According to the research, O_2 transfer increases with longer exposure times on the steps (i.e., reduced flow rate), higher temperatures, and larger total exposed areas of the steps (Moulick et al., 2010). The circular stepped cascade (CSC), combined with a pump, is the core component of the main aeration unit commonly used in wastewater treatment plants and aquaculture ponds. Its design, featuring only static components, simplifies maintenance and reduces energy consumption. This system consists of circular steps that allow water to flow from the upper step to the lower one and finally to the bottom step in a uniformly distributed manner. Key geometric parameters for optimal performance include the overall step number (N), total height to the bottom radius ratio of the cascade (H/R_b), and volume of water under aeration to the bottom radius ratio of the cascade (V/R_b^3) (Nguyen et al., 2024).

Compared to stepped and CSC aerators, a pooled circular stepped cascade (PCSC) is employed as the primary aeration unit in aquaculture ponds. Combined with a pump, this type of gravity aerator allows water to flow radially over the steps. By adding zigzag-patterned obstructions at the periphery of each step, the exposure time of the water is increased, thereby enhancing O_2 transfer. As a simple and effective aeration device, the PCSC is believed to significantly reduce operational and maintenance costs. It addresses several issues simultaneously, including low DO levels, high concentrations of VOCs, excessive nitrogenous compounds, and water stratification (Kumar et al., 2013).

Design characteristics of a PCSC aeration system can be explained starting with the two-film theory, where the difference form of O_2 transfer is reported as follows (Stenstrom et al., 2006):

$$\frac{dC}{dt} = K_L a_T \left(C_\infty^* - C\right) \tag{1}$$

where dC/dt is the O_2 concentration change rate in the liquid media. $K_L a_T$ is the coefficient for the total volumetric O_2 transfer at specific T (°C), and C_∞^* is the O_2 concentration in liquid media at equilibrium. C is the real O_2 concentration in the liquid media. The incorporated form of this equation can be reported as follows:

$$C = C_\infty^* - \left(C_\infty^* - C_0\right)e^{K_L a_T t} \tag{2}$$

C_0 is the initial O_2 concentration, and $K_L a_T$ is best estimated, considering C_0, C_∞^* and $K_L a_T$ parameters concurrently using a non-linear method. The value of $K_L a_T$ is converted to a value at a temperature of 20 °C as $K_L a_{20}$; for comparison of results, the following equation can be used:

$$K_L a_{20} = \frac{K_L a_T}{\theta^{(T-20)}} \tag{3}$$

where θ is the temperature correction factor, 1.024 for pure water.

The mass of O_2 that aerating equipment can introduce into a reservoir of water per unit of time under standard conditions is known as the standard O_2 transfer rate (SOTR). An aeration system's SOTR may be calculated as follows:

$$\begin{aligned} SOTR(kgO_2/h) &= K_L a_{20} \times \left(C^* - C_0\right) \times V \\ &= K_L a_{20} \times 9.07 \times V \times 10^{-3} \end{aligned} \tag{4}$$

Moreover, the standard aeration efficiency (SAE) is an important parameter, defined as SOTR per power unit (Lawson and Merry, 1993).

$$SAE(kgO_2/kWh) = SOTR(kgO_2/h)/P(kW) \quad (5)$$

where P is the energy consumption of the aerator.

The O_2 transfer rate is another important parameter in aeration applications. The physical properties of air and water crucial for O_2 transfer include the mass density of air (ρ_a), the mass density of water (ρ_w), the kinematic viscosity of water (ν_w) and surface tension of water (σ_w).

The schematic diagram of the PCSC depicted in Fig. 5.6 outlines several geometric parameters essential for its characterization. These parameters encompass the radius of the bottommost circular step (R_b), the number of cascade steps (N), the cascade height (H), the volume of water (V), the percentage coverage of each step's circumference by enclosures (P_e), and the number of enclosures within each step (N_e). The steps are assumed to uniformly decrease in radius from the bottommost to the topmost, with $R_i = [R_b - (R_b/N) \times (i-1)]$. Consequently, the difference in radius between consecutive steps equals Rb/N, equivalent to the radius of the topmost circular step, denoted as R_t. In addition, the height of each step (h) can be derived from H and N, where $N \times h = H$. Hence, h is not

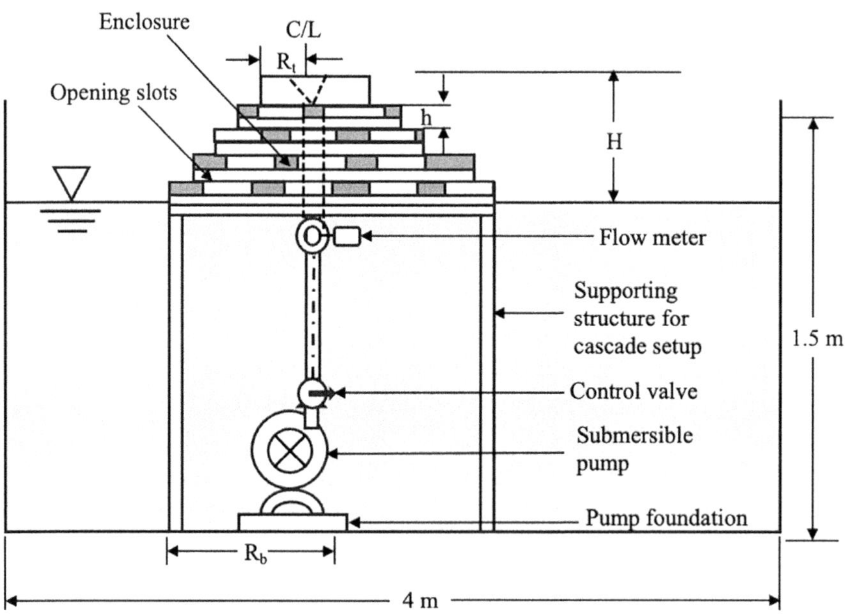

Fig. 5.6 Schematic view of pooled circular stepped cascade system aerator. Adapted with permission from Kumar, A., Moulick, S., Singh, B.K., Mal, B.C. 2013. Design characteristics of pooled circular stepped cascade aeration system. Aquac. Eng. 56, 51–58. https://doi.org/10.1016/j.aquaeng.2013.04.004. 01448609, Elsevier (license number: 5844630126679).

considered a variable in the analysis, while the enclosure height matches each step's. The flow parameters encompass the acceleration due to gravity (g) and the pump discharge (Q).

Weir aerator

The weir aeration technique, introduced by Gameson (1957), is a popular gravity-based system used in water treatment and aquaculture. This method involves water flowing over a weir with a splashboard, which fragments the water into droplets that pass through various screens. The falling water jet from the weir mixes with the downstream water, introducing air bubbles that improve aeration. Compared to other systems like surface aeration, weir aeration is more cost-effective, even when considering the energy required for water pumping (Baylar and Bağatur, 2000).

Tray aerator

In contrast to cascade and spray aerators, tray aerators have the advantage of requiring significantly less installation space. These aeration systems make use of the head in the incoming water to lift it above a distribution tray. From this elevated position, the water flows downward due to gravity across a series of horizontally aligned perforated plates, as depicted in Fig. 5.7, which also details the main parameters of tray aerators. A tray aerator generally features multiple trays (N) stacked vertically with a space (SP) between each. Each tray is constructed from a sheet with several holes (n) of diameter (d). As the water traverses the trays, a thin water film with a height (h_w) forms on the surface before passing through the holes. To prevent overflow, the trays are designed with sides of height (h_s) greater than the height of the water film. Tray aerators are primarily used for stripping processes, such as removing unwanted gases from water, including iron and manganese in water treatment, carbon dioxide in clean water, and sulfides from groundwater (El-Zahaby and El-Gendy, 2016).

During the aeration of water, the constituents naturally move from high to low concentration regions until equilibrium is reached, resulting in similar concentrations in both zones. The rate of O_2 transfer to water across an air-water interface can be described by Eq. (6):

$$V\frac{\partial C}{\partial t} = K_L A(C_s - C) \tag{6}$$

where V is the volume of water over which the concentration of O_2 in water (C) and the interface area (A) are measured. t is the time, K_L is the coefficient for liquid mass transfer, and C_S is the saturation concentration of the gas in water.

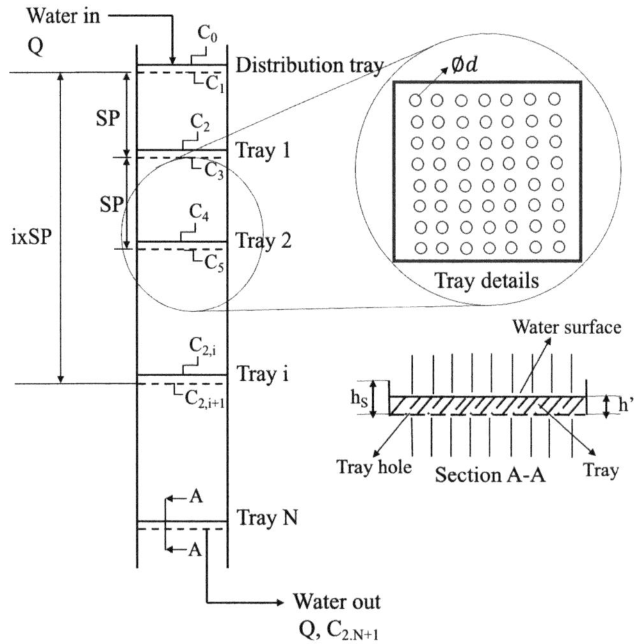

Fig. 5.7 Tray aerator setup. Adapted with permission from El-Zahaby, A.M., El-Gendy, A.S. 2016. Passive aeration of wastewater treated by an anaerobic process—a design approach. J. Environ. Chem. Eng. 4 (4), 4565–4573. https://doi.org/10.1016/j.jece.2016.10.025, 22133437. Elsevier, Egypt. Available from: http://www.journals.elsevier.com/journal-of-environmental-chemical-engineering/. Elsevier (license number: 5844630489953).

For the tray aerator system, Eq. (6) can be rearranged and integrated over the time interval from zero to t and the concentration interval from the initial to time t concentration, resulting in the form shown in Eq. (7).

$$ln\left(\frac{C_S - C_t}{C_S - C_0}\right) = -K_L at \tag{7}$$

where a is the specific area equal to the ratio between the interface area and the liquid volume.

The DO content for wastewater entering the aeration unit is usually zero when it comes from an anaerobic treatment process, which makes it an uncontrollable factor during the design phase. Three widely accepted theories describe the rate of gas transfer: the two-film theory proposed by Lewis and Whitman (1924), the penetration theory by Higbie (1935), and the surface renewal theory by Danckwerts and Lannus (1970). The primary distinction among these theories lies in their physical explanations of the diffusion process. The two-film theory is based on a rigid interface between air and liquid, making it more suitable for still water, such as in estuaries (Wang and Langemann, 1994). In contrast, the penetration and surface renewal theories address liquid displacement at the interface, which is relevant for flowing rivers and falling water jets or droplets. Another key difference lies in the techniques for estimating K_L, which indicates the gas transfer rate per unit concentration gradient and is measured in velocity units. The methods for calculating K_L are demonstrated in Eqs. (8)–(10) for the two-film, penetration, and surface renewal theories, respectively, and these equations help estimate K_L based on the hydraulic conditions of the flow regime.

$$K_L = \frac{D_L}{\delta} \tag{8}$$

$$K_L = 2\sqrt{\frac{D_L}{\pi t_e}} \tag{9}$$

$$K_L = \sqrt{D_L S} \tag{10}$$

In these theories, D_L denotes the diffusivity of air in water, while δ represents the thickness of a thin laminar water film through which mass transfer occurs, assuming the remainder of the water is turbulently mixed, according to the thin film theory. The parameter t_e refers to the exposure time of this laminar film to the gas, with the film being uniformly replenished, as outlined in the penetration theory. S signifies the rate at which the liquid surface is renewed at a random pace, as described in the surface renewal theory. When deriving a mathematical expression for the DO concentration in the effluent from tray aerators, El-Zahaby and El-Gendy (2016) examined a system featuring one aeration tray alongside a distribution tray. Water entering the distribution tray undergoes two flow modes before reaching the aeration tray: a thin film above the distribution tray and free-falling water between the two trays.

Water exiting nozzles or holes can display various flow regimes, such as dripping or jetting. In the dripping regime, spherical drops of constant mass detach from the nozzle or hole at a regular frequency (periodic dripping), or variable mass drops form irregularly (dripping faucet). This regime occurs when the Weber number (*We*), a dimensionless quantity, is below a critical Weber number (We_c), as defined by Clanet and Lasheras for any Newtonian fluid, as detailed in Eqs. (11)–(13).

$$We = \frac{\rho v_1^2 D_1}{\sigma} \tag{11}$$

$$We_c = 4\frac{BO_o}{BO}\left[1 + kBO_oBO - \left((1 + kBO_oBO)^2 - 1\right)^{0.5}\right]^2 \tag{12}$$

$$BO = \left[\frac{\rho g d^2}{2\sigma}\right]^{0.5} \tag{13}$$

where the dimensionless *We* compare the inertia force with the surface tension force in the context of water exiting from nozzles or holes. ρ denotes the liquid density, v_1 represents the initial velocity upon exit from the nozzle, and D_1 signifies the initial diameter of the water stream. σ stands for the surface tension of the liquid, while BO_o corresponds to the Bond Number relative to the outside diameter (wetting diameter) of the nozzle, and BO pertains to the Bond Number relative to the inside diameter of the nozzle. The constant k equals 0.37 for water injected into the air, and g denotes the acceleration due to gravity.

When free-falling water exits tray holes under gravitational forces, the holes function like nozzles. Under steady-state flow conditions, a thin water film of height h forms over the tray, as depicted in Fig. 5.7. The initial exit velocity v_1 from the nozzle can be estimated based on h using the nozzle equation, as given in Eq. (14).

$$v_1 = C_v\sqrt{2gh} \tag{14}$$

where C_V is the coefficient of velocity through nozzles that is assumed to be 0.99.

In the jetting regime, the water jet exiting the hole has an initial diameter of D_1, estimated as indicated in Eq. (15). The number of holes per tray required to achieve a desired film height is estimated as indicated in Eq. (16).

$$D_1 = d\sqrt{\frac{C_d}{C_v}} \tag{15}$$

$$n = \frac{Q}{\frac{\pi}{4} v_1 D_1^2} \tag{16}$$

where C_d is the coefficient of discharge through nozzles assumed to be 0.6, and Q is the total flow rate.

Since the number of holes must be an integer, the value derived from Eq. (16) is rounded to the nearest integer. Subsequently, the initial water velocity and the film height are recalculated as indicated in Eqs. (17), (18).

$$v_1' = \frac{Q_{hole}}{A_j} = \frac{Q}{n' \frac{\pi}{4} D_1^2} \tag{17}$$

$$h' = \frac{\left(v_1'/C_v\right)^2}{2g} \tag{18}$$

where v_1' denotes the corrected initial exit velocity from the nozzle; Q_{hole} represents the flow rate per hole; A_j is the cross-sectional area of the water jet; n′ stands for the corrected number of holes per tray; and h' signifies the corrected film height.

To prevent water overflow from the tray sides, if the corrected film height h′ exceeds the tray side height h_s, the corrected number of holes n′ is increased by one hole iteratively until h′ is less than h_s. This iterative process for n′, v_1', and h′ continues until all conditions are satisfied. However, due to these adjustments, the final film height may differ from the initially desired film height. The velocity of the jet increases as the water descends under gravitational acceleration, causing the jet diameter to decrease to maintain continuity. The parameters describing the water jet are calculated according to Eqs. (19)–(26).

$$L_j = SP - h' \tag{19}$$

$$v_2 = \sqrt{\left(v_1'\right)^2 + \left(2gL_j\right)} \tag{20}$$

$$D_2 = D_1\sqrt{\frac{v_1'}{v_2}} \tag{21}$$

$$D_j = \frac{D_1 + D_2}{2} \tag{22}$$

$$v_j = \frac{v_1' + v_2}{2} \tag{23}$$

where L_j is the jet length, v_2 is the jet velocity reaching the subsequent tray, D_2 is the jet diameter reaching the subsequent tray, D_j is the mean jet diameter, and v_j is the mean jet velocity.

The thin film is assumed to be uniform over the tray area, with a volume equal to the tray area multiplied by the film height. The specific area of the thin film (a_f), which is the ratio of the interface area to the liquid volume, is calculated as shown in Eq. (24).

$$a_f = \frac{A}{V_f} = \frac{A}{Ah'} = \frac{1}{h'} \tag{24}$$

where A represents the interface area of the thin film, equal to the tray area, and V_f denotes the liquid volume over the tray, which equals the tray area multiplied by the film height h'.

For the aeration time, the duration for water to be aerated as it travels from point C_0 to point C_1, shown in Fig. 5.7 and separated by a distance h′, can be estimated using Eq. (25).

$$t_f = \frac{V_f}{Q} = \frac{Ah'}{Q} = \frac{h'}{HLR} \tag{25}$$

where t_f is the thin film aeration time, and HLR is the hydraulic loading rate, defined as the flow rate per unit area.

Using Eq. (7) and substituting *a* and *t* from Eqs. (24) and (25), respectively, the DO concentration of water at the exit from the tray, indicated as C1 in Fig. 5.7, can be calculated as shown in Eq. (26).

$$\begin{aligned} C_1 &= C_S - (C_S - C_0)\exp\left(-K_L \times \frac{1}{h'} \times \frac{h'}{HLR}\right) \\ &= C_S - (C_S - C_0)\exp\left(-\frac{K_L}{HLR}\right) \end{aligned} \tag{26}$$

Using the same approach for estimating the specific area and aeration time for the thin film, the specific area of the water jet is estimated as shown in Eq. (27). According to Eq. (21), the jet diameter decreases as it travels downward, resulting in an inverted cone shape rather than a perfect cylinder. However, for simplification, the jet is modeled as a cylinder with a diameter D_j having a specific area of a_j and traveling at a velocity v_j. The aeration time is estimated by Eq. (28). Hence, the DO concentration at the point reaching the next aeration tray (C_2) is calculated by Eq. (29), considering the DO concentration at the exit point from the distribution tray (C_1) and by substituting a by a_j in Eq. (7).

$$a_j = \frac{A_j}{V_j} = \frac{\pi D_j L_j}{\frac{\pi}{4} D_j^2 L_j} = \frac{4}{D_j} \tag{27}$$

$$t_j = \frac{L_j}{v_j} \tag{28}$$

$$C_2 = C_s - (C_s - C_1)\exp(-K_L a_j t_j) \tag{29}$$

To estimate the overall aeration occurring over a single tray, Eqs. (26) and (29) are combined to account for the two aeration regimes: the thin film above the tray and the water jetting from the tray, as in Eq. (30).

$$C_2 = C_S - (C_S - C_o)\,exp\left[-K_L\left(\frac{1}{HLR} + a_j t_j\right)\right] \tag{30}$$

Given that a system with N consecutive trays arranged vertically beneath the distribution tray, each with a constant spacing SP as in Fig. 5.7, the influent DO to the ith tray is denoted as C_{2i}, where i ranges from 1 to N, representing the number of trays. Water reaching the ith tray undergoes i – 1 thin film regimes above the tray and i – 1 falling water regimes. Therefore, Eq. (30) can be generalized to account for the tray number, as indicated in Eq. (31).

$$C_{2i} = C_S - (C_S - C_0)\,exp\left[-K_L(i-1)\left(\frac{1}{HLR} + a_j t_j\right)\right] \tag{31}$$

The overall system height (H) is calculated after determining the total number of trays (N) as follows:

$$H = SP \times N \tag{32}$$

5.3.2 Diffused aeration system

Fine-bubble diffused air (FBDA) systems are more commonly implemented in wastewater treatment facilities serving large populations than mechanical SA systems. FBDA systems are more frequently utilized to produce high-quality, nitrified effluents than surface aeration (SA) systems. The FBDA system is highly efficient in low-rate activated sludge systems. However, this aeration technique suffers from poor O_2 transfer efficiency and diffuser clogging under increased loading conditions, such as those near the tank inlet in plug-flow systems. In contrast, mechanical surface aeration systems, while seemingly less efficient, do not encounter these issues. Therefore, the optimal configuration for an aeration tank to achieve maximum efficiency includes an inlet anoxic zone, a completely mixed surface aerated zone, and a plug-flow, fine-bubble diffused air zone with tapered aeration leading to clarification tanks. This design is predicted to achieve an overall aeration efficiency of 1.5–2.0 kg/kWh for nitrifying systems and extend diffuser lifetime compared to conventional FBDA tanks. This design approach is exemplified using parameters from a large wastewater treatment plant (Thomas et al., 1989).

The optimal design for an aeration tank appears to be a plug-flow system with an anoxic zone, aerated by fine-bubble diffusers. However, fine-bubble systems face challenges such as the accumulation of surfactant material at the air/water interface, which reduces O_2 transfer efficiency, particularly near the inlet of plug-flow systems. This effect diminishes as treatment progresses. In addition, diffusers tend to clog due to slime and sludge attachment at the inlet of plug-flow tanks, leading to increased pressure drop and higher aeration energy costs. In contrast, vertical-shaft mechanical surface aerators do not suffer from these issues and are thus better suited for high-load environments. Therefore, an optimally efficient aeration tank designed to produce fully nitrified effluent would have an inlet anoxic zone for denitrification, followed by a hybrid aeration system. This system would include a fully mixed, mechanically surface-aerated section and a plug-flow, fine-bubble diffused air section with tapered aeration. This configuration is expected to achieve an overall aeration efficiency of 1.5–1.7 kg/kWh and extend the operational lifetime of the diffusers.

Incorporating inlet anoxic zones for denitrification in nitrifying activated sludge plants offers numerous well-documented advantages. There is no known process disadvantages associated with well-designed anoxic zones. Therefore, any new nitrifying activated sludge plant should integrate denitrification in inlet anoxic zones. The benefits of this approach include a reduction in overall energy requirement because utilizing nitrate O_2 can meet 10–20% of the total O_2 requirement for treating typical mixed domestic/industrial sewage. It has improved sludge settleability by incorporating a non-aerated zone at the tank inlet. Reduction of BOD (biological oxygen demand) load is another benefit of this approach, as the BOD load imposed on the aerated section of the activated sludge plant is decreased. Mitigating rising sludge in clarification tanks reduces or eliminates denitrification, preventing the formation of rising sludge in final clarification tanks. The pH decrease caused by nitrification in soft water is reduced by pH stabilization.

Mixing characteristics within a reactor are often described using the Dispersion Number (D/uL), as defined by Levenspiel (1962). This parameter is essential for understanding and optimizing the mixing and flow patterns within the reactor, contributing to the overall efficiency and effectiveness of the treatment process.

For design purposes, it is useful to conceptualize longitudinal mixing using the "tanks-in-series" model. This model posits that any system can be represented by a series of N completely mixed tanks connected in sequence. Although a precise comparison between different mixing models is not available, the following relationship provides an adequately accurate approximation:

$$Dispersionnumber, \frac{D}{uL} = \frac{1}{2N} \tag{33}$$

where D is the dispersion coefficient (m^2/s), u is the average fluid velocity (m/s), L is the total aeration tank length (m), and N is the number of tanks in series.

The degree of longitudinal mixing within an aeration tank is primarily determined by the tank's geometry. Generally, the mixing characteristics are a function of the linear velocity of a fluid element traveling from the tank inlet to the outlet. For conventional diffused air systems without spiral roll, the degree of longitudinal mixing can be estimated, with an accuracy of $\pm 15\%$, using the following empirically derived equation:

$$N = \frac{7.4}{W \times H} L \times q_s(1 + r) \tag{34}$$

where W is the aeration width (m), H is the aeration tank depth (m), q_s is the average sewage flow rate (m^3/s), and r is the recycle ratio.

The equation is valid for diffused air systems, provided that the variables fall within the following ranges: $28 < L < 500$ (m), $2 < W < 20$ (m), $2.4 < H < 6$ (m), $0.7 < r < 1.5$ and $1.3 < t < 8$ (h) where t is defined as $t = V/(3600q_s/1 + r))$. Note that V is the aeration tank volume (m^3). A representation of the activated sludge process, as utilized in the model, is illustrated in Fig. 5.8. The optimal setup for an activated sludge plant, designed to produce high-quality, fully nitrified effluent, consists of an anoxic zone, a completely mixed mechanical SA section, a plug-flow FBDA section with tapered aeration and secondary clarification.

To find DO, a mass balance for DO in the arbitrary jth tank within the system can be expressed as follows:

$$\begin{aligned} \frac{V}{N}\frac{dC_j}{dt} &= q_s(1 + r)C_{j-1} - q_s(1 + r)C_j \\ &+ (k_L a)_j \frac{V}{N}(C_S - C_j) - R_j \end{aligned} \tag{35}$$

where C_j and C_{j-1} are the concentration of DO in the jth and (j−1) th tanks, respectively. $(k_L a)_j$ is the overall mass transfer coefficient for O_2 in the jth tank, and R_j is the DO consumption rate of microorganisms. $(V/N)dC_j/dt$ is the term for the accumulation of O_2 in

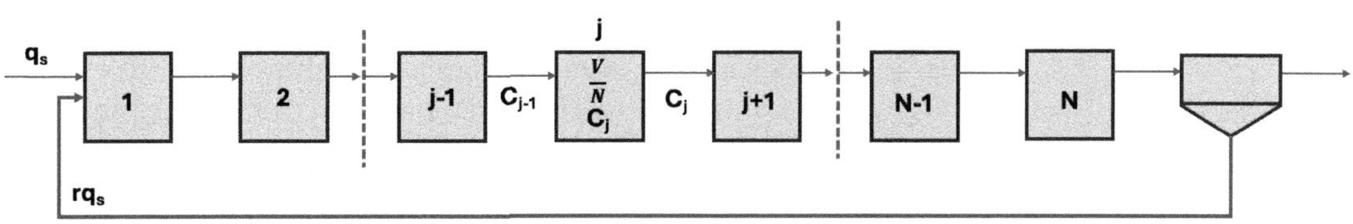

Fig. 5.8 Flow diagram of activated sludge model.

the tank, $q_s(1+r)C_{j-1}$ is the bulk flow of O_2 into the tank, $q_s(1+r)C_j$ is the bulk flow of O_2 leaving the tank, $(k_La)_j(V/N)(C_S-C_j)$ denotes for the O_2 transferred to liquid from air, and R_j is the O_2 consumed by microorganisms.

5.3.3 Surface aeration system

Surface aeration systems used in activated sludge plants are the most energy-consuming components and generally represent a significant portion of the total energy usage of the treatment facility (Rao et al., 2009). Surface aeration involves the entrainment of gas at the gas-liquid interface. Reactors designed for this type of gas-liquid interaction are called surface aerators. Their primary functions are to enhance the O_2 transfer rate, ensure thorough mixing of the liquid phase for consistent O_2 distribution throughout the contactor, and keep microorganisms suspended. Surface aerators are preferred for their simplicity and efficient O_2 transfer rate relative to their input power under practical conditions. Most bioreactors used in biotechnology are stirred-tank bioreactors, as stirring effectively distributes gas and nutrients to the growing organisms. Surface aeration systems are utilized in aerobic biological reactors to supply O_2 to microorganisms and mix the fermentation broth (Rao and Kumar, 2007).

Fig. 5.9 illustrates a typical surface aerator. The aerator's geometric dimensions include a (cross-sectional area), H (water depth), h (distance between the tank's horizontal bottom and the top of the blades), and D (rotor diameter). The rotor has six symmetrically and evenly spaced flat blades, each with linear dimensions b and l. The rotor shaft is connected to a DC motor, allowing it to operate at the desired speeds. The rotor is operated to generate turbulence in the water, facilitating aeration at the interface between atmospheric air and the water surface. The O_2 transfer rate is influenced by several factors, including the turbulence intensity, which is determined by the rotor's speed, the size, shape, and number of blades, the rotor's diameter and immersion depth, the size and shape of the aeration tank, and the physical, chemical, and biological properties of the water. The two fundamental parameters essential for the design and scale-up of surface aerators are O_2 transfer and the corresponding power requirement (Rao and Kumar, 2007). To evaluate O_2 transfer into the water body per unit time and the power requirement to run the aerator to transfer enough amount of O_2 from the atmospheric air to water, SOTR (Eq. (4)) and SAE (Eq. (5)) equations are also valid for surface aeration systems (Tanveer et al., 2018).

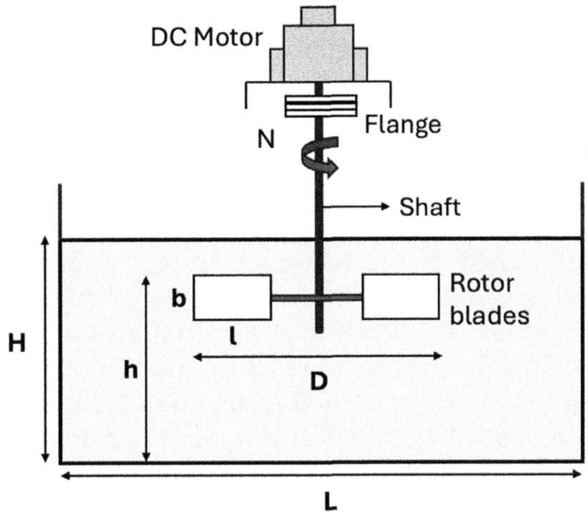

Fig. 5.9 Schematic diagram of a surface aerator.

5.4 Aeration systems: Laboratory scale experiments

Selecting the appropriate aeration device is crucial for optimizing O_2 transfer efficiency and minimizing operational costs in wastewater treatment. Fine bubble diffusers, with an O_2 transfer rate of 2.0–2.5 kg O_2/kWh, are widely utilized in treatment facilities due to their ability to produce small bubbles that dissolve efficiently in wastewater, enhancing aeration efficiency and reducing energy consumption. They avoid mechanical disruption of flocs, which results in lower VOC emissions. On the other hand, the coarse bubble diffusers, with an O_2 transfer rate of 0.8–1.2 kg O_2/kWh, are ideal for environments with high solids concentration, such as sludge stabilization, but require more air and energy consumption than fine bubble diffusers. Vertical shaft aerators, offering up to 2.0 kg O_2/kWh, are used in vertical shaft bioreactors due to their low operating and maintenance costs and high O_2 transfer efficiency. They can function as fixed or floating devices without tank drainage, although they may cause mechanical disruption of flocs and require precise adjustments in rotation speed and impeller diameter ratio to optimize performance. Horizontal shaft aerators, capable of up to 2.0 kg O_2/kWh, are preferred in shallow, wide channels due to their effective mixing and high O_2 transfer efficiency. These aerators are fully accessible for maintenance without draining tanks, but their mixing performance is limited to their immediate vicinity, and they are prone to failure in cold conditions due to splashing water and exposure to freezing temperatures. Therefore, selecting aeration methods should consider the specific requirements of the wastewater treatment process, including O_2 transfer efficiency, energy consumption, maintenance needs, and the operational environment, to ensure optimal performance and cost-efficiency (Drewnowski et al., 2019).

Fast et al. (1999) evaluated the O_2 transfer efficiencies of two types of paddlewheel aerators, such as single impeller (0.37 kW) and two impellers (0.75 kW), at three different salinity levels. In freshwater (0‰ salinity), the mean SAE values for single impellers with the 0.37-kW aerator ranged from 1.6 to 2.1 kg O_2/kWh, with an overall average of 1.93 kg O_2/kWh. At 11‰ salinity, SAE values spanned from 3.1 to 3.4 kg O_2/kWh, averaging 3.22 kg O_2/kWh. At a salinity of 22‰, the SAE values varied between 3.3 and 3.7 kg O_2/kWh, resulting in an overall average of 3.46 kg O_2/kWh. These results showed that salinity is a crucial factor that significantly impacts the O_2 transfer performance of aerators, thereby influencing the sizing of aeration systems (Fast et al., 1999). Moreover, Tanveer et al. (2018) reported a wide range of SAEs among different paddle wheel aerator types, from as low as 1 kg O_2/kWh for the spiral type to as high as 3.46 kg O_2/kWh for the two-impeller design (Tanveer et al., 2018).

As aerators are in different configurations than typical ones, Roy et al. (2017) assessed the performance of a spiral aerator to evaluate its suitability in aquaculture ponds. They concluded that the O_2 transfer rate increases with the number of handles and the rotational speed of a spiral aerator. At the same time, SAE decreases as the rotational speed increases from 70 to 250 rpm. At the rotational speed of 70 rpm, the highest SAE value of 1.0 kg O_2/kWh is achieved (Roy et al., 2017). The performance of the perforated tray aerator (PTA) was assessed through experiments conducted in a tank measuring $4\,m \times 4\,m \times 1.5\,m$ to enhance the efficiency by optimizing the geometric variables. The flow rate (Q) of $0.013\,m^3/s$ and the number of trays (n) set at 3 were constants. The optimal performance of the PTA was achieved with $W/H=0.665$, $d/H=1.85 \times 10^{-4}$, and $V/H^3=312.50$, resulting in a maximum non-dimensional standard aeration efficiency (NDSAE) of 35.58×10^{-3} and a SAE of 1.45 kg O_2/kWh (Roy et al., 2020a). Furthermore, an enhanced design based on the CSC and PCSC aerators, known as the perforated pooled circular stepped cascade (PPCSC) aerator, has been conceptualized and tested for its suitability as an aeration solution for small intensive aquaculture ponds. Simulation equations were developed to predict the SOTR, SAE, and power consumption for PPCSC aerators, applicable within the

$0.00141 \leq Fr \leq 0.01441$ range. Based on brake power, the SAE values for the PPCSC aerators ranged from 3.36 to 4.98 kg O_2/kWh, with an average of 4.45 ± 0.741 kg O_2/kWh (Roy et al., 2020b). The showering aeration system (SAS) was designed, and its performance was evaluated through aeration experiments conducted in a tank measuring 2 m × 4 m × 1.5 m. The results indicate that the maximum NDSAE of 16.98×10^6 was achieved from the SAS performance at $n=80$, H/l = 2, diameter of shower hole/l = 4, and $V/l^3 = 448$ where the flowrate was 0.0015 m^3/s the length of shower arm (l) and distance between showers (D) were 0.5 m (Roshan et al., 2022).

Falahi et al. (2021) assessed a pilot-scale vertical subsurface flow constructed wetland (VSSFCW) with *Scirpus grossus* and an aeration system for removing ibuprofen, chemical oxygen demand (COD), and nutrients (NH_3-N, NO_3-N, and PO_4-P) from domestic wastewater. Using 500 L platforms filled with gravel and sand, the system operated continuously for 21 days. Three hydraulic retention times (HRTs) of 3, 4, and 5 days and aeration rates of 0, 1, and 2 L/min were tested. Statistical analysis showed that aeration and HRT significantly influenced ibuprofen and COD removal, with longer HRT improving but not significantly affecting nutrient removal. At 2 L/min aeration and a 5-day HRT, the VSSFCW achieved 99.3% ibuprofen, 88.2% COD, 99.1% ammonia, 72.9% nitrate-nitrogen, and 83.2% orthophosphate removal, meeting Malaysian effluent standards. Thus, the VSSFCW with *S. grossus* effectively treats pharmaceutical-laden wastewater (Falahi et al., 2021). Teoh et al. (2020) examined the effects of floating plants, aeration, and circuit connection on wastewater treatment using an upflow-constructed wetland microbial fuel cell (UFCW-MFC). The study consisted of four stages: stage I without plants or aeration, stage II with *Eichhornia crassipes*, stage III with supplementary aeration, and stage IV with an open circuit. Supplementary aeration had the most significant impact on UFCW-MFC performance. Stage III, with both floating plants and aeration, showed optimal wastewater treatment and electricity production, achieving 99% COD, 96% NH_4^+, and 44% NO_3^- removal. *Eichhornia crassipes* and aeration improved DO levels at the cathode and enhanced NH_4^+ removal by 10% (stage II) and 67% (stage III) compared to stage I (Teoh et al., 2020). A bench-scale activated sludge treatment system was established with an aeration tank containing 5000 mg/L initial biomass and a clarifier chamber for biomass settling. The system acclimatized for 3 weeks using real domestic wastewater. Subsequently, a mixture of 20% industrial pulp and 80% domestic wastewater was fed. Throughout the study, concentrations of organic and nutrient parameters were tested from influent and effluent samples for data analysis. The system achieved COD and total suspended solid (TSS) removal rates of 83% and 90%, respectively, with an average BOD value of 74.6% in the treated wastewater. The study concluded that the bench-scale extended aeration activated sludge (EAAS) effectively treated BOD and TSS to standard levels. However, modifications such as adding a biocarrier or activated carbon are suggested to meet COD removal standards and enhance treatment efficiency, even with higher wastewater concentrations (Jagaba et al., 2021).

Overall, the case studies on designing aeration systems in wastewater treatment emphasize the importance of selecting the right device to optimize efficiency and minimize costs. Fine bubble diffusers, with their high O_2 transfer rates of 2.0–2.5 kg O_2/kWh, are highly effective for general applications due to their efficiency and energy savings, though they require frequent maintenance due to fouling. Coarse bubble diffusers, while less efficient (0.8–1.2 kg O_2/kWh), are advantageous in environments with high solids concentrations, such as sludge stabilization, but their higher air and energy consumption is a drawback. Vertical shaft aerators offer a balance of up to 2.0 kg O_2/kWh, providing high efficiency and low maintenance costs, though they can disrupt flocs mechanically. Horizontal shaft aerators, also up to 2.0 kg O_2/kWh, are ideal for shallow, wide channels due to effective mixing, but they are limited by cold weather conditions and their localized mixing capacity. However, advanced designs such as paddlewheel aerators show variable performance with salinity, achieving SAE values significantly influenced by salinity levels. The perforated pooled circular stepped cascade aerator stands out with superior SAE values (3.36 to 4.98 kg O_2/kWh), outperforming traditional cascade aerators and highlighting the potential for innovative designs to enhance efficiency. These insights underscore that a tailored approach, considering specific operational requirements and leveraging the latest advancements, is essential for optimizing aeration processes in wastewater treatment, leading to economic and environmental benefits.

5.5 Aeration systems in municipal wastewater treatment facilities

The aeration methods have been performed in wastewater treatment (WWT) facilities in various treatment units. For instance, natural aeration is performed in the waste stabilization ponds, and stabilization ponds are artificial ponds that can be utilized in series or as a single step in a longer treatment chain at any point in the wastewater treatment process. Furthermore, facultative ponds use both anaerobic and aerobic processes as secondary treatment. Facultative ponds use two levels of stratification to work with influent: the top layer, which is created by atmospheric reaeration and algal respiration, contains DO, and the sludge, which settles at the bottom and creates an anaerobic environment where decomposition takes place, needs to be regularly extracted. A facultative pond functions under a lower organic loading than an anaerobic pond because it often gets settling water from an anaerobic pond. Furthermore, aerobic ponds use UV disinfection, sedimentation, and natural aeration as tertiary treatment methods for wastewater. This procedure mimics the natural cleansing in a body of water. Natural oxygenation promotes organic breakdown and nutrient removal through atmospheric reaeration and algal respiration. Wastewater enters the pond regularly, and because of its shallow depth, sunlight can reach every part of it, mixing with O_2 to aid in eliminating pathogens. All these processes facilitate healthy fecal bacterial clearance. When photosynthetic algae consume carbon dioxide created by bacterial action, they release O_2 into the water. In WWT facilities, the anaerobic, facultative, and aerobic units are utilized as the sequential waste stabilization pond (Fig. 5.10).

Natural aeration also takes place in a secondary aerobic treatment process using constructed wetlands. To benefit from the natural processes that wetlands use to clean wastewater, constructed wetlands are artificial regions designed to resemble natural wetlands. They comprise a planted bed and a porous layer of rock, gravel, or sand. Some suspended materials are trapped, and the porous layer performs filtering functions. A portion of the pollutants are absorbed by the planted bed, which also encourages the growth of bacteria and invertebrates that further purify the water as it passes through by breaking down the organic contaminants. These wetlands can be designed in different types, such as vertical flow constructed, horizontal flow constructed, and free water surface constructed wetlands (Fig. 5.11). The vertical flow-built wetlands require less space than horizontal-flow wetlands, considering that wastewater flows downward and is equally loaded from the top to allow for oxygenation. Using a pump or siphon, intermittent loading raises the oxygenation and anaerobic phase even more. On the other hand, wastewater in wetlands with horizontal flow construction moves across the basin and filters as it is pushed by fresh influent to the opposite side of the wetland. Aerobic bacteria can colonize the region by transferring a small quantity of O_2 to the root zone through the plants. In free-water surface-constructed wetlands, water flows above ground, and plants are either floating in the water or rooted in the sediment

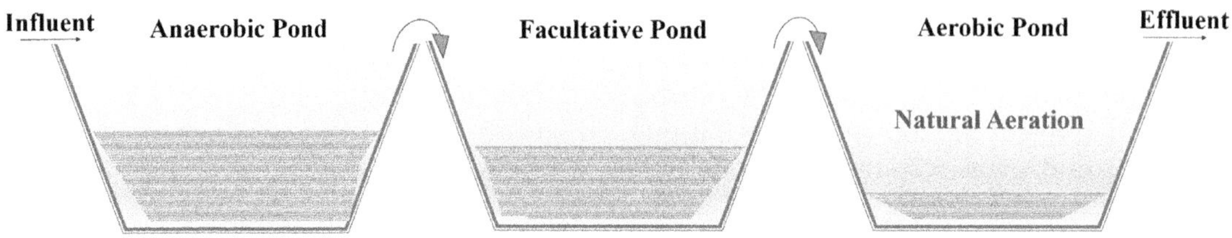

Fig. 5.10 Sequential waste stabilization pond with natural aeration unit.

layer at the base of the basin. In contrast to other wetlands, submerged and floating plants can be found there. These systems expose the wetland's water surface to the atmosphere, which transfers the O_2 naturally into the water and provides oxygenation.

Moreover, in secondary aerobic treatment, the artificial pond that receives mechanical aeration is also referred to as an aerated lagoon or aerated pond (Fig. 5.12). This process, which is facilitated by dispersed or mechanical aeration, is similar to the natural treatment that takes place in a river body. The O_2 promotes nutrition extraction and organic decomposition. The aeration preserves the particles in suspension as the wastewater flows in constantly. The suspended solids and DO concentrations of the basin are kept constant. Moreover, a facultative aerated lagoon is a term given to a pond if only the upper layer remains aerated. In that case, some floating materials sink to the bottom of the basin and begin anaerobic breakdown there. The suspended materials agglomerate into sludge during the settling stage, which needs to be periodically removed.

In addition, aerated lagoons can be constructed in a series for optimal treatment, with variable aeration throughout the sequence in WWT facilities. The effluent is cleaned in an aerated or high-performance aerated lagoon after passing through the facultative lagoon. In addition, communities where industrial or agricultural activities significantly alter the organic pollutant content of wastewater are especially well-suited for the aerated lagoon. Lagoon processes are known for their great buffering ability against changes in organic or hydraulic loads because they have a substantially longer hydraulic retention period than other processes.

In secondary aerobic treatment options, extended aeration can also be utilized as an established form of the activated sludge method. In contrast to other complex examples of that procedure, extended aeration is based on its simplicity. Since there are no primary sedimentation tanks and the waste-activated sludge is kept

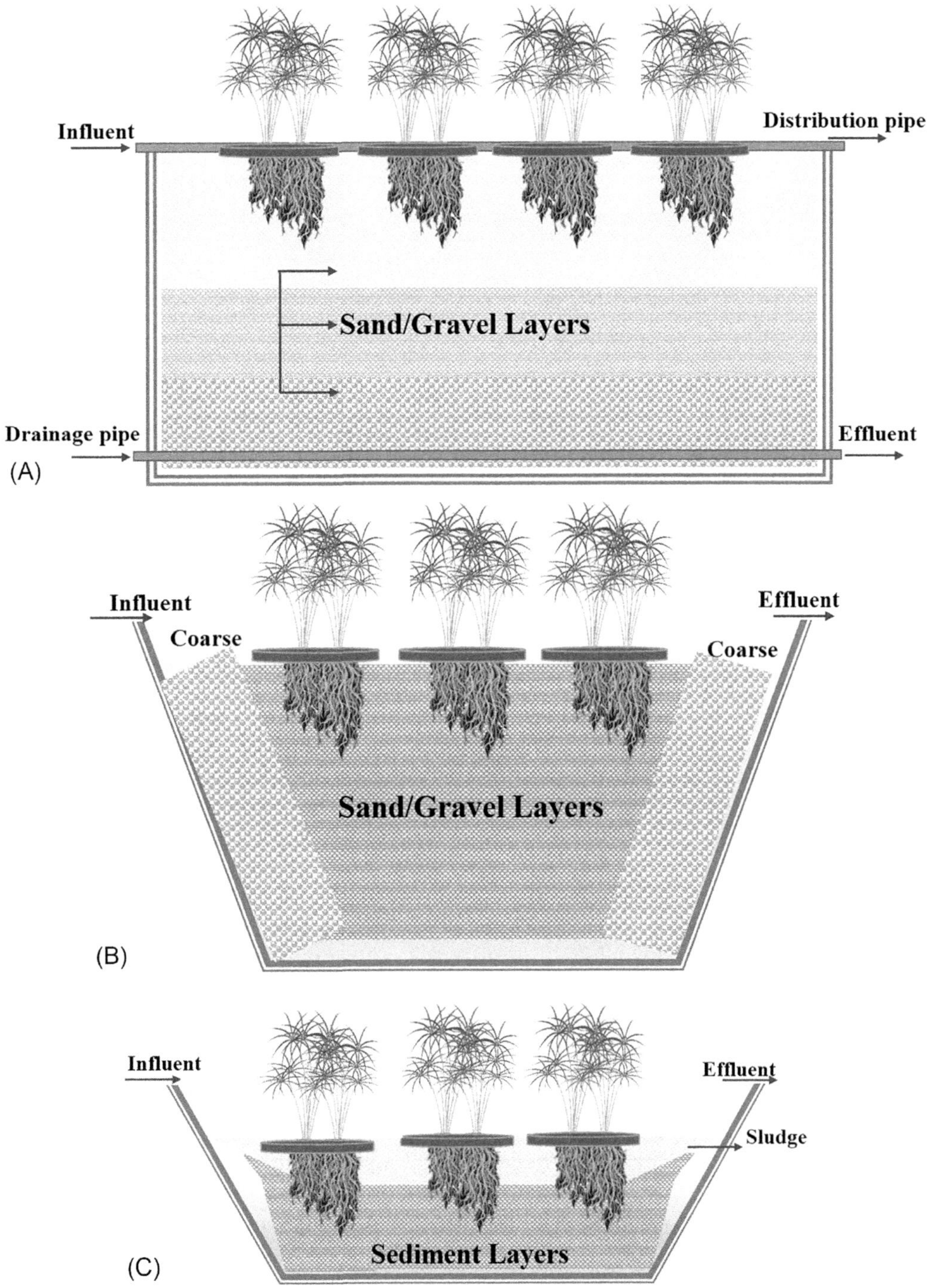

Fig. 5.11 Natural aeration in (A) vertical flow, (B) horizontal flow, and (C) free water surface constructed wetlands.

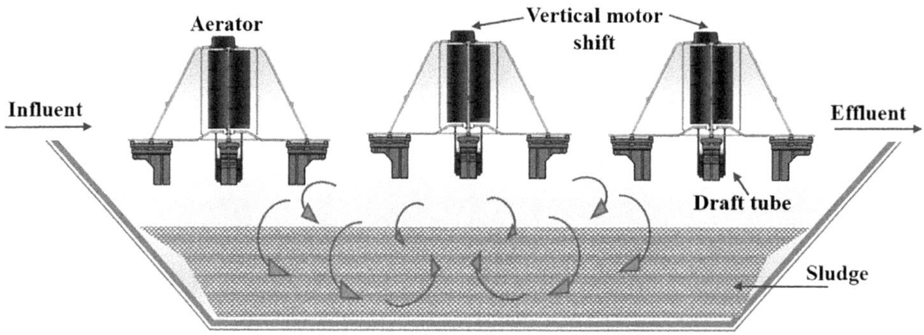

Fig. 5.12 Aeration lagoon in a wastewater treatment facility with mechanical aerators.

in the aeration tanks for extended periods, sludge digestion and stabilization may be accomplished without sludge digesters. That is, waste sludge extracted from the tanks may be thickened and dewatered immediately. There are several forms of extended aeration; however, the most popular ones are oxidation ditch, plug-flow, and sequencing batch reactor-type extended aeration systems (Fig. 5.13). The flow conditions are completely mixed since the aeration tank is built as a closed loop channel in an oxidation ditch configuration. In addition to providing microbes with the O_2 they require, the aerators also provide horizontal force, which improves mixing conditions. The sludge flocs are allowed to settle by gravity at the bottom of a subsequent secondary sedimentation tank, after which the sludge is pumped back to the aeration tank (Fig. 5.13A). Moreover, in plug-flow extended aeration configuration, the flow enters at one end and exits at the other because of how the aeration tanks are built. The main goal of these steps is to increase efficiency. Pressurized aeration is used, water depth is increased to typically 5–7 m, aerated and non-aerated zones are installed intermittently, and smart automation systems, which include effluent quality control sensors and frequency-controlled blowers, are introduced to control the air supply. On the other hand, one specific type of activated sludge is the sequencing batch reactor, which is especially useful for extended aeration in small cities. Since all operations take place in the same clarifier, which results in high sludge age, aeration works best with lower waste loads and necessitates longer mixing durations (Fig. 5.13B). The following steps are usually included in the batch mode process sequence: filling, mixing with aeration, decantation, and supernatant removal.

Moreover, the rotating biological systems are also utilized for the aeration in the WWT facilities. In this unit, discs act as the

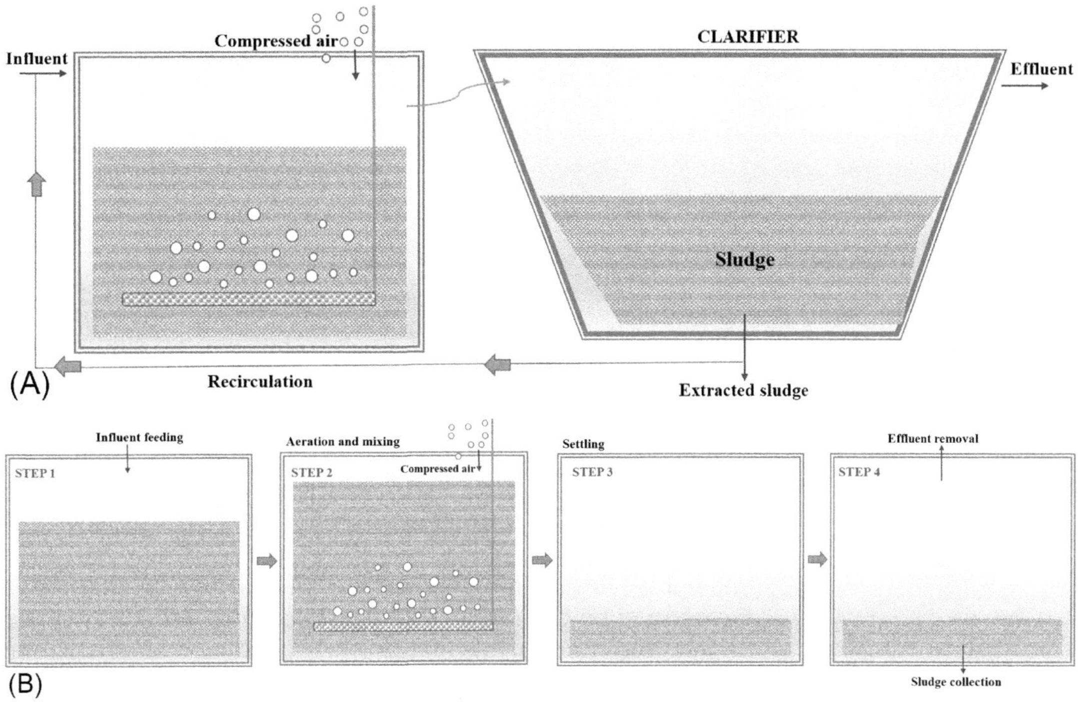

Fig. 5.13 Extended aeration configurations: (A) mechanical and (B) sequencing batch reactor-type aeration systems.

growth supports for bacteria. To guarantee mixing and aeration, they are propelled by a circular movement along a horizontal axis and partially submerged in the effluent (Fig. 5.14). The immersion condition of biomass is alternated by the rotation, enabling both oxygenation and the absorption of organic materials. Depending on the amount of organic material present in the wastewater, the rotating speed, which regulates both the rate of aeration and the degree of contact between the biomass and the wastewater, can be changed.

Overall, different types of aeration systems have been widely utilized in municipal wastewater treatment (MWWT) facilities to enhance the DO level in water media and to improve the treatment of wastewater resources. For instance, one of the biggest municipal wastewater treatment plants in Asia is the Bai Long Gang MWWTP, which was built in Shanghai, China. It serves a population of 7.2 million people and treats 2.8 million m^3/day on average, making it around 40% of Shanghai's treatment capability. During a 2018 reconstruction, the facility was modified to treat 3.5 million m^3/day while following strict effluent quality standards. During the 2007–08 Bai Long Gang MWWT facility

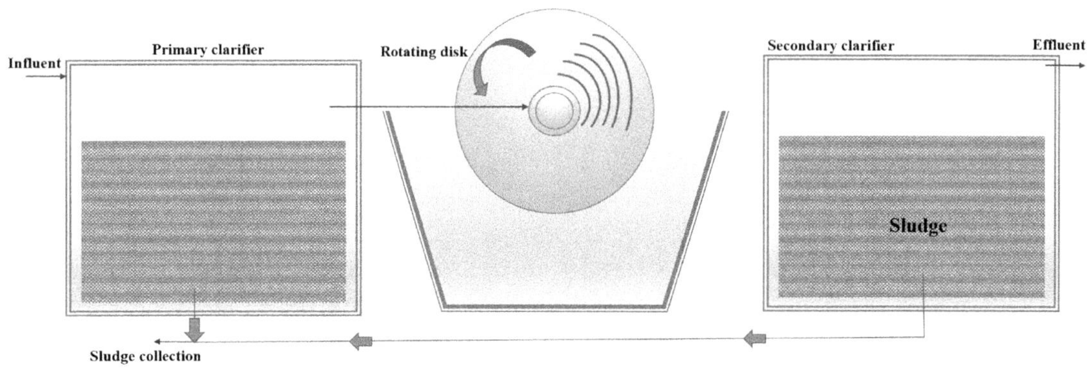

Fig. 5.14 Rotating biological reactor with aeration and mixing.

reconstruction, Environmental Dynamics International (EDI) developed the idea for the aeration system and supplied 22,000 fine bubble flexible membrane tube diffusers to Shanghai WINda Environmental Co. Ltd. (WINda). WINda provided and installed the aeration components for the project with committed EDI support. This project used tube diffusers, which have a 91 mm diameter by 1000 mm length in the aeration lagoon (Fig. 5.15). These diffusers have several advantages: their configuration is ideal for high-density floor coverage; they have the greatest potential for fine-bubble aeration improvements; the diffuser building improves chemical, temperature, and UV resistance; and high-quality membrane materials minimize fouling and maintenance needs. Moreover, a 10% increase in O_2 transfer efficiency was achieved with the aeration system change, which was made feasible by the high-performance diffusers from EDI. Moreover, the aeration design's robust diffuser construction and steady operating performance have reduced the need for regular aeration tank draining, resulting in further cost savings.

5.6 Conclusions and future trends

Aeration systems play a crucial role in environmental engineering, particularly in the treatment of wastewater and the enhancement of aquatic environments. These systems are essential for maintaining the O_2 levels required for aerobic microorganisms to grow, thereby facilitating the breakdown of organic pollutants and improving water quality. Based on the

Fig. 5.15 Aeration unit with fine bubble flexible membrane tube diffusers in China. Adapted from https://blog.wastewater.com/bai-long-gang.

comprehensive literature survey, from conventional mechanical aerators to advanced diffused aeration technologies, each system offers unique benefits and challenges. Namely, mechanical aeration systems, such as surface aerators and turbine aerators, provide high O_2 transfer rates and are relatively simple to operate. However, they can be energy-intensive and may cause significant disturbance to the water surface. On the other hand, diffused aeration systems, including fine and coarse bubble diffusers, offer higher energy efficiency and less surface disruption, but they require more complex installation and maintenance procedures.

Moreover, selecting an appropriate aeration system depends on various factors, including the specific requirements of the water body, the nature of the pollutants, energy consumption considerations, and maintenance capabilities. Technological advances, such as integrating automation and smart monitoring systems, continually improve the efficiency and effectiveness of aeration processes. Scientific research and field studies have demonstrated the critical impact of proper aeration on water quality, ecosystem health, and overall environmental sustainability. Nevertheless, continued innovation and optimization of aeration systems are essential for addressing the growing challenges of water pollution and for supporting sustainable development goals. In conclusion, aeration systems are indispensable tools in environmental management, offering diverse solutions for oxygenating water bodies. Their effective implementation requires a careful balance of technological, environmental, and operational factors to achieve optimal outcomes in water treatment and environmental protection in a sustainable manner.

The extent to which environmentally friendly and cutting-edge sustainable aeration technologies achieve commercial maturity will determine the role that aeration methods will perform in the future. To achieve sustainable aeration, it is necessary to establish sustainable water aeration systems that link bioenergy and renewable sources to market end users (Fig. 5.16). Moreover, detailed future recommendations are reported as follows:

- Implement Internet of Things (IoT) and Artificial Intelligence (AI) technologies for real-time monitoring and adaptive control. These systems can optimize aeration parameters based on real-time data, potentially reducing energy consumption by up to 30% compared to traditional systems.
- Develop advanced blower technologies and fine-bubble diffusers to maximize O_2 transfer efficiency. Fine-bubble diffusers can achieve an O_2 transfer efficiency of up to 35%, compared to 10–12% for coarse-bubble systems, thereby reducing energy requirements.
- Integrate renewable energy sources, such as solar photovoltaic (PV) panels or wind turbines, to power aeration systems. A typical wastewater treatment plant could offset up to 50% of its energy consumption by utilizing on-site renewable energy installations.
- Incorporate energy recovery systems, such as MFCs, to generate electricity from wastewater. MFCs can potentially produce 0.5–2 kWh of electricity per cubic meter of wastewater treated, contributing to the energy-neutral operation of treatment plants.
- Utilize nanomaterials and advanced polymers to create more durable and efficient aeration components. These materials can enhance the lifespan of diffusers by up to 50% and improve O_2 transfer rates by 10–15%.

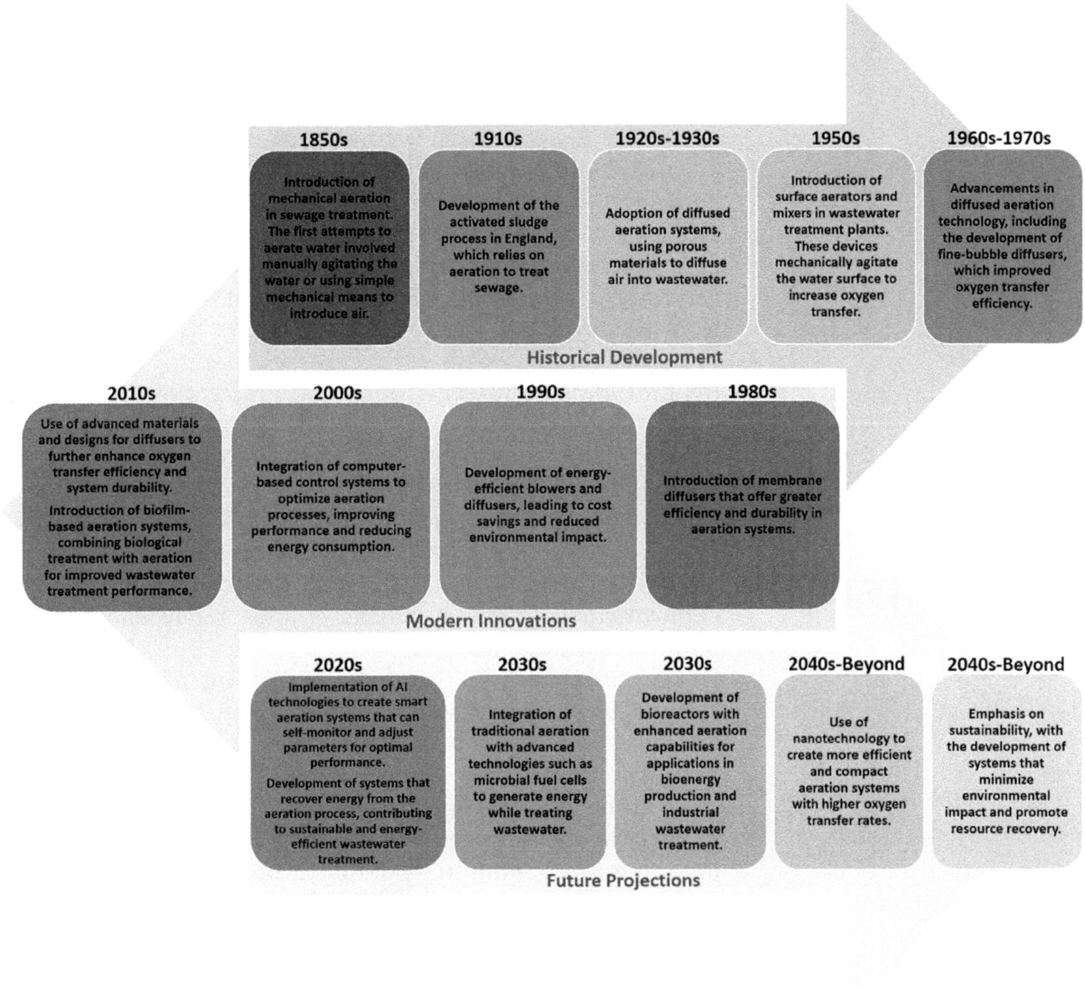

Fig. 5.16 Historical development and some pathways to reach sustainable and effective aeration approaches.

- Develop hybrid aeration systems that combine conventional aeration with bioelectrochemical systems, such as bioelectrochemical reactors, to enhance treatment efficiency and energy recovery. Hybrid systems can increase the overall energy efficiency of treatment processes by 20–30%.
- Integrate nutrient recovery processes, such as precipitation, to recover phosphorus and nitrogen during aeration. This can reduce the need for chemical fertilizers and improve the sustainability of agricultural practices.

- Promote the deployment of small-scale, decentralized aeration systems to reduce infrastructure costs and improve water treatment accessibility in remote or underserved areas. Decentralized systems can reduce the energy footprint by up to 40% compared to centralized treatment plants.
- To complement aeration processes, incorporate green infrastructure, such as constructed wetlands and vegetated swales. These systems can enhance pollutant removal and provide additional ecosystem services. Constructed wetlands can remove up to 90% of BOD and 85% of total nitrogen (TN) from wastewater.
- Conduct comprehensive lifecycle assessments (LCAs) to identify and mitigate the environmental impacts of aeration systems from production to disposal. LCAs can inform the development of systems with lower carbon footprints and reduced environmental impacts.
- Advocate for policies and incentives supporting the development and adoption of sustainable aeration technologies. Incentives such as tax credits and grants can accelerate the deployment of energy-efficient and renewable energy-powered aeration systems.
- Foster collaborations between public institutions, private companies, and academic institutions to drive innovation and investment in sustainable aeration solutions. These partnerships can leverage funding and expertise to develop and scale up new technologies.
- Engage local communities in planning, developing, and implementing aeration systems to ensure they meet local needs and encourage sustainable practices. Community-based approaches can improve system acceptance and long-term sustainability.

Overall, we believe that these future recommendations provide a comprehensive and scientific roadmap for advancing aeration systems toward greater sustainability, efficiency, and environmental compatibility.

AI disclosure

During the preparation of this work, the author(s) used ChatGPT to assist with language-related tasks such as paraphrasing, grammar checking, and improving sentence clarity. The tool was used solely to enhance the linguistic quality of the manuscript, and no part of the scientific content or interpretation was generated by AI. Following the use of this tool, the author(s) thoroughly reviewed and edited the content to ensure accuracy, coherence, and originality. The author(s) take full responsibility for the final content of the publication.

References

Al-Ahmady, K.K., 2006. Analysis of oxygen transfer performance on sub-surface aeration systems. Int. J. Environ. Res. Public Health 3 (3), 301–308. 16604601. MDPI, Iraq. Available from: http://www.mdpi.org/ijerph/papers/ijerph2006030037.pdf. https://doi.org/10.3390/ijerph2006030037.

Alkhalidi, A.A.T., Al Ba'ba'a, H.B., Amano, R.S., 2016. Wave generation in subsurface aeration system: a new approach to enhance mixing in aeration tank in wastewater treatment. Desalin. Water Treat. 57 (56), 27144–27151. 19443986. Taylor and Francis Inc., Jordan. Available from: http://www.tandfonline.com/toc/tdwt20/current. https://doi.org/10.1080/19443994.2016.1172263.

Baylar, A., Bağatur, T., 2000. Study of aeration efficiency at weirs. Turk. J. Eng. Environ. Sci. 24 (4), 255–264. 13000160.

Bencsik, D., Takács, I., Rosso, D., 2022. Dynamic alpha factors: prediction in time and evolution along reactors. Water Res. 216, 118339. https://doi.org/10.1016/j.watres.2022.118339. 00431354.

Boyd, C.E., Moore, J.M., 1993. Factors affecting the performance of diffused-air aeration systems for aquaculture. J. Appl. Aquac. 2 (2), 1–12. https://doi.org/10.1300/J028v02n02_01. 15450805.

Cardoso, B.J., Rodrigues, E., Gaspar, A.R., Gomes, Á., 2021. Energy performance factors in wastewater treatment plants: a review. J. Clean. Prod. 322. 09596526. Elsevier Ltd, Portugal. Available from: https://www.journals.elsevier.com/journal-of-cleaner-production. https://doi.org/10.1016/j.jclepro.2021.129107.

Chanson, H., 2009. Turbulent air–water flows in hydraulic structures: dynamic similarity and scale effects. Environ. Fluid Mech. 9 (2), 125–142. https://doi.org/10.1007/s10652-008-9078-3. 1567-7419.

Chen, G.-H., van Loosdrecht, M.C.M., Ekama, G.A., Brdjanovic, D., 2020. Biological Wastewater Treatment: Principles, Modeling and Design. IWA Publishing.

Cravotta, C.A., 2007. Passive aerobic treatment of net-alkaline, iron-laden drainage from a flooded underground anthracite mine, Pennsylvania, USA. Mine Water Environ. 26 (3), 128–149. https://doi.org/10.1007/s10230-007-0002-8. 16161068.

Danckwerts, P.V., Lannus, A., 1970. Gas-liquid reactions. J. Electrochem. Soc. 117 (10), 369C. https://doi.org/10.1149/1.2407312. 00134651.

Dawen, G., Nabi, M., 2024. Novel Approaches Towards Wastewater Treatment: Effective Strategies and Techniques Springer Water. Springer Nature Switzerland, https://doi.org/10.1007/978-3-031-55189-5.

Drewnowski, J., Remiszewska-Skwarek, A., Duda, S., Łagód, G., 2019. Aeration process in bioreactors as the main energy consumer in a wastewater treatment plant. Review of solutions and methods of process optimization. Processes 7 (5), 311. https://doi.org/10.3390/pr7050311. 2227-9717.

Eddy, M., Abu-Orf, M., Bowden, G., Burton, F.L., Pfrang, W., Stensel, H.D., Tchobanoglous, G., Tsuchihashi, R., 2014. Wastewater Engineering: Treatment and Resource Recovery. McGraw Hill Education.

El-Zahaby, A.M., El-Gendy, A.S., 2016. Passive aeration of wastewater treated by an anaerobic process—a design approach. J. Environ. Chem. Eng. 4 (4), 4565–4573. 22133437. Elsevier Ltd, Egypt. Available from: http://www.journals.elsevier.com/journal-of-environmental-chemical-engineering/. https://doi.org/10.1016/j.jece.2016.10.025.

Falahi, O.A.A.L., Abdullah, S.R.S., Hasan, H.A., Othman, A.R., Ewadh, H.M., Al-Baldawi, I.A., Kurniawan, S.B., Imron, M.F., Ismail, N.I., 2021. Simultaneous removal of ibuprofen, organic material, and nutrients from domestic wastewater through a pilot-scale vertical sub-surface flow constructed wetland with aeration system. J. Water Process Eng. 43. 22147144. Elsevier Ltd, Malaysia Available from: http://www.journals.elsevier.com/journal-of-water-process-engineering/. https://doi.org/10.1016/j.jwpe.2021.102214.

Fast, A.W., Tan, E.C., Stevens, D.F., Olson, J.C., Qin, J., Barclay, D.K., 1999. Paddle-wheel aerator oxygen transfer efficiencies at three salinities. Aquac. Eng. 19 (2), 99–103. https://doi.org/10.1016/S0144-8609(98)00044-2.

Fischer, H.B., List, E.J., Koh, R.C.Y., Imberger, J., Brooks, N.H., 1979. Mixing in Inland and Coastal Waters. Acaemic Press Inc., San Diego, CA.

Freeman, A.I., Surridge, B.W.J., Matthews, M., Stewart, M., Haygarth, P.M., 2018. New approaches to enhance pollutant removal in artificially aerated wastewater treatment systems. Sci. Total Environ. 627, 1182–1194. 18791026. Elsevier B.V., United Kingdom. Available from: www.elsevier.com/locate/scitotenv. https://doi.org/10.1016/j.scitotenv.2018.01.261.

Gameson, A.L.H., 1957. Weirs and the aeration of rivers. J. Inst. Water Eng. 11, 477–490.

Geroni, J.N., Cravotta, C.A., Sapsford, D.J., 2012. Evolution of the chemistry of Fe bearing waters during CO2 degassing. Appl. Geochem. 27 (12), 2335–2347. https://doi.org/10.1016/j.apgeochem.2012.07.017. 08832927.

Gu, Y., Li, Y., Yuan, F., Yang, Q., 2023. Optimization and control strategies of aeration in WWTPs: a review. J. Clean. Prod. 418, 138008. https://doi.org/10.1016/j.jclepro.2023.138008. 09596526.

Higbie, R., 1935. The rate of absorption of pure gas into a still liquid during short periods of exposure. Trans. Am. Inst. Chem. Eng. 31, 365–389.

Jagaba, A.H., Kutty, S.R.M., Fauzi, M.A.H.M., Razali, M.A., Hafiz, M.F.U.M., Noor, A., 2021. Organic and nutrient removal from pulp and paper industry wastewater by extended aeration activated sludge system. IOP Conf. Ser. Earth Environ. Sci. 842 (1), 012021. https://doi.org/10.1088/1755-1315/842/1/012021. 1755-1307.

John, A., Brookes, A., Carra, I., Jefferson, B., Jarvis, P., 2022. Microbubbles and their application to ozonation in water treatment: a critical review exploring their benefit and future application. Crit. Rev. Environ. Sci. Technol. 52 (9), 1561–1603. 15476537. Taylor and Francis Ltd., United Kingdom. Available from: www.tandf.co.uk/journals/titles/10643389.asp. https://doi.org/10.1080/10643389.2020.1860406.

John, A., Carra, I., Jefferson, B., Bertolaso, L., Brookes, A., Jarvis, P., 2024. Enhancement of ozonation using microbubbles—micropollutant removal, mass transfer and bromate formation. Chem. Eng. Sci. 283, 119369. https://doi.org/10.1016/j.ces.2023.119369. 00092509.

Kumar, A., Moulick, S., Singh, B.K., Mal, B.C., 2013. Design characteristics of pooled circular stepped cascade aeration system. Aquac. Eng. 56, 51–58. https://doi.org/10.1016/j.aquaeng.2013.04.004. 01448609.

Lawson, T.B., Merry, G.E., 1993. Procedures for evaluating low-power surface aerators under field conditions. In: Techniques for Modern Aquaculture. Proceedings of an Aquacultural Engineering Conference. Spokane, Washington, USA, 21-23 June. ASAE Publication, pp. 511–518.

Levenspiel, O., 1962. Mixed models to represent flow of fluids through vessels. Can. J. Chem. Eng. 40, 135–138.

Lewis, W.K., Whitman, W.G., 1924. Principles of gas absorption. Ind. Eng. Chem. 16 (12), 1215–1220. https://doi.org/10.1021/ie50180a002. 0019-7866.

Liew, K.C.S., Rasdi, A., Budhijanto, W., Yusoff, M.H.M., Bilad, M.R., Shamsuddin, N., Nordin, N.A.H.M., Putra, Z.A., 2020. Porous venturi-orifice microbubble generator for oxygen dissolution in water. Processes 8 (10), 1–15. 22279717. MDPI AG, Malaysia. Available from: https://www.mdpi.com/2227-9717/8/10/1266/pdf. https://doi.org/10.3390/pr8101266.

Mara, D., Pearson, H., 1998. Design Manual for Waste Stabilization Ponds in Mediterranean Countries. Lagoon Technology International, Leeds.

Marcelino, K.R., Wongkiew, S., Shitanaka, T., Surendra, K.C., Song, B., Khanal, S.K., 2023. Micronanobubble aeration enhances plant yield and nitrification in aquaponic systems. ACS ES&T Eng. 3 (11), 2081–2096. 26900645. American Chemical Society, United States. Available from: https://pubs.acs.org/page/aeecco/about.html. https://doi.org/10.1021/acsestengg.3c00344.

Moulick, S., Tambada, N.V., Singh, B.K., Mal, B.C., 2010. Aeration characteristics of a rectangular stepped cascade system. Water Sci. Technol. 61 (2), 415–420. 02731223. IWA Publishing, India. Available from: http://www.iwaponline.com/wst/toc.htm. https://doi.org/10.2166/wst.2010.828.

Musvoto, E., Ikumi, D., 2016. Energy Use Reduction in Biological Nutrient Removal Wastewater Treatment Plants. Water Research Commission.

Nguyen, N.T., Vo, T.S., Tran-Nguyen, P.L., Nguyen, M.N., Pham, V.H., Matsuhashi, R., Kim, K., Vo, T.T.B.C., 2024. A comprehensive review of aeration and wastewater treatment. Aquaculture 591. 00448486. Elsevier B.V., Vietnam. Available from: https://www.sciencedirect.com/science/journal/00448486. https://doi.org/10.1016/j.aquaculture.2024.741113.

Oh, C., Ji, S., Cheong, Y., Yim, G., Hong, J.-H., 2016. Evaluation of design factors for a cascade aerator to enhance the efficiency of an oxidation pond for ferruginous mine drainage. Environ. Technol. 37 (19), 2483–2493. https://doi.org/10.1080/09593330.2016.1153154. 0959-3330.

Oprina, G., Bunea, F., Pincovschi, I., Mândrea, L., 2010. Aspects of hydrodynamics and mass transfer in diffused aeration systems. Environ. Eng. Manag. J. 9 (4), 565–571. 15829596. Gheorghe Asachi Technical University of Iasi, Romania. Available from: www.eemj.eu. 10.30638/eemj.2010.079.

Padrón-Páez, J.I., Almaraz, S.D.L., Román-Martínez, A., 2016. Design of sustainable municipal wastewater treatment plants. Comput. Aided Chem. Eng. 38, 1509–1514. Elsevier B.V., Mexico. Available from: http://www.elsevier.com/wps/find/bookdescription.cws_home/BS_CCE/description#description. https://doi.org/10.1016/B978-0-444-63428-3.50256-3.

Radkevich, M., Abdukodirova, M., Shipilova, K., Abdullaev, B., 2021. Determination of the optimal parameters of the jet aeration. IOP Conf. Ser. Earth Environ. Sci.

Rao, A.R., Kumar, B., 2007. Scale-up criteria of square tank surface aerator. Biotechnol. Bioeng. 96 (3), 464–470. https://doi.org/10.1002/bit.21149. 10970290.

Rao, A.R., Patel, A.K., Kumar, B., 2009. Oxygen transfer in circular surface aeration tanks. Environ. Technol. 30 (7), 747–753. 1479487X. Taylor and Francis Ltd., India. Available from: http://www.tandf.co.uk/journals/titles/09593330.asp. https://doi.org/10.1080/09593330902911705.

Roshan, R.U., Mohammed, T., Roy, S.M., Rajendran, R., 2022. Design characteristics of showering aeration system. Aqua Water Infrastruct. Ecosyst. Soc. 71 (1), 139–153.

Rosso, D., Stenstrom, M.K., 2005. Comparative economic analysis of the impacts of mean cell retention time and denitrification on aeration systems. Water Res. 39 (16), 3773–3780. 00431354. Elsevier Ltd, United States. Available from: https://www.sciencedirect.com/science/journal/00431354. https://doi.org/10.1016/j.watres.2005.07.002.

Rosso, D., Stenstrom, M.K., 2006. Surfactant effects on α-factors in aeration systems. Water Res. 40 (7), 1397–1404. 00431354. Elsevier Ltd, United States. Available from: https://www.sciencedirect.com/science/journal/00431354. https://doi.org/10.1016/j.watres.2006.01.044.

Rosso, D., Iranpour, R., Stenstrom, M.K., 2005. Fifteen years of offgas transfer efficiency measurements on fine-pore aerators: key role of sludge age and normalized air flux. Water Environ. Res. 77 (3), 266–273. https://doi.org/10.2175/106143005x41843. 1061-4303.

Rosso, D., Larson, L.E., Stenstrom, M.K., 2006. Surfactant effects on alpha factors in full-scale wastewater aeration systems. Water Sci. Technol. 54 (10), 143–153. https://doi.org/10.2166/wst.2006.768. 0273-1223.

Rosso, D., Larson, L.E., Stenstrom, M.K., 2008. Aeration of large-scale municipal wastewater treatment plants: state of the art. Water Sci. Technol. 57 (7), 973–978. https://doi.org/10.2166/wst.2008.218. 02731223.

Rosso, D., Stenstrom, M.K., Garrido-Baserba, M., 2018. Aeration Fundamentals, Performance and Monitoring. IWA Publishing, https://doi.org/10.2166/9781780407845_0031.

Rosso, D., Volcke, E.I.P., Garrido-Baserba, M., Pretorius, C., Stenstrom, M.K., 2023. Aeration and Mixing. IWA Publishing, pp. 317–340, https://doi.org/10.2166/9781789062304_0317.

Roy, S.M., Moulick, S., Mal, B.C., 2017. Design characteristics of spiral aerator. J. World Aquacult. Soc. 48 (6), 898–908. 17497345. Blackwell Publishing Inc., India. Available from: http://www.blackwellpublishing.com/aims.asp?ref=0893-8849. https://doi.org/10.1111/jwas.12410.

Roy, S.M., Tanveer, M., Mukherjee, C.K., Mal, B.C., 2020a. Design characteristics of perforated tray aerator. Water Supply 20 (5), 1643–1652. https://doi.org/10.2166/ws.2020.069. 1606-9749.

Roy, S.M., Moulick, S., Mukherjee, C.K., 2020b. Design characteristics of perforated pooled circular stepped cascade (PPCSC) aeration system. Water Supply 20 (5), 1692–1705. https://doi.org/10.2166/ws.2020.078. 1606-9749.

Roy, S.M., Tanveer, M., Gupta, D., Pareek, C.M., Mal, B.C., 2021. Prediction of standard aeration efficiency of a propeller diffused aeration system using response surface methodology and an artificial neural network. Water Supply 21 (8), 4534–4547. https://doi.org/10.2166/ws.2021.199.

Roy, S.M., Machavaram, R., Moulick, S., Mukherjee, C.K., 2022a. Economic feasibility study of aerators in aquaculture using life cycle costing (LCC) approach. J. Environ. Manag. 302, 114037. https://doi.org/10.1016/j.jenvman.2021.114037. 03014797.

Roy, S.M., Tanveer, M., Machavaram, R., 2022b. Applications of gravity aeration system in aquaculture—a systematic review. Aquac. Int. 30 (3), 1593–1621. 1573143X. Springer Science and Business Media Deutschland GmbH, India. Available from: www.wkap.nl/journalhome.htm/0967-6120. https://doi.org/10.1007/s10499-022-00851-5.

Rozainy, M.A.Z.M.R., Jamil, R., Adlan, M.N., 2015. A review of removal iron and manganese by using cascade aeration systems. J. Teknol. 74 (11), 69–76. 21803722. Penerbit UTM Press, Malaysia. Available from: http://www.jurnalteknologi.utm.my/index.php/jurnalteknologi/article/download/4875/3398.

Shukla, B.K., Sharma, P.K., Goel, A., 2020. Study on oxygenation performance of solid jet aerator having circular opening corresponding to variable jet length and flow area. J. Phys. Conf. Ser. 1531 (1), 012117. https://doi.org/10.1088/1742-6596/1531/1/012117. 1742-6588.

Singh, D., Kumar, M., 2022. Study on aeration performance of different types of piano key weir. Water Supply 22 (5), 4810–4821. https://doi.org/10.2166/ws.2022.131. 1606-9749.

Stensel, H.D., Makinia, J., 2014. Chapter 3: activated sludge process development. In: Jenkins, D., Wanner, J. (Eds.), Activated Sludge-100 Years and Counting. IWA Publishing, pp. 33–52.

Stenstrom, M.K., (Ben) Leu, S.-Y., Jiang, P., 2006. Theory to practice: oxygen transfer and the new ASCE Standard. Proc. Water Environ. Fed. 2006 (7), 4838–4852. https://doi.org/10.2175/193864706783762931. 1938-6478.

Tanveer, M., Roy, S.M., Vikneswaran, M., Renganathan, P., Balasubramanian, S., 2018. Surface aeration systems for application in aquaculture: a review. Int. J. Fish Aquat. Stud. 6, 342–347.

Tchobanoglus, G., Burton, F., Stensel, H.D., 2003. Wastewater Engineering: Treatment and Reuse. vol. 95 American Water Works Association.

Teoh, T.P., Ong, S.A., Ho, L.N., Wong, Y.S., Oon, Y.L., Oon, Y.S., Tan, S.M., Thung, W.E., 2020. Up-flow constructed wetland-microbial fuel cell: influence of floating plant, aeration and circuit connection on wastewater treatment performance and bioelectricity generation. J. Water Process. Eng. 36. 22147144. Elsevier Ltd, Malaysia. Available from: http://www.journals.elsevier.com/journal-of-water-process-engineering/. https://doi.org/10.1016/j.jwpe.2020.101371.

Thomas, V.K., Chambers, B., Dunn, W., 1989. Optimisation of aeration efficiency: a design procedure for secondary treatment using a hybrid aeration system. Water Sci. Technol. 21 (10–11), 1403–1419. https://doi.org/10.2166/wst.1989.0337. 0273-1223.

Van Loosdrecht, M.C.M., Ekama, G.A., Wentzel, M.C., Hooijmans, C.M., Lopez-Vazquez, C.M., Meijer, S.C.F., Brdjanovic, D., 2015. Introduction to modelling of activated sludge processes. In: Applications of Activated Sludge Models. vol. 1.

Wang, J., Langemann, H., 1994. Unsteady two-film model for mass transfer. Chem. Eng. Technol. 17 (4), 280–284. https://doi.org/10.1002/ceat.270170410. 15214125.

Wanner, J., Jobbágy, A., Jenkins, D., 2014. Chapter 10: activated sludge solids separation. In: Jenkins, D., Wanner, J. (Eds.), Activated Sludge-100 Years and Counting. IWA Publishing, pp. 171–194.

Yim, G.J., Cheong, Y.W., Hong, J.H., Hur, W., 2014. The role of each compartment in a two-compartment vertical flow reactor for ferruginous mine water treatment. Water Res. 62, 11–19. https://doi.org/10.1016/j.watres.2014.05.016. 00431354.

6

Moving bed biofilm reactor

Nawaf S. Alhajeri[a], Ahmed Tawfik[a], Mohamed Elsamadony[b,c], and Fahad M. Al-Fadhli[d]

[a]Department of Environmental Sciences, College of Life Sciences, Kuwait University, Safat, Kuwait. [b]Department of Mechanical Engineering, King Fahd University of Petroleum & Minerals, Dhahran, Saudi Arabia. [c]Interdisciplinary Research Center for Refining & Advanced Chemicals, King Fahd University of Petroleum & Minerals, Dhahran, Saudi Arabia. [d]Department of Chemical Engineering, College of Engineering and Petroleum, Kuwait University, Safat, Kuwait

Abbreviations

A/O	anoxic/oxic
AOB	ammonia oxidizing bacteria
C/N	carbon/nitrogen
COD	chemical oxygen demand
COD/N	chemical oxygen demand/nitrogen
DO	dissolved oxygen
EDC	endocrine disrupting compounds
EPS	extracellular polymeric substances
FNA	free nitrous acid
HRT	hydraulic retention time
MBBRs	moving bed biofilm reactors
MBR	membrane bioreactor
MMBBR	membrane moving bed biofilm reactor
NOB	nitrite-oxidizing bacteria
PN/A	partial nitritation/anammox
RBCs	rotating biological contactors
SBR	sequencing batch reactor
SMZ	sulfamethoxazole
SRT	sludge retention time
TFs	trickling filters
TN	total nitrogen
TP	total phosphorous

6.1 Introduction

Rapid industrialization and urbanization represent two primary contributors to wastewater generation, driven by the significant increase in population in both developing and advanced

Municipal Wastewater Treatment. https://doi.org/10.1016/B978-0-443-24826-9.00006-1

countries (Elsheikh et al., 2013; Elnmer et al., 2019). Effective management of wastewater is imperative, necessitating a transition from a linear to a circular approach, wherein pollutants can be repurposed as sources of energy or agricultural inputs, thereby achieving sustainable environmental objectives (Elsamadony et al., 2015; Mostafa et al., 2023). Discharging wastewater into the environment poses serious health risks, particularly in impoverished and low-income nations. Additionally, water scarcity in these regions incentivizes the reuse of wastewater to support livelihoods and economies (Tawfik and El-Qelish, 2014; Elsamadony et al., 2022).

Biofilm reactors are predominantly utilized for the removal of organics, nutrients (such as nitrogen and phosphorus), viruses, parasites, and pathogenic bacteria from wastewater (Elsamadony et al., 2021; Allam et al., 2015). As shown in Fig. 6.1, the biofilm formation process begins with the attachment of microorganisms to a surface. This is followed by the development of microcolonies and the formation of a protective matrix, which collectively enhances the stability and resilience of the biofilm. The mature biofilm stage is crucial for efficient wastewater treatment because it harbors a diverse microbial community capable of degrading a wide range of pollutants. However, detachment of biofilm fragments can pose challenges, such as reducing pollutant removal efficiency and necessitating regular maintenance. Various technologies, including trickling filters (TFs) (Tawfik et al., 2006b), fluidized-bed reactors, aerated submerged fixed film biofilm systems, and rotating biological contactors (RBCs), have been employed for organic degradation and nutrient removal (Tawfik et al., 2004). However, these treatments face certain limitations. For instance, TFs exhibit reduced effectiveness due to limited working volume and insufficient oxygen for

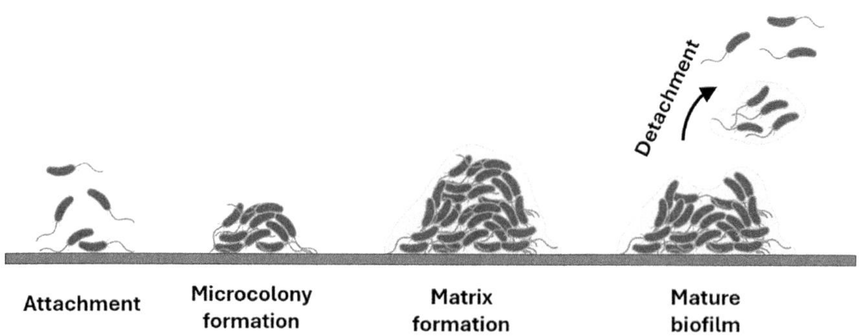

Fig. 6.1 Stages of biofilm formation and maturation in bioreactors.

nitrification (Ismail et al., 2021b; Ismail and Tawfik, 2016a). RBC applications are susceptible to mechanical failures and biofilm detachment, thereby compromising pollutant removal efficiency (Tawfik et al., 2006a, 2011). Hydraulic instability and challenges in biofilm distribution hinder the widespread adoption of aerated submerged fixed film biofilm systems and fluidized bed reactors (Mahmoud et al., 2009). Furthermore, biofilm systems often encounter diffusional limitations, resulting in decreased removal efficiency of particulate organic matter and nitrification (Tawfik et al., 2022b,c).

In response to these limitations, Norwegian authorities in the late 1980s and early 1990s developed the moving bed biofilm reactor (MBBR) for sewage treatment, with a focus on its practical application and implementation in small communities ranging from 20 to 2000 people. The MBBR integrates both attached and suspended growth systems, where bacteria adhere to moving carrier materials within the reaction tank (Tawfik and Salem, 2012; Mostafa et al., 2017). It can be employed to enhance existing wastewater treatment plants and is particularly effective for treating heavily polluted municipal and industrial wastewater (Tawfik et al., 2021; Mostafa et al., 2020). The MBBR possesses a compact design, simplified operation, and eliminates the need for returning sludge, unlike conventional activated sludge units. Moreover, the MBBR could be utilized in different configurations, i.e., anoxic/anaerobic systems. The concept involves employing free-floating moving carriers with a lower density than water, providing high surface area to avoid substrate diffusion limitations and allowing for easy facilitation of high mass transfer. MBBR systems generate minimal excess sludge due to longer sludge residence times compared to activated sludge treatment processes. Sludge recirculation is unnecessary since the biomass is retained or attached as a biofilm on packing carriers, and backwashing is not required due to reduced clogging compared to traditional fixed-film systems. MBBRs exhibit high resistance to fluctuations in influent wastewater characteristics, including shock loads, temperature variations, organic loading rates, pH levels, and toxic compounds. However, the main drawback of MBBR technology lies in aeration and mixing costs (Ødegaard, 2016; Tawfik and Elsamadony, 2018; Elsamadony and Guo, 2019).

The COD/N ratio of influent wastewater significantly affects the microbial composition, structure, and dynamic behaviors in MBBRs. Dominance of heterotrophic bacteria occurs at high COD/N ratios. Coexistence of heterotrophic and autotrophic bacteria is observed at medium COD/N ratios. Sensitive autotrophic bacteria prefer low COD/N ratio conditions in the MBBR system.

The COD/N ratio mainly affects microbial activity, particularly in the production of ammonia monooxygenase, in MBBR systems (Torresi et al., 2016). The effect of carrier material type on microbial community and nitrification efficiency in MBBRs was investigated by Tawfik et al. (2006b) and Hu et al. (Zhao et al., 2022). Biofilm development on carriers with grids of 50 μm in height was delayed, resulting in the formation of thin biofilms. The carpet-like structure of carriers had the lowest α-diversity. Thick biofilms and abundant microbial colonies were formed on carriers with grids of 200 and 400 μm in height. However, Nitrospiria was dominant (19.3–41.4%) in all carriers, with Alphaproteobacteria and Gammaproteobacteria being the abundant classes. Nitrospira, Nitrobacter (nitrite-oxidizers), and Nitrosomonas (ammonia-oxidizers with >5% abundance) were present in all MBBRs. The abundance of Ideonella (heterotrophs) accounted for 13.6%.

Recently, MBBR has become known as an efficient and economical biofilm bioreactor, attracting significant attention for upgrading traditional wastewater treatment processes. However, biofilm formation on the carrier materials and the stable performance of MBBRs at low temperatures during the winter season are susceptible to negatively affecting substrate utilization rate, diffusion, and microbial community activity (Hu et al., 2017). This also delays biofilm formation and requires a long start-up period. Thus, bioaugmented MBBR with Quorum Sensing bacteria (*Sphingomonas rubra* BH3T) achieved a high removal efficiency of 93% for COD and 75% for ammonia (Fu et al., 2022). Operating the sequencing batch moving bed biofilm reactor in mode promoted microbial activity and provided ammonia and COD removal efficiencies of 95.6% and 77.1%, respectively. The module achieved a removal efficiency of 47.4% for TN and 50.7% for TP (Yuan et al., 2023). Nitrifying bacteria such as Nitrospira (2.9%) and Nitrosomonas (1.3%), and denitrifying bacteria like Terrimonas (1.8%) and Denitratisoma (0.7%), were abundant in the sequencing batch moving bed biofilm reactor. Heterotrophic denitrifying bacteria including Ferruginibacter, Terrimonas, Dechloromonas, and f_Burkholderiaceae_G_IS were responsible for COD and TN removal in the MBBR reactor (Hetz and Horn, 2021). Decolorization of acid orange 7 by 95% was achieved in an MBBR packed with polyhydroxyalkanoate/activated carbon (Tong et al., 2024). Heterotrophic nitrification/aerobic denitrification bacteria were inoculated onto the MBBR to enhance ammonia removal from wastewater (Zhang et al., 2024). Biofilm formation took place within 20 days, achieving high ammonia and TN removal efficacies. Alcaligenes was responsible for nitrogen removal, resulting in an increase in enzymes hydroxylamine reductase, which provided a new nitrogen removal pathway as follows,

Nitrogen reaction pathway in MBBR:

$$NH_4^+ \rightarrow NH_2OH \rightarrow NO_3^- \rightarrow NO_2^- \rightarrow NO \rightarrow N_2O \rightarrow N_2$$

The objectives of this chapter are to assess the feasibility of replacing/upgrading the existing wastewater treatment plants by MBBR for organics and nitrogen removal. The emphasis was afforded for characterization (packing carriers, mixing and filling ratios), and operational parameters (dissolved oxygen, organic loading rate, hydraulic retention time and sludge residence time, temperature and pH) of moving bed biofilm reactors. The incorporation of moving Bed biofilm reactors with other technologies was discussed.

6.2 Municipal wastewater treatment using moving bed biofilm reactors

Municipal wastewater is the mixture of water with particulate, soluble organic and inorganic substances. Organic matters mainly include carbohydrates, proteins, soaps, fats, lignin, synthetic detergents and their decomposition by-products, as well as synthetic and natural organic chemicals generated from the industries processes. 96.1% of COD was removed from the municipal wastewater at influent concentration of 576.5 mg/L using activated sludge at 12 h HRT (Tandukar et al., 2007). The Sequencing batch reactor (SBR) treating domestic sewage with COD of 290.1 ± 7.4 mg/L achieved removal efficiency of 88.3 ± 7.4% HRT of 8 h (Liu et al., 2018). A higher removal efficiency of COD 77–99% and 98.6% was obtained using membrane bioreactor (MBR) treating municipal wastewater However, clogging of the membrane bioreactor was occurred and backwashing was needed (Barreto et al., 2017; Zhang et al., 2020). Low removal efficiency of recalcitrant organics i.e. various toxic emerging and micropollutants from municipal wastewater is the main concern with these biological treatment systems. The ciprofloxacin removal efficiency reached to 27.1% in aerobic MBBR module inoculated with *Achromobacter* sp. JL9 (Liang et al., 2022). Ciprofloxacin removal was optimized during the biological nitrification process in the MBBR, probably due to the co-metabolic microbial activity of ammonia oxidizing bacteria (Xu et al., 2016). These ammonia oxidizing bacteria (AOB) secrets monooxygenase enzyme that catalyze phenolic, aromatic and hydrocarbons compounds (Khunjar et al., 2011). Heterotrophic bacteria in the MBBR module highly induce metabolic biodegradation of pharmaceuticals (De Gusseme et al., 2011). The pharmaceuticals removals was highly promoted in the presence of the nitrifying situated in MBBR treating wastewater (Torresi et al., 2016). Cometabolic biodegradation is mainly occurred by ammonium oxidizing bacteria membrane-aerated biofilm reactors and had a major contribution for biodegradation of cephalexin and sulfadiazine (Wang et al., 2021). The ciprofloxacin were removed by value of 55% using enriched nitrifying MBBR (Xu et al., 2023). Better adsorption of ciprofloxacin onto the biofilm in the presence of ammonium occurred due to AOB that induced co-metabolism in the biodegradation of ciprofloxacin.

Anoxic/oxic (A/O) MBBR treating wastewater containing tetracycline (0.05–0.2 mg/L) at HRT of 8 h provided 52% removal efficiency (Chen et al., 2018). Sulfamethoxazole (SMZ) antibiotics were removed from mariculture wastewater at initial concentration of 5 mg/L using A/O MBBR (Chen et al., 2022). The major portion of the SMZ was biodegraded in anoxic stage of MBBR where the nitrate was utilized as electron acceptor. This was not the case for (Kang et al., 2018) who found that aerobic degradation pathway was dominance for SMZ due to the amino functional group's oxidation of the antibiotic by autotrophic nitrifying bacteria. In another study, the SMZ was hardly removed using multistage nitrifying and denitrifying MBBR. This was mainly due to metabolites deconjugation by products using bacterial active enzymes. The presence of antibiotics in the wastewater at high concentrations highly affected the nitrifiers; thus, reducing the nitrification efficiency and ammonia oxidation process (Du et al., 2019). The effects of sulfamethoxazole (SMZ), on the efficiency of A/O MBBR, was investigated by Saidulu et al. (2024). The SMZ was removed by 85% in the anoxic zone by denitrifying heterotrophic microbes that were highly resistance to antibiotics loads. This was due to an increase of protein content of extracellular polymeric substances in the anoxic tank. Proteobacteria and Actinobacteria were dominant for degradation of SMZ in A/O MBBR.

6.3 Critical operational parameters of the moving bed biofilm reactor

The performance of an MBBR module is greatly influenced by factors such as hydraulic retention time, carrier filling ratio, wastewater characteristics, aeration mode, and biofilm thickness. These operational conditions play an unusual role in facilitating the operation and design of the module for wastewater treatment.

6.3.1 Characterization of packing carriers, mixing, and filling ratios

Carriers form the primary component of the MBBR module, where biological activities for organic degradation, nitrification, and denitrification take place in the treatment of polluted water (Tawfik et al., 2012). Increasing the surface area of the carrier materials enhances the bacterial community, prolongs the residence time of bacteria, and subsequently improves the removal efficiency of pollutants (Yuan et al., 2023). The accumulation of bacteria in the carrier helps resolve conflicts between nitrifiers and heterotrophs for oxygen and space, as reported earlier (Tawfik et al., 2010). The carrier provides an excellent environment for the formation of aerobic, anoxic, and anaerobic biofilm zones, which play a vital role in the treatment process (Tawfik et al., 2010). The packing of carriers in the MBBR should have a density of 0.94–0.98 g/cm^3, which is less than water to enable suspension in the reactor tank (Zhang et al., 2024). Aerators, with fine and coarse bubble air distribution systems, primarily serve as mixers and provide the necessary oxygen for bacteria to oxidize organics and ammonia in aerobic MBBR systems. Mechanical mixing or stirrers are employed in anaerobic and anoxic MBBR systems to facilitate contact between substrates and organisms degrading pollutants as shown in Fig. 6.2 (Elsharkawy et al., 2020; Soltan et al., 2019). AnoxKaldnes® biocarriers, developed in Norway by researcher Ødegaard, were originally manufactured from polyethylene with a density of 0.95 g/cm^3 and various

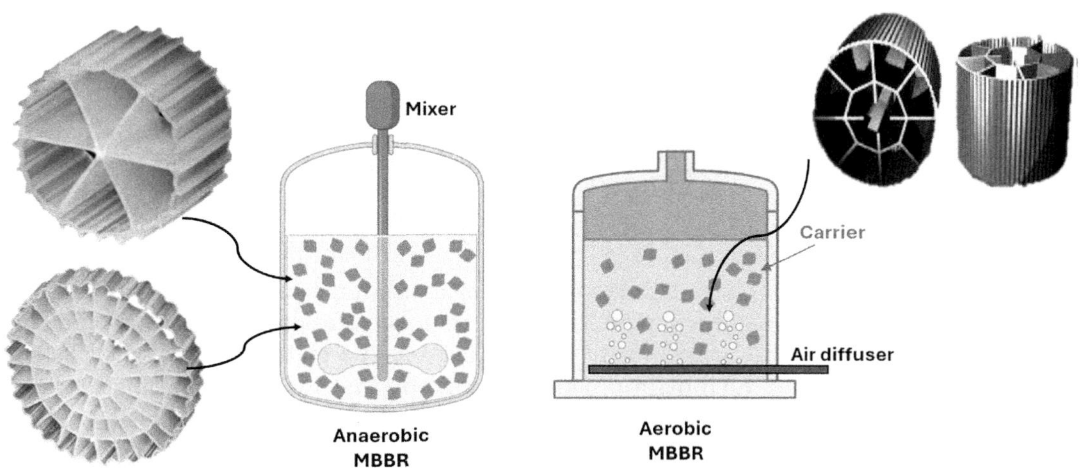

Fig. 6.2 Anaerobic and aerobic configurations of moving bed biofilm reactor (MBBR).

dimensions featuring circular and longitudinal exterior (Fu et al., 2022; Salvetti et al., 2006). Biocarriers in MBBR processes treating wastewater should possess the following characteristics: facilitation of mass transfer between bulk liquid and the active fraction of the biofilm, sufficient surface area for bacterial growth, slightly lower density than water, rough surface for easy attachment of microorganisms, adequate air distribution to prevent biomass sloughing and resistance to turbulence, and resistance to shock loads while remaining cost-effective.

Cuboid and plastic carriers with different shapes and sizes are commercially available i.e. polymeric reticulated foam that are utilized in MBBR processes. Cuboid-shaped polyurethane/polyether/polyethylene foam-based carriers with dimensions of $1 \times 1.2\,cm^2$ or $2.5 \times 2.5 \times 1.2\,cm^3$ (side dimensions) and a specific surface area of 1000–5000 m^2/m^3 are commonly used for wastewater treatment. Plastic-based biocarriers are preferred for MBBR design due to their shape and size versatility, offering a high range of surface areas (Zhang et al., 2024). These carriers, typically made of recycled high-density polyethylene and polypropylene, have a specific surface area of 200–1200 m^2/m^3 and are highly durable, resistant to degradation from intensive ultraviolet irradiation, and do not require replacement throughout the MBBR operation (Ahmadi et al., 2023). Biofilm growth on biocarriers varies from 5 to 30 mg solids/m^2 due to variations in organic load and dissolved oxygen (Forrest et al., 2016). Typical plastic-based carriers, primarily Kaldnes® types, include K1 (500 m^2/m^3), K2 (350 m^2/m^3), K3 (500 m^2/m^3), K5 (800 m^2/m^3), Natrix C2 (220 m^2/m^3), Natrix F3 (200 m^2/m^3), BiofilmChip M (1200 m^2/m^3), and BiofilmChip P (900 m^2/m^3), designed to prevent carrier clogging (Torresi et al., 2016; Ren et al., 2021). Additionally, a saddle-shaped grid height Z-carrier (Z-200 and Z-400) was developed, providing a biofilm thickness of 200–400 μm. The biofilm is thicker on the outer surface, making the carrier surface area nearly equal to the active biofilm, and this configuration prevents clogging due to the desirable biofilm thickness, the impact of voids on the value of the surface area specific to bulk is represented in Fig. 6.3 (Torresi et al., 2016).

Materials with less negative charge and greater hydrophobicity offer maximum organic removal rates and nitrification efficiency. Carriers made from negatively charged cellulosic materials with high hydrophilicity exhibit good nitrification and biofilm formation (Sousa et al., 1997; Kim et al., 1997). Biofilm adhesion on sepiolite and red clay is lower than on plastic materials due to the higher hydrophobicity of plastic carrier materials (Bonastre and Paris, 1988; Teixeira and Oliveira, 1999). Polyethylene,

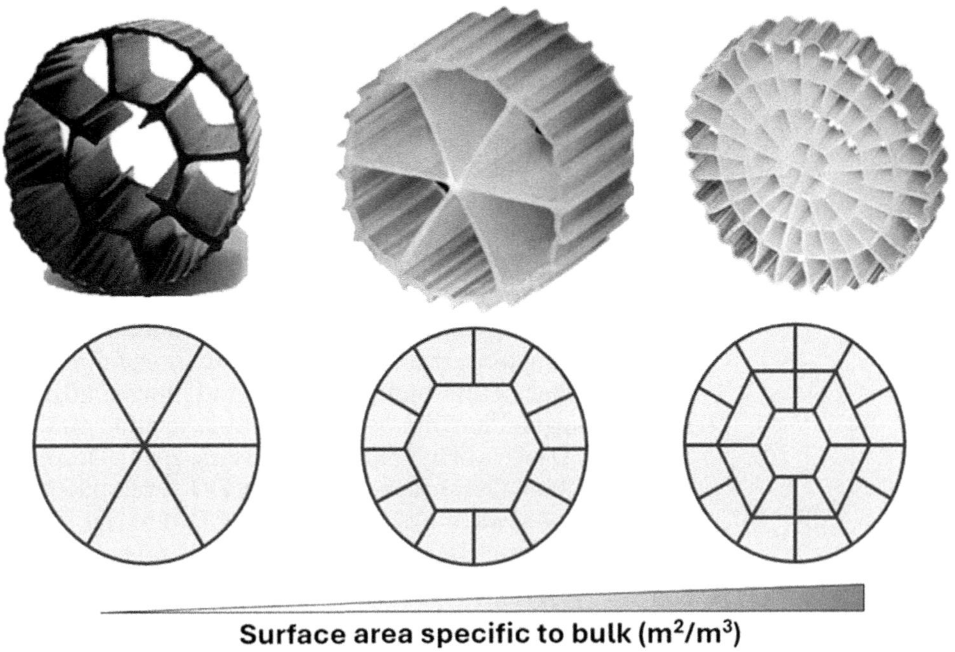

Fig. 6.3 Higher voids increase the surface area specific to bulk.

polypropylene, polyurethane foam-based biocarriers have shown the highest denitrification efficiency in wastewater rich in nitrate (Yuan et al., 2015). Further, the filling ratio of biocarriers in the MBBR is a crucial parameter for nitrification and denitrification processes. A filling ratio below 70% is preferable to avoid stagnant conditions and allow carriers to move freely in the bulk liquid (Salvetti et al., 2006). Sokół (2003) found that a filling ratio of 55% in the MBBR achieved a 90% COD removal from refinery wastewater. Carriers with a specific surface area of 400 m^2/m^3 require a 60% filling ratio to provide a specific surface area of 240 m^2/m^3 for biofilm growth (Tawfik and Klapwijk, 2010). A higher filling ratio is desirable for the removal of nutrients and organics. However, if the filling fraction exceeds 70%, it negatively impacts the performance of the MBBR module.

6.3.2 Dissolved oxygen

The aerobic or anaerobic/anoxic MBBR modules can be easily utilized for removal of pollutants from wastewater. The movement of the biocarriers is mechanically mixing in an anaerobic MBBR system. Aerators induce carrier movement in the aerobic system, to ensure adequate dissolved oxygen (DO) levels for microbial oxygenation. Consequently, an additional oxygen supply is necessary, leading to increased operational costs (Cao et al., 2017). A minimum of 2 mg/L of dissolved oxygen is required for the biodegradation process (Ali et al., 2017; Arora et al., 2021). In aerobic MBBR processes, higher levels of DO are necessary to oxygenate microbes degrading pollutants to maintain carrier suspension in the bulk liquid and in the diffused layer of the biofilm. Hence, the size of air bubbles should be adequate to facilitate oxygen transfer but not too large to prevent a decrease in oxygen transferability and biofilm sloughing. Although smaller air bubbles are less effective in moving biocarrier, they can quickly transfer oxygen in the diffused layer of the biofilm (Ismail and Tawfik, 2016b). Some studies suggest that the aerobic bacteria in the MBBR module requires DO levels of 2–4 mg/L (de Oliveira, 2014). Despite a lower filling fraction (20%), achieving high DO levels (5–6 mg/L) for treatment of sewage has resulted in COD removal efficiency of 99.6%. Adequate DO levels or aeration positively influence organic removal and nitrification (Yuan et al., 2023).

Additionally, in aerobic systems, carriers are uniformly distributed due to high enforced aeration compared to mechanical mixing anaerobic/anoxic systems. Further, oxygen transfer to the liquid is more efficient in aerobic systems. Heterotrophic bacteria in aerobic biofilm systems, utilize organic matter as a carbon source, facilitating the growth of nitrifying bacteria. Additional carbon sources rich acids or carbohydrates are added to support heterotroph growth in anoxic reactors (Tawfik et al., 2004). Carbon sources are also essential for cultivating denitrifying communities, particularly in post-denitrification processes within anoxic systems (Tawfik et al., 2006b). Moreover, anaerobic/anoxic systems produce toxic gases such as H_2S and CH_4, which can have adverse environmental effects (Wazeri et al., 2018).

6.3.3 Hydraulic retention time, organic loading rate and sludge residence time

Removal efficiency of pollutants in the MBBR process, mostly relies on surface loading. The surface loading rate is defined as the flow rate multiplied by organic content (COD influent) divided by the total surface area of the carriers. Soluble organic matter encounters biofilm, undergoing hydrolysis and degradation (Yuan et al., 2023). Studies indicate that COD removal efficiency is enhanced at high imposed organic loading rate due to increased biofilm thickness but negatively impacted the nitrification efficiency (Bassin et al., 2016). Consequently, higher organic loads necessitate increased aeration to prevent biomass accumulation in the biofilm and ensure adequate oxygen for heterotrophic bacteria and nitrifying organisms (Bassin et al., 2016). Liang et al. (2022) employed MBBR technology for dairy wastewater treatment containing COD (189–6219 mg/L), achieving 80.5–97.3% for COD removal. Additionally, Bassin et al. (2016) reported that increasing COD thickens the biofilm and initiates biofilm sloughing, leading to increased suspended solids. Kamilya et al. (2023) also indicate that increasing the organic C/N ratio boosts both COD and BOD removal efficiencies (~2%) but decreases nutrient removal efficiency (~4–8%).

Researchers note that SRT has a minimal impact compared to hydraulic retention time (HRT) in the MBBR module, as the biofilm remains attached to the biocarriers except for occasional sloughing incidents (Zhu et al., 2023). Unlike conventional biological processes where biomass exists with effluent and settles in secondary clarifiers, in the MBBR system, most biomass remains, leading to a high SRT (Liu et al., 2022). Consequently, HRT significantly influences removal efficiency, with higher HRT correlating with increased COD removal (Chen et al., 2023). For instance, MBBR treating pulp and paper mill industry wastewater at an HRT of 3.3 h resulted in 65% COD removal (de Oliveira, 2014; Leyva-Díaz et al., 2013). Increasing the HRT up to 7.0 h improved the nitrification efficiency in MBBR module (Ismail and Tawfik, 2016a). Mannina et al. (2018) demonstrated that an HRT of 14 h yielded maximum nitrification rates and over 90% ammonia removal. Generally, HRT in MBBR ranges from 4 to 24 h (Saidulu et al., 2021). A study by Amin et al. (2018) varied HRT and SRT from 4 to 16 h. and 11 to 46 days, respectively, for endocrine disrupting compounds (EDC) removal using MBBR system, observing over 90% removal rates at long SRTs. Furthermore, increasing HRT positively influences organics and nutrient removal, with removal efficiencies significantly improved at HRT above 6 h.

6.3.4 Temperature

The microbial community in the biofilm of the MBBR highly depends on the temperature. The biofilm formation is quite slow at low temperature due to the slow of growth of microbial community particularly heterotrophs and nitrifiers. Further, the stability of the MBBR performance for wastewater treatment at low temperature is necessary for application. A temperature fluctuation would affect the performance of the MBBR reactor particularly nitrifying bacteria (Mannina et al., 2018; Young et al., 2017).

The temperature could highly affect the organic utilization rate, adsorption of particulate matter, oxygen diffusion onto the deep layers of the biofilm, controlling the biofilm thickness and microbial community structure. The startup of MBBR module and acceleration of the biofilm formation at low temperature of 5 °C was occurred by adding Quorum Sensing bacteria namely *Sphingomonas rubra* BH3T (Fu et al., 2022). The addition of *Sphingomonas rubra* BH3T improved the efficiency of the COD removal (93%) and ammonia oxidation (75%) due to increase the biofilm thickness (60.2%), and activities. The extracellular polymeric substances (EPS) were increased up to 8.1–9.1 mg/carrier enhancing biofilm mechanical stability, formation, and surface adhesion. Further, the nitrifying bacteria (Nitrospira), Pseudomonas, Arcobacter, Dechloromonas, Aeromonas, and Flavobacterium were abundant at low temperatures. The ammonia removal rate by MBBR treating domestic wastewater was increased from 0.11 to 0.26 kgN/m^3 d at increasing the temperature from 1 to 20 °C respectively (Almomani et al., 2014). The intermittent aerated MBBRs treating anaerobic effluent achieved 50–60% removal for total nitrogen and 40–70% for COD at operational temperature of 10 °C (Luostarinen et al., 2006). The nitrification was complete (100%) due to sufficient dissolved oxygen, nitrifiers and low imposed loading rate that enhance the ammonia oxidation. The denitrification was quite low due to the deficiency of carbon source for denitrifies. The nitrification rates by sequencing batch moving bed biofilm reactor treating wastewater were 0.24 and 0.25 mg NO_X-N/m^2 d, at temperatures of 20 and 15 °C. This nitrification efficiency was highly reduced to 0.18 mg NO_X-N/m^2 d at lower temperature of 10 °C (Dulkadiroglu et al., 2005). The nitrification rate in MBBR treating wastewater at temperature of 5 °C was 64–66% and decreased to 18.7–15.7% at temperature of 1 °C (Hoang et al., 2014). The detachment and sloughing of biomass were occurred at low temperature that deteriorated the nitrification efficiency. Nitritation/anammox MBBR at temperatures of 13–19 °C was stable for nitrogen removal and deteriorated at 10 °C (Persson et al., 2014). *Brocadia* sp. and *Brocadia anammox* populations were dominance in the biofilm of carrier materials incorporation with nitrifiers (Nitrosomonas, Nitrobacter, Nitrospira). Partial nitritation was successfully obtained at low temperature of 10 °C in MBBR treating wastewater which is beneficial for nitrogen removal via annamox bacteria (Kowalski et al., 2019). The average NH_4-N; NO_2-N and NO_3-N concentrations were 16.0, 15.7 and 2.6 mg/L in the treated effluent of MBBR at influent of 40 mgN/L. This nitritation process was achieved at ratio of dissolved oxygen to total ammonium nitrogen of 0.06 and free ammonia concentration of 1.1 mg NH_3-N/L. *Nitrosomonas oligotropha* and *Nitrosomonas nitrosa* was dominant for ammonia oxidation at high temperature of 35–45 °C imposed to MBBR module resulting nitrification efficiency of 90% (Shore et al., 2012).

6.4 Moving bed biofilm reactors incorporated with other technologies

The MBBR technology is mainly employed for carbon and nitrogen removal from wastewaters. However, standalone MBBR unit cannot completely remove of nitrogen and phosphorous from wastewater. Thus, the integration of MBBR with other technologies and or treatment processes are needed to achieve the required effluent quality for safely discharge.

6.4.1 Moving bed biofilm reactor coupled with membrane bioreactor

The MBBR process is primarily utilized as a standalone unit for the simultaneous removal of carbon and nitrogen from municipal wastewaters. However, the treated wastewater often fails to meet the required standards for discharge into sewage networks or for agricultural reuse. Therefore, coupling the MBBR with other technologies, such as the coagulation process (Ahmad et al., 2022) or membrane filtration technology, could significantly improve the effluent quality of the MBBR system (Tawfik et al., 2022a). The coagulation process presents itself as a viable treatment option for removing particulate organic matter, suspended pathogenic bacteria, and color from municipal wastewater before entering the MBBR module (Mahmoud et al., 2009). Nonetheless, chemical coagulation consumes energy and chemicals, generates a substantial amount of chemical sludge as a secondary pollutant, and necessitates proper management. On the other hand, membrane filtration processes are highly efficient, offering flexibility, ease of operation, and effective removal of both particulate and dissolved organics from wastewater (Alhajeri et al., 2024). Various polymeric membrane technologies, such as microfiltration, ultrafiltration, nanofiltration, and reverse osmosis membranes, can be coupled with the MBBR system for the treatment of municipal and industrial wastewater (Jabornig and Favero, 2013). The coupling of MBBR with different membrane units depends on the specific application of the treated effluent, whether for industrial or agricultural purposes. For instance, the membrane moving bed biofilm reactor (MMBBR) followed by reverse osmosis removed

COD and total nitrogen by 97% and 93%, respectively. This resulted in an effluent quality with COD and TN concentrations of 8.2 and 2.3 mg/L, respectively, which comply with the requirements for reuse in industry and agriculture. The attached biofilm accounted for a suspended sludge concentration of 75 mg VS/L, with a reported stable permeating flux of 30 L/m^2 h (Sun et al., 2021). Simultaneous nitrification and denitrification of municipal wastewater was investigated in MBBR supplied with iron(II) sulfate heptahydrate (Wang et al., 2006). Total nitrogen was removed by 89.9% and 90.6% was accounted for total phosphorous (TP) elimination at dissolved oxygen (DO) level of 2 mg/L. Microbial communities of the biofilms were seasonally identified in full-scale MBBR treating municipal wastewater. Aerobic Gammaproteobacteria (52%) was initially dominance for organics degradations. However, Deltaproteobacteria and Clostridiales were observed in the biofilm indicating the presence of anaerobic conditions in the biofilm layer (Biswas et al., 2014). Soluble COD, total nitrogen and phosphorus were removed by values of 96.9%, 84.6% and 95.8% from municipal wastewater using integrated MBBR (anaerobic/anoxic/aerobic) (Kermani et al., 2008). MBBR treating municipal wastewater at organic loading rate of 1.58 $kgCOD/m^3$ d provided removal efficiency of 94.6% for BOD and 92.30% for COD (Javid et al., 2013).

Gray water underwent efficient treatment through a combination of MBBR and membrane filtration with a pore size of 0.2 μm. The permeated flux reached 12.9 L/m^2 h, with an estimated rejection efficiency of approximately 80% and an energy consumption of 1.3 kWh/m^3 during operation (Jabornig and Favero, 2013). In another study, two anaerobic and aerobic MBBRs, integrated with membrane filtration, were utilized for treating printing and dyeing wastewater at hydraulic retention times of 11 and 5 h, respectively. The filling materials ratio in the anaerobic and aerobic MBBRs was maintained at 35% (v/v). This integrated system achieved significant removal rates, with 90% removal for color, 85% for COD, and 84% for suspended solids (Dong et al., 2014). Furthermore, Jalali et al. (2023) investigated the treatment of industrial estate wastewater using MBBR followed by polyvinylidene fluoride membrane. The study varied HRT from 5 to 15 h, air flow rate (AFR) from 1 to 4 L/min, and filling carrier ratio (FR) from 30% to 70%. Soluble COD and total nitrogen (TN) were removed by 93.4% and 27.6%, respectively. To determine the attached biomass on a carrier in the MBBR module treating wastewater, Eq. (1) is applied. Carriers with biofilm are randomly selected from the MBBR module at different intervals and locations to ensure accuracy. The harvested carriers with biomass undergo placement in an oven at 100 °C for 1 h, followed by washing to separate the biomass. The attached biomass concentration is then calculated according to the provided equation (Tian et al., 2023).

$$\text{Attached biomass concentration } \frac{\text{mg}}{\text{L}} = \left(\frac{(A-B)*C}{V}\right)x1000 \tag{1}$$

where, A, B, C and V (g) are the weight of carrier with attached biofilm.

6.4.2 Microbial community for nitrogen removal in anoxic/oxic-moving bed biofilm reactor process

Conventional methods of nitrification and denitrification have long been employed for nitrogen removal from wastewater, particularly in activated sludge systems. These processes rely on factors such as hydraulic retention time, biomass concentration, and the flow rate of return sludge. However, they often require a large footprint and pose challenges in excess sludge disposal management. In contrast, the application of MBBR for nitrogen removal offers several advantages, including a smaller footprint and reduced production of excess sludge. Additionally, the flexibility to adjust the sludge retention time enables efficient nitrification and denitrification processes. The anoxic/oxic (A/O) configuration of MBBR is commonly utilized for total nitrogen removal. Despite its potential, A/O-MBBR has faced limitations in achieving high nitrogen removal efficiency due to constraints in internal recycling ratios (Wang et al., 2018). For instance, Zhou et al. (2022a) demonstrated that a two-stage A/O-MBBR system achieved a total nitrogen removal efficiency of 91.8% at a hydraulic retention time of 11.2h. In a similar study, a two-stage A/O MBBR system maintained low concentrations of COD, NH_4^+-N, and TN, with removal efficiencies of 91.68% for COD, 99.11% for ammonia, and 74.03% for TN (Zhou et al., 2022b). This remarkable efficiency can be attributed to the abundance of denitrifiers and nitrifiers within the biofilm, including genera such as Methylotenera (denitrifiers), Nitrosomonas, and Nitrospira (nitrifiers). The distribution of Proteobacteria and Actinobacteriota phyla in the post-anoxic and oxic MBBR biofilms, respectively, further supports the nitrogen removal process.

Proteobacteria phylum, closely associated with denitrification, predominates in the post-anoxic compartments of the MBBR unit

(Zhao et al., 2019), while Actinobacteriota phyla, comprising heterotrophic bacteria, facilitate organic matter degradation in the oxic compartments (Wang et al., 2018). Acidobacteriota, Bacteroidota, and Chloroflexi phyla play crucial roles in nitrogen removal, with Acidobacteriota particularly significant in this process (Kristensen et al., 2021). Bacteroidota phylum is primarily involved in the hydrolysis of macromolecular organics under anaerobic or anoxic conditions, while Chloroflexi phylum contributes to organic degradation in wastewater treatment (Zhao et al., 2019). Furthermore, specific genera such as Gemmobacter, Rhodobacter, Thermomonas, Pseudorhodobacter, Simplicispira, and Thiothrix dominate the denitrification process in anoxic MBBR units (Chao et al., 2020). Methylotenera genera are known for their role in denitrification within anoxic treatment systems, contributing to the overall efficiency of nitrogen removal (Herlambang et al., 2021). The *Mycobacterium* genus was found to be abundant in the post anoxic MBBR compartments, while the pre-anoxic MBBR compartments exhibited abundance in genera such as Devosia, Simplicispira, Defluviimonas, Mesorhizobium, Trichococcus, Hydrogenophaga, CL500-29 marine group, Dokdonella, Arenimonas, Thermomonas, Ferruginibacter, Sphingobium, Rhodobacter, and Lautropia. Additionally, certain genera like Methylotenera, Trichococcus, Denitratisoma, Hyphomicrobium, Flavobacterium, Terrimonas, and Brevundimonas were found to be abundant in both pre-anoxic and post-anoxic MBBR compartments (Zhou et al., 2022b). The denitrifying bacterial communities in the pre-anoxic MBBR compartments exhibited diversity due to the presence of sufficient organic matter, which serves as a substrate for denitrifiers. Notably, nitrifying bacteria, such as Nitrosomonas (ammonia-oxidizing genus) and Nitrospira (nitrite-oxidizing genus), were dominant in MBBR systems treating wastewater (Yan et al., 2018). Moreover, the presence of Candidatus Brocadia (anammox genus) in the pre-anoxic compartment of MBBR enables the utilization of nitrite and ammonia for nitrogen removal (Zhang et al., 2019). The abundance and diversity of bacterial communities within MBBR biofilms offer unique advantages for efficient removal of organic compounds and nitrogen from wastewater.

6.4.3 Endogenous carbon source for nitrogen and phosphorous removal

The carbon-to-nitrogen (C/N) ratio sometimes becomes inadequate in the influent wastewater, which can compromise conventional treatment processes due to the scarcity of available

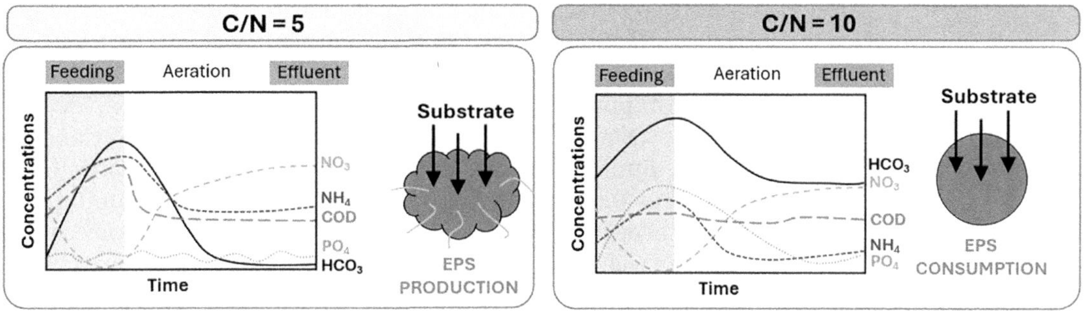

Fig. 6.4 Impact of carbon-to-nitrogen ratios on substrate dynamics and EPS production in moving bed biofilm reactors.

carbon sources (Elsharkawy et al., 2019; Ahmad et al., 2022). Therefore, simultaneous nitrification/denitrifying phosphorus removal from wastewater at low C/N ratios in standalone moving bed biofilm reactors (MBBRs) has garnered increasing interest due to its high efficiency, low footprints, and investment costs. In Fig. 6.4, at a C/N ratio of 5, there is significant production of extracellular polymeric substances (EPS), which helps in the formation and maintenance of biofilms. This results in better retention and removal of nitrogen and phosphorus. Conversely, at a higher C/N ratio of 10, the consumption of EPS is more pronounced, leading to different substrate dynamics. The higher C/N ratio may support higher microbial activity but can also result in less stable biofilms due to reduced EPS production. In an intermittent aeration MBBR module, simultaneous nitrification-denitrifying phosphorus achieved removal efficiencies of 62% for total inorganic nitrogen and 75% for phosphorus (Iannacone et al., 2020). Similarly, an anaerobic/intermittently aerated MBBR at a C/N ratio of 3 provided TN and TP removal efficiencies of 90% and 74%, respectively, for short aeration periods. However, these efficiencies dropped to 88% for TN and 59% for TP with long aeration due to biofilm detachment and low secretion of extracellular polymeric substances (Luan et al., 2024). The functional genera for nitrogen removal included Nitrosomonas, Terrimonas, Ferruginibacter, Nitrospira, and Tetrasphaera. Anoxic conditions prevailing inside the biofilm of the carrier materials enriched aerobic denitrifiers (Osmani et al., 2021). The alternating aerobic/anoxic environment in the MBBR, created by intermittent aeration, fostered the growth of facultative bacteria, particularly aerobic denitrifiers, which can utilize nitrate (NO_3-N) as an electron acceptor to complete denitrification alongside the nitrification process. Notably, the aerobic denitrifier *Lysobacter* genus exhibited the capability for nitrogen removal at low C/N ratios.

At C/N ratios exceeding 5, Thiothrix and Flavobacterium (anoxic denitrifiers) were the dominant genera for nitrogen removal. Furthermore, at a C/N ratio of 5, Acinetobacter was dominant for phosphorus uptake, utilizing NO_2^--N as an electron acceptor (Xie et al., 2021). Reducing the C/N ratio to 3, Candidatus_Accumulibacter, Rhodobacteraceae, and Candidatus_Competibacter became dominant, exhibiting efficient simultaneous phosphorus uptake and denitrification using endogenous carbon sources (Yuan et al., 2020). Most of the organics in the wastewater were converted into endogenous carbon sources for phosphorus and nitrogen removal in the anaerobic MBBR stage. Candidatus_Accumulibacter, Rhodobacteraceae, and Candidatus_Competibacter utilized NO_2^--N and NO_3-N as electron acceptors and completed denitrification using endogenous carbon sources (Zaman et al., 2021). Furthermore, phosphorus uptake from wastewater could be achieved simultaneously by denitrifying phosphorus-accumulating organisms. While the endogenous denitrification and removal of phosphorus in the MBBR cannot efficiently utilize carbon sources, they reduce energy consumption for aeration. The regulation of intermittent aeration strategies in the MBBR module can enrich the microorganisms required for nitrogen and phosphorous removal.

6.4.4 Anammox moving bed biofilm reactor for nitrogen removal from acidic partial nitrification

Bacteria autotrophically convert ammonium ions and nitrite into dinitrogen gas, a process known as anaerobic ammonium oxidation (anammox) (Eq. 2). Partial nitritation combined with anammox (PN/A) for deammonification of wastewater is a promising option due to a 60% reduction in aeration and a 100% decrease in carbon source usage compared to the classical nitrification/denitrification process as highlighted in Fig. 6.5 (Ismail et al., 2021a). The PN/A process occurs in two biological steps (Eqs. 2 and 3).

$$\text{Anammox } NH_4^+ + 1.32NO_2^- \rightarrow 1.02N_2 + 0.26NO_3^- \quad (2)$$

$$\text{Partial nitritation(PA)} = NH_4^+ \rightarrow NO_2^- \quad (3)$$

The suppression of nitrite-oxidizing bacteria (NOB), which converts nitrite into nitrate, poses a challenge and can lead to the failure of the PN/A process. Acid-tolerant ammonia-oxidizing

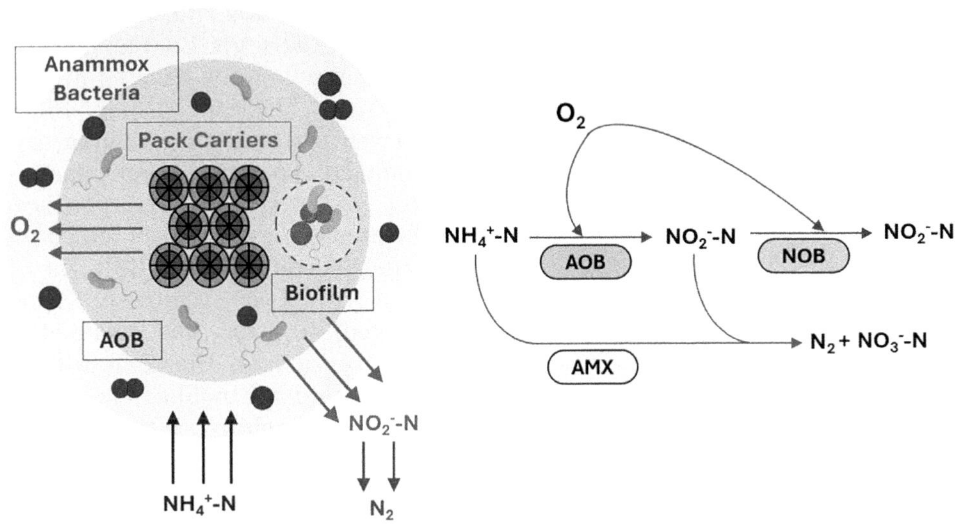

Fig. 6.5 Coupled anammox and partial nitrification in AnMBBR.

bacteria (AOB) such as Candidatus Nitrosoglobus can lower the pH of the wastewater to 4.5 through aerobic oxidation of ammonium (Eq. 4).

$$NH_4^+ + 1.5O_2 \rightarrow NO_2^- + 2H^+ + H_2O \quad (4)$$

This process produces a nitrite concentration exceeding 25 mg N/L at a pH of 4.5 and a free nitrous acid (FNA) concentration of 1.0 mg N/L, resulting in conditions conducive to NOB suppression (Ahmad et al., 2022; Zhao et al., 2023). Anammox bacteria are efficient at neutral pH values; thus, neutralization of the treated effluent of PN is needed to achieve total nitrogen removal (Ismail et al., 2021b; Ismail and Tawfik, 2016a). Fortunately, biofilms on the carrier materials act as a biological barrier, protecting anammox bacteria inside the biomass against the prevailing low pH value in the PN effluent.

The suppression of NOB in biofilm systems is significant at low pH because NOB live inside biofilms and are protected from acidic pH stress in the bulk liquid (Zhang et al., 2023; Van Tendeloo et al., 2021). It is hypothesized that biofilms play a key role in protecting anammox bacteria from acidic conditions in the bulk liquid. Thus, biofilm systems have the capability to treat acidic effluent from the PN process. A lab-scale anammox MBBR was utilized for the treatment of acidic PN effluent at a pH of 4.6

(Hu et al., 2023). Total nitrogen removal efficiency exceeded 80% at a loading rate of up to 149.7±3.9 mgN/Ld. Candidatus Brocadia (16.3%) was the dominant anammox bacteria in the MBBR module at a nitrite concentration of 10 mgN/L in the PN effluent. These anammox bacteria were situated in the inner layer of thick biofilms of the MBBR and were not present in the surface layer. The NO_2^- removed/NH_4^+ removed and NO_3^- produced/NH_4^+ removed ratios were accounted for 1.24±0.06 and 0.20±0.04, respectively, in the anammox bacteria in the MBBR. These values are comparable to those of 1.15 for NO_2^- removed/NH_4^+ removed and 0.16 for NO_3^- produced/NH_4^+ removed obtained by Lotti et al. (2014) using anammox bacteria. The composition of the microbial community of the anammox biofilm in the MBBR treating acidic PN included Planctomycetota (29.3%), Verrucomicrobiota (0.9%), Chloroflexi (21.2%), Proteobacteria (14.4%), Patescibacteria (4.6%), Cercozoa (3.3%), Bacteroidota (12.8%), Acidobacteriota (5.6%), Nitrospirota (1.7%), and Firmicutes (1.0%). It is known that anammox species mainly belong to the Planctomycetota phylum (Pradhan et al., 2020). The denitrifiers (Denitratisoma and Pseudomonas) were present in the anammox biofilms of the MBBR (Hu et al., 2023).

6.4.5 Surfactant enhancing the performance moving bed biofilm reactor

Biofilm formation under anaerobic conditions in MBBR is a slow, complex, and time-consuming process. The startup of anaerobic biofilm MBBR reactors treating wastewater requires a long time because bacterial growth is quite slow, limiting its potential applications. The addition of biosurfactants changes the structure of bacterial cell membranes, reducing the toxicity of pollutants and enhancing biofilm formation (Zhang et al., 2017). Biosurfactants are biodegradable and possess excellent chemical surface properties that enhance biofilm formation as described in Fig. 6.6 (Ismail et al., 2022). Rhamnolipid, a type of biosurfactant secreted by *Pseudomonas aeruginosa*, significantly promotes the deposition and adhesion of biofilm while reducing interfacial surface tension (Peng et al., 2021). Rhamnolipid interacts with existing functional groups, such as hydroxyl, carboxyl, and amine groups, on the cell surface of the biofilm, thereby altering the biomass compositions. This biosurfactant has great potential to enhance and promote biofilm characteristics, formation, attachment, and growth, thereby accelerating the startup of MBBR for wastewater treatment. For instance, the addition of 40 mg/L rhamnolipid to a microbial fuel cell accelerated biofilm

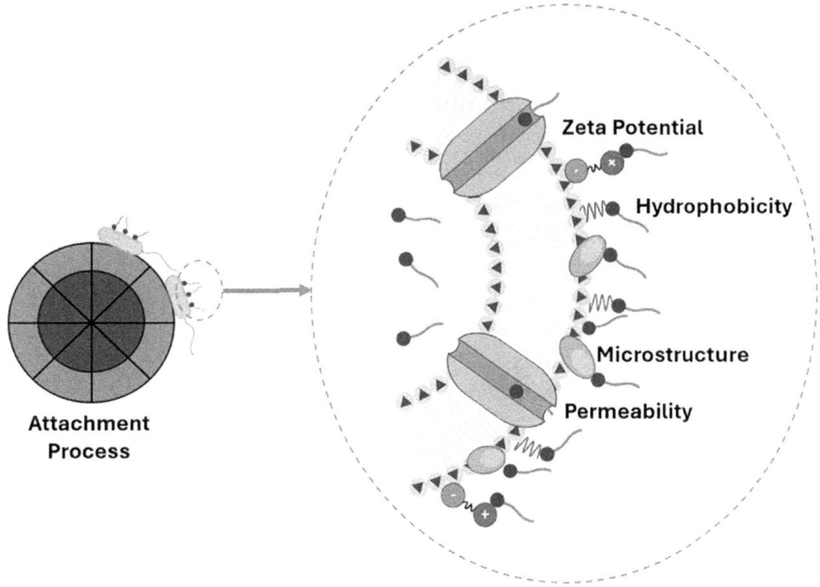

Fig. 6.6 Proposed mechanisms of the effect of biosurfactant (rhamnolipid) on biofilm formation in MBBR.

formation on the anode and shortened the startup period by 48 h (Zhang et al., 2017). Rhamnolipid (20–50 mg/L) reduces interfacial hydration rejection and induces high deposition of microorganisms on the surface of biocarriers (Huang et al., 2018).

The hydrophilic/hydrophobic groups of rhamnolipid interact with macromolecular substances, promoting the biofilm deposition and adhesion process (Peng et al., 2021). Addition of rhamnolipid reduced the startup period of anaerobic moving bed biofilm reactor (AnMBBR) by 50%, to within 25 days compared to the control unit. This reduction was attributed to an increase in attached biomass concentrations on carriers by 1.5 times and the removal of organic pollutants (Song et al., 2022). Furthermore, rhamnolipid enhanced internal microbial activities such as metabolic and enzyme activity, as well as extracellular polymeric substances, thereby regulating biofilm growth. Moreover, it increased the abundance of microbial community structure, including Thauera, Methylomonas, Azoarcus, Leptonema, and Desulfomicrobium. The addition of rhamnolipid promoted the denitrification performance of AnMBBR and shortened the startup period to 20 days (Peng et al., 2021). Rhamnolipid contains hydrophobic groups that combine with cell membrane functional groups, facilitating the transport of bacteria from bulk liquid to the surface of biocarriers. The polar structure of the cell surface interacts with rhamnolipid through hydrogen bonds, dipoles, and/or electrostatic forces, resulting in a hydrophobic cell surface (Zeng et al., 2018). The addition of rhamnolipid reduces the surface tension between the biocarrier and microorganisms, thereby reducing the energy between them and promoting biofilm formation (Zeng et al., 2018).

Furthermore, rhamnolipid adsorbs onto the cell membrane of microorganisms, interacting with their active functional groups (hydroxyl, carboxyl, etc.), resulting in the modification of the microorganism cell surface. This significantly decreases the zeta potential and increases biofilm hydrophobicity (Liu et al., 2012). Microbial adhesion is subsequently increased to the biofilm surface. Moreover, rhamnolipid has a positive effect on microbial growth, metabolic activity, and enzymatic activity. Rhamnolipid improves the permeability of the bacterial cell membrane, facilitating the exchange and transmission of substrates/substances inside and outside the cell membrane of the microorganisms, resulting in high microbial and enzymatic activities (Shao et al., 2017). Microbial activities are promoted, improving metabolic processes and enzyme secretion. Rhamnolipid addition stimulates the secretion of extracellular polymeric substances on biocarriers, which helps to agglomerate thick biofilm.

6.5 Conclusions and future trends

The application of MBBR for wastewater treatment particularly for organics and nitrogen removal represents a promising approach. Further, the MBBR can be easily applied for upgrading the existing wastewater treatment plants. The technology can be easily applied for treatment of domestic, municipal, and industrial wastewater due to its flexibility for forming various consortia and microbes for removal toxic and non-toxic pollutants. Additionally, the flexibility to adjust the sludge retention time enables efficient nitrification and denitrification processes. Partial nitritation combined with anammox (PN/A) for deammonification of wastewater using MBBR is a promising option due to a 60% reduction in aeration and a 100% decrease in carbon source usage compared to the classical nitrification/denitrification process. Further, the filling ratio of biocarriers in the MBBR is a crucial parameter for nitrification and denitrification processes. A filling ratio below 70% is preferable to avoid stagnant conditions and allow carriers to move freely in the bulk liquid. The filling ratio of 55% in the MBBR achieved a 90% COD removal from refinery wastewater. Carriers with a specific surface area of 400 m^2/m^3 require a 60% filling ratio to provide a specific surface area of

$240\,m^2/m^3$ for biofilm growth. A higher filling ratio is desirable for the removal of nutrients and organics. However, if the filling fraction exceeds 70%, it negatively impacts the performance of the MBBR module.

Further research is needed for the development of MBBR technology i.e. combination with aerobic membrane biofilm to reduce the aeration and mixing costs as well as increase the removal efficiency of organics and nutrients. Low removal efficiency of recalcitrant organics i.e. various toxic emerging and micropollutants from municipal wastewater is the main concern with these biological treatment systems. The biosurfactant/biochar addition has great potential to enhance and promote biofilm characteristics, formation, attachment, and growth, thereby accelerating the startup of MBBR for wastewater treatment. Furthermore, phosphorus uptake from wastewater could be achieved simultaneously by denitrifying phosphorus-accumulating organisms. While the endogenous denitrification and removal of phosphorus in the MBBR cannot efficiently utilize carbon sources, they reduce energy consumption for aeration. The regulation of intermittent aeration strategies in the MBBR module can enrich the microorganisms required for nitrogen and phosphorous removal.

Acknowledgments

This work was financially supported by College of Life Sciences, Department of Environmental Sciences, Kuwait University.

References

Ahmad, H.A., Ahmad, S., Cui, Q., Wang, Z., Wei, H., Chen, X., Ni, S.Q., Ismail, S., Awad, H.M., Tawfik, A., 2022. The environmental distribution and removal of emerging pollutants, highlighting the importance of using microbes as a potential degrader: a review. Sci. Total Environ. 809. 18791026. Elsevier B.V., China. Available from: www.elsevier.com/locate/scitotenv. https://doi.org/10.1016/j.scitotenv.2021.151926.

Ahmadi, N., Abbasi, M., Torabian, A., van Loosdrecht, M.C.M., Ducoste, J., 2023. Biotransformation of micropollutants in moving bed biofilm reactors under heterotrophic and autotrophic conditions. J. Hazard. Mater. 460. 18733336. Elsevier B.V., Iran, Available from: www.elsevier.com/locate/jhazmat. https://doi.org/10.1016/j.jhazmat.2023.132232.

Alhajeri, N.S., Tawfik, A., Elsamadony, M., Al-Fadhli, F.M., Meng, F., 2024. Synergistic algal/bacterial interaction in membrane bioreactor for detoxification of 1,2-dichloroethane-rich petroleum wastewater. J. Hazard. Mater. 470. https://doi.org/10.1016/j.jhazmat.2024.134125. 18733336. Elsevier B.V., Kuwait. Available from: https://www.sciencedirect.com/science/journal/03043894.

Ali, M., Farghaly, A., Roux, S.L., Peu, P., Dabert, P., Tawfik, A., 2017. Potential of using non-inoculated self-aerated immobilized biomass reactor for post-treatment of upflow anaerobic staged reactor treating high strength industrial wastewater. J. Chem. Technol. Biotechnol. 92 (5), 1065–1075. 10974660. John Wiley and Sons Ltd, Egypt. Available from: http://onlinelibrary.wiley.com/journal/10.1002/(ISSN)1097-4660. https://doi.org/10.1002/jctb.5082.

Allam, A., Fleifle, A., Tawfik, A., Yoshimura, C., El-Saadi, A., 2015. A simulation-based suitability index of the quality and quantity of agricultural drainage water for reuse in irrigation. Sci. Total Environ. 536, 79–90. 18791026. Elsevier, Egypt. Available from: www.elsevier.com/locate/scitotenv. https://doi.org/10.1016/j.scitotenv.2015.07.029.

Almomani, F.A., Delatolla, R., Örmeci, B., 2014. Field study of moving bed biofilm reactor technology for post-treatment of wastewater lagoon effluent at 1°C. Can. Environ. Technol. 35 (13), 1596–1604. 1479487X. Taylor and Francis Ltd. Available from: http://www.tandf.co.uk/journals/titles/09593330.asp. https://doi.org/10.1080/09593330.2013.874500.

Amin, M.M., Bina, B., Ebrahim, K., Yavari, Z., Mohammadi, F., 2018. Biodegradation of natural and synthetic estrogens in moving bed bioreactor. Chin. J. Chem. Eng. 26, 393–399. https://doi.org/10.1016/j.cjche.2017.06.006.

Arora, A.S., Nawaz, A., Qyyum, M.A., Ismail, S., Aslam, M., Tawfik, A., Yun, C.M., Lee, M., 2021. Energy saving anammox technology-based nitrogen removal and bioenergy recovery from wastewater: inhibition mechanisms, state-of-the-art control strategies, and prospects. Renew. Sust. Energ. Rev. 135. 18790690. Elsevier Ltd, South Korea. Available from: https://www.journals.elsevier.com/renewable-and-sustainable-energy-reviews. https://doi.org/10.1016/j.rser.2020.110126.

Barreto, C.M., Garcia, H.A., Hooijmans, C.M., Herrera, A., Brdjanovic, D., 2017. Assessing the performance of an MBR operated at high biomass concentrations. Int. Biodeterior. Biodegrad. 119, 528–537. 09648305. Elsevier Ltd, Netherlands. Available from: www.elsevier.com/inca/publications/store/4/0/5/8/9/9. https://doi.org/10.1016/j.ibiod.2016.10.006.

Bassin, J.P., Dias, I.N., Cao, S.M.S., Senra, E., Laranjeira, Y., Dezotti, M., 2016. Effect of increasing organic loading rates on the performance of moving-bed biofilm reactors filled with different support media: assessing the activity of suspended and attached biomass fractions. Process. Saf. Environ. Prot. 100, 131–141. https://doi.org/10.1016/j.psep.2016.01.007. 09575820.

Biswas, K., Taylor, M.W., Turner, S.J., 2014. Successional development of biofilms in moving bed biofilm reactor (MBBR) systems treating municipal wastewater. Appl. Microbiol. Biotechnol. 98 (3), 1429–1440. 14320614. Springer Verlag, New Zealand. Available from: link.springer.de/link/service/journals/00253/index.htm https://doi.org/10.1007/s00253-013-5082-8.

Bonastre, N., Paris, J.M., 1988. Colonisation and stimulation/inhibition properties of different supports used in anaerobic fixed-film reactors. Environ. Technol. Lett. 9 (8), 763–768. https://doi.org/10.1080/09593338809384630. 0143-2060.

Cao, Y., Zhang, C., Rong, H., Zheng, G., Zhao, L., 2017. The effect of dissolved oxygen concentration (DO) on oxygen diffusion and bacterial community structure in moving bed sequencing batch reactor (MBSBR). Water Res. 108, 86–94. 18792448. Elsevier Ltd, China. Available from: www.elsevier.com/locate/watres. https://doi.org/10.1016/j.watres.2016.10.063.

Chao, C., Zhao, Y., Keskar, J., Ji, M., Wang, Z., Li, X., 2020. Simultaneous removal of COD, nitrogen and phosphorus and the tridimensional microbial response in a sequencing batch biofilm reactor: with varying C/N/P ratios. Biochem. Eng. J. 154. 1873295X. Elsevier B.V., China. Available from: www.elsevier.com/locate/bej. https://doi.org/10.1016/j.bej.2019.04.017.

Chen, H.Y., Liu, Y.D., Dong, B., 2018. Biodegradation of tetracycline antibiotics in A/O moving-bed biofilm reactor systems. Bioprocess Biosyst. Eng. 41 (1), 47–56. 16157605. Springer Verlag, China. Available from: link.springer.de/link/service/journals/00449/index.htm https://doi.org/10.1007/s00449-017-1842-7.

Chen, Y., Wang, J., Zhao, Y.-G., Maqbool, F., Gao, M., Guo, L., Ji, J., Zhao, X., Zhang, M., 2022. Sulfamethoxazole removal from mariculture wastewater in moving bed biofilm reactor and insight into the changes of antibiotic and resistance genes. Chemosphere 298, 134327. https://doi.org/10.1016/j.chemosphere.2022.134327. 00456535.

Chen, P., Chen, D., Yang, B., Zhang, K., Li, S., 2023. Enhancing the sewage treatment effect, reducing membrane fouling, and increasing microbial community diversity by a novel biofiller-carrying functional bacteria *Rhodococcus* sp. CPZ24 in a moving bed biofilm reactor. J. Water Process Eng. 56, 104576. https://doi.org/10.1016/j.jwpe.2023.104576. 22147144.

De Gusseme, B., Vanhaecke, L., Verstraete, W., Boon, N., 2011. Degradation of acetaminophen by *Delftia tsuruhatensis* and *Pseudomonas aeruginosa* in a membrane bioreactor. Water Res. 45 (4), 1829–1837. 00431354. Elsevier Ltd, Belgium. Available from: https://www.sciencedirect.com/science/journal/00431354. https://doi.org/10.1016/j.watres.2010.11.040.

de Oliveira, D.V.M., 2014. Evaluation of a MBBR (moving bed biofilm reactor) pilot plant for treatment of pulp and paper mill wastewater. Int. J. Environ. Monitor. Anal. 2 (4), 220. https://doi.org/10.11648/j.ijema.20140204.15. 2328-7659.

Dong, B., Chen, H., Yang, Y., He, Q., Dai, X., 2014. Treatment of printing and dyeing wastewater using MBBR followed by membrane separation process. Desalin. Water Treat. 52 (22–24), 4562–4567. 19443986. Taylor and Francis Inc., China. Available from: http://www.tandfonline.com/toc/tdwt20/current. https://doi.org/10.1080/19443994.2013.803780.

Du, B., Yang, Q., Li, X., Yuan, W., Chen, Y., Wang, R., 2019. Impacts of long-term exposure to tetracycline and sulfamethoxazole on the sludge granules in an anoxic-aerobic wastewater treatment system. Sci. Total Environ. 684, 67–77. 18791026. Elsevier B.V., China. Available from: www.elsevier.com/locate/scitotenv. https://doi.org/10.1016/j.scitotenv.2019.05.313.

Dulkadiroglu, H., Cokgor, E.U., Artan, N., Orhon, D., 2005. The effect of temperature and sludge age on COD removal and nitrification in a moving bed sequencing batch biofilm reactor. Water Sci. Technol. 51 (11), 95–103. https://doi.org/10.2166/wst.2005.0395. 0273-1223.

Elnmer, A., Khadr, M., Kanae, S., Tawfik, A., 2019. Mapping daily and seasonally evapotranspiration using remote sensing techniques over the Nile delta. Agric. Water Manag. 213, 682–692. https://doi.org/10.1016/j.agwat.2018.11.009. 03783774.

Elsamadony, M., Guo, Y., 2019. Enrich waste activated sludge digestibility via natural enzyme supplementation. E3S Web Conf. 83. https://doi.org/10.1051/e3sconf/20198301012. 01012. 2267-1242.

Elsamadony, M., Tawfik, A., Danial, A., Suzuki, M., 2015. Use of Carica papaya enzymes for enhancement of H2 production and degradation of glucose, protein, and lipids. Energy Procedia 75, 975–980. https://doi.org/10.1016/j.egypro.2015.07.308. 18766102.

Elsamadony, M., Elreedy, A., Mostafa, A., Fujii, M., Gescher, J., Shakeri Yekta, S., Schnürer, A., Gaillard, J.F., Pant, D., 2021. Perspectives on potential applications of nanometal derivatives in gaseous bioenergy pathways: mechanisms, life cycle, and toxicity. ACS Sustain. Chem. Eng. 9 (29), 9563–9589. 21680485.

American Chemical Society, Japan. Available from: http://pubs.acs.org/journal/ascecg. https://doi.org/10.1021/acssuschemeng.1c02260.

Elsamadony, M., Fujii, M., Ryo, M., Nerini, F.F., Kakinuma, K., Kanae, S., 2022. Preliminary quantitative assessment of the multidimensional impact of the COVID-19 pandemic on Sustainable Development Goals. J. Clean. Prod. 372. 09596526. Elsevier Ltd, Japan. Available from: https://www.journals.elsevier.com/journal-of-cleaner-production. https://doi.org/10.1016/j.jclepro.2022.133812.

Elsharkawy, K., Elsamadony, M., Afify, H., 2019. EDP sciences Egypt comparative analysis of common full scale reactors for dry anaerobic digestion process. E3S Web Conf. 83. https://doi.org/10.1051/e3sconf/20198301011. 22671242 www.e3s-conferences.org/.

Elsharkawy, K., Gar Alalm, M., Fujii, M., Afify, H., Tawfik, A., Elsamadony, M., 2020. Paperboard mill wastewater treatment via combined dark and LED-mediated fermentation in the absence of external chemical addition. Bioresour. Technol. 295. 18732976. Elsevier Ltd, Egypt. Available from: www.elsevier.com/locate/biortech. https://doi.org/10.1016/j.biortech.2019.122312.

Elsheikh, M.A., Saleh, H.I., Rashwan, I.M., El-Samadoni, M.M., 2013. Hydraulic modelling of water supply distribution for improving its quantity and quality. Sustainable Environ. Res. 23 (6), 403–411. 24682039. Chinese Institute of Environmental Engineering, Egypt. Available from: https://link.springer.com/journal/42834.

Forrest, D., Delatolla, R., Kennedy, K., 2016. Carrier effects on tertiary nitrifying moving bed biofilm reactor: an examination of performance, biofilm and biologically produced solids. Environ. Technol. 37 (6), 662–671. 1479487X. Taylor and Francis Ltd., Canada. Available from: http://www.tandf.co.uk/journals/titles/09593330.asp. https://doi.org/10.1080/09593330.2015.1077272.

Fu, H.-M., Wang, J., Ren, H., Ding, L., 2022. Acceleration of start-up of moving bed biofilm reactor at low temperature by adding specialized quorum sensing bacteria. Bioresour. Technol. 358, 127249. https://doi.org/10.1016/j.biortech.2022.127249. 09608524.

Herlambang, A., Murwantoko, M., Istiqomah, I., 2021. Dynamic change in bacterial communities in the integrated rice–fish farming system in Sleman, Yogyakarta, Indonesia. Aquac. Res. 52 (11), 5566–5578. 13652109. John Wiley and Sons Inc, Indonesia. Available from: http://onlinelibrary.wiley.com/journal/10.1111/(ISSN)1365-2109. https://doi.org/10.1111/are.15432.

Hetz, S.A., Horn, M.A., 2021. Burkholderiaceae are key acetate assimilators during complete denitrification in acidic cryoturbated peat circles of the Arctic Tundra. Front. Microbiol. 12. 1664302X. Frontiers Media S.A., Germany. Available from: https://www.frontiersin.org/journals/microbiology. https://doi.org/10.3389/fmicb.2021.628269.

Hoang, V., Delatolla, R., Laflamme, E., Gadbois, A., 2014. An investigation of moving bed biofilm reactor nitrification during long-term exposure to cold temperatures. Water Environ. Res. 86 (1), 36–42. 15547531. Water Environment Federation, Canada. Available from: http://docserver.ingentaconnect.com/deliver/connect/wef/10614303/v86n1/s5.pdf?expires=1389829413%26id=76898304%26titleid=11548%26accname=Elsevier+BV%26checksum=B25D4A8A003D9EAA33E26074DD6C1525. https://doi.org/10.2175/106143013X13807328848694.

Hu, H., He, J., Yu, H., Liu, J., Zhang, J., 2017. A strategy to speed up formation and strengthen activity of biofilms at low temperature. RSC Adv. 7 (37), 22788–22796. 20462069. Royal Society of Chemistry, China. Available from: http://pubs.rsc.org/en/journals/journalissues. https://doi.org/10.1039/c7ra02223a.

Hu, Z., Liu, T., Zicheng, S., Zhao, J., Guo, J., Shihu, H., Yuan, Z., Zheng, M., 2023. Adaptation of anammox process for nitrogen removal from acidic nitritation effluent in a low pH moving bed biofilm reactor. Water Res. 243, 120370. https://doi.org/10.1016/j.watres.2023.120370. 00431354.

Huang, H., Lin, Y., Peng, P., Geng, J., Xu, K., Zhang, Y., Ding, L., Ren, H., 2018. Calcium ion- and rhamnolipid-mediated deposition of soluble matters on biocarriers. Water Res. 133, 37–46. 18792448. Elsevier Ltd, China. Available from: www.elsevier.com/locate/watres. https://doi.org/10.1016/j.watres.2018.01.010.

Iannacone, F., Di Capua, F., Granata, F., Gargano, R., Esposito, G., 2020. Simultaneous nitrification, denitrification and phosphorus removal in a continuous-flow moving bed biofilm reactor alternating microaerobic and aerobic conditions. Bioresour. Technol. 310, 123453. https://doi.org/10.1016/j.biortech.2020.123453. 09608524.

Ismail, S., Tawfik, A., 2016a. Performance of passive aerated immobilized biomass reactor coupled with Fenton process for treatment of landfill leachate. Int. Biodeterior. Biodegrad. 111, 22–30. 09648305. Elsevier Ltd, Egypt. Available from: www.elsevier.com/inca/publications/store/4/0/5/8/9/9. https://doi.org/10.1016/j.ibiod.2016.04.010.

Ismail, S., Tawfik, A., 2016b. Treatment of hazardous landfill leachate using Fenton process followed by a combined (UASB/DHS) system. Water Sci. Technol. 73 (7), 1700–1708. 02731223. IWA Publishing, Egypt. Available from: http://wst.iwaponline.com/content/ppiwawst/73/7/1700.full.pdf. https://doi.org/10.2166/wst.2015.655.

Ismail, S., Elreedy, A., Fujii, M., Ni, S.Q., Tawfik, A., Elsamadony, M., 2021a. Fatigue of anammox consortia under long-term 1,4-dioxane exposure and recovery potential: N-kinetics and microbial dynamics. J. Hazard. Mater. 414. 18733336. Elsevier B.V., China. Available from: www.elsevier.com/locate/jhazmat. https://doi.org/10.1016/j.jhazmat.2021.125533.

Ismail, S., Elreedy, A., Elsamadony, M., Abdelrazek, E., Fujii, M., Ni, S.-Q., Tawfik, A., 2021b. Response of anammox bacteria to short-term exposure of 1,4-dioxane: bacterial activity and community dynamics. Sep. Purif. Technol. 266, 118539. https://doi.org/10.1016/j.seppur.2021.118539. 13835866.

Ismail, S., Elsamadony, M., Abdalla, M., Ni, S.-Q., Tawfik, A., 2022. Stimulating the fermentation process of industrial food waste via nonionic surfactant/graphene nanosheet combined supplementation. ACS ES&T Eng. 2 (11), 2043–2057. https://doi.org/10.1021/acsestengg.2c00140. 2690-0645.

Jabornig, S., Favero, E., 2013. Single household greywater treatment with a moving bed biofilm membrane reactor (MBBMR). J. Membr. Sci. 446, 277–285. https://doi.org/10.1016/j.memsci.2013.06.049. 18733123.

Jalali, F., Zinatizadeh, A.A., Asadi, A., Zinadini, S., 2023. A moving bed biofilm reactor coupled with an upgraded nanocomposite polyvinylidene fluoride membrane to treat an industrial estate wastewater. Chem. Eng. J. 470, 144128. https://doi.org/10.1016/j.cej.2023.144128. 13858947.

Javid, A.H., Hassani, A.H., Ghanbari, B., Yaghmaeian, K., 2013. Feasibility of utilizing moving bed biofilm reactor to upgrade and retrofit municipal wastewater treatment plants. Int. J. Environ. Res. 7 (4), 963–972. 17356865. Iran. Available from: http://ijer.ut.ac.ir/?_action=showPDF%26article=679%26_ob=de8379c10c19f7bcb7ef510f72e8064f%26fileName=full_text.pdf.

Kamilya, T., Majumder, A., Saidulu, D., Tripathy, S., Gupta, A.K., 2023. Optimization of a continuous hybrid moving bed biofilm reactor and constructed wetland system for the treatment of paracetamol-spiked domestic wastewater. Chem. Eng. J. 477, 147139. https://doi.org/10.1016/j.cej.2023.147139. 13858947.

Kang, A.J., Brown, A.K., Wong, C.S., Yuan, Q., 2018. Removal of antibiotic sulfamethoxazole by anoxic/anaerobic/oxic granular and suspended activated sludge processes. Bioresour. Technol. 251, 151–157. 18732976. Elsevier Ltd, Canada. Available from: www.elsevier.com/locate/biortech. https://doi.org/10.1016/j.biortech.2017.12.021.

Kermani, M., Bina, B., Movahedian, H., Amin, M.M., Nikaein, M., 2008. Application of moving bed biofilm process for biological organics and nutrients from municipal wastewater. Am. J. Environ. Sci. 4 (6), 675–682. https://doi.org/10.3844/ajessp.2008.675.682. 1553345X.

Khunjar, W.O., MacKintosh, S.A., Skotnicka-Pitak, J., Baik, S., Aga, D.S., Love, N.G., 2011. Elucidating the relative roles of ammonia oxidizing and heterotrophic bacteria during the biotransformation of 17α-ethinylestradiol and trimethoprim. Environ. Sci. Technol. 45 (8), 3605–3612. https://doi.org/10.1021/es1037035. 15205851.

Kim, Y.H., Cho, J.H., Lee, Y.W., Lee, W.K., 1997. Development of a carrier for adhesion of nitrifying bacteria using a thermodynamic approach. Biotechnol. Tech. 11 (11), 773–776. https://doi.org/10.1023/A:1018460805328. 0951208X.

Kowalski, M.S., Devlin, T.R., di Biase, A., Oleszkiewicz, J.A., 2019. Controlling cold temperature partial nitritation in moving bed biofilm reactor. Chemosphere 227, 216–224. 18791298. Elsevier Ltd, Canada. Available from: www.elsevier.com/locate/chemosphere. https://doi.org/10.1016/j.chemosphere.2019.04.025.

Kristensen, J.M., Singleton, C., Clegg, L.A., Petriglieri, F., Nielsen, P.H., 2021. High diversity and functional potential of undescribed "Acidobacteriota" in Danish Wastewater Treatment Plants. Front. Microbiol. 12. 1664302X. Frontiers Media S.A., Denmark. Available from: https://www.frontiersin.org/journals/microbiology#. https://doi.org/10.3389/fmicb.2021.643950.

Leyva-Díaz, J.C., Calderón, K., Rodríguez, F.A., González-López, J., Hontoria, E., Poyatos, J.M., 2013. Comparative kinetic study between moving bed biofilm reactor-membrane bioreactor and membrane bioreactor systems and their influence on organic matter and nutrients removal. Biochem. Eng. J. 77, 28–40. https://doi.org/10.1016/j.bej.2013.04.023. 1369703X.

Liang, D., Yongyou, H., Huang, R., Cheng, J., Chen, Y., 2022. Effects of various antibiotics on aerobic nitrogen removal and antibiotic degradation performance: mechanism, degradation pathways, and microbial community evolution. J. Hazard. Mater. 422, 126818. https://doi.org/10.1016/j.jhazmat.2021.126818. 03043894.

Liu, Z., Zeng, Z., Zeng, G., Li, J., Zhong, H., Yuan, X., Liu, Y., Zhang, J., Chen, M., Liu, Y., Xie, G., 2012. Influence of rhamnolipids and Triton X-100 on adsorption of phenol by *Penicillium simplicissimum*. Bioresour. Technol. 110, 468–473. https://doi.org/10.1016/j.biortech.2012.01.092. 18732976.

Liu, Y., Li, J., Guo, W., Ngo, H.H., Hu, J., Gao, M.t., 2018. Use of magnetic powder to effectively improve the performance of sequencing batch reactors (SBRs) in municipal wastewater treatment. Bioresour. Technol. 248, 135–139. 18732976. Elsevier Ltd, China. Available from: www.elsevier.com/locate/biortech. https://doi.org/10.1016/j.biortech.2017.06.069.

Liu, M., Meng, P., Chen, G., Guan, Y., Liu, G., 2022. Correlation of structural extracellular polymeric substances in the mesh biofilms with solids retention time and biofilm hydraulic resistance in dynamic membrane bioreactors. Sci. Total Environ. 832, 155000. https://doi.org/10.1016/j.scitotenv.2022.155000. 00489697.

Lotti, T., Kleerebezem, R., Lubello, C., van Loosdrecht, M.C.M., 2014. Physiological and kinetic characterization of a suspended cell anammox culture. Water Res. 60, 1–14. https://doi.org/10.1016/j.watres.2014.04.017. 00431354.

Luan, Y.-n., Yin, Y., Guo, Z., Yang, J., Wang, G., Zhang, F., Xiao, Y., Liu, C., 2024. Achieving simultaneous nitrification and endogenous denitrifying phosphorus removal in anaerobic/intermittently-aerated moving bed biofilm reactor for low carbon-to-nitrogen ratio wastewater treatment. Bioresour. Technol. 394, 130178. https://doi.org/10.1016/j.biortech.2023.130178. 09608524.

Luostarinen, S., Luste, S., Valentín, L., Rintala, J., 2006. Nitrogen removal from on-site treated anaerobic effluents using intermittently aerated moving bed biofilm reactors at low temperatures. Water Res. 40 (8), 1607–1615. https://doi.org/10.1016/j.watres.2006.02.022. 00431354.

Mahmoud, M., Tawfik, A., Samhan, F., El-Gohary, F., 2009. Sewage treatment using an integrated system consisting of anaerobic hybrid reactor (AHR) and downflow hanging sponge (DHS). Desalin. Water Treat. 4 (1–3), 168–176. https://doi.org/10.5004/dwt.2009.372. 19443986.

Mannina, G., Capodici, M., Cosenza, A., Di Trapani, D., Ekama, G.A., 2018. The effect of the solids and hydraulic retention time on moving bed membrane bioreactor performance. J. Clean. Prod. 170, 1305–1315. https://doi.org/10.1016/j.jclepro.2017.09.200. 09596526.

Mostafa, A., Elsamadony, M., El-Dissouky, A., Elhusseiny, A., Tawfik, A., 2017. Biological H2 potential harvested from complex gelatinaceous wastewater via attached versus suspended growth culture anaerobes. Bioresour. Technol. 231, 9–18. 18732976. Elsevier Ltd, Egypt. Available from: www.elsevier.com/locate/biortech. https://doi.org/10.1016/j.biortech.2017.01.062.

Mostafa, A., Tolba, A., Gar Alalm, M., Fujii, M., Afify, H., Elsamadony, M., 2020. Application of magnetic multi-wall carbon nanotube composite into fermentative treatment process of ultrasonicated waste activated sludge. Bioresour. Technol. 306. 18732976. Elsevier Ltd, South Korea. Available from: www.elsevier.com/locate/biortech. https://doi.org/10.1016/j.biortech.2020.123186.

Mostafa, A., Elsamadony, M., Khalil, T.E., Elhusseiny, A.F., Tawfik, A., Fujii, M., El-dissouky, A., Pant, D., 2023. Bioelectrochemical system for enhancing anaerobic digestion of pharmaceutical-containing domestic wastewater. Chemosphere 339. 18791298. Elsevier Ltd, Egypt. Available from: www.elsevier.com/locate/chemosphere. https://doi.org/10.1016/j.chemosphere.2023.139766.

Ødegaard, H., 2016. A road-map for energy-neutral wastewater treatment plants of the future based on compact technologies (including MBBR). Front. Environ. Sci. Eng. 10 (4). 2095221X. Higher Education Press, Norway. Available from: http://www.springerlink.com/content/2095-2201/. https://doi.org/10.1007/s11783-016-0835-0.

Osmani, S.A., Rajpal, A., Kazmi, A.A., 2021. Upgradation of conventional MBBR into aerobic/anoxic/aerobic configuration: a case study of carbon and nitrogen removal based sewage treatment plant. J. Water Process Eng. 40, 101921. https://doi.org/10.1016/j.jwpe.2021.101921. 22147144.

Peng, C., Fan, X., Yujin, X., Ren, H., Huang, H., 2021. Microscopic analysis towards rhamnolipid-mediated adhesion of *Thiobacillus denitrificans*: a QCM-D study. Chemosphere 271, 129539. https://doi.org/10.1016/j.chemosphere.2021.129539. 00456535.

Persson, F., Sultana, R., Suarez, C., Hermansson, M., Plaza, E., Wilén, B.M., 2014. Structure and composition of biofilm communities in a moving bed biofilm reactor for nitritation-anammox at low temperatures. Bioresour. Technol. 154, 267–273. 18732976. Elsevier Ltd, Sweden. Available from: www.elsevier.com/locate/biortech. https://doi.org/10.1016/j.biortech.2013.12.062.

Pradhan, N., Swa Thi, S., Wuertz, S., 2020. Inhibition factors and kinetic model for anaerobic ammonia oxidation in a granular sludge bioreactor with Candidatus Brocadia. Chem. Eng. J. 389, 123618. https://doi.org/10.1016/j.cej.2019.123618. 13858947.

Ren, L., Chen, M., Zheng, J., Li, Z., Tian, C., Wang, Q., Wang, Z., 2021. Efficacy of a novel electrochemical membrane-aerated biofilm reactor for removal of antibiotics from micro-polluted surface water and suppression of antibiotic resistance genes. Bioresour. Technol. 338, 125527. https://doi.org/10.1016/j.biortech.2021.125527. 09608524.

Saidulu, D., Gupta, B., Gupta, A.K., Ghosal, P.S., 2021. A review on occurrences, ecotoxic effects, and remediation of emerging contaminants from wastewater: special emphasis on biological treatment based hybrid systems. J. Environ. Chem. Eng. 9 (4). 22133437. Elsevier Ltd, India. Available from: http://www.journals.elsevier.com/journal-of-environmental-chemical-engineering/. https://doi.org/10.1016/j.jece.2021.105282.

Saidulu, D., Agrawal, S., Bhatnagar, A., Gupta, A.K., 2024. Sulfamethoxazole removal from wastewater via anoxic/oxic moving bed biofilm reactor: degradation pathways and toxicity assessment. Bioresour. Technol. 392, 129998. https://doi.org/10.1016/j.biortech.2023.129998. 09608524.

Salvetti, R., Azzellino, A., Canziani, R., Bonomo, L., 2006. Effects of temperature on tertiary nitrification in moving-bed biofilm reactors. Water Res. 40 (15), 2981–2993. Elsevier Ltd, Italy. Available from: https://www.sciencedirect.com/science/journal/00431354. https://doi.org/10.1016/j.watres.2006.05.013. 00431354.

Shao, B., Liu, Z., Zhong, H., Zeng, G., Liu, G., Yu, M., Liu, Y., Yang, X., Li, Z., Fang, Z., Zhang, J., Zhao, C., 2017. Effects of rhamnolipids on microorganism characteristics and applications in composting: a review. Microbiol. Res. 200, 33–44. 09445013. Elsevier GmbH, China. Available from: www.urbanfischer.de/journals/microbiolres/microbio.htm. https://doi.org/10.1016/j.micres.2017.04.005.

Shore, J.L., M'Coy, W.S., Gunsch, C.K., Deshusses, M.A., 2012. Application of a moving bed biofilm reactor for tertiary ammonia treatment in high temperature industrial wastewater. Bioresour. Technol. 112, 51–60. https://doi.org/10.1016/j.biortech.2012.02.045. 18732976.

Sokół, W., 2003. Treatment of refinery wastewater in a three-phase fluidised bed bioreactor with a low density biomass support. Biochem. Eng. J. 15 (1), 1–10. https://doi.org/10.1016/s1369-703x(02)00174-2. 1369703X.

Soltan, M., Elsamadony, M., Mostafa, A., Awad, H., Tawfik, A., 2019. Harvesting zero waste from co-digested fruit and vegetable peels via integrated fermentation and pyrolysis processes. Environ. Sci. Pollut. Res. 26 (10), 10429–10438. 16147499. Springer Verlag, Egypt. Available from: http://www.springerlink.com/content/0944-1344. https://doi.org/10.1007/s11356-019-04647-8.

Song, Z., Liao, R., Wang, M., Zhang, J., Dong, Z., Wen, Z., Dong, W., Sun, F., 2022. Biosurfactant-driven synergic roles to accelerate biofilm formation in an anaerobic moving bed biofilm reactor (MBBR): a comprehensive analysis on meso and micro scale. J. Clean. Prod. 373, 133930. https://doi.org/10.1016/j.jclepro.2022.133930. 09596526.

Sousa, M., Azeredo, J., Feijó, J., Oliveira, R., 1997. Polymeric supports for the adhesion of a consortium of autotrophic nitrifying bacteria. Biotechnol. Tech. 11 (10), 751–754. https://doi.org/10.1023/A:1018400619440. 0951208X.

Sun, H., Liu, H., Zhang, M., Liu, Y., 2021. A novel single-stage ceramic membrane moving bed biofilm reactor coupled with reverse osmosis for reclamation of municipal wastewater to NEWater-like product water. Chemosphere 268, 128836. https://doi.org/10.1016/j.chemosphere.2020.128836. 00456535.

Tandukar, M., Ohashi, A., Harada, H., 2007. Performance comparison of a pilot-scale UASB and DHS system and activated sludge process for the treatment of municipal wastewater. Water Res. 41 (12), 2697–2705. 00431354. Elsevier Ltd., Japan. Available from: https://www.sciencedirect.com/science/journal/00431354. https://doi.org/10.1016/j.watres.2007.02.027.

Tawfik, A., El-Qelish, M., 2014. Key factors affecting on bio-hydrogen production from co-digestion of organic fraction of municipal solid waste and kitchen wastewater. Bioresour. Technol. 168, 106–111. 18732976. Elsevier Ltd., Egypt. Available from: www.elsevier.com/locate/biortech. https://doi.org/10.1016/j.biortech.2014.02.127.

Tawfik, A., Elsamadony, M., 2018. Development of dry anaerobic technologies of bio-waste and unlock the barriers for valorization. In: Optimization and Applicability of Bioprocesses. Springer Singapore, Egypt, pp. 267–282, https://doi.org/10.1007/978-981-10-6863-8_13. Available from: http://www.springer.com/in/book/9789811068621.

Tawfik, A., Klapwijk, A., 2010. Polyurethane rotating disc system for post-treatment of anaerobically pre-treated sewage. J. Environ. Manag. 91 (5), 1183–1192. https://doi.org/10.1016/j.jenvman.2010.01.010. 03014797.

Tawfik, A., Salem, A., 2012. The effect of organic loading rate on bio-hydrogen production from pre-treated rice straw waste via mesophilic up-flow anaerobic reactor. Bioresour. Technol. 107, 186–190. https://doi.org/10.1016/j.biortech.2011.11.086. 09608524.

Tawfik, A., Klapwijk, B., Van Buuren, J., El-Gohary, F., Lettinga, G., 2004. Physico-chemical factors affecting the *E. coli* removal in a rotating biological contactor (RBC) treating UASB effluent. Water Res. 38 (5), 1081–1088. 00431354. Elsevier Ltd., Egypt. Available from: https://www.sciencedirect.com/science/journal/00431354. https://doi.org/10.1016/S0043-1354(03)00345-2.

Tawfik, A., Temmink, H., Zeeman, G., Klapwijk, B., 2006a. Sewage treatment in a rotating biological contactor (RBC) system. Water Air Soil Pollut. 0049-6979. 175 (1–4), 275–289. https://doi.org/10.1007/s11270-006-9138-6.

Tawfik, A., El-Gohary, F., Ohashi, A., Harada, H., 2006b. The influence of physical–chemical and biological factors on the removal of faecal coliform through down-flow hanging sponge (DHS) system treating UASB reactor effluent. Water Res. 40 (9), 1877–1883. https://doi.org/10.1016/j.watres.2006.02.038. 00431354.

Tawfik, A., El-Gohary, F., Temmink, H., 2010. Treatment of domestic wastewater in an up-flow anaerobic sludge blanket reactor followed by moving bed biofilm reactor. Bioprocess Biosyst. Eng. 33 (2), 267–276. https://doi.org/10.1007/s00449-009-0321-1. 1615-7591.

Tawfik, A., Wahab, R.A., Al-Asmer, A., Matary, F., 2011. Effect of hydraulic retention time on the performance of down-flow hanging sponge system treating grey wastewater. Bioprocess Biosyst. Eng. 34 (6), 767–776. 16157605. Springer Verlag, Egypt. Available from: https://rd.springer.com/journal/449. https://doi.org/10.1007/s00449-011-0528-9.

Tawfik, A., Badr, N., Taleb, E.A., El-Senousy, W., 2012. Sewage treatment in an up-flow anaerobic sponge reactor followed by moving bed biofilm reactor based on polyurethane carrier material. Desalin. Water Treat. 37 (1–3), 350–358. 19443986. Taylor and Francis Inc., Egypt. Available from: http://www.deswater.com/readpaper.php?id=1616. https://doi.org/10.1080/19443994.2012.661291.

Tawfik, A., Hassan, G.K., Awad, H., Hassan, M., Rojas, P., Sanz, J.L., Elsamadony, M., Pant, D., Fujii, M., 2021. Strengthen “the sustainable farm” concept via efficacious conversion of farm wastes into methane. Bioresour. Technol. 341. 18732976. Elsevier Ltd., Egypt. Available from: www.elsevier.com/locate/biortech. https://doi.org/10.1016/j.biortech.2021.125838.

Tawfik, A., Al-Sayed, A., Hassan, G.K., Nasr, M., El-Shafai, S.A., Alhajeri, N.S., Khan, M.S., Akhtar, M.S., Ahmad, Z., Rojas, P., Sanz, J.L., 2022a. Electron donor addition for stimulating the microbial degradation of 1,4 dioxane by sequential batch membrane bioreactor: a techno-economic approach. Chemosphere 306. 18791298. Elsevier Ltd., Egypt. Available from: www.elsevier.com/locate/chemosphere. https://doi.org/10.1016/j.chemosphere.2022.135580.

Tawfik, A., Hasanan, K., Abdullah, M., Badr, O.A., Awad, H.M., Elsamadony, M., El-Dissouky, A., Qyyum, M.A., Nizami, A.S., 2022b. Graphene enhanced detoxification of wastewater rich 4-nitrophenol in multistage anaerobic reactor followed by baffled high-rate algal pond. J. Hazard. Mater. 424. 18733336. Elsevier B.V., Egypt. Available from: www.elsevier.com/locate/jhazmat. https://doi.org/10.1016/j.jhazmat.2021.127395.

Tawfik, A., Tan, X., Elsamadony, M., Qyyum, M.A., Azzam, A.M., Mubashir, M., Ng, H.S., Akhtar, M.S., Khoo, K.S., 2022c. Graphene/hydroxyapatite nano-composite for enhancement of hydrogen productivity from delignified duckweed. Fuel 330. 00162361. Elsevier Ltd., Egypt. Available from: http://www.journals.elsevier.com/fuel/. https://doi.org/10.1016/j.fuel.2022.125537.

Teixeira, P., Oliveira, R., 1999. Influence of surface characteristics on the adhesion of *Alcaligenes denitrificans* to polymeric substrates. J. Adhes. Sci. Technol. 13 (11), 1287–1294. https://doi.org/10.1163/156856199x00190. 0169-4243.

Tian, H., Li, Y., Chen, H., Zhang, J., Hui, M., Xu, X., Qingxian, S., Smets, B.F., 2023. Aerobic biodegradation of quinoline under denitrifying conditions in membrane-aerated biofilm reactor. Environ. Pollut. 326, 121507. https://doi.org/10.1016/j.envpol.2023.121507. 02697491.

Tong, W.S., Oo, C.W., Sudesh, K., Ng, S.L., 2024. Decolorization of acid orange 7 in moving bed biofilm reactor packed with polyhydroxyalkanoate/activated carbon. J. Water Process Eng. 58. 22147144. Elsevier Ltd., Malaysia. Available from: https://www.sciencedirect.com/science/journal/22147144. https://doi.org/10.1016/j.jwpe.2023.104750.

Torresi, E., Fowler, S.J., Polesel, F., Bester, K., Andersen, H.R., Smets, B.F., Plósz, B.G., Christensson, M., 2016. Biofilm thickness influences biodiversity in nitrifying MBBRs—implications on micropollutant removal. Environ. Sci. Technol. 50 (17), 9279–9288. 15205851. American Chemical Society, Denmark. Available from: http://pubs.acs.org/journal/esthag. https://doi.org/10.1021/acs.est.6b02007.

Van Tendeloo, M., Xie, Y., Van Beeck, W., Zhu, W., Lebeer, S., Vlaeminck, S.E., 2021. Oxygen control and stressor treatments for complete and long-term suppression of nitrite-oxidizing bacteria in biofilm-based partial nitritation/anammox. Bioresour. Technol. 342, 125996. https://doi.org/10.1016/j.biortech.2021.125996. 09608524.

Wang, X.J., Xia, S.Q., Chen, L., Zhao, J.F., Renault, N.J., Chovelon, J.M., 2006. Nutrients removal from municipal wastewater by chemical precipitation in a moving bed biofilm reactor. Process Biochem. 41 (4), 824–828. https://doi.org/10.1016/j.procbio.2005.10.015. 13595113.

Wang, X., Bi, X., Hem, L.J., Ratnaweera, H., 2018. Microbial community composition of a multi-stage moving bed biofilm reactor and its interaction with kinetic model parameters estimation. J. Environ. Manag. 218, 340–347. 10958630. Academic Press, Norway. Available from: https://www.sciencedirect.com/journal/journal-of-environmental-management. https://doi.org/10.1016/j.jenvman.2018.04.015.

Wang, B., Li, H., Liu, T., Guo, J., 2021. Enhanced removal of cephalexin and sulfadiazine in nitrifying membrane-aerated biofilm reactors. Chemosphere 263, 128224. https://doi.org/10.1016/j.chemosphere.2020.128224. 00456535.

Wazeri, A., Elsamadony, M., Tawfik, A., 2018. Carbon emissions reduction by catalyzing H2 gas harvested from water hyacinth fermentation process using metallic salts. Energy Procedia 152, 1254–1259. https://doi.org/10.1016/j.

egypro.2018.09.178. 18766102, Elsevier Ltd. Egypt. Available from: http://www.sciencedirect.com/science/journal/18766102.

Xie, S., Zhao, J., Zhang, Q., Zhao, J., Lei, S., Ma, X., Yan, C., 2021. Improvement of the performance of simultaneous nitrification denitrification and phosphorus removal (SNDPR) system by nitrite stress. Sci. Total Environ. 788, 147825. https://doi.org/10.1016/j.scitotenv.2021.147825. 00489697.

Xu, Y., Yuan, Z., Ni, B.J., 2016. Biotransformation of pharmaceuticals by ammonia oxidizing bacteria in wastewater treatment processes. Sci. Total Environ. 566-567, 796–805. 18791026. Elsevier B.V., Australia. Available from: www.elsevier.com/locate/scitotenv. https://doi.org/10.1016/j.scitotenv.2016.05.118.

Xu, Y., Ying, G., Peng, L., Wang, N., Chen, S., Liang, C., Liu, Y., Ni, B.-J., 2023. Unravelling ciprofloxacin removal in a nitrifying moving bed biofilm reactor: biodegradation mechanisms and pathways. Chemosphere 320, 138099. https://doi.org/10.1016/j.chemosphere.2023.138099. 00456535.

Yan, X., Zhu, C., Huang, B., Yan, Q., Zhang, G., 2018. Enhanced nitrogen removal from electroplating tail wastewater through two-staged anoxic-oxic (A/O) process. Bioresour. Technol. 247, 157–164. Elsevier Ltd., China. Available from: www.elsevier.com/locate/biortech. https://doi.org/10.1016/j.biortech.2017.09.084. 18732976.

Young, B., Delatolla, R., Kennedy, K., Laflamme, E., Stintzi, A., 2017. Low temperature MBBR nitrification: microbiome analysis. Water Res. 111, 224–233. 18792448. Elsevier Ltd., Canada. Available from: www.elsevier.com/locate/watres. https://doi.org/10.1016/j.watres.2016.12.050.

Yuan, Q., Wang, H., Hang, Q., Deng, Y., Liu, K., Li, C., Zheng, S., 2015. Comparison of the MBBR denitrification carriers for advanced nitrogen removal of wastewater treatment plant effluent. Environ. Sci. Pollut. Res. 22 (18), 13970–13979. 16147499. Springer Verlag, China. Available from: http://www.springerlink.com/content/0944-1344. https://doi.org/10.1007/s11356-015-4546-z.

Yuan, C., Wang, B., Peng, Y., Li, X., Zhang, Q., Hu, T., 2020. Enhanced nutrient removal of simultaneous partial nitrification, denitrification and phosphorus removal (SPNDPR) in a single-stage anaerobic/micro-aerobic sequencing batch reactor for treating real sewage with low carbon/nitrogen. Chemosphere 257, 127097. https://doi.org/10.1016/j.chemosphere.2020.127097. 00456535.

Yuan, S., Guo, S., Tan, Y., Li, M., Lu, Y., Xu, R., Tawfik, A., Zhou, Z., Chen, J., Liu, W., Meng, F., 2023. Deciphering community assembly and succession in sequencing batch moving bed biofilm reactor: differentiation between attached and suspended communities. Sci. Total Environ. 873. 18791026, Elsevier B.V., China. Available from: www.elsevier.com/locate/scitotenv. https://doi.org/10.1016/j.scitotenv.2023.162448.

Zaman, M., Kim, M., Nakhla, G., 2021. Simultaneous nitrification-denitrifying phosphorus removal (SNDPR) at low DO for treating carbon-limited municipal wastewater. Sci. Total Environ. 760, 143387. https://doi.org/10.1016/j.scitotenv.2020.143387. 00489697.

Zeng, Z., Liu, Y., Zhong, H., Xiao, R., Zeng, G., Liu, Z., Cheng, M., Lai, C., Zhang, C., Liu, G., Qin, L., 2018. Mechanisms for rhamnolipids-mediated biodegradation of hydrophobic organic compounds. Sci. Total Environ. 634, 1–11. 18791026. Elsevier B.V., China. Available from: www.elsevier.com/locate/scitotenv. https://doi.org/10.1016/j.scitotenv.2018.03.349.

Zhang, Y., Jiang, J., Zhao, Q., Gao, Y.Z., Wang, K., Ding, J., Yu, H., Yao, Y., 2017. Accelerating anodic biofilms formation and electron transfer in microbial fuel cells: role of anionic biosurfactants and mechanism. Bioelectrochemistry 117, 48–56. 1878562X. Elsevier B.V., China. Available from: www.elsevier.com/locate/bioelechem. https://doi.org/10.1016/j.bioelechem.2017.06.002.

Zhang, J., Liang, Z., Miao, Y., Sun, Y., Zhang, Q., Wu, L., Peng, Y., 2019. Enhancing sewage nitrogen removal via anammox and endogenous denitrification:

significance of anaerobic/oxic/anoxic operation mode. Bioresour. Technol. 289, 121665. https://doi.org/10.1016/j.biortech.2019.121665. 09608524.

Zhang, Z., Zhang, Y., Chen, Y., 2020. Recent advances in partial denitrification in biological nitrogen removal: from enrichment to application. Bioresour. Technol. 298. 18732976. Elsevier Ltd., China. Available from: www.elsevier.com/locate/biortech. https://doi.org/10.1016/j.biortech.2019.122444.

Zhang, X., Feng, C., Xu, Z., Yang, W., Tong, K., Wang, Y., Liu, X., 2023. Coupling of partial nitrification and aerated vertical flow constructed wetland for enhancing nitrite removal and reducing nitrous oxide. J. Environ. Chem. Eng. 11 (1), 109114. https://doi.org/10.1016/j.jece.2022.109114. 22133437.

Zhang, Q., Liu, H., Liang, S., Wang, C., Tan, S., Yang, C., Qin, S., Long, K., 2024. Comparison of moving bed biofilm reactor and bio-contact oxidation reactor start-up with heterotrophic nitrification-aerobic denitrification bacteria and activated sludge inoculation under high ammonia nitrogen conditions. Bioresour. Technol. 395, 130408. https://doi.org/10.1016/j.biortech.2024.130408. 09608524.

Zhao, W., Peng, Y., Wang, M., Yu, H., Li, X., 2019. Nutrient removal and microbial community structure variation in the two-sludge system treating low carbon/nitrogen domestic wastewater. Bioresour. Technol. 294, 122161. https://doi.org/10.1016/j.biortech.2019.122161. 09608524.

Zhao, J., Ni, G., Piculell, M., Li, J., Hu, Z., Wang, Z., Guo, J., Yuan, Z., Zheng, M., Hu, S., 2022. Characterizing and comparing microbial community and biofilm structure in three nitrifying moving bed biofilm reactors. J. Environ. Manag. 320, 115883. https://doi.org/10.1016/j.jenvman.2022.115883. 03014797.

Zhao, Z.C., Fan, S.Q., Lu, Y., Dang, C.C., Wang, X., Liu, B.F., Xing, D.F., Ma, J., Ren, N.Q., Wang, Q., Xie, G.J., 2023. Reactivated biofilm coupling n-DAMO with anammox achieved high-rate nitrogen removal in membrane aerated moving bed biofilm reactor. Environ. Res. 220. 10960953. Academic Press Inc., China. Available from: https://www.sciencedirect.com/journal/environmental-research. https://doi.org/10.1016/j.envres.2022.115184.

Zhou, X., Bi, X., Fan, X., Yang, T., Wang, X., Chen, S., Cheng, L.H., Zhang, Y., Zhao, W., Zhao, F., Nie, S., Deng, X., 2022a. Performance and bacterial community analysis of a two-stage A/O-MBBR system with multiple chambers for biological nitrogen removal. Chemosphere 303 (3), 135195. https://doi.org/10.1016/j.chemosphere.2022.135195.

Zhou, X., Zhang, Y., Li, Z., Sun, P., Hui, X., Fan, X., Bi, X., Yang, T., Huang, S., Cheng, L., Ratnaweera, H., 2022b. A novel two-stage anoxic/oxic-moving bed biofilm reactor process for biological nitrogen removal in a full-scale municipal WWTP: performance and bacterial community analysis. J. Water Process Eng. 50, 103224. https://doi.org/10.1016/j.jwpe.2022.103224. 22147144.

Zhu, L., Ma, J., Yuan, H., Lin, D., Shi, Z., He, Q., Ke, S., 2023. Effects of successional sulfadiazine exposure on biofilm in moving bed biofilm reactor: secretion of extracellular polymeric substances, community activity and functional gene expression. Bioresour. Technol. 380, 129092. https://doi.org/10.1016/j.biortech.2023.129092. 09608524.

7

Membrane based processes for municipal wastewater reuse

Tara Torabi[a], **Kamran Ghasemzadeh**[b,c], **Alfredo Cassano**[d], **and Angelo Basile**[e,f]

[a]*Faculty of Chemical Engineering, Urmia University of Technology, Urmia, Iran.* [b]*Chemical Engineering Department, Urmia University of Technology, Urmia, Iran.* [c]*Institutes for Materials and Processes, University of Edinburgh, Scotland, United Kingdom.* [d]*Institute of Membrane Technology (ITM) of the Italian National Research Council (CNR), Rende, Italy.* [e]*Institute of Membrane Technology (ITM) of the Italian National Research Council (CNR), Cosenza, Italy - Retired.* [f]*General TAG, Ligornetto, Switzerland*

Abbreviations

AnMBR	anaerobic membrane bioreactor
ASP	activated sludge process
BOD_5	biological oxygen demand after 5 days
CA	cellulose acetate
CNTs	carbon nanotubes
CVD	chemical vapor deposition
EPS	extracellular polymeric substances
FS	flat sheet
GO	graphene oxide
HF	hollow fiber
HRT	hydraulic retention time
iMBR	immersed membrane bioreactor
IUPAC	international union of pure and applied chemistry
MBR	membrane bioreactor
MF	microfiltration
MLSS	mixed liquor suspended solids
MMM	mixed matrix membrane
MT	multi-tubular
NF	nanofiltration
PA	polyamide
PAN	polyacrylonitrile
PE	polyethylene
PEG	polyethylene glycol
PES	polyether sulfone
PESA	polyether sulfone amide
PP	polypropylene
PS	polysulfone
PTFE	polytetrafluoroethylene

Municipal Wastewater Treatment. https://doi.org/10.1016/B978-0-443-24826-9.00007-3

PVA	polyvinyl acetate
PVDF	polyvinylidene fluoride
RO	reverse osmosis
sMBR	side-stream membrane bioreactor
SRT	solids retention time
TFC	thin-film composite
TFN	thin-film nanocomposite
TSS	total suspended solids
UF	ultrafiltration

7.1 Introduction

Water is fundamental to human life and critical for sustenance and societal advancement. However, the global challenges of water scarcity, resource depletion, climate change, and widespread water pollution have led to a growing emphasis on wastewater treatment and reuse. Among various types of wastewater, municipal wastewater stands out as particularly significant. Treated wastewater from municipal plants holds promise not only for producing clean water but also as a potential renewable energy source (Aslam et al., 2022; Wu et al., 2021; Sikosana et al., 2019). The volume of municipal wastewater directly correlates with residents' water usage patterns and municipal services such as irrigation and sanitation operations (Thomas et al., 2007). Municipal wastewater typically comprises water containing relatively small concentrations of suspended solids and dissolved organic and inorganic materials. Some of these substances are toxic, posing serious health and environmental risks. Polluted wastewater may contain significant amounts of toxic organic compounds, ammonia, heavy metals, and other complex components, necessitating proper treatment before safe reuse or discharge into the environment (Pereira et al., 2014). Moreover, organic pollutants present in raw municipal wastewater contain chemically bound energy, which can be harnessed to produce useful energy carriers alongside other recyclable and reusable byproducts (Sikosana et al., 2019).

Various wastewater treatment methods exist, encompassing physical processes like adsorption and separation, chemical processes such as coagulation and oxidation, and biological processes like activated sludge processes (ASPs) and sequencing batch reactors. Selection of the appropriate treatment method depends on factors such as cost, energy usage, and available space. Enhanced and cost-effective wastewater treatment is imperative to meet escalating global demands and combat pollution (Jalilnejad et al., 2019). In recent decades, membrane technology has garnered significant attention for wastewater treatment. Originating in the 1960s with the advent of the first reverse osmosis desalination plants, membrane-based treatment swiftly emerged as a cost-effective and widely adopted method. Research indicates that membrane technology offers advantages over other processes due to its low energy consumption, continuous operation, and scalability to higher capacities. As membrane systems evolve to become smaller, more energy-efficient, and less costly, they present numerous opportunities in wastewater treatment, bridging the gap between economics and sustainability. Membrane processes, characterized by minimal or no chemical usage, environmental friendliness, and high accessibility, have seen increased favorability in wastewater treatment due to reduced equipment footprint, energy usage, and capital costs (Ezugbe and Rathilal, 2020; Bera et al., 2022).

Different membrane processes, including microfiltration (MF), ultrafiltration (UF), nanofiltration (NF), and reverse osmosis (RO), can be employed independently at various stages from pre-treatment to post-treatment of wastewater. Alternatively, they can be integrated into hybrid form with other chemical, physical, or biological processes such as coagulation, ion exchange, adsorption, additional membranes, or ASPs. Hybrid membrane processes aim to achieve superior performance compared to individual processes by leveraging the complementary nature of their components, thereby overcoming limitations and enhancing treated water quality (Ezugbe and Rathilal, 2020; Yang et al., 2020b).

Membrane bioreactors (MBRs) represent a hybrid membrane system that combines biological treatment with membrane filtration. Unlike traditional methods where membrane filtration follows conventional biological treatment, MBR integrates both processes, enabling more effective treatment of municipal wastewater. Conventional ASPs for municipal wastewater treatment typically involve three stages: separation of large solids through settling, aerobic degradation of organics by biomass, and removal of biomass through sedimentation. MBR replaces the two separation stages by filtering biomass through a membrane, resulting in higher-quality treated water compared to conventional treatment and eliminating the need for further disinfection (Ezugbe and Rathilal, 2020; Judd, 2008). Consequently, compared to conventional ASPs, MBRs offer several advantages, including quicker activation, reduced floor space requirements, elimination of the need for a secondary clarifier, operation with higher concentrations of mixed liquor suspended solids (MLSS) in the bioreactor tankage, and overall reduced waste sludge production (Zanetti et al., 2010).

The primary aim of this chapter is to underscore the necessity for municipal wastewater reuse. The chapter will provide a comprehensive overview of membrane materials and MBR technology for wastewater treatment, highlighting their advantages as well as limitations.

7.2 Necessity of municipal wastewater reusing

While the Earth's surface is predominantly covered by water, only a small fraction, approximately 3%, is freshwater suitable for human use. Moreover, a significant portion of this freshwater, around 2.5%, remains inaccessible as it is locked in ice caps and glaciers. Consequently, a mere 0.5% of accessible freshwater, derived from rivers, lakes, and groundwater sources, must suffice to meet the water demands of the global population. However, this finite resource faces mounting pressures due to factors such as population growth, rising living standards, urbanization, and the adverse effects of climate change, including phenomena like floods and droughts, exacerbating water scarcity issues, particularly evident in regions like the Middle East and North Africa (Gude, 2017; Garcia and Pargament, 2015; Lavrnić et al., 2017).

Concurrently, the escalating demand for freshwater, coupled with its unsustainable extraction at rates surpassing natural replenishment, exerts immense strain on underground reservoirs, leading to their depletion and the onset of critical issues such as land subsidence. Additionally, heightened water usage results in increased wastewater generation and uncontrolled discharge into the environment poses grave ecological and public health risks (Wu et al., 2021; Jalilnejad et al., 2020).

Municipal wastewater comprises a complex mixture of human waste, biological contaminants, and chemical pollutants, with its composition varying across countries due to diverse factors like climate, socioeconomics, and technological standards. They mainly consist of water alongside suspended and dissolved organic and inorganic solids; organic substances encompass a range of compounds including carbohydrates, fats, proteins, and synthetic chemicals. Inorganic constituents originate from both domestic and industrial sources, with potentially hazardous elements like arsenic, lead, and mercury posing significant health and environmental hazards. Furthermore, pathogenic microorganisms such as viruses, bacteria, protozoa, and helminths may persist in raw municipal wastewater, posing ongoing risks. Consequently, the treatment of municipal wastewater is imperative to avert severe adverse impacts (Assayie et al., 2017; Shevah, 2019).

In response to these challenges, wastewater reuse has emerged as a viable alternative water source, offering a consistent supply unaffected by climatic variations. Wastewater reuse holds potential environmental benefits, including reductions in freshwater extraction from delicate ecosystems and curtailment of wastewater discharge into natural water bodies. Furthermore, it can support the preservation of wetlands and riparian habitats. With further treatment to meet stringent water quality standards, reclaimed wastewater could serve diverse industrial and urban purposes, including manufacturing, irrigation, landscaping, and even potable water supply (Wu et al., 2021; Lavrnić et al., 2017).

Economically, wastewater reuse carries significance, particularly when utilized for direct applications like irrigation, where it assumes the characteristics of a private good with inherent market value. Overall, wastewater reuse presents a promising strategy for bolstering local economies and ensuring socioeconomic and political stability, particularly in developing nations (Garcia and Pargament, 2015). For wastewater to be suitable for reuse, it must satisfy specific criteria, with key parameters including pH, total suspended solids (TSS), biological oxygen demand after 5 days (BOD_5), turbidity, and levels of total and fecal coliforms. Wastewater treatment encompasses various technologies categorized broadly into preliminary, secondary, and tertiary processes. Preliminary treatment targets physical contaminants through screening, sedimentation, and biological processes, while primary treatment focuses on the removal of solids and biodegradable materials. Secondary treatment further reduces organic matter, albeit with limited reuse potential due to residual health and environmental concerns. Tertiary treatment, employing advanced techniques such as ion exchange, ozonation, and membrane filtration, becomes necessary if primary or secondary treatment fails to meet reuse standards (Yang et al., 2020b; Assayie et al., 2017).

In recent years, membrane filtration has emerged as a highly efficient technology for municipal wastewater reuse. Advanced membrane processes can effectively eliminate suspended solids, colloids, bacteria, and viruses by up to 99%, rendering the treated water suitable for reuse. Despite its effectiveness, conventional wastewater treatment methods possess drawbacks such as the generation of substantial amounts of sludge necessitating costly disposal, susceptibility to influent quality variations, high energy requirements, and significant operational and maintenance demands. Hence, there is a pressing need for advanced treatment methods tailored to be cost-effective, especially in regions with limited energy resources per capita (Yang et al., 2020b; Assayie et al., 2017).

7.3 Membrane technology for municipal wastewater reuse

Membranes serve as barriers facilitating the selective passage of substances under various driving forces such as concentration, pressure, temperature, and electric potential. In the realm of water and wastewater treatment, membranes play a crucial role in selectively allowing certain components to pass through while blocking others (Nikolaidis and Poullikkas, 2017; Sadr and Saroj, 2015).

For municipal wastewater reuse applications, pressure-driven membrane processes have demonstrated efficacy. These processes can be classified based on membrane pore size, the size and charge of retained particles/molecules, and the pressure applied to the membrane. As depicted in Fig. 7.1, common classifications utilized in wastewater treatment include MF, UF, NF, and RO (Van Der Bruggen et al., 2003).

Furthermore, Table 7.1 offers a succinct overview of the characteristics of these membrane processes. MF employs membranes with relatively large pores and high permeability, facilitating adequate water flux at low pressures. Larger components exceeding the pore size are effectively removed through sieving, making MF particularly effective in eliminating suspended solids, including larger microorganisms like protozoa and bacteria, which could potentially cause issues in subsequent treatment processes.

UF, on the other hand, utilizes membranes with smaller pores and lower permeability compared to MF, necessitating higher pressures. UF is instrumental in removing viruses and organic macromolecules with sizes around 20 nm, offering enhanced purification capabilities beyond MF.

NF membranes possess even smaller pores, approximately 1 nm in size. This characteristic enables NF to effectively eliminate relatively small organic molecules, micropollutants, multivalent ions, and color from surface or groundwater, as well as degradation products present in biologically treated wastewater effluent.

RO membranes, characterized by their dense structure devoid of defined pores, rely on solution-diffusion mechanisms rather than sieving for permeation. RO stands out as a highly efficient

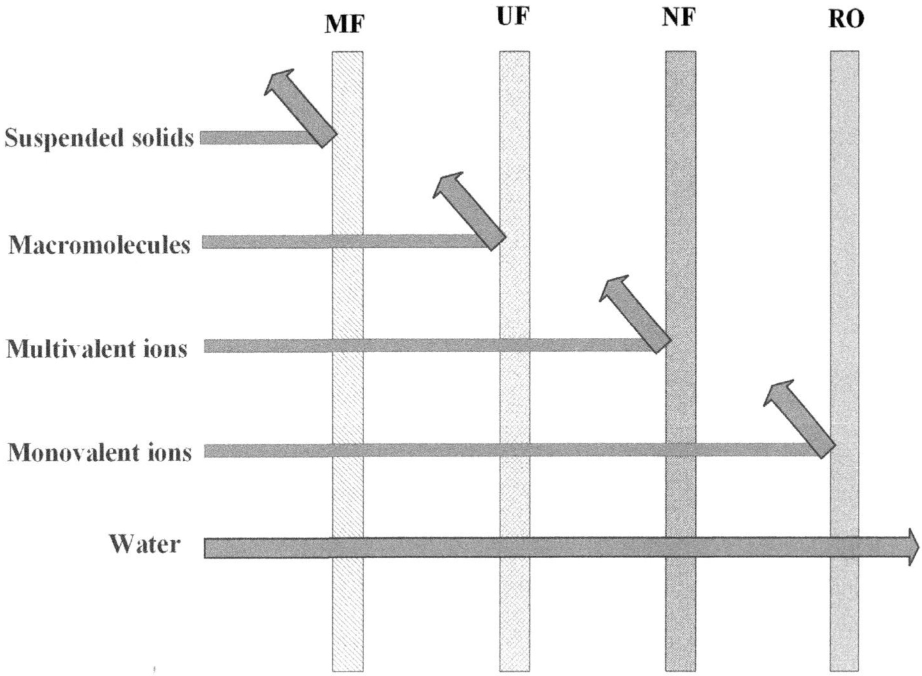

Fig. 7.1 Pressure-driven membrane processes for wastewater treatment.

Table 7.1 Characteristics of different pressure-driven membranes (Ezugbe and Rathilal, 2020; Madhura et al., 2018; Zeynali et al., 2020).

Membrane	Microfiltration	Ultrafiltration	Nanofiltration	Reverse osmosis
Pore size (nm)	100–10,000	2–100	0.5–2	<0.5
Pressure (bar)	0.1–2	1–5	5–20	10–100
Membrane type	Porous isotropic	Porous asymmetric	Semi porous asymmetric	Nonporous asymmetric
Mechanism	Sieving	Sieving	Sieving (Donnan-exclusion)	Solution-diffusion
Applications	Pre-treatment, sterilizing, removal of bacteria, fat, oil, colloids, organics, micro-particles	Removal of macromolecule, bacteria, viruses, proteins, oils, sugar, organics, microplastics	Removal of multivalent ions, sulfates, lactose, sucrose, sodium chloride	Production of ultrapure water, removal of salts and micro-solutes from water through desalination

pressure-driven membrane process capable of separating small particles such as bacteria and ions from water. Widely employed in desalination and pure water production, RO operates by applying pressure to overcome natural osmotic pressures, thereby forcing water molecules to pass through a semi-permeable membrane while retaining contaminants. The purified water is collected on the permeate side, with RO demonstrating proficiency in eliminating monovalent ions like sodium and all dissolved species (Ezugbe and Rathilal, 2020; Van Der Bruggen et al., 2003; Wintgens et al., 2005). These processes can be employed independently, combined, or integrated with conventional treatments (Assayie et al., 2017). For example, MBR technology integrates ASPs or anaerobic biodegradation processes with MF or UF. Furthermore, MF or UF is often integrated to function as a post-treatment step in the tertiary treatment of municipal wastewater, aiming to achieve high-quality effluent (Sadr and Saroj, 2015).

Research has demonstrated that membrane-based processes offer several advantages over traditional methods for wastewater reuse. Membrane processes in water treatment enable physical or physicochemical separations without involving phase changes, rendering them more energy-efficient compared to distillation methods. Moreover, they operate continuously under steady-state conditions, contrasting with adsorptive processes that necessitate regeneration cycles. Furthermore, minimal or no chemical addition is required, resulting in reduced sludge production compared to conventional clarification methods employing coagulants/flocculants (Assayie et al., 2017).

The morphology or structure of membranes can be categorized into symmetric membranes (or isotropic) and asymmetric membranes (or anisotropic), as illustrated in Fig. 7.2.

Symmetric membranes can be further delineated into:

- Symmetric porous membranes: These membranes feature a rigid, highly voided structure with interconnected and randomly distributed pores ranging from 0.01 to 10 μm in diameter. They are proficient in separating molecules that exhibit significant differences in size, such as in UF and MF processes.
- Symmetric nonporous or dense membranes: Comprising a dense film, these membranes facilitate permeate transport through diffusion under the driving force of pressure, concentration, or electrical potential gradients. The separation of mixture components is directly linked to their relative transport rates within the membrane, determined by their diffusivity and solubility in the membrane material, as observed in the RO process.

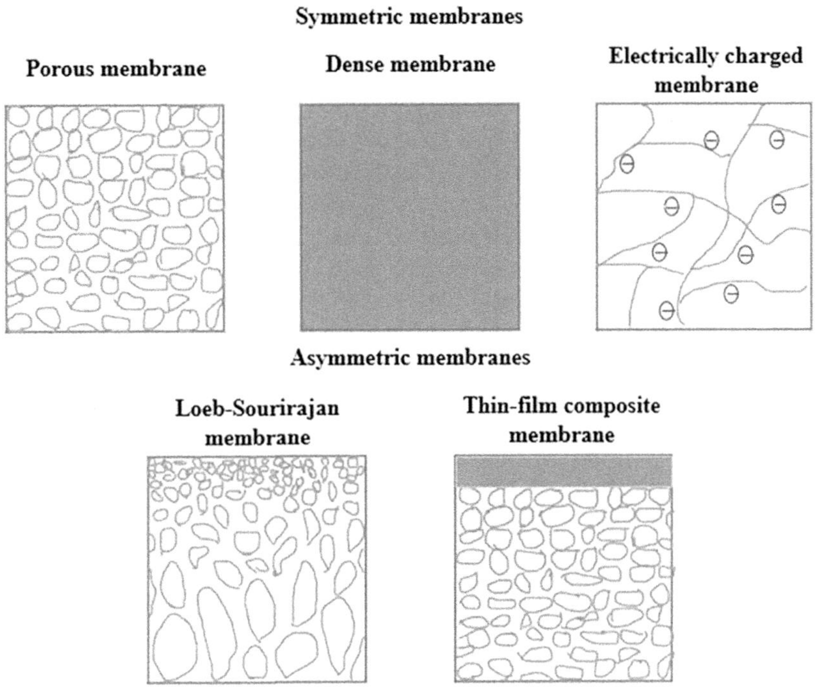

Fig. 7.2 Symmetric and asymmetric membranes.

- Symmetric electrically charged membranes: also referred to as anion-exchange or cation-exchange membranes, these may be dense or porous, but are typically finely microporous, with the pore walls carrying fixed positive or negative charged ions. The separation process is influenced by the charge and concentration of ions in the solution.

In contrast, asymmetric membranes encompass layered structures with varying porosity and pore size across the membrane cross-section. These membranes consist of an extremely thin surface layer supported by a much thicker, porous substructure. The surface layer and its substructure can be formed in a single operation or separately. The thin layer serves as the selective layer for separation, while the thickness of this layer contributes to high membrane fluxes.

Asymmetric membranes can be categorized into *Loeb-Sourirajan* and *composite* membrane types. In the former, the porosity and pore size gradually change throughout the cross-section, with a selective thin layer on the top surface formed on a progressively looser layer. The thin skin layer and porous substrate are coupled together as they are formed simultaneously during the phase inversion process (Abdullah et al., 2018). Anisotropic thin-film composite membranes typically consist of layers made from different polymers. The separation properties and permeation rates of the membrane are determined exclusively by the surface layer, while the substructure functions as a mechanical support. The advantages of the higher fluxes provided by anisotropic membranes are so significant that almost all commercial processes employ such membranes (Purkait et al., 2018; Baker, 2023).

To achieve high-quality treated wastewater suitable for reuse, it is essential to determine the appropriate membrane characteristics, such as hydrophobicity/hydrophilicity, permeability, nominal pore sizes, porosity, surface charges, and roughness. The membrane materials must exhibit chemical and thermal resilience within specific parameters, as they are subjected to varying temperatures, pH levels, and oxidant concentrations during chemical cleaning procedures. Additionally, mechanical durability is essential as particles come into contact with the membrane surface.

Membranes can be classified into two main categories based on their materials: polymeric or organic membranes, and ceramic or inorganic membranes. To be functional, the membrane material must be structured or configured in a manner that permits water to pass through it. A variety of different polymeric and ceramic materials are employed in the fabrication of membranes (Jalilnejad et al., 2019; Cui et al., 2010; Hoek and Tarabara, 2013).

7.3.1 Organic membrane applications

Organic materials utilized in membrane technology are primarily derived from a range of polymers, including cellulose acetate (CA), polyamide (PA), polysulfone (PS), polyethylene (PE), polytetrafluoroethylene (PTFE), polyethersulfone (PES), polyvinylidene fluoride (PVDF), and polypropylene (PP) (Cui et al., 2010; Jose et al., 2018; Lee et al., 2016). These polymers offer diverse structures and properties, rendering them suitable for membrane applications. Polymeric membranes are favored due to their ease of preparation, cost-effectiveness, efficient removal of contaminants, compact size, lower energy consumption, easy manufacture, availability in various pore sizes, and flexibility in configuration. Consequently, they are extensively utilized in wastewater treatment processes.

Polymeric membranes serve crucial roles across different filtration processes, with the choice of polymer dependent on the specific membrane application. For MF and UF systems operating based on a pore-flow mechanism, PES and PS are commonly preferred choices. CA finds applications in both MF and RO membranes, while cross-linked aromatic PA is well-suited for NF and RO membranes. Additionally, materials like polyether sulfone amide (PESA) and polyacrylonitrile (PAN) can be employed in synthesizing various membrane types (Jalilnejad et al., 2019). Despite the widespread adoption of polymeric membranes in wastewater reuse applications, they are associated with several drawbacks that necessitate further modifications and improvements. These drawbacks include issues related to hydrophobicity, membrane fouling, limited operating conditions, and low resistance to chlorine. Addressing these challenges is imperative to enhance the performance and durability of polymeric membranes for sustainable wastewater treatment.

7.3.1.1 Challenges and limitations

While polymeric membranes are extensively utilized in municipal wastewater reuse processes, they encounter several challenges such as hydrophobicity, membrane fouling, limited lifespan, and constraints in operating conditions including pH, temperature, pressure, and chlorine tolerance, which can hinder their effectiveness (Cui et al., 2010; Warsinger et al., 2018). For instance, CA, a conventional material for producing skinned membranes, suffers from drawbacks such as low-temperature tolerance (30–40 °C), a narrow pH range (2–8, preferably 2–6), and low chlorine tolerance (below 1 mg/L free chlorine).

The predominant issue with most polymeric membranes is their inherent hydrophobic nature, leading to reduced water flux during operation due to the accumulation of organic compounds that facilitate the attachment and proliferation of microorganisms on the membrane surface. This phenomenon results in membrane fouling, ultimately leading to membrane failure (Lee et al., 2016; Giwa et al., 2018). Fouling escalates operational costs, diminishes membrane lifespan, causes inconsistent separations, reduces selectivity, and limits applications. It occurs when colloidal or particulate matter accumulates within the membrane's pores or on its surface, altering the membrane's transport properties. As a result, the flux diminishes, necessitating frequent membrane cleaning, periodic replacement, and increased energy input to sustain high flux rates (Geise et al., 2010).

Moreover, polymeric membranes exhibit poor resistance to chlorine, a commonly used disinfectant in water treatment processes, which can react with electron-rich functional groups in the membranes. For example, PA membranes are often hindered by continuous exposure to chlorine and other oxidizing agents during desalination processes. The amide groups in PA membranes are susceptible to chlorine attacks during chemical cleaning, leading to membrane degradation. To address this issue, an additional de-chlorination step is required to reduce chlorine concentration and prevent PA membrane degradation (Giwa et al., 2018; Wen et al., 2019). Additionally, polymeric membranes face challenges in achieving a balance between permeability and selectivity, making it difficult to enhance one without compromising the other in current polymeric membrane designs (Wen et al., 2019).

7.3.1.2 Recent advances

To address the limitations and enhance the performance of polymeric materials in wastewater treatment applications, extensive efforts have been made, including the development of thin-film composite (TFC) membranes and modification techniques (Giwa et al., 2018; Wen et al., 2019). TFC membranes are widely utilized in NF and RO applications, representing a significant advancement in polymeric membrane technology. These membranes consist of three layers: a semipermeable active layer, a microporous polymer layer, and a nonwoven support layer, each contributing to improved performance and durability. Among various polymer combinations, PA-based TFC membranes stand out for their superior properties, revolutionizing wastewater treatment and desalination processes (Zeynali et al., 2020; Maletskyi, 2020).

Polymeric membrane modifications aim to increase hydrophilicity and enhance fouling resistance through methods such as polymer blending, plasma treatment, UV exposure, chemical treatment, and nanoparticle incorporation (Le and Nunes, 2016; Khulbe et al., 2010). Nanotechnology offers unique opportunities for membrane improvement, with nanoparticles like silver, titanium dioxide (TiO_2), graphene oxide (GO), and carbon nanotubes (CNTs) exhibiting remarkable antifouling properties, particularly against microorganisms (Remanan et al., 2018).

Mixed matrix membranes (MMMs), incorporating hydrophilic nanomaterials into the polymer matrix, combat membrane fouling effectively. These membranes, consisting of a polymer matrix embedded with inorganic fillers, offer enhanced hydrophilicity, mechanical strength, thermal stability, and adsorption capabilities for pollutant removal (Abdullah et al., 2023; Lim et al., 2021). Thin-film nanocomposite (TFN) membranes, incorporating nanomaterials into the PA rejection layer, aim to improve water permeability and antifouling properties compared to traditional TFC membranes. Research in this area has led to successful commercialization for desalination and wastewater treatment applications (Zeynali et al., 2020; Yang et al., 2020a). Modifying polymeric materials is crucial for enhancing membrane anti-fouling capabilities while treating wastewater, improving hydrophilicity, mechanical strength, water permeability, rejection rate, and resistance to fouling (Lee et al., 2016; Miller et al., 2017). Strategies include the use of polymers with robust backbones to enhance chlorine resistance, such as polysulfones, polyimides, and polybenzimidazoles, along with the attachment of hydrophilic ionic groups to improve water permeability (Geise et al., 2010).

7.3.2 Inorganic membrane applications

The development of polymeric membranes has advanced steadily, and these membranes dominate the majority of the membrane market due to their affordability. However, polymeric membranes face several issues such as poor mechanical resilience and susceptibility to fouling. In contrast, inorganic membranes offer superior attributes, including exceptional chemical and physical properties, thermal and mechanical stability, high tunability, and reusability (Chen et al., 2017; Goh and Ismail, 2018; Fard et al., 2018). These qualities make inorganic membranes well-suited for applications in demanding environments, such as corrosive and high-temperature conditions, making them ideal candidates for harsh water purification processes, such as wastewater treatment, while exhibiting high selectivity and permeability for specific molecules (Chen et al., 2017; Goh and Ismail, 2018; Fard et al., 2018). Inorganic membranes comprise ceramic, silica, zeolite, and carbon-based materials, and several synthesis methods, including slip casting, the sol-gel approach, chemical vapor deposition (CVD), and pyrolysis, have been proposed as prevalent techniques for crafting these membranes (Fard et al., 2018).

Inorganic filtration membranes exhibit chemical resistance to organic solvents, chlorine, and other chemicals, making them more resistant than organic membranes. This chemical resistivity is beneficial for water treatment applications where water is dosed with chlorine and/or other disinfectants, as organic membranes could fail. Additionally, this property allows inorganic membranes to be cleaned and washed regularly with different solvents and anti-scaling agents. Furthermore, inorganic membranes are not vulnerable to microbial attack and are mechanically robust (Fard et al., 2018).

Inorganic membranes are classified according to their pore size into dense (<0.5 nm), microporous (0.5–2 nm), mesoporous (2–50 nm), and macroporous (>50 nm) structures. Dense and microporous membranes are impermeable to liquids but can allow single-molecule transport, while mesoporous and macroporous membranes facilitate liquid flow via viscous mechanisms. Selective separations can occur at any pore scale through size exclusion or other mechanisms. Dense membranes ideally separate single molecular species, while microporous membranes can differentiate between various molecules based on feed composition, material, and operating conditions. The structure and morphology of inorganic membranes, whether porous or dense, symmetric or asymmetric, significantly influence their separation performance in wastewater treatment applications (Fard et al., 2018; Verweij, 2012).

Among different types of inorganic membranes, ceramic membranes are increasingly favored in the drinking water and wastewater treatment sectors due to their superior resistance to fouling and chemical stability compared to prevalent polymeric membranes (Lee et al., 2015, 2016). Ceramic membranes have a multi-layered structure, comprising a macroporous support layer, an intermediate layer, and a selective microporous top layer (Arumugham et al., 2021). They are composed of materials such as alumina (Al_2O_3), titania (TiO_2), silicon carbide (SiC), zirconia (ZrO_2), and zinc oxide (ZnO), with alternative materials like zeolites, fly ash, kaolinite, and mullite being investigated for preparing ceramic membranes (Warsinger et al., 2018; Lee et al., 2015; Arumugham et al., 2021; Benfer et al., 2001; Hakami et al., 2020). Ceramic membranes have been evaluated across various pore sizes for treating secondary municipal wastewater effluent. While most commercially available ceramic membranes lack pore sizes small enough for virus removal, they exhibit a high tolerance for chlorine, oxidants, and coagulants, potentially enabling efficient virus removal. Moreover, ceramic UF and NF membranes serve as effective pretreatment for RO, requiring less frequent cleaning and preventing biofouling of downstream RO membranes. With their ability to withstand extreme conditions and sustainable cleaning methods, ceramic membranes present a compelling alternative for enhancing municipal wastewater treatment processes, ensuring extended service life and efficient filtration performance (Warsinger et al., 2018; Lee et al., 2015; Arumugham et al., 2021; Benfer et al., 2001; Hakami et al., 2020).

Although ceramic membranes offer several advantages such as high chemical stability, mechanical strength, ease of cleaning, and long lifetimes over organic membranes, their applications are limited for municipal wastewater treatment due to some disadvantages that need to be addressed (Le and Nunes, 2016).

7.3.2.1 Challenges and limitations

One of the primary disadvantages of ceramic membranes is their significantly higher manufacturing cost, often reaching levels approximately ten times greater than that of organic membranes (Cui et al., 2010; Le and Nunes, 2016; Qiu et al., 2018; Samaei et al., 2018; Tolkou et al., 2014). Beyond cost considerations, ceramic membranes also suffer from inherent brittleness, rendering them susceptible to damage from mechanical shocks or excessive vibrations, necessitating meticulous handling and operation procedures. Moreover, their rigid composition can lead to potential complications, such as thermal expansion mismatch between the membrane and module housing, posing challenges in both design and operation phases (Cui et al., 2010; Le and Nunes, 2016; Qiu et al., 2018; Samaei et al., 2018; Tolkou et al., 2014). Presently, ceramic membranes are predominantly available in UF and MF types, commonly utilized in municipal wastewater treatment processes. However, their performance may suffer due to fouling issues. Additionally, ensuring consistent product quality throughout the ceramic membrane manufacturing process remains a persistent challenge (Cui et al., 2010; Le and Nunes, 2016; Qiu et al., 2018; Samaei et al., 2018; Tolkou et al., 2014).

Similar to polymeric membranes, ceramic membranes also experience fouling during typical separation processes. Interestingly, reversible fouling is more prominent in ceramic membranes, whereas irreversible fouling is more significant when polymeric membranes are employed. Most ceramic membranes are inherently hydrophilic, possessing a lower fouling tendency. Cleaning methods such as physical cleaning and chemical cleaning have proven to be more effective in the case of ceramic membranes compared with polymeric ones (Arumugham et al., 2021). Membrane fouling not only impairs wastewater treatment performance but also increases operational expenses. While ceramic membranes for municipal wastewater treatment have gained more attention, research on fouling mechanisms has been relatively limited (Qiu et al., 2018).

The high cost associated with ceramic membranes has been a significant barrier to their widespread adoption, particularly in cost-sensitive environmental treatment processes. Conventional ceramic membranes, typically crafted from high-purity refractory oxides, often incur unnecessary expenses, especially for applications like municipal or industrial wastewater treatment, where lower-cost materials could suffice. The complex multilayer structure of ceramic membranes poses additional challenges, contributing to elevated fabrication costs. Each layer requires high-temperature sintering, leading to substantial energy consumption, which accounts for the majority of fabrication expenses (Mestre et al., 2019; Zou and Fan, 2021). Previous studies have shown that sintering energy consumption alone can represent up to 60% of the total cost. Consequently, efforts to enhance the competitiveness of ceramic membranes in the market have predominantly focused on reducing fabrication expenses by mitigating energy consumption (Mestre et al., 2019; Zou and Fan, 2021). While ceramic MF and UF membranes are commercially available and have been extensively tested for wastewater treatment at laboratory and full-scale levels, their widespread use in water treatment is limited due to high capital costs. Much research has focused on decreasing fabrication costs, such as using clay as the membrane material, to encourage larger-scale application of ceramic membranes. However, further research on ceramic MF and UF membranes is needed to develop cheaper manufacturing processes and produce higher-quality membranes for continuous operation (Li et al., 2020).

7.3.2.2 Recent advances

Porous ceramic membranes initially employed for MF and UF applications are now being expanded to NF applications for removing organics and salts (Arumugham et al., 2021). While porous ceramic NF membranes are still in the intermediate stage between laboratory research and commercial implementation, current research efforts are focused on in-depth investigations to alleviate membrane fouling, improve packing density, augment flux and selectivity, and explore novel low-cost materials for developing application-specific porous membranes (Arumugham et al., 2021).

Efforts to enhance the cost-effectiveness and versatility of ceramic membranes have led to the development of low-cost ceramic membranes through various innovative strategies. These membranes capitalize on affordably sourced raw materials commonly found in traditional ceramic products like tiles and porcelain, employing traditional ceramic industry techniques such as pressing and extrusion for substrate processing. By utilizing lower sintering temperatures compared to commercial ceramic membranes, significant reductions in processing costs are achieved, resulting in a more economical product overall (Mestre et al., 2019; Zou and Fan, 2021). Leveraging conventional ceramic materials and processing techniques, low-cost ceramic membranes offer a promising solution for broader adoption in cost-sensitive environmental applications, while still maintaining robust operational capabilities. Recent research has focused intensely on designing new ceramic membrane materials based on raw materials typical of traditional silicate ceramics, to lower costs while preserving performance. These membranes exhibit characteristics typical of ceramic materials such as chemical stability, high mechanical resistance, and thermal insulation, making them suitable for diverse applications including oil and heavy metal removal, wastewater treatment, and various production processes. However, further industrial research is necessary to validate their performance in real-world application environments (Mestre et al., 2019; Zou and Fan, 2021).

To reduce the energy consumption associated with sintering during the production of ceramic membranes, several strategies have been proven efficient (Zou and Fan, 2021). These strategies include: (1) lowering the sintering temperature of ceramic membranes, achieved through the utilization of low-melting-point materials or the implementation of processes that involve low, cold, or no sintering; (2) enhancing the sintering rate for ceramic membranes to expedite the process; (3) minimizing sintering durations by employing co-sintering techniques or optimizing membrane structures to streamline the process.

Combining highly selective materials, fabricating thin defect-free layers on supported structures, characterizing these TFC, and improving cost-effectiveness through rapid manufacturing methods and inorganic/polymeric hybrids can also be used to improve ceramic membrane characteristics (Fard et al., 2018; Souza and Quadri, 2013). The structure and morphology of inorganic membranes, whether porous or non-porous, symmetric or asymmetric, significantly influence their separation performance in diverse applications ranging from gas separation and pervaporation to harsh wastewater treatment.

Nanotechnology's rapid advances from academic research into a wide array of commercial products over the past few decades have led to the focus of numerous studies on incorporating nanomaterials into ceramic MF and UF membranes to improve permeability, selectivity, and mitigate fouling in wastewater treatment (Li et al., 2020; Madaeni et al., 2015). Incorporating inorganic nanoparticles into polymeric materials results in the formation of nanocomposite (nano-enhanced) membranes with improved mechanical and physicochemical properties. A variety of nanomaterials have been employed as fillers in these membranes, including CNTs, nanoclay, nanosilver, and nanosized particles of TiO_2, ZnO, Al_2O_3, Fe_3O_4, SiO_2, and ZrO_2. By dispersing these nanofillers within the polymer matrix, researchers aim to create membranes that combine the advantages of polymeric materials, such as processability and low cost, with the unique properties offered by nanomaterials, like high surface area, improved thermal stability, and enhanced selectivity (Li et al., 2020; Madaeni et al., 2015). This nanocomposite approach represents a promising avenue for advancing membrane technology and overcoming some of the limitations of conventional polymeric membranes.

MMMs are a type of membrane that combines inorganic nanoparticle materials dispersed within a polymer matrix (Fard et al., 2018). The goal of creating MMMs is to combine the advantages of inorganic fillers, such as high mechanical strength, superior processability, and low costs associated with polymeric materials. This aims to overcome some limitations of using only polymeric or inorganic membranes alone. Several fabrication techniques can be used to produce MMMs, including phase inversion and surface coating methods that incorporate the nanoparticle filler into the polymer membrane structure.

7.3.3 Membrane bioreactor applications

When combined with conventional biological treatment processes, membrane processes can form robust membrane technologies that offer a higher level of treatment. A prime example is MBR technology, which uniquely integrates a conventional biological treatment system with physical liquid-solid separation by membrane filtration into a single hybrid system for municipal wastewater treatment (Sadr and Saroj, 2015; Iorhemen et al., 2016). MBR technology was first utilized in 1969, employing a UF membrane to segregate ASP from the final effluent of a biological wastewater treatment system, with the sludge then recycled back into the aeration tank (Ng and Kim, 2007).

MBRs can be categorized into *side-stream* or *submerged/immersed* configurations for solid-liquid separation based on how the membrane module is incorporated. In side-stream MBRs (sMBRs) the membrane module is placed externally to the main bioreactor, with separation occurring via pressure-driven filtration. Mixed liquor from the bioreactor is pumped into the membrane module, where a concentrated stream is retained by the membrane and returned to the bioreactor (Fig. 7.3A) (Assayie et al., 2017; Sadr and Saroj, 2015).

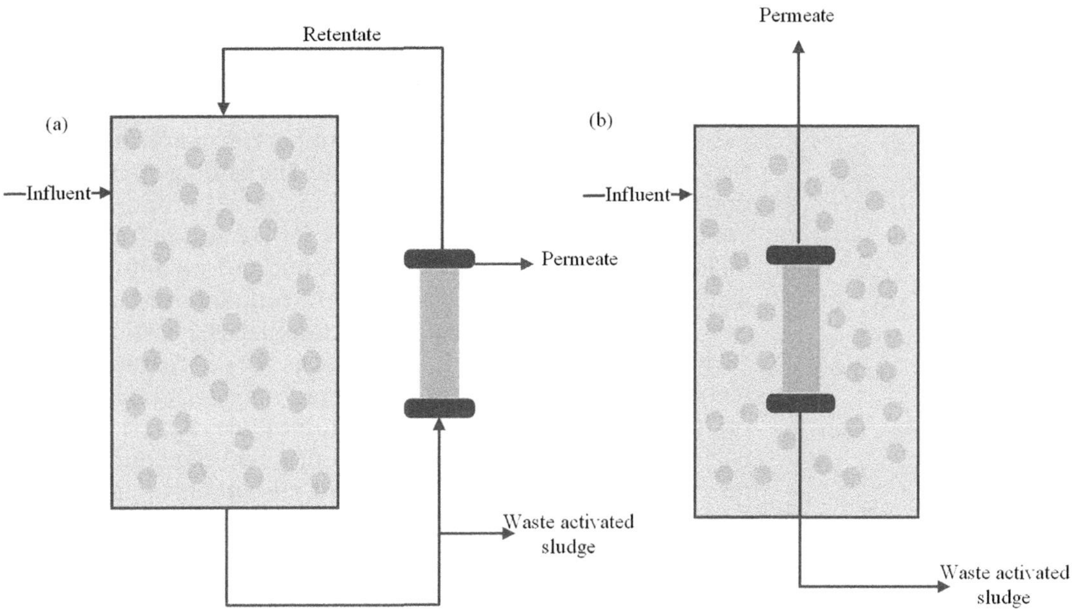

Fig. 7.3 (A) Side-stream MBR; (B) submerged MBR.

Submerged or Immersed MBRs (iMBRs) were introduced the first time in 1989: this configuration aims to reduce the energy required for recirculation pumping in side-stream systems. Here, the membrane module is directly immersed into the bioreactor itself. In an iMBR, the permeate is extracted either under suction pressure on the permeate side and/or through the static head of pressure on the retentate side (Fig. 7.3B) (Sadr and Saroj, 2015; Judd, 2016).

The submerged setup, being less energy-intensive and offering a built-in cleaning mechanism to mitigate membrane fouling, is more prevalent than the recirculated configuration, prompting greater focus on fouling models tailored to this setup. In general, the more benign conditions of the iMBR configuration make it more suitable for municipal wastewater (Ng and Kim, 2007; Judd, 2016).

In large-scale membrane processes such as those utilized in industrial or commercial applications, substantial membrane areas are necessary. These significant areas are efficiently packaged into what is known as membrane modules, which come in three main types: flat sheets (FS), hollow fiber (HF), and multi-tubular (MT) modules (Ezugbe and Rathilal, 2020; Jalilnejad et al., 2020; Al-Asheh et al., 2021; Giorno et al., 2017). FS modules consist of membrane sheets arranged alternately with feed and permeate spacers, all enclosed within a rigid frame. MT modules house semipermeable membranes inside perforated tubes within an outer shell, allowing feed to flow through the membrane tubes while permeate exits through perforations into the shell side for collection. This design is adapt at handling high-solids feeds. HF modules contain bundles of hollow fiber membranes within a shell. These fibers have a porous support layer and an ultra-thin selective layer forming the membrane. Depending on the application, a feed can either flow through the fiber bores or across the outer shell side. Critical dimensions for HF and MT membranes include outer diameter, inner diameter, and membrane length, while for flat sheet membranes, significant dimensions are thickness, length, and width (Ezugbe and Rathilal, 2020; Jalilnejad et al., 2020; Al-Asheh et al., 2021; Giorno et al., 2017). A crucial consideration in MBR plant design and operations is selecting the appropriate system configuration. Each module type has unique characteristics well-suited to different operating conditions and feed stream properties when selecting an optimal membrane system design. FS membranes offer the advantages of simple cleaning and replacement methods, as well as good efficiency for commercial use. However, their main drawback is their high cost, prompting researchers to explore more cost-effective alternatives. In contrast, HF membranes are appealing due to their high surface area per unit volume, cost-effective fabrication process, and ability to withstand severe operating conditions. Nonetheless, they are susceptible to fouling, necessitating frequent washing. MT membranes are also attractive due to their high mechanical strength, low fouling rate, long lifespan, and straightforward cleaning and replacement processes. However, their low packing density, as well as high capital and operational costs, hinder their widespread adoption on a large scale. Overall, understanding the unique advantages and drawbacks of each membrane module type is crucial in designing efficient and effective membrane processes for large-scale applications (Ezugbe and Rathilal, 2020; Jalilnejad et al., 2020; Al-Asheh et al., 2021; Giorno et al., 2017).

The range of membrane materials employed commercially for MBR technologies is fairly limited, and rated pore sizes are generally in the range between 0.03 and 0.4 μm. Around half of the polymeric MBR membranes are based on PVDF; most of the others are either polyolefins (PP, or PE) or PES. There is also a growing number of ceramic products, with FS ceramic membranes representing the most recent innovation (Warsinger et al., 2018; Judd, 2016).

MBRs offer numerous advantages over traditional ASP systems, particularly in municipal wastewater reuse applications (Warsinger et al., 2018; Jyoti et al., 2013; Melin et al., 2006). One notable advantage is the replacement of the secondary clarifier with a compact membrane module, ensuring effluent quality remains consistent regardless of sludge settling properties. With the ability to achieve higher biomass concentrations, MBRs enable more thorough biodegradation, thereby increasing treatment efficiency beyond the limitations imposed by poor settling in ASP processes. Additionally, the elevated MLSS levels in MBRs allow for a reduction in aeration tank volume. Moreover, MBRs significantly decrease sludge production, leading to reduced disposal costs. Furthermore, the technology facilitates direct in-situ treatment and water reuse, thereby cutting down on manufacturing costs associated with water consumption. Another significant advantage lies in MBRs' capability to control solids retention time (SRT) independently from hydraulic retention time (HRT), thereby enhancing organic load application and pollutant degradation. These advantages, including the ability to achieve high MLSS concentrations in a small footprint and produce superior disinfected effluent, are key drivers for the increasing adoption of MBRs over conventional ASP processes in municipal wastewater reuse endeavors (Warsinger et al., 2018; Jyoti et al., 2013; Melin et al., 2006).

7.3.3.1 Challenges and limitations

Despite the advantages of MBR technology, widespread application is constrained by the challenge of higher energy costs, the need to control persistent membrane fouling issues, and potentially high costs for periodic membrane replacement (Iorhemen et al., 2016; Wisniewski, 2007).

Recent reviews on membrane applications in bioreactors have shown that, as with other membrane separation processes, membrane fouling is the most serious issue affecting system performance (Iorhemen et al., 2016; Chang et al., 2002). Fouling leads to a decline in permeate flux, necessitating more frequent membrane cleaning and replacement, which increases operating costs. Consequently, most MBR studies aim to identify, investigate, control, and model membrane fouling. According to the International Union of Pure and Applied Chemistry (IUPAC), membrane fouling refers to the process where a membrane's performance declines due to the accumulation of suspended or dissolved substances on its external surfaces, at the pore openings, or within the pores themselves. The foulants can include particulates like microorganisms and cell debris, colloids, and solutes present in the MLSS. Physicochemical interactions between these foulants and the membrane material lead to fouling. Uncontrolled membrane fouling in MBRs can potentially cause failure to treat the required design flows (Iorhemen et al., 2016; Chang et al., 2002).

Membrane fouling can be subdivided into three groups based on the effectiveness of cleaning methods, characteristics of foulants, and location of fouling.

Membrane fouling can be categorized into reversible and irreversible fouling based on the effectiveness of cleaning methods. Reversible fouling is caused by the formation of a cake layer, which is a porous layer of rejected solids accumulated on the membrane surface. This type of fouling can typically be mitigated through physical cleaning techniques such as backwashing or relaxation cycles. In contrast, long-term flux decline that cannot be recovered by simple hydrodynamic techniques indicates irreversible fouling, often attributable to colloidal deposition or soluble compound adsorption onto the membrane. Chemical cleaning is necessary to eliminate such fouling. On the other hand, the deposition of particulates is considered reversible fouling and can be mitigated by appropriate filtration conditions (Wisniewski, 2007; Maaz et al., 2019; Drews, 2010).

Membrane fouling mechanisms can also be differentiated based on the size of solutes and micro-colloids relative to the membrane pores. When the foulants are larger than the average membrane pore size, external fouling occurs. External fouling consists of two layers: a reversible cake layer fouling caused by the agglomeration of gross solids on the membrane surface, and an irreversible gel layer fouling formed by the precipitation of soluble macromolecules, colloids, and inorganic matter. Conversely, internal fouling happens when fine particles, solutes, and undissolved matter are smaller than the average pore size. In such cases, these foulants penetrate and accumulate within the membrane pores, causing pore narrowing and blocking. This type of internal fouling is generally irreversible (Maaz et al., 2019; Helmi and Gallucci, 2020). During continuous filtration operations, the direct flow path of foulants towards the membrane is higher than the backflow diffusion to the bulk solution before the steady state is reached. This leads to concentration polarization, where solutes or particles accumulate in a thin liquid layer adjacent to the membrane surface. While concentration polarization is a reversible and inherent phenomenon in membrane filtration processes, it can increase resistance to liquid flow and reduce permeate flux (Maaz et al., 2019; Wang et al., 2014).

Membrane foulants in MBRs can be categorized into three groups based on their biological and chemical nature: biofoulants, organic foulants, and inorganic foulants.

Biofoulants refer to bacteria or flocs that cause fouling through deposition, growth, and metabolism on the membrane surface. For MF and UF membranes treating wastewater, biofoulants are a major issue because most foulants are significantly larger than the membrane pores. Initially, a single bacterial cell may attach to the membrane surface or within its pores. Over time, this cell multiplies into a cluster, leading to the formation of a biocake that reduces permeability. Both the bacteria themselves and their metabolic products contribute to biofouling. Essentially, membrane biofouling is a two-step process, beginning with early bacterial attachment followed by the multiplication of bacteria on the membrane surface (Iorhemen et al., 2016; Jabbari et al., 2019; Meng et al., 2009). Organic foulants in MBRs are biopolymers, such as polysaccharides and proteins, whose deposition on the membrane results in a decline in permeability. These foulants are found in the metabolic products of bacteria, collectively known as extracellular polymeric substances (EPS). Compared to larger particles like sludge flocs, the deposition of organic foulants on the membrane surface is more difficult to remove. Due to their small size, biopolymers can easily be deposited onto membranes by the permeate flow. However, compared to larger particles like colloids and sludge flocs, biopolymers exhibit a lower back-transport velocity off the membranes because the lifting forces acting on them are weaker (Iorhemen et al., 2016; Meng et al., 2009). Inorganic foulants are a group of inorganic substances that precipitate onto the membrane surface or into the pores, causing fouling. Examples include cations and anions such as Ca^{2+}, Mg^{2+}, Fe^{3+}, Al^{3+}, CO_3^{2-}, SO_4^{2-}, PO_4^{3-}, and OH^-. These species precipitate onto the membrane due to processes like hydrolysis leading to pH changes and oxidation reactions (Iorhemen et al., 2016; Wang et al., 2014).

7.3.3.2 Recent advances

Numerous efforts and notable advancements have been dedicated to enhancing the performance of MBRs in municipal wastewater treatment. Some strategies concentrate on altering membrane properties to increase resistance to fouling by modifying surface characteristics. Others aim to optimize operational parameters to establish conditions less prone to fouling. Pretreatment methods endeavor to modify characteristics of the mixed liquor, such as particle size distribution or EPS concentrations, which contribute to fouling. Additionally, research has delved into exploring diverse configurations and designs for MBRs to augment wastewater treatment efficiency (Le-Clech et al., 2006).

Since membrane fouling represents the primary limitation for MBRs, a significant portion of research and development efforts in membrane materials and processes has been dedicated to characterizing and mitigating fouling. Resulting from advances in understanding fouling, as well as the pressing economic and ecological problems associated with it, several novel strategies for fouling mitigation have recently emerged. These strategies exploit a variety of biochemical, mechanical, or hydrodynamic means, such as the development of anti-fouling membranes, the use of additives, sponge-like carriers or other circulating abrasive particles, sludge granulation, pre-settling of biomass from the bulk, membrane surface modification, and others. Additionally, traditional methods such as back pulsing or bubbling, as well as geometric design features, are being systematically optimized. Over the past years, significant numbers of research have also focused on advanced control to improve the efficiency of traditional and novel antifouling methods and minimize energy expenditure (Drews, 2010; Kraume and Drews, 2010).

Membrane cleaning aims to restore permeate flux that is lost due to fouling. It involves removing deposited materials from the membrane surface to allow permeate flow. Cleaning methods can be broadly classified as physical, chemical, physicochemical, and biological/biochemical cleaning. The approach depends on the fouling removal mechanism or cleaning agent used. Cleaning can occur in situ with the membrane module inside the reactor, or ex-situ after removing the module (Wang et al., 2014; Ozgun et al., 2013; Lefebvre et al., 2019). Physical cleaning utilizes mechanical forces to dislodge foulants. Physical cleaning techniques aim to remove reversible foulants that accumulate on the membrane surface and within the pores. Common techniques include periodic backflushing, air scouring, ultrasonic vibrations, and sponge ball cleaning of tubular membranes. Backflushing applies reverse pressure to lift off surface deposits, though it cannot remove internal pore fouling. Air scouring creates shear forces without chemical use, but requires air pumping energy. Ultrasound generates cavitation bubbles that create turbulence at the membrane surface. While physical cleaning allows membrane systems to maintain relatively constant flux without introducing additional contaminants, it requires frequent application to be effective. This necessity for recurring physical cleaning cycles increases the operational costs associated with membrane processes. However, physical cleaning avoids the use of chemical agents and the potential for secondary contamination issues they can introduce. Despite the added operational expense, physical cleaning remains an attractive option as it extends membrane lifespan by controlling the buildup of reversible surface fouling. Optimizing physical cleaning regimes allows membrane systems to operate productively for longer periods between more intensive chemical cleaning steps. Overall, physical cleaning trades off higher frequency and costs for simpler, chemically benign membrane maintenance (Ezugbe and Rathilal, 2020; Ozgun et al., 2013; Du et al., 2020).

As operation time increases, the effectiveness of membrane physical cleaning tends to diminish due to accumulation of irreversible fouling on the membrane surface. Therefore, in addition to physical cleaning strategies, various types and intensities of chemical cleaning may also be recommended for MBRs (Wintgens et al., 2003). Chemical cleaning uses acids, alkalis, chelants, enzymes, surfactants, or disinfectants to dissolve and remove different foulant types. Acids like hydrochloric acid target inorganic scales, while alkalis like sodium hydroxide are used for organic fouling (Ezugbe and Rathilal, 2020). The chemical cleaning procedure generally involves six fundamental stages: initial reaction of the cleaning agent in large quantities; conveyance to the membrane interface; infiltration into fouling layers; dissolution and separation of contaminants; return transport of the utilized agent carrying separated contaminants; and disposal of residual material back into the primary solution or cleaning agent (Porcelli and Judd, 2010).

While physical and chemical cleaning methods are capable of removing fouling, they pose risks of membrane damage, the release of harmful chemical substances, and the production of cleaning byproducts that can be harmful to membranes, the environment, and microbial communities. In response to these concerns, biological and biochemical approaches have proven to be useful. Biological cleaning harnesses enzymes or biochemicals in a more eco-friendly process compared to harsh chemicals. Enzymatic cleaning, energy uncoupling, and quorum quenching are some of the biological cleaning examples. These techniques have the potential to reduce biofilm formation with fewer negative effects on microbial activity, microbial communities, and membrane properties compared to physical and chemical cleaning methods (Ezugbe and Rathilal, 2020; Deng et al., 2016).

In addition, physicochemical methods combine physical and chemical actions, such as chemically enhanced backwashing or ultrasound-aided chemical cleaning to improve flux recovery.

The appropriate cleaning approach depends on the foulant-membrane interactions, foulant composition, and the need to avoid membrane damage while maximizing flux recovery (Ezugbe and Rathilal, 2020).

Furthermore, in line with extensive efforts to find more efficient, cost-effective, and environmentally friendly MBR configurations for wastewater reuse, considerable research has been devoted to anaerobic membrane bioreactors (AnMBR).

Aerobic wastewater treatment methods, though long-established, require over 2.0 mg/L of dissolved oxygen for aerobic degradation of organic matter resulting in high energy demands, generating excessive sludge, demanding substantial space for plant operations, and involving high maintenance costs, offsetting the advantages of wastewater.

7.4 Conclusion and future trends

In conclusion, membrane technologies stand as indispensable tools in the realm of municipal wastewater treatment, offering unparalleled treatment capabilities and environmental advantages. Pressure-driven membrane processes like MF, UF, NF, and RO have demonstrated their effectiveness in this domain. While polymeric membranes lead the market due to their affordability and versatility, inorganic membranes present a promising alternative for applications necessitating superior thermal and chemical resistance. Despite their merits, the widespread embrace of membrane technologies faces hurdles such as membrane fouling and high manufacturing costs. However, recent strides in fouling mitigation strategies, including the development of fouling-resistant membrane materials and the implementation of efficient pretreatment and cleaning techniques, offer prospective solutions to combat this challenge. Moreover, ongoing research endeavors in cost-effective membrane materials and manufacturing processes aim to alleviate the overall expenses associated with membrane technologies.

MBRs emerge as potent tools for enhanced treatment efficiency, rendering them attractive for sustainable wastewater treatment practices. Nonetheless, they grapple with issues such as high energy consumption and membrane fouling. Recent advancements in fouling mitigation strategies, such as advanced membrane material integration and operating condition optimization, present encouraging outcomes in tackling these obstacles. Furthermore, the advent of AnMBR opens doors to sustainable wastewater treatment practices by curbing energy consumption and facilitating resource recovery.

Looking ahead, sustained research and development endeavors are imperative to further enhance membrane technologies, curtail costs, and foster sustainable water management practices on a global scale. These efforts should prioritize the development of more efficient and cost-effective membrane materials, the optimization of membrane processes tailored for specific applications, and the integration of membrane technologies with other treatment processes to craft comprehensive and sustainable wastewater treatment solutions. By harnessing innovation and collaboration, the future of membrane technologies in wastewater treatment holds promise for addressing pressing environmental challenges and advancing toward a more sustainable future.

References

Abdullah, N., Rahman, M.A., Othman, M.H.D., Jaafar, J., Ismail, A.F., 2018. Membranes and Membrane Processes: Fundamentals. Current Trends and Future Developments on (Bio-) Membranes: Photocatalytic Membranes and Photocatalytic Membrane Reactors. Elsevier Inc., Malaysia, pp. 45–70. Available from: https://www.sciencedirect.com/book/9780128135495/current-trends-and-future-developments-on-bio-membranes. https://doi.org/10.1016/B978-0-12-813549-5.00002-5.

Abdullah, R., Astira, D., Widyanto, A.R., Dharma, H.N.C., Hidayat, A.R.P., Santoso, L., Sulistiono, D.O., Rahmawati, Z., Gunawan, T., Jaafar, J., Kusumawati, Y., Othman, M.H.D., Fansuri, H., 2023. Recent development of mixed matrix membrane as a membrane bioreactor for wastewater treatment: A review. Case Stud. Chem. Environ. Eng. 8. Elsevier Ltd, Indonesia 26660164. Available from: www.journals.elsevier.com/case-studies-in-chemical-and-environmental-engineering/. https://doi.org/10.1016/j.cscee.2023.100485.

Al-Asheh, S., Bagheri, M., Aidan, A., 2021. Membrane bioreactor for wastewater treatment: A review. Case Stud. Chem. Environ. Eng. 4, 100109. https://doi.org/10.1016/j.cscee.2021.100109. 26660164.

Arumugham, T., Kaleekkal, N.J., Gopal, S., Nambikkattu, J., Rambabu, K., Aboulella, A.M., Wickramasinghe, S.R., Banat, F., 2021. Recent developments in porous ceramic membranes for wastewater treatment and desalination: A review. J. Environ. Manage. 293. Academic Press, United Arab Emirates, 10958630. Available from: https://www.sciencedirect.com/journal/journal-of-environmental-management. https://doi.org/10.1016/j.jenvman.2021.112925.

Aslam, A., Khan, S.J., Shahzad, H.M.A., 2022. Anaerobic membrane bioreactors (AnMBRs) for municipal wastewater treatment- potential benefits, constraints, and future perspectives: An updated review. Sci. Total Environ. 802. Elsevier B.V., Pakistan, 18791026. Available from: www.elsevier.com/locate/scitotenv. https://doi.org/10.1016/j.scitotenv.2021.149612.

Assayie, A.A., Gebreyohannes, A.Y., Giorno, L., 2017. Municipal Wastewater Treatment by Membrane Bioreactors. Springer Science and Business Media LLC, pp. 265–294, https://doi.org/10.1007/978-981-10-5623-9_10.

Baker, R.W., 2023. Membrane Technology and Applications. John Wiley & Sons.

Benfer, S., Popp, U., Richter, H., Siewert, C., Tomandl, G., 2001. Development and characterization of ceramic nanofiltration membranes. Sep. Purif. Technol. 22-23, 231–237. https://doi.org/10.1016/S1383-5866(00)00133-7. 13835866.

Bera, S.P., Godhaniya, M., Kothari, C., 2022. Emerging and advanced membrane technology for wastewater treatment: a review. J. Basic Microbiol. 62 (3-4), 245–259. John Wiley and Sons Inc, India, 15214028. Available from: http://onlinelibrary.wiley.com/journal/10.1002/(ISSN)1521-4028. https://doi.org/10.1002/jobm.202100259.

Chang, I.S., Clech, P.L., Jefferson, B., Judd, S., 2002. Membrane fouling in membrane bioreactors for wastewater treatment. J. Environ. Eng. 128 (11), 1018–1029. https://doi.org/10.1061/(ASCE)0733-9372(2002)128:11(1018). 07339372.

Chen, L.-H., Chen, Y.-R., Chou, C.-Y., Chen, C.-H., Ko, C.-C., Tung, K.-L., 2017. Inorganic Membranes in Water and Wastewater Treatment. Springer Science and Business Media LLC, pp. 121–154, https://doi.org/10.1007/978-981-10-5623-9_5.

Cui, Z.F., Jiang, Y., Field, R.W., 2010. Fundamentals of Pressure-Driven Membrane Separation Processes Membrane Technology. Elsevier Ltd, United Kingdom, pp. 1–18. Available from: http://www.sciencedirect.com/science/book/9781856176323. https://doi.org/10.1016/B978-1-85617-632-3.00001-X.

Deng, L., Guo, W., Ngo, H.H., Zhang, H., Wang, J., Li, J., Xia, S., Wu, Y., 2016. Biofouling and control approaches in membrane bioreactors. Bioresour. Technol. 221, 656–665. Elsevier Ltd, China, 18732976. Available from: www.elsevier.com/locate/biortech. https://doi.org/10.1016/j.biortech.2016.09.105.

Drews, A., 2010. Membrane fouling in membrane bioreactors—characterisation, contradictions, cause and cures. J. Membr. Sci. 363 (1-2), 1–28. https://doi.org/10.1016/j.memsci.2010.06.046. 03767388.

Du, X., Shi, Y., Jegatheesan, V., Haq, I.U., 2020. A review on the mechanism, impacts and control methods of membrane fouling in MBR system. Membranes 10 (2), 24. https://doi.org/10.3390/membranes10020024. 2077-0375.

Ezugbe, E.O., Rathilal, S., 2020. Membrane technologies in wastewater treatment: a review. Membranes 10 (5). MDPI AG, South Africa, 20770375. Available from: https://www.mdpi.com/2077-0375/10/5/89/pdf. https://doi.org/10.3390/membranes10050089.

Fard, A.K., McKay, G., Buekenhoudt, A., Al Sulaiti, H., Motmans, F., Khraisheh, M., Atieh, M., 2018. Inorganic membranes: preparation and application for water treatment and desalination. Materials 11 (1), 74. https://doi.org/10.3390/ma11010074. 1996-1944.

Garcia, X., Pargament, D., 2015. Reusing wastewater to cope with water scarcity: economic, social and environmental considerations for decision-making. Resour. Conserv. Recycl. 101, 154–166. https://doi.org/10.1016/j.resconrec.2015.05.015. 09213449.

Geise, G.M., Lee, H.S., Miller, D.J., Freeman, B.D., McGrath, J.E., Paul, D.R., 2010. Water purification by membranes: the role of polymer science. J. Polym. Sci. B Polym. Phys. 48 (15), 1685–1718. United States, 10990488. Available from: http://www3.interscience.wiley.com/cgi-bin/fulltext/123563298/PDFSTART. https://doi.org/10.1002/polb.22037.

Giorno, L., Gebreyohannes, A.Y., Drioli, E., Mazzei, R., 2017. Biocatalytic Membranes and Membrane Bioreactors.

Giwa, A., Ahmed, M., Hasan, S.W., 2018. Polymers for Membrane Filtration in Water Purification. Springer Science and Business Media LLC, pp. 167–190, https://doi.org/10.1007/978-3-030-00743-0_8.

Goh, P.S., Ismail, A.F., 2018. A review on inorganic membranes for desalination and wastewater treatment. Desalination 434, 60–80. https://doi.org/10.1016/j.desal.2017.07.023. 00119164.

Gude, V.G., 2017. Desalination and water reuse to address global water scarcity. Rev. Environ. Sci. Biotechnol. 16 (4), 591–609. Springer Netherlands, United States, 15729826. Available from: http://www.kluweronline.com/issn/1569-1705. https://doi.org/10.1007/s11157-017-9449-7.

Hakami, M.W., Alkhudhiri, A., Al-Batty, S., Zacharof, M.P., Maddy, J., Hilal, N., 2020. Ceramic microfiltration membranes in wastewater treatment: filtration behavior, fouling and prevention. Membranes 10 (9), 1–34. MDPI AG, Saudi Arabia, 20770375. Available from: https://www.mdpi.com/2077-0375/10/9/248/pdf. https://doi.org/10.3390/membranes10090248.

Helmi, A., Gallucci, F., 2020. Latest developments in membrane (bio)reactors. Processes 8 (10), 1239. https://doi.org/10.3390/pr8101239. 2227-9717.

Hoek, E.M.V., Tarabara, V.V., 2013. Membrane Bioreactors, Applications to Wastewater Treatment and Reuse Encyclopedia of Membrane Science and Technology. Wiley, pp. 1–17, https://doi.org/10.1002/9781118522318.emst150.

Iorhemen, O.T., Hamza, R.A., Tay, J.H., 2016. Membrane bioreactor (MBR) technology for wastewater treatment and reclamation: membrane fouling. Membranes 6 (2). MDPI AG, Canada, 20770375. Available from: http://www.mdpi.com/2077-0375/6/2/33/pdf. https://doi.org/10.3390/membranes6020033.

Jabbari, B., Jalilnejad, E., Ghasemzadeh, K., Iulianelli, A., 2019. Recent progresses in application of membrane bioreactors in production of biohydrogen. Membranes 9 (8), 100. https://doi.org/10.3390/membranes9080100. 2077-0375.

Jalilnejad, E., Sadeghpour, P., Ghasemzadeh, K., 2019. Advances in membrane bioreactor technology. In: Current Trends and Future Developments on (Bio-) Membranes: Ceramic Membrane Bioreactors. Elsevier, Iran, pp. 1–29. Available from: http://www.sciencedirect.com/science/book/9780128168226. https://doi.org/10.1016/B978-0-12-816822-6.00001-X.

Jalilnejad, E., Sadeghpour, P., Ghasemzadeh, K., 2020. Achievements in membrane bioreactors for wastewater and water treatment. In: Current Trends and Future Developments on (Bio-) Membranes: Recent Achievements in Wastewater and Water Treatments. Elsevier, Iran, pp. 187–220. Available from: https://www.elsevier.com/books/current-trends-and-future-developments-on-bio-membranes/basile/978-0-12-817378-7. https://doi.org/10.1016/B978-0-12-817378-7.00007-0.

Jose, A.J., Kappen, J., Alagar, M., 2018. Polymeric membranes: classification, preparation, structure physiochemical, and transport mechanisms. In: Fundamental Biomaterials: Polymers. Elsevier Inc., India, pp. 21–35. Available from: https://www.sciencedirect.com/science/book/9780081021941. https://doi.org/10.1016/B978-0-08-102194-1.00002-5.

Judd, S., 2008. The status of membrane bioreactor technology. Trends Biotechnol. 26 (2), 109–116. https://doi.org/10.1016/j.tibtech.2007.11.005, 01677799.

Judd, S.J., 2016. The status of industrial and municipal effluent treatment with membrane bioreactor technology. Chem. Eng. J. 305, 37–45. https://doi.org/10.1016/j.cej.2015.08.141. 13858947.

Jyoti, J., Alka, D., Kumar, S.J., 2013. Application of membrane-bio-reactor in wastewater treatment: a review. Int. J. Chem. Chem. Eng. 3 (2), 115–122.

Khulbe, K.C., Feng, C., Matsuura, T., 2010. The art of surface modification of synthetic polymeric membranes. J. Appl. Polym. Sci. 115 (2), 855–895. https://doi.org/10.1002/app.31108. 0021-8995.

Kraume, M., Drews, A., 2010. Membrane bioreactors in waste water treatment—status and trends. Chem. Eng. Technol. 33 (8), 1251–1259. https://doi.org/10.1002/ceat.201000104. 0930-7516.

Lavrnić, S., Zapater-Pereyra, M., Mancini, M.L., 2017. Water scarcity and wastewater reuse standards in Southern Europe: focus on agriculture. Water Air Soil Pollut. 228 (7). https://doi.org/10.1007/s11270-017-3425-2. 0049-6979.

Le, N.L., Nunes, S.P., 2016. Materials and membrane technologies for water and energy sustainability. Sustain. Mater. Technol. 7, 1–28. https://doi.org/10.1016/j.susmat.2016.02.001. 22149937.

Le-Clech, P., Chen, V., Fane, T.A.G., 2006. Fouling in membrane bioreactors used in wastewater treatment. J. Membr. Sci. 284 (1-2), 17–53. https://doi.org/10.1016/j.memsci.2006.08.019. 03767388.

Lee, M., Wu, Z., Li, K., 2015. Advances in Ceramic Membranes for Water Treatment. Elsevier BV, pp. 43–82, https://doi.org/10.1016/b978-1-78242-121-4.00002-2.

Lee, A., Elam, J.W., Darling, S.B., 2016. Membrane materials for water purification: design, development, and application. Environ. Sci. Water Res. Technol. 2 (1),

17–42. Royal Society of Chemistry, United States, 20531419. Available from: http://pubs.rsc.org/en/journals/journalissues/ew#!recentarticles%26adv. https://doi.org/10.1039/c5ew00159e.

Lefebvre, O., Ng, K.K., Tang, K.Y., Ng, T.C.A., Ng, H.Y., 2019. Hybrid Processes, New Generation Membranes and Novel MBR Designs.

Li, C., Sun, W., Zedong, L., Ao, X., Li, S., 2020. Ceramic nanocomposite membranes and membrane fouling: a review. Water Res. 175, 115674. https://doi.org/10.1016/j.watres.2020.115674. 00431354.

Lim, Y.J., Lee, S.M., Wang, R., Lee, J., 2021. Emerging materials to prepare mixed matrix membranes for pollutant removal in water. Membranes 11 (7). MDPI AG, Singapore, 20770375. Available from: http://www.mdpi.com/journal/membranes. https://doi.org/10.3390/membranes11070508.

Maaz, M., Yasin, M., Aslam, M., Kumar, G., Atabani, A.E., Idrees, M., Anjum, F., Jamil, F., Ahmad, R., Khan, A.L., Lesage, G., Heran, M., Kim, J., 2019. Anaerobic membrane bioreactors for wastewater treatment: novel configurations, fouling control and energy considerations. Bioresour. Technol. 283, 358–372. Elsevier Ltd, Pakistan, 18732976. Available from: www.elsevier.com/locate/biortech. https://doi.org/10.1016/j.biortech.2019.03.061.

Madaeni, S.S., Ghaemi, N., Rajabi, H., 2015. Advances in Polymeric Membranes for Water Treatment. Elsevier BV, pp. 3–41, https://doi.org/10.1016/b978-1-78242-121-4.00001-0.

Madhura, L., Kanchi, S., Sabela, M.I., Singh, S., Bisetty, K., Inamuddin, 2018. Membrane technology for water purification. Environ. Chem. Lett. 16 (2), 343–365. Springer Verlag, South Africa, 16103661. Available from: https://www.springer.com/journal/10311. https://doi.org/10.1007/s10311-017-0699-y.

Maletskyi, Z., 2020. Advances in Membrane Materials and Processes for Water and Wastewater Treatment. ACS Symposium Series, vol. 1348, pp. 3–35. American Chemical Society, Norway. Available from: http://global.oup.com/academic/content/series/a/acs-symposium-series-acsss/?cc=in%26lang=en. https://doi.org/10.1021/bk-2020-1348.ch001.

Melin, T., Jefferson, B., Bixio, D., Thoeye, C., De Wilde, W., De Koning, J., van der Graaf, J., Wintgens, T., 2006. Membrane bioreactor technology for wastewater treatment and reuse. Desalination 187 (1-3), 271–282. https://doi.org/10.1016/j.desal.2005.04.086. 00119164.

Meng, F., Chae, S.R., Drews, A., Kraume, M., Shin, H.S., Yang, F., 2009. Recent advances in membrane bioreactors (MBRs): membrane fouling and membrane material. Water Res. 43 (6), 1489–1512. Elsevier Ltd, China, 00431354. Available from: www.elsevier.com/locate/watres. https://doi.org/10.1016/j.watres.2008.12.044.

Mestre, S., Gozalbo, A., Lorente-Ayza, M.M., Sánchez, E., 2019. Low-cost ceramic membranes: a research opportunity for industrial application. J. Eur. Ceram. Soc. 39 (12), 3392–3407. https://doi.org/10.1016/j.jeurceramsoc.2019.03.054. 09552219.

Miller, D.J., Dreyer, D.R., Bielawski, C.W., Paul, D.R., Freeman, B.D., 2017. Surface modification of water purification membranes. Angew. Chem. Int. Ed. 56 (17), 4662–4711. 15213773. Available from: http://onlinelibrary.wiley.com/journal/10.1002/(ISSN)1521-3773. https://doi.org/10.1002/anie.201601509.

Ng, A.N.L., Kim, A.S., 2007. A mini-review of modeling studies on membrane bioreactor (MBR) treatment for municipal wastewaters. Desalination 212 (1-3), 261–281. https://doi.org/10.1016/j.desal.2006.10.013. 00119164.

Nikolaidis, P., Poullikkas, A., 2017. A comparative overview of hydrogen production processes. Renew. Sustain. Energy Rev. 67, 597–611. Elsevier Ltd, Cyprus,

18790690. Available from: https://www.journals.elsevier.com/renewable-and-sustainable-energy-reviews. https://doi.org/10.1016/j.rser.2016.09.044.

Ozgun, H., Dereli, R.K., Ersahin, M.E., Kinaci, C., Spanjers, H., Van Lier, J.B., 2013. A review of anaerobic membrane bioreactors for municipal wastewater treatment: integration options, limitations and expectations. Sep. Purif. Technol. 118, 89–104. https://doi.org/10.1016/j.seppur.2013.06.036. 13835866.

Pereira, L.S., Duarte, E., Fragoso, R., 2014. Water Use: Recycling and Desalination for Agriculture. Elsevier BV, pp. 407–424, https://doi.org/10.1016/b978-0-444-52512-3.00084-x.

Porcelli, N., Judd, S., 2010. Chemical cleaning of potable water membranes: a review. Sep. Purif. Technol. 71 (2), 137–143. https://doi.org/10.1016/j.seppur.2009.12.007. 13835866.

Purkait, M.K., Sinha, M.K., Mondal, P., Singh, R., 2018. Introduction to membranes. Interface Sci. Technol. 25, 1–37. Elsevier B.V., undefined. Available from: http://www.elsevier.com/wps/find/bookdescription.cws_home/BS_3710/description#description. https://doi.org/10.1016/B978-0-12-813961-5.00001-2.

Qiu, Q., Zhao, B.W., Qiu, L.P., 2018. Development of flat ceramic membrane technology on municipal wastewater treatment. IOP Conf. Ser. Mater. Sci. Eng. 392 (2), 022036. https://doi.org/10.1088/1757-899x/392/2/022036. 1757-899X.

Remanan, S., Sharma, M., Bose, S., Das, N.C., 2018. Recent advances in preparation of porous polymeric membranes by unique techniques and mitigation of fouling through surface modification. ChemistrySelect 3 (2), 609–633. Wiley-Blackwell, India, 23656549. Available from: http://onlinelibrary.wiley.com/journal/10.1002/(ISSN)2365-6549. https://doi.org/10.1002/slct.201702503.

Sadr, S.M.K., Saroj, D.P., 2015. Membrane technologies for municipal wastewater treatment. In: Advances in Membrane Technologies for Water Treatment: Materials, Processes and Applications. Elsevier Inc., United Kingdom, pp. 443–463. Available from: http://www.sciencedirect.com/science/book/9781782421214. https://doi.org/10.1016/B978-1-78242-121-4.00014-9.

Samaei, S.M., Gato-Trinidad, S., Altaee, A., 2018. The application of pressure-driven ceramic membrane technology for the treatment of industrial wastewaters—a review. Sep. Purif. Technol. 200, 198–220. Elsevier B.V., Australia, 18733794. Available from: http://www.journals.elsevier.com/separation-and-purification-technology/. https://doi.org/10.1016/j.seppur.2018.02.041.

Shevah, Y., 2019. Impact of Persistent Droughts on the Quality of the Middle East Water Resources. vol. 11 Elsevier BV, pp. 51–84, https://doi.org/10.1016/b978-0-12-815730-5.00003-x.

Sikosana, M.L., Sikhwivhilu, K., Moutloali, R., Madyira, D.M., 2019. Municipal wastewater treatment technologies: A review. Procedia Manuf. 35, 1018–1024. Elsevier B.V. South Africa, 23519789 http://www.journals.elsevier.com/procedia-manufacturing. https://doi.org/10.1016/j.promfg.2019.06.051.

Souza, V.C., Quadri, M.G.N., 2013. Organic-inorganic hybrid membranes in separation processes: a 10-year review. Braz. J. Chem. Eng. 30 (4), 683–700. https://doi.org/10.1590/s0104-66322013000400001. 0104-6632.

Thomas, O., Theraulaz, F., Vaillant, S., Pouet, M.F., 2007. Urban wastewater. In: Techniques and Instrumentation in Analytical Chemistry. vol. 27, pp. 189–216, https://doi.org/10.1016/S0167-9244(07)80010-2. 01679244 (Chapter 8).

Tolkou, A., Zouboulis, A., Samaras, P., 2014. The incorporation of ceramic membranes in MBR systems for wastewater treatment: advantages and patented new developments. Recent Pat. Eng. 8 (1), 24–32. Bentham Science Publishers B.V., Greece, 18722121. Available from: http://www.benthamdirect.org/pages/all_b_bypublication.php. https://doi.org/10.2174/1872212107666131126234626.

Van Der Bruggen, B., Vandecasteele, C., Van Gestel, T., Doyen, W., Leysen, R., 2003. A review of pressure-driven membrane processes in wastewater treatment and drinking water production. Environ. Prog. 22 (1), 46–56. https://doi.org/10.1002/ep.670220116.

Verweij, H., 2012. Inorganic membranes. Curr. Opin. Chem. Eng. 1 (2), 156–162. Elsevier Ltd, United States, 22113398. Available from: http://www.elsevier.com/wps/find/journaldescription.cws_home/725837/description#description. https://doi.org/10.1016/j.coche.2012.03.006.

Wang, Z., Ma, J., Tang, C.Y., Kimura, K., Wang, Q., Han, X., 2014. Membrane cleaning in membrane bioreactors: a review. J. Membr. Sci. 468, 276–307. Elsevier, China, 18733123. Available from: www.elsevier.com/locate/memsci. https://doi.org/10.1016/j.memsci.2014.05.060.

Warsinger, D.M., Chakraborty, S., Tow, E.W., Plumlee, M.H., Bellona, C., Loutatidou, S., Karimi, L., Mikelonis, A.M., Achilli, A., Ghassemi, A., Padhye, L.P., Snyder, S.A., Curcio, S., Vecitis, C.D., Arafat, H.A., Lienhard, J.H., 2018. A review of polymeric membranes and processes for potable water reuse. Prog. Polym. Sci. 81, 209–237. Elsevier Ltd, United States, 00796700. Available from: http://www.sciencedirect.com/science/journal/00796700. https://doi.org/10.1016/j.progpolymsci.2018.01.004.

Wen, Y., Yuan, J., Ma, X., Wang, S., Liu, Y., 2019. Polymeric nanocomposite membranes for water treatment: a review. Environ. Chem. Lett. 17 (4), 1539–1551. Springer Verlag, United States, 16103661. Available from: https://www.springer.com/journal/10311. https://doi.org/10.1007/s10311-019-00895-9.

Wintgens, T., Rosen, J., Melin, T., Brepols, C., Drensla, K., Engelhardt, N., 2003. Modelling of a membrane bioreactor system for municipal wastewater treatment. J. Membr. Sci. 216 (1-2), 55–65. Elsevier, Germany, 03767388. Available from: www.elsevier.com/locate/memsci. https://doi.org/10.1016/S0376-7388(03)00046-2.

Wintgens, T., Melin, T., Schäfer, A., Khan, S., Muston, M., Bixio, D., Thoeye, C., 2005. The role of membrane processes in municipal wastewater reclamation and reuse. Desalination 178 (1-3), 1–11. https://doi.org/10.1016/j.desal.2004.12.014. 00119164.

Wisniewski, C., 2007. Membrane bioreactor for water reuse. Desalination 203 (1-3), 15–19. https://doi.org/10.1016/j.desal.2006.05.002. 00119164.

Wu, J., Zhang, Y., Wang, J., Zheng, X., Chen, Y., 2021. Municipal wastewater reclamation and reuse using membrane-based technologies: a review. Desalin. Water Treat. 224, 65–82. Desalination Publications, China, 19443986. Available from: https://www.deswater.com/readfulltextopenaccess.php?id=RFdUX2FydGljbGVzL3ZvbF8yMjRfcGFwZXJzLzIyNF8yMDIxXzY1LnBkZg. https://doi.org/10.5004/dwt.2021.27175.

Yang, Z., Sun, P.F., Li, X., Gan, B., Wang, L., Song, X., Park, H.D., Tang, C.Y., 2020a. A critical review on thin-film nanocomposite membranes with interlayered structure: mechanisms, recent developments, and environmental applications. Environ. Sci. Technol. 54 (24), 15563–15583. American Chemical Society, Hong Kong, 15205851. Available from: http://pubs.acs.org/journal/esthag. https://doi.org/10.1021/acs.est.0c05377.

Yang, J., Monnot, M., Ercolei, L., Moulin, P., 2020b. Membrane-based processes used in municipal wastewater treatment for water reuse: state-of-the-art and performance analysis. Membranes 10 (6), 1–56. MDPI AG, France, 20770375. Available from: https://www.mdpi.com/2077-0375/10/6/131/pdf. https://doi.org/10.3390/membranes10060131.

Zanetti, F., De Luca, G., Sacchetti, R., 2010. Performance of a full-scale membrane bioreactor system in treating municipal wastewater for reuse purposes. Bioresour. Technol. 101 (10), 3768–3771. https://doi.org/10.1016/j.biortech.2009.12.091. 09608524.

Zeynali, R., Akbari, M., Ghasemzadeh, K., Jalilnejad, E., Basile, A., 2020. Achievements in high pressure membrane processes NF and RO for wastewater and water treatment. In: Current Trends and Future Developments on (Bio-) Membranes: Recent Achievements in Wastewater and Water Treatments. Elsevier, Iran, pp. 109–126. Available from: https://www.elsevier.com/books/current-trends-and-future-developments-on-bio-membranes/basile/978-0-12-817378-7. https://doi.org/10.1016/B978-0-12-817378-7.00004-5.

Zou, D., Fan, Y., 2021. State-of-the-art developments in fabricating ceramic membranes with low energy consumption. Ceram. Int. 47 (11), 14966–14987. Elsevier Ltd, China, 02728842. Available from: https://www.journals.elsevier.com/ceramics-international. https://doi.org/10.1016/j.ceramint.2021.02.195.

8

Membrane aerated biofilm reactors

Ahmed Tawfik[a], Nawaf S. Alhajeri[a], Fahad M. Al-Fadhli[b], and Mohamed Elsamadony[c,d]

[a]Department of Environmental Sciences, College of Life Sciences, Kuwait University, Safat, Kuwait. [b]Department of Chemical Engineering, College of Engineering and Petroleum, Kuwait University, Safat, Kuwait. [c]Department of Mechanical Engineering, King Fahd University of Petroleum & Minerals, Dhahran, Saudi Arabia. [d]Interdisciplinary Research Center for Refining & Advanced Chemicals, King Fahd University of Petroleum & Minerals, Dhahran, Saudi Arabia

Abbreviations

A^2O	anaerobic/anoxic/oxic
C/N	carbon to nitrogen ratio
COD	chemical oxygen demand
EPS	extracellular polymeric substances
HRT	hydraulic retention time
L-DOPA	levodopa
MABR	membrane aerated biofilm reactor
PDMS	polydimethylsiloxane
PP	polypropylene
PTFE	polytetrafluoroethylene
PVDF	polyvinylidene fluoride
TN	total nitrogen
TOC	total organic carbon

8.1 Introduction

Traditional activated sludge, moving bed biofilm reactors, and sequencing batch reactor technologies are commonly used for various types of wastewaters (Allam et al., 2015). These technologies are efficient for organic degradation (Tawfik et al., 2023; Ismail et al., 2021a; Elsheikh et al., 2013). However, the treated effluent often contains high residual levels of nitrogen and phosphorus, which have negative effects on the environment

Municipal Wastewater Treatment. https://doi.org/10.1016/B978-0-443-24826-9.00008-5

(Tawfik et al., 2021; Wazeri et al., 2018). Conversely, membrane bioreactors are highly efficient for organic and nutrient removal and can potentially replace conventional treatment processes (Tawfik and Elsamadony, 2018). The main drawbacks of applying membrane technology for full-scale wastewater treatment are the operational costs and fouling (Tawfik et al., 2023; Elsharkawy et al., 2020). Classical wastewater treatment technologies consume energy and lead to secondary pollution issues, such as sludge containing micropollutants and hazardous chemicals (Elsamadony et al., 2015). There's an urgent need for additional funding for sludge bioremediation and management to safeguard our environment (Alhajeri et al., 2024; Mostafa et al., 2023). The MABR presents a promising alternative technology for effective wastewater treatment, particularly focusing on nutrient removal without generating excess sludge (Mei et al., 2020).

The operational mode of MABR differs significantly from classical membrane technologies, where the permeable membrane functions to supply oxygen necessary for the oxidation of organics and ammonia by heterotrophs and nitrifiers within the attached biofilm (Tawfik et al., 2022b). Oxygen transfer efficiencies in MABR can reach 100%. Unlike in classical membranes, the permeable membrane in MABR serves as a carrier for the biofilm containing active bacteria and enzymes responsible for pollutant removal from wastewater (Syron et al., 2015). In MABR, the membrane acts as a bubble-free aeration system, evenly diffusing oxygen throughout the entire reaction tank, thus preventing the formation of dead zones (Renuka et al., 2018; Mei et al., 2020). Moreover, MABR exhibits low energy consumption, efficiency in organic and nutrient removal from wastewater, and generates no secondary pollution (Ismail et al., 2021b). Gas-permeable hollow fiber membranes are commonly used in MABR applications to remove organics and nutrients from wastewater. They facilitate oxygen transfer to the attached biofilm by controlling air bubbling (Lin et al., 2016; Tawfik et al., 2006). This unique feature of MABR enables simultaneous nitrification and denitrification processes, allowing nitrifiers and denitrifiers to coexist in the biofilm for nutrient removal at low temperatures with less energy supply and space requirements (Lan et al., 2018). Other advantages of MABR over conventional wastewater treatment systems include a 70–90% reduction in aeration energy, high oxygen transfer rates to bacteria in the attached biofilm, a high specific surface area of hollow fiber membranes for biofilm formation, and the absence of secondary pollution due to air diffusion and stripping. The stability of MABR operation is attributed to the dense membrane biofilm and hydrophobic micropores (Wei et al., 2012b).

MABR has been utilized for removing refractory pollutants from wastewater, including nutrients, formaldehyde, perchloroethylene, acetonitrile, pharmaceuticals, chemical oxygen demand (COD), organic matter, and 1,2-dichloroethane (Li et al., 2016b; Li and Zhang, 2018; Wang et al., 2019). Li and Zhang (2018) employed a 470 L polyvinylidene fluoride (PVDF) microporous MABR with a specific surface area of 54.1 m^2/m^3 for contaminated water treatment, achieving removal efficiencies of 80% for total organic carbon (TOC), 92.2% for NH_4-N, and 60.3% for total nitrogen (TN) within 12–48 h, indicating simultaneous nitrification/denitrification. Additionally, the MABR system achieved a 92–95% removal efficiency for perchloroethylene from wastewater with an initial influent concentration of 5.5 mg/L and an hydraulic retention time (HRT) of 44 days (Ohandja and Stuckey, 2010). Li et al. (2008a,b) successfully eliminated acetonitrile, 83.3% of TN, and 98.6% of TOC from the wastewater industry using an up-flow fluid velocity of 12 cm/s, a loading rate of 11.3 g/m^2 day, and an HRT of 30 h. A pilot scale MABR system removed 80–99% of NH_4-N and 80–83% of chemical oxygen demand (COD) from landfill leachate at oxygen transfer rate of 35 g/m^2 day oxygen and influent 500–2500 mg/L for ammonia and 1000–3000 mg/L for COD (Xia et al., 2024). Terada et al. (2003) treated swine wastewater with the MABR system, achieving 96% removal for TOC and 83% for TN. The reactor also removed 83% of ammonia and 65% of tetrabromobisphenol A from contaminated water (Potvin et al., 2012). The chapter aims to analyze MABR technology's effectiveness in treating wastewater, with a focus on nutrients, refractory pollutants, and organics. It reviews the operational principles, characteristics, and influencing factors of MABR, while also evaluating microbial processes for nitrogen removal, including partial nitritation and anammox (Ismail et al., 2021b).

8.2 Membrane materials (microporous, composite, and dense)

The membrane material is the key component in MABR technology, facilitating oxygen diffusion into biological reactions and serving as carriers for biofilm. MABR membranes are typically classified as microporous, composite, or dense materials. Organic polymers like PVDF, polypropylene (PP), and polytetrafluoroethylene (PTFE) are commonly used to produce microporous membranes due to their easy assembly and affordability. These membranes, both hydrophobic and hydrophilic, enable bubble-free aeration for the biofilm community (Hou et al., 2013; Lu et al., 2022). Standalone MABR systems made from microporous carbon tubes combined with non-woven porous fabric have shown enhanced autotrophic nitrogen removal, owing to their high biomass retention time (Gong et al., 2008; Liu et al., 2008). The membranes, initially derived from carbon source materials, are relatively new, offering affordability and high chemical and thermal stability (Liu et al., 2008). However, they suffer from poor radial stress resistance and a low surface area. To address this, low-cost nylon silk carriers were integrated into MABR, facilitating bubble-free aeration for removing pollutants from wastewater (Ma et al., 2022). Blockage of the membrane pores could occur due to biofilm growth, secretion of extracellular polymeric substances, and cell debris, inevitably reducing oxygen transfer flux in the microporous membrane during operation (Casey et al., 2000). Consequently, microporous membranes are suitable for small-scale applications only and require further research to mitigate fouling processes.

Dense membranes were utilized in the MABR for treating wastewater. The mass transfer mechanism in these dense membranes relies on solution diffusion (Nagy, 2004). Two types of dense membranes include silicone and polydimethylsiloxane, which facilitate excellent membrane interface pressures. However, dense membranes are characterized by thick walls, high resistance to oxygen gas transfer onto the materials, and consequently, low aeration flux. Recently, there has been significant attention on composite membranes to address the challenges of fouling in microporous modules and the low oxygen gas flux in dense membrane units. The coating or immobilization of a polymer layer onto a microporous membrane module increases the pressure of bubble-free aeration, ensuring sufficient oxygen saturation for the removal of pollutants from wastewater (Ahmed et al., 2004; Cerqueira et al., 2013). The preparation of a composite membrane from L-3,3-dihydroxyphenylalanine (L-DOPA) represents a novel approach where the biological pollutant affinity and adhesive properties surpass those of other commercial materials (Hou et al., 2013; Lu et al., 2022). The PVDF membrane composite coated with levodopa (L-DOPA) exhibits a high oxygen gas flux and excellent bacterial adhesion. L-DOPA facilitates biofilm formation more effectively than the classical PVDF membrane (Hou et al., 2013). A major concern regarding the application of composite membranes is the challenge of controlling and achieving an accurately thick polymer coating. While no specific membrane materials tailored for MABR wastewater treatment currently exist, commercial membrane modules are available with their own set of advantages and drawbacks (Uri-Carreño et al., 2021).

OxyMem MABR, made from silicone-dense hollow fiber membranes with high oxygen gas permeability, is illustrated in Fig. 8.1A. In this configuration, air is fed into the module, and exhaust gas exits after oxygen transfer. The feed flow (liquid stream) enters and permeate flow (treated liquid) exits, demonstrating the process flow of both gaseous and liquid streams (Castrillo et al., 2019). Hollow fiber curtain membrane units, depicted in Fig. 8.1B, include ZeeLung MABR modules (Siddiqui et al., 2022; Chen et al., 2022) and HZH-MABR (Wei et al., 2012a; Lan et al., 2018). These units feature a significant surface area, promoting microorganism attachment and biofilm growth, which enhances the removal of recalcitrant and micropollutants from wastewater. Air enters at the top, and exhaust gas exits at the bottom, while the wastewater feed flows through the module and exits as permeate. Composite membranes, such as the Fluence MABR module, composed of spirally wound flat sheets of

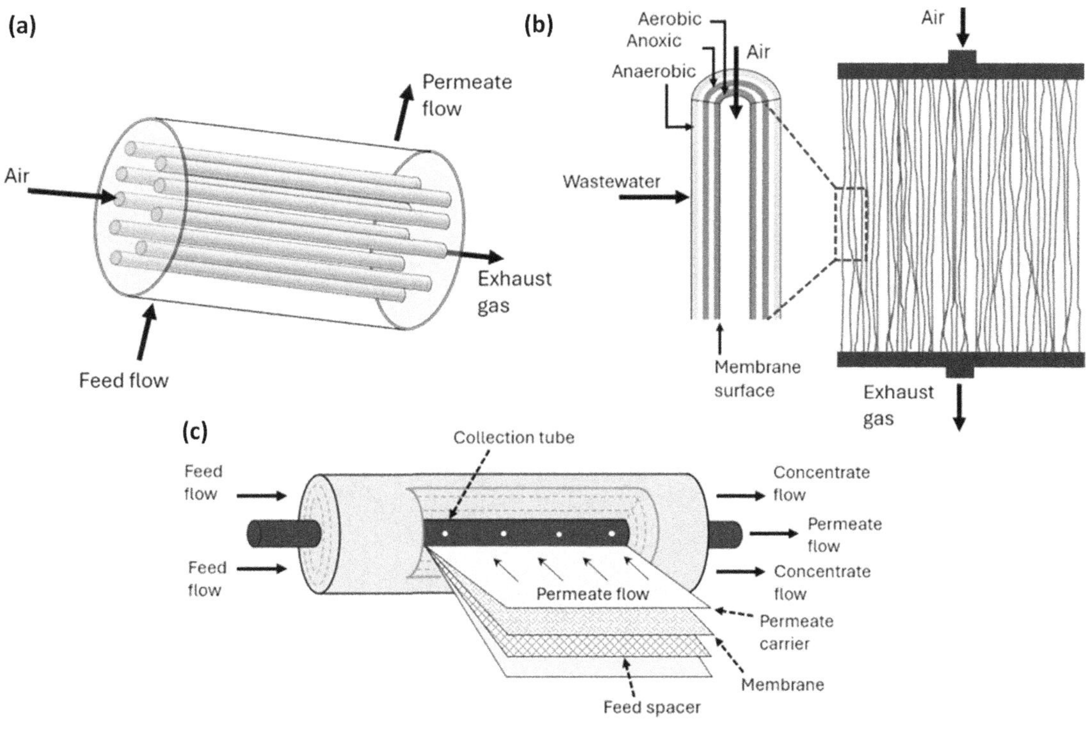

Fig. 8.1 Schematic representation of the MABR configurations (A) Hollow fiber bundles, (B) curtain shape hollow fiber module, and (C) spiral-wound flat membrane module.

polymer, are illustrated in Fig. 8.1C. Here, the bulk liquid containing wastewater is circulated through an upward spiral sheet, with airflow stirring from the bottom. Permeate flow and concentrate flow are indicated, showing the separation of treated water and concentrated waste. Oxygen transfer in these modules is facilitated through an open-end supply, enhancing mass transfer efficiency and preventing water vapor condensation. A hybrid MABR system, combining MABR and activated sludge, can significantly upgrade existing wastewater treatment plants, particularly for in situ applications.

8.3 Membrane aerated biofilm reactor operation mode and microbial community

A gas-permeable hydrophobic membrane in the MABR configuration is depicted in Fig. 8.2. This membrane allows oxygen to diffuse from the lumen side into the biofilm, where it is utilized by microbial degrading bacteria as an electron acceptor for removing pollutants, particularly organics and ammonia, from wastewater (Casey et al., 2000). The hollow fiber membrane unit effectively separates the reactor into two zones: the gas phase (lumen side) and the liquid phase (shell side) (Xia et al., 2024). Biofilm growth occurs on the membrane unit where air is diffused through the filter to the biofilm (Bunse et al., 2020). Stratification of the biofilm into aerobic, anoxic, and anaerobic zones allows for sequential processes such as nitrification (ammonia oxidation

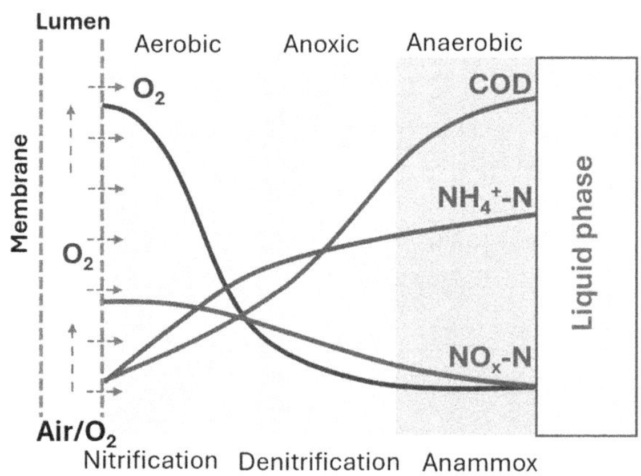

Fig. 8.2 Schematic of ideal oxygen and pollution delivery model in MABRs. *Solid lines* indicate the conceptual concentration profiles of COD, oxygen, NH_4^+-N, and NO_x-N inside the biofilm.

in the aerobic zone), denitrification (reduction of nitrogen oxides in the anoxic zone), and anammox (anaerobic ammonium oxidation). This stratification is crucial for achieving efficient organics removal and simultaneous nitrification/denitrification (Lin et al., 2016). In MABR systems, the dissolved oxygen diffuses through the membrane and is utilized in the aerobic zone for the oxidation of COD and NH_4^+-N. As oxygen concentration decreases, the biofilm transitions to anoxic conditions, facilitating denitrification where NO_x-N is reduced. In the deeper anaerobic zone, further biological processes occur, contributing to comprehensive wastewater treatment. The standalone MABR module can be used for wastewater treatment systems and can also be combined with ozonation or activated carbon for enhanced treatment (Li et al., 2015).

The biofilm within the MABR module can develop naturally within 3 days after adding the substrate during the startup period (Li et al., 2016b; Tawfik and Salem, 2012) or by introducing oxic bacteria from the aeration tank of an activated sludge unit treating wastewater (Wang et al., 2012). Bacterial populations in the MABR modules form on the surface due to adhesion and attraction forces, with dominance observed by Ciliates, Vorticella, and Rotatoria (Li et al., 2016b). Acid orange 7 decolorization was achieved using the MABR system under an oxygen supply pressure of 0.03 MPa, resulting in the removal of 98% of color and 80% of COD within 6 h, with dominance by Shewanella sp. (Wang et al., 2012). The MABR effectively removed fluoroacetate from wastewater by 91.5% using *P. fluorescens* DSM8341. Additionally, MABR with a biofilm rich in *P. knackmussii* B13 strains was utilized for the removal of 4-fluorobenzoate, although the degradation efficiency decreased over time due to fluorine accumulation (Li et al., 2016b). Nitrosospira dominated in intermittently aerated MABRs at low oxygen supply, ensuring stable nitrification under changing conditions (Elad et al., 2023).

8.4 Removal of hazardous pollutants from municipal wastewater

The high persistence of hazardous pollutants in municipal wastewater represents a serious environmental and health problems. The conventional treatment process fails to completely remove hazardous pollutants, especially xenobiotic compounds from municipal wastewater. Biofilm reactors are favored due to their extended sludge residence time and resilience to shock load conditions (Mei et al., 2019; Tawfik et al., 2022a).

The MABR was very efficient for removal of hazardous pollutants from municipal wastewater. Nevertheless, the efficiency of MABR depends on the bacterial cell density and biofilm thickness as reported by Sanchez-Huerta et al. (2023). The module removed COD and ammonia by values exceeding 50%. Furthermore, acetaminophen and triclosan was removed by values of 70% and 99% at biofilm thickness (<0.33 mm) and cell density of 2.9×10^5 cell/mL. Ninety percent of nonpolar, hydrophobic and hydrophilic organic micropollutants were removed at biofilm thickness of 1.02 mm and cell density 2.2×10^6 cell/mL due to the abundance of Zoogloea, Aqua-bacterium, Runella, Paludilbaculum and Leucobacter in the biofilm. Biodegradation of hazardous micropollutants i.e. acetaminophen is the main removal mechanism in the MABR (Sanchez-Huerta et al., 2023). Sorption of lipophilic compounds like triclosan was dominance in MABR due to increase of extracellular polymeric substances (EPS) protein fraction. A two-stage MABR was employed for treatment of municipal wastewater containing o-aminophenol (Tian et al., 2019). TN was removed by 90% with glucose addition and Pseudomonas and Nitrosomonas were the dominance species. Ninety-nine percent of o-aminophenol was removed by unclassified Sphingomonadaceae, Lysobacter, Cupriavidus, Thauera, and Azotobacter. Acetonitrile in the municipal wastewater was removed by MABR at hydraulic retention times of 4, 6, 9, and 12 h and surface loading rates (Kunlasubpreedee and Visvanathan, 2020). Mechanical aeration caused stripping of 49% for acetonitrile. The removal of acetonitrile (98%), total organic carbon (87%), chemical oxygen demand (COD) (84%), and total nitrogen (69%) was optimized at an HRT of 6.0 h.

Complete degradation of formaldehyde-rich municipal wastewater occurred in an MABR system (Fig. 8.3). Formaldehyde undergoes oxidation into formic acid within the aerobic biofilm zone, followed by reduction into methanol by bacteria in both aerobic and anoxic biofilms. Subsequently, formaldehyde undergoes polymerization in the anaerobic biofilm zone, forming long-chain polymers. These polymers are then degraded into 2 and 3-carbon volatile fatty acids by acid-producing bacteria. The resulting intermediate by-products are mineralized into carbon dioxide and water by acidophilus bacteria within the aerobic biofilm (Oliveira et al., 2004; Ismail and Tawfik, 2016). In an MABR operating at a formaldehyde influent concentration of 116.3 mg/L, with an HRT of 4.0 h and an aeration pressure of 0.01 MPa, removal efficiencies were achieved: 99.9% for formaldehyde, 81.5% for COD, and 97.14% for methanol (Mei et al., 2019).

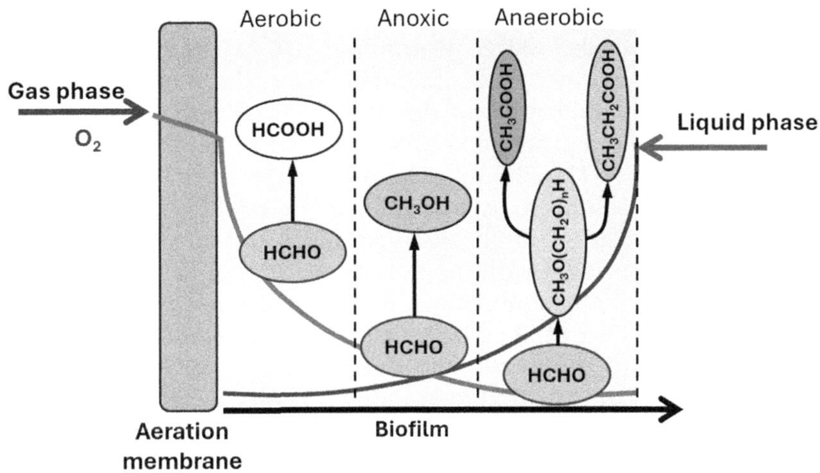

Fig. 8.3 Degradation pathways of formaldehyde in MABR system (Mei et al., 2019).

The investigation by Tian et al. (2019) examined an integrated system for treating pharmaceutical wastewater, comprising ozone/MABR/coagulation/flocculation. It achieved a removal of 95% for COD and 92% for TN, starting at initial concentrations of 2058 mg/L for COD and 62.6 mg/L for TN. In the MABR process, hollow fiber membranes with a surface area of 287 m^2/m^3 attained removal efficiencies of 90% for COD and 98% for ammonia, with organic loading rates of 1311 g COD/m^3 day and 48.2 g NH_4^+-N/m^3 day, respectively (Wei et al., 2012b). Similarly, Li et al. (2016b) utilized MABR/ozone/activated carbon to treat oil field wastewater, achieving removal rates of 90.6% for COD, 78.1% for TN, and 87.9% for oil.

8.5 Membrane aerated biofilm reactor pores clogging

The classical membrane bioreactor encounters fouling, creating a barrier to its application due to the need for backwashing and restarting biological activities. The membrane separates treated wastewater from suspended growth bacteria, with aeration occurring in this suspended growth. The MABR was proposed to overcome such obstacles, especially the fouling process, and enhance the quality of the treated effluent. Biofilms form on the membrane surface of MABR, with bacterial cells producing extracellular polymeric substances (EPS) to maintain mechanical stability and counter detachment/sloughing phenomena (Wang et al., 2022; Flemming and Wingender, 2010). Additionally, matured bacteria detach from the biofilm into the bulk liquid, stabilizing and digesting themselves, releasing polysaccharides and proteins as EPS (LaMotta, 1976).

The EPS migrate to the outer layer of the membrane surface, leading to pore clogging in the MABR system. Regrettably, divalent and/or multivalent cations in the bulk liquid can form carboxyl bridges with the polysaccharides of EPS, significantly increasing the accumulation of these enzymes on the membrane surface (Xin et al., 2016). EPS accumulate/adsorb on the aerated layer of the membrane due to the presence of terminal hydrophilic and hydrophobic groups. Under water pressure, EPS enter the membrane pores through an adsorption process, causing membrane clogging by attracting hydrophilic and hydrophobic organic substances. This process changes the originally hydrophobic membrane into hydrophilic pores (Wang et al., 2018). The adsorption capacity of organics from wastewater increases due to the enhanced hydrophilicity of the aerated membrane surface (Xiao et al., 2011). Consequently, this leads to a decrease in oxygen transfer and subsequent deterioration of reactor efficiency.

8.6 Membrane aerated biofilm reactor fouling

The application of MABR for municipal wastewater treatment introduces a unique process for removal of pollutants. This module configuration overcome most of the serious fouling problem in classical membrane bioreactor. Nevertheless, biofilm formation and microbial secretion i.e. EPS caused a fouling of MABR and clogging the pores of membrane resulting a low oxygen transfer for bacterial degradation process (Wang et al., 2022). Organic matter particularly colloidal matter in the wastewater could penetrate the pores of membrane and adsorbed on the walls resulting an accumulation of foulants between the aeration membrane and the biofilm. Algae and quorum quenching minimized membrane fouling due to the reduction of bacterial soluble metabolites by-products and polysaccharides production (Sun et al., 2020). Coupling of biochar and voltage reduced protein/polysaccharide ratio in EPS, minimizing the EPS adsorption performance on the hydrophobic membrane (Jiang et al., 2020). The microbial interactions with pollutants at the MABRs are necessary to establish the stability of the biofilms to accomplishment biological activities. The biofilm thickness in MABRs is the main parameter affecting the efficiency and performance of this technology. Controlling the thickness of biofilm will minimize the membrane fouling causing a success of MABRs treating wastewater. A thick intensive biofilm and enzymes activities will highly reduce the dissolved oxygen penetration distance in the biofilm resulting a reduction in oxygen transfer rate and subsequently low substrate removal rate. Likely, insufficient biofilm formation (thin biofilm thickness) in the MABRs treating wastewater reduces the substrate removal due to the bacterial and biomass limitations (Martin et al., 2013). The hydrophobic membranes are mainly employed in MABRs to minimize the fouling and microbial attachment due to a lack of hydrogen bonding interactions at the boundary layer between the liquid interface and membrane. The water molecules are mainly repulsed from the membrane hydrophobic surface allowing microbial cells or the flount from bulk liquid to adsorb to the surface of membrane dominating in the boundary layer. Membrane biofouling was reduced using hydrophilic surfaces where the water molecules adsorption is occurred (Matar et al., 2016). However, Li and Zhang (2018) found that the fouling of MABRs is mainly affected by the interaction between membrane surface charge and foulant particles. A minor effect of membrane hydrophilicity on the membrane fouling was reported by the authors. Song et al. (2015) found that the surface charge determined the adhesion force between the microbes and membrane surface. Surface positive charge membrane promoted negative charge microbial adhesion at neutral pH. The surface negatively charged membrane prevent microbial and bacterial attachment to membranes (Song et al., 2015). Surface membrane roughness is an important factor affecting the biofilm formation in MABRs (Zhao et al., 2016).

The membrane fouling, biofilm formation and microbial adhesion mechanism are illustrated in Fig. 8.4. The bacterial cells

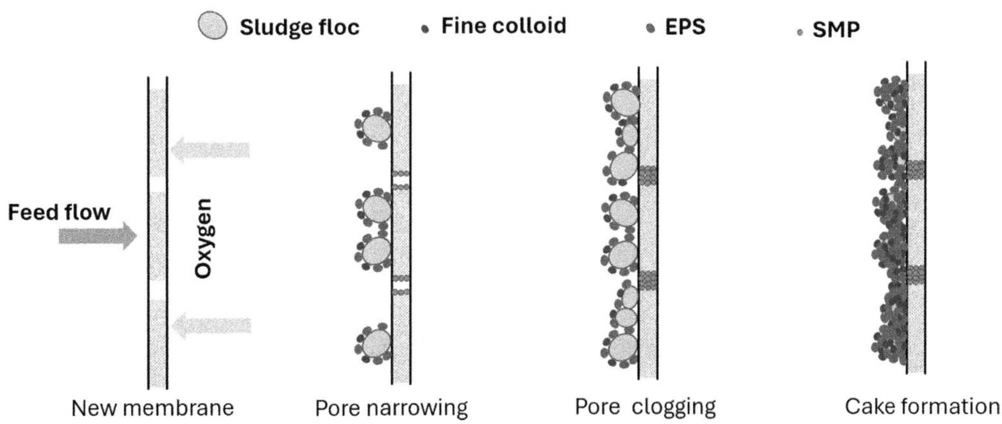

Fig. 8.4 MABR fouling; biofilm formation and pore clogging in microporous gas permeable membrane and biofilm formation in dense gas permeable membrane.

adhesion is initially taken place at the surface of gas permeable membranes forming thin biofilm. This followed by biofilm growing (thickness) with the oxygen and substrates supplying. The biofilm may be detached or sloughed from the membrane surface particularly under shear stress conditions, and eventually attained a stable biofilm thickness. Back diffusion of water vapor into the lumen side of the membrane creates suitable conditions for the growth of biofilm inside the pores resulting in pore clogging. Pore clogging of the membrane is minor for dense gas permeable membranes, and only biofilm formation is occurred at the membrane surface. EPS and soluble microbial products play an important role not only for membrane foulants but also in biofilm formation in MABRs. Periodically intensive air scouring minimizes the fouling and controls biofilm thickness in MABRs (Wang et al., 2022). Microbial predation can control the biofouling in MABRs. Klein et al. (2016) found that the use of Aelosoma hemprichi and Plectus aquatilis (metazoans) increased the membrane flux by values of 50% and 119–164% respectively, compared to the control (no predators). The predators at low COD concentration provided large fraction (45.4%) of voids in MABR compared (4%) of voids with high COD concentration in the MABR (Aybar et al., 2019).

8.7 Factors affecting the performance of membrane aerated biofilm reactor

The design parameters for application of MABR for treatment of municipal wastewater is important. Hence, the effect of hydraulic retention time, biocarrier materials, carbon to nitrogen (C/N) ratio, predation, substrate and nutrients limitations on the performance of MABR are comprehensively reviewed here.

8.7.1 Effect of hydraulic retention time on the organics and nutrients removal

The hydraulic retention time (HRT) represents the reactor's volumetric capacity (L) divided by the imposed flow rate (L/h) (Mostafa et al., 2020; Elsharkawy et al., 2019). This parameter is crucial for microbial and metabolic activities, impacting the degradation efficiency of organics and the removal of nutrients from wastewater (Elsamadony et al., 2021; Soltan et al., 2019). Despite advancements, scaling up MABR technology for wastewater treatment remains challenging due to limited nitrogen removal efficiency at short HRT, primarily caused by low denitrification efficiency (Premarathna and Visvanathan, 2019). While COD and ammonia removal are effective in pilot-scale MABR systems, total nitrogen (TN) removal ranges from 40% to 66% at low HRTs of 2.4–7.5 h (Li and Zhang, 2018). TN removal efficiency surpasses 60% at longer HRTs of 15.0 h (Li and Zhang, 2018; Downing et al., 2010). Wei et al. (2012b) reported 45–60% of TN removal efficiencies at longer HRT of 36–50 h., respectively. Wei et al. (2012b), Li et al. (2016a), and Downing et al. (2010) achieved TN removal efficiencies of 70–75%, at mentation an HRT exceeding 15 h. However, a longer HRT makes the application of MABR technology less commercially attractive to stakeholders due to the large reactor volume requirement. To address this significant issue, researchers implemented a pilot-scale membrane aerated biofilm reactor (MABR) for treating domestic wastewater. They introduced immobilized denitrifying bacteria on polyvinyl alcohol gel bio-carriers to enhance total nitrogen removal (Siriweera et al., 2021). The researchers operated the modified MABR reactor at HRTs of 6, 9, and 12 h, observing that the addition of bio-carriers increased nitrification efficiency and TN removal efficiencies by 11% and 18%, respectively, at an HRT of 12 h. Using the modified MABR, they achieved 88% nitrification efficiency, 79% TN removal, and 90% organic removal due to the high metabolic activities of heterotrophs, nitrifiers, and denitrifiers. Introducing bio-carriers to the MABR extended the sludge residence time and reduced the required HRT for pollutant removal from wastewater. Denitrifying bacteria on the polyvinyl alcohol gel bio-carriers consumed most organics in the bulk liquid, reducing biofilm thickness on the membrane. This facilitated ammonia penetration to the accumulated nitrifiers on the membrane surface for efficient nitrification. Additionally, the synergistic effect of nitrification/denitrification was regulated by the alkalinity of the bulk liquid, where denitrifiers produced the required alkalinity for nitrifiers to oxidize ammonia into nitrate fractions. Thant et al. (2023) discovered that HRT of 12.5 h is necessary for the successful formation of biofilm in an MABR reactor and for aerobic autotrophs to achieve maximum nitrification efficiency. Table 8.1 summarizes the recent results for MABR treating municipal wastewater at different operational conditions.

8.7.2 Effect of addition biocarrier materials (hybrid MABR performance)

The primary obstacle to the application of MABR technology for nutrient removal is the insufficient elimination of total nitrogen (TN) by denitrifiers (Ismail et al., 2019). An anoxic unit

Table 8.1 Results for MABR treating municipal wastewater.

Configuration	Membrane type	HRT (h)	Carbon removal (%)	References
One-stage MABR	Dense polydimethylsiloxane (PDMS) membrane	108–156	Effluent COD soluble: 200–500 mg/L	Syron et al. (2015)
	OxyMem MABR PDMS membrane	5.2–18	26–64% COD soluble removal	He et al. (2021)
	ZeeLung MABR dense membrane	0.38	N/A	Houweling et al. (2017)
	Fluence MABR membrane	N/A	N/A	He et al. (2021)
Hybrid MABR in the anoxic zone in the A^2O process	Fluence MABR membrane	1.3–5.5, average: 3.1	N/A	Sanchez-Huerta et al. (2023)
	ZeeLung dense MABR membrane	1.2–2.6	77.5% COD soluble removal	Peeters et al. (2017)
		Average: 1.6 Peak: 3	42% filtered BOD_5 removal	Carlson et al. (2021)
		2.36	Not reported	Li et al. (2008a)
		2.1, up to 3.1	Not reported	Peeters et al. (2017)
		0.37–5.93	56.66% COD soluble removed	Li et al. (2008a)
3-Stage hybrid MABRs in the anoxic zone in the MLE process	ZeeLung MABR dense membrane	1–3	N/A	Syron et al. (2015), Wang et al. (2022), and Carlson et al. (2021)

addition is necessary to achieve complete nitrogen removal, leading to increased costs and footprint requirements. Classical MABR systems only achieved a TN removal efficiency of 50%, which is unsatisfactory from an environmental perspective (Fu et al., 2009). Denitrification relies on nitrate produced by nitrifiers, highlighting the need for their presence. Enhancing nitrification/denitrification in MABR can be achieved by incorporating biocarrier materials for denitrifiers, while nitrifiers can easily accumulate on the membrane surface to facilitate ammonia oxidation into nitrate. Immobilizing denitrifiers on biocarriers is an alternative method for enhancing TN removal with MABR technology. Polyvinyl alcohol gel biocarriers are non-toxic, have a micro-porous structure, high surface area, a lifespan exceeding 10 years, and are cost-effective (Chaikasem et al., 2015). These biocarriers have demonstrated effectiveness in improving TN removal from wastewater (Premarathna and Visvanathan, 2019).

Incorporating polyvinyl alcohol gel biocarriers into the MABR system led to a 14% improvement in nitrification and a 13.4% enhancement in denitrification efficiency (Premarathna and Visvanathan, 2019). The hybrid denitrifiers biocarrier/MABR achieved maximum nitrification, denitrification, and organics removal efficiencies of 88%, 79%, and 90%, respectively (Siriweera et al., 2021). Adding biocarriers with denitrifiers increased alkalinity, necessary for nitrifiers to oxidize ammonia. Moreover, immobilized denitrifiers on biocarriers consumed organics from wastewater, reducing biofilm formation on the membrane surface. This facilitated ammonia penetration through the thin biofilm to the inner surface layer, creating favorable conditions for the nitrification-denitrification process.

8.7.3 Effect of C/N ratio

Fig. 8.5 illustrates the response of aerobic denitrifying bacteria to varying C/N ratios from a metabolic division perspective. At a decreased C/N ratio, there is enhanced microbial assimilation characterized by improved nitrate removal, increased carbon validity, higher cell yield, greater intracellular-N proportion, and elevated transcriptional expression levels of assimilation genes. Conversely, an increased C/N ratio fosters enhanced microbial respiration, leading to reduced nitrate removal, decreased carbon validity, lower cell yield, a reduced intracellular-N proportion, and heightened transcriptional expression levels of respiration genes. The turning region, with a C/N ratio of 5.0–7.5, marks the transition between these two metabolic states. In this context, organics and nutrients are essential elements for microbial growth and biofilm formation. Bacterial metabolism deteriorates due to an imbalance in nutrients and organics. The organic fraction in wastewater consumes the most oxygen and promotes the growth of heterotrophic organisms. Nitrogen is necessary for the growth of slow-growing nitrifiers, which compete with heterotrophs for space and oxygen in wastewater-fed reactors. MABR technology may contain both heterotrophs and nitrifiers, but their efficiency depends greatly on the C/N ratio imposed. The COD/N ratio in domestic wastewater varies from 5 to 8 (Sun et al., 2010;

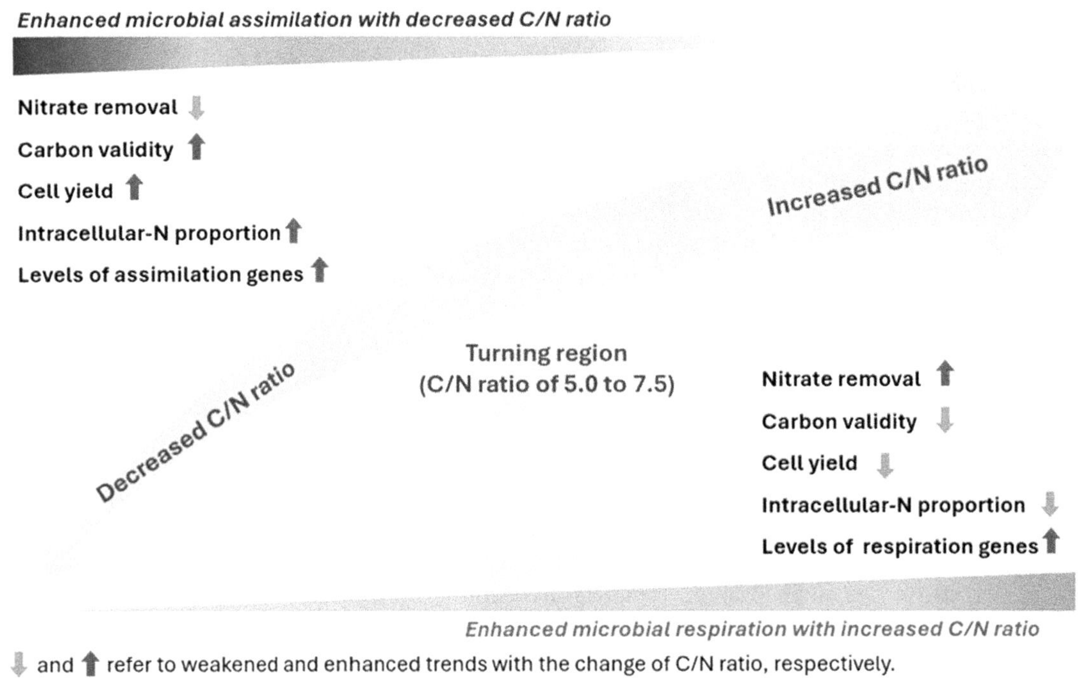

Fig. 8.5 Impact of carbon-to-nitrogen ratios on aerobic denitrifying bacteria in MABR.

Semmens et al., 2003), with variations based on the wastewater source and seasonal changes.

An optimum COD/N ratio of 10 for biocarrier/MABR achieved an average nitrification efficiency of 82% and TN removal efficiency of 63% at an HRT of 6 h (Siriweera et al., 2021). High COD/N ratios favor organic-consuming bacteria like aerobic heterotrophs but inhibit the growth and metabolic activities of nitrifiers. Low imposed COD/N ratios encourage the growth of slow nitrifier organisms by limiting electron donors (organics) to heterotrophs and denitrifiers. A COD/N ratio exceeding 20 in black water treated in MABR at an HRT of 39–50 h achieved ammonia removal efficiency of 98% and TN elimination of 70% (Wei et al., 2012b).

8.7.4 Effect of predation on heterotrophic MABR (predation changing the biofilm characteristics)

The composition of the biofilm, which includes active bacteria and a microbial community, plays a crucial role in removing pollutants using MABR. Oxygen penetration into the biofilm primarily depends on its thickness and adhesive layers. Predation significantly impacts biofilm formation and, consequently, the efficiency of MABR in treating low-strength wastewater, where protozoa can thrive (Aybar et al., 2019). Protozoa exhibit high growth rates under aerobic conditions, utilizing carbon and nitrogen as energy sources. However, their presence in the bulk liquid of MABR can diminish their activity by grazing the outer layer of the biofilm (anoxic/anaerobic) and the base of the aerobic layer, thus reducing pollutant removal efficiency. Amoebae, ciliates, and flagellates are the primary predatory protozoa in biofilm systems, playing a crucial role in the treatment process (Madoni, 2011). Amoebae influence the growth pattern and structure of the biofilm, while flagellates thrive in the outer biofilm and consume its attached biomass (Dopheide et al., 2011). Ciliates effectively remove the biofilm by creating holes and channels using their cilia (Klein et al., 2016).

Predation initiates within the aerobic base layer of the biofilm in the MABR treating wastewater, where oxygen is sufficient for microbial metabolic activities. This process leads to void layers in the aerobic biofilm, primarily due to inadequate penetration of electron donors for microbial activities. Enhanced penetration of electron donors reduces biofilm predation, as microbial growth surpasses predation, countering void formation. Void formation within the biofilm promotes biomass sloughing, thereby diminishing MABR efficiency. Kim et al. (2020) noted that predation in MABR increased the internal void ratio of the biofilm from 6% to 50%, with a biofilm loss of 44%, reduced to 7% in the absence of protozoans. This observation introduces a new characteristic of the MABR biofilm, including increased porosity, reduced mechanical strength, and enhanced potential for sloughing/detachment. Understanding this phenomenon can aid in developing more effective strategies for biofilm control in MABRs. Within the MABR biofilm, the deepest layer is the aerobic layer, where insufficient penetration of organics (electron donors) and nutrients can facilitate easy predation. This leads to the formation of voids/spaces and detachment of biomass, consequently decreasing the efficiency of MABR in treating wastewater (Nerenberg, 2016). Insufficient organics and limited nutrients in the aerobic layer of the biofilm decrease microbial growth and impair MABR efficiency. Additionally, this oxygen-rich aerobic layer fosters predator growth, reducing biofilm thickness and increasing voids. These findings are corroborated by other researchers. Aybar et al. (2019) observed that inadequate COD substrate (38 mg/L) increased MABR biofilm void ratios to 45.4%. Void ratios decreased at higher COD concentrations. Nutrient limitations escalate biofilm predation rates by protozoa (flagellates and ciliates) (Nygaard and Tobiesen, 1993). Constraints on organic substrate and nutrients at the aerobic biofilm base layer intensify planktonic microorganism growth within existing voids, augmenting protozoa (flagellates and ciliates) growth and predation.

8.7.5 Effect of substrate and nutrients limitations on the detachment of biofilm

The implementation of MABR for wastewater treatment requires a thorough understanding of biofilm behavior under both ideal and stressful conditions. Unlike conventional biofilm systems, where the outer layer is aerobic and directly linked to the bulk liquid containing available substrate and nutrients, MABR presents a different scenario. In MABR, the deepest layer consists of aerobic microbial biofilm, leading to challenges such as limited substrate and nutrient penetration, resulting in biomass sloughing/detachment (Nerenberg, 2016). However, the concentration of dissolved oxygen (electron acceptor) in the MABR is notably high, while the penetration of the electron donor (substrate) is relatively low, which can impact the efficiency of wastewater treatment. In conventional biofilm systems, the interaction between electron donors and bacteria mainly occurs in the outer layer of the biofilm, where microbial metabolic activities are predominant. Aybar et al. (2019) observed that in MABR, the biofilm formed rapidly at a low COD concentration of 58 mg/L, without predators, resulting in a thickness of 220 mm and a void fraction of 7%. Over time, the void fraction increased to 45% after 80 h due to predation conditions and limitations of COD for the microbial community in the deepest layer of the biofilm. Conversely, the high oxygen levels in the deepest aerobic layer of the membrane promote the growth of predators at an initial COD value of 60 mg/L. Voids form in the inner layer of the biofilm due to high dissolved oxygen and insufficient COD for bacterial growth, enhancing the growth of protozoa and increasing biomass detachment. Similarly, voids may be present in the outer layer of the biofilm due to limitations in dissolved oxygen and excess substrate availability. Increasing the influent COD up to 267 mg/L in the MABR resulted in a biofilm thickness of 223 mm and a lower void fraction of 3.9% in the presence of protozoa. The reduced void fractions in the biofilm were attributed to higher COD (substrate) penetration into the inner layer of the biofilm, resulting in increased microbial metabolic activity at high dissolved oxygen levels.

Hunt et al. (2004) discovered that voids formed at the base of the MABR biofilm when the COD (substrate) was limited, leading to the degradation of EPS and detachment of the biofilm due to predation and starvation. Amoebae consume the biofilm content, resulting in the loss of attached active biomass (Weitere et al., 2005). These organisms can easily penetrate the inner surface of MABR biofilms, creating more void fractions and disrupting the biofilm structure (Raghupathi et al., 2018). Moreover, the high dissolved oxygen in the inner biofilm layer attracts amoebae to the aerobic base of the MABR (Fenchel, 2014). Ciliates also contribute to biofilm detachment in MABR by feeding on the suspended fraction of attached biomass, creating large internal voids (Böhme et al., 2009).

8.7.6 Effect of substrate conditions on the microbial community

The biofilm content mainly represents the primary biological and metabolic activities in the MABR. Microbial composition within the biofilm is influenced by substrate concentration and aeration conditions. The Proteobacteria phylum predominates in MABR biofilms, with its relative abundance ranging from 54% at a low initial COD of 60 mg/L to 65% at a higher COD of 267 mg/L (Aybar et al., 2019). Betaproteobacteria (Dechloromonas and Zoogloea genera) and Bacteroidetes (Flavobacterium genus) are prevalent in MABR biofilms, exhibiting high activity in the presence of electron donors (organics) and dissolved oxygen. At a high initial COD of 267 mg/L, amoebae (Naegleria spp.) constitute 64% of the biofilm, which decreases to 42% at a low COD value of 60 mg/L. Additionally, other amoeba species, namely 5.0% (Vermamoeba spp.) and 8.0% (Acanthamoeba spp.), were detected in high-COD biofilms. The limited penetration of low COD through the biofilm enhances the growth of Discicristata flagellates by 20% (Aybar et al., 2019). This active group of flagellates, along with amoebae, significantly contributes to the creation of voids in the attached biofilm by feeding on the biomass (Parry, 2004).

The porosity of the biofilm in the MABR is increased fourfold compared to the control due to the relative abundance of Spumella spp., a flagellated protozoan, in the reactor (Böhme et al., 2009). The pores created in the MABR biofilm are distributed within the layers of its matrix. This differs from the findings of Aybar et al. (2019), who observed voids and porosity occurring on the inner surface of the attached biofilm. Protozoan abundance was noted in the internal void fractions of the MABR biofilm under low COD influent. The outer layer of the MABR biofilm experiences anaerobic conditions with high COD availability, minimizing biofilm predation and detachment. Although the thin biofilm facilitates COD penetration to the inner bacteria, it reduces detachment. However, the sludge residence time may be insufficient for heterotrophs and nitrifiers to effectively remove organics and nutrients. Further research is necessary for the advanced application of this technology in wastewater treatment.

In summary, the biofilm formation, thickness, and oxygen transfer rate are the most parameters affecting the efficiency of MABR treating municipal wastewater. Controlling the biofilm formation and fouling process will facilitate the application of the technology at full scale replacing the conventional treatment processes. Predation of the biofilm reduces the fouling process but negatively affected the treatment efficiency and could be utilized only for low strength wastewater. COD/N ratio is the major parameter affecting the nitrification/denitrification process and shifting the microbial activities. This parameter should be controlled to achieve a high nitrogen removal and enhance the growth of the nitrifiers and heterotrophs. The dissolved oxygen is the main parameter affecting the performance of the MABR treating wastewater where the nitrifiers and heterotrophs removed nitrogen and organics.

8.8 Conclusions and future trends

MABR technology offers a distinctive feature by integrating the benefits of biofilm systems and oxygen bubble-free aeration processes. It serves as an energy-saving and efficient solution for wastewater treatment, particularly focusing on nutrient and organics removal. MABR, when combined with innovative bacterial heterotrophic biofilm for organics/micropollutants removal and autotrophic microbial community for nitrogen removal, contributes to reducing the carbon footprint of wastewater treatment plants. However, the cost of membrane material needs further reduction to enable its widespread application in low-income countries for wastewater treatment. There is a demand for new membrane configurations, materials, and unique components to enhance the bio-affinity of microorganisms, increase surface area, and mitigate biofilm membrane fouling.

MABRs have demonstrated great potentials for refractory pollutants, TN and COD removal from municipal wastewaters. However, several challenges remain, including biofilm thickness and pore clogging control, availability of low cost and efficient membrane materials. In-depth fundamental understanding and mechanism of biofilm formation and stability is needed for application of MABRs in wastewater treatment.

Predation within the biofilm took place in the deep aerobic layer, prompted by the challenge of substrate and nutrient penetration to microbial communities, resulting in void spaces within the heterotrophic MABR. This issue could be significantly mitigated by elevating the organic loading rate and diminishing the oxygen supply. Further investigation is required in nitrifying MABR to evaluate predator behavior in the presence of slowly growing nitrifiers. The role of hydrophobicity, surface roughness, surface charge in the fouling and biofilm formation in gas permeable membranes of MABR needs further studies. Controlling biofilm thickness, soluble microbial by-products and EPS is essential for successful application of MABRs full-scale.

Acknowledgments

This work was financially supported by College of Life Sciences, Department of Environmental Sciences, Kuwait University.

References

Ahmed, T., Semmens, M.J., Voss, M.A., 2004. Oxygen transfer characteristics of hollow-fiber, composite membranes. Adv. Environ. Res. 8 (3–4), 637–646. https://doi.org/10.1016/S1093-0191(03)00036-4.

Alhajeri, N.S., Tawfik, A., Elsamadony, M., Al-Fadhli, F.M., Meng, F., 2024. Synergistic algal/bacterial interaction in membrane bioreactor for detoxification of 1,2-dichloroethane-rich petroleum wastewater. J. Hazard. Mater. 470. Elsevier B.V., Kuwait, 18733336. Available from: https://www.sciencedirect.com/science/journal/03043894. https://doi.org/10.1016/j.jhazmat.2024.134125.

Allam, A., Fleifle, A., Tawfik, A., Yoshimura, C., El-Saadi, A., 2015. A simulation-based suitability index of the quality and quantity of agricultural drainage water for reuse in irrigation. Sci. Total Environ. 536, 79–90. https://doi.org/10.1016/j.scitotenv.2015.07.029. 00489697.

Aybar, M., Perez-Calleja, P., Li, M., Pavissich, J.P., Nerenberg, R., 2019. Predation creates unique void layer in membrane-aerated biofilms. Water Res. 149, 232–242. https://doi.org/10.1016/j.watres.2018.10.084. 00431354.

Böhme, A., Risse-Buhl, U., Küsel, K., 2009. Protists with different feeding modes change biofilm morphology. FEMS Microbiol. Ecol. 69 (2), 158–169. https://doi.org/10.1111/j.1574-6941.2009.00710.x. 15746941.

Bunse, P., Orschler, L., Agrawal, S., Lackner, S., 2020. Membrane aerated biofilm reactors for mainstream partial nitritation/anammox: experiences using real municipal wastewater. Water Res. X 9, 100066. https://doi.org/10.1016/j.wroa.2020.100066. 25899147.

Carlson, A.L., He, H., Yang, C., Daigger, G.T., 2021. Comparison of hybrid membrane aerated biofilm reactor (MABR)/suspended growth and conventional biological nutrient removal processes. Water Sci. Technol. 83 (6), 1418–1428. IWA Publishing, United States, 19969732. Available from: https://watermark.silverchair.com/wst083061418.pdf. https://doi.org/10.2166/wst.2021.062.

Casey, E., Glennon, B., Hamer, G., 2000. Biofilm development in a membrane-aerated biofilm reactor: effect of flow velocity on performance. Biotechnol. Bioeng. 67 (4), 476–486. https://doi.org/10.1002/(sici)1097-0290(20000220)67:4<476::aid-bit11>3.0.co;2-2. 0006-3592.

Castrillo, M., Díez-Montero, R., Esteban-García, A.L., Tejero, I., 2019. Mass transfer enhancement and improved nitrification in MABR through specific membrane configuration. Water Res. 152, 1–11. https://doi.org/10.1016/j.watres.2019.01.001. 00431354.

Cerqueira, A.C., Nobrega, R., Sant'Anna, G.L., Dezotti, M., 2013. Oxygen air enrichment through composite membrane: application to an aerated biofilm reactor. Braz. J. Chem. Eng. 30 (4), 771–779. Brazil, 01046632. Available from: http://www.scielo.br/pdf/bjce/v30n4/a09v30n4.pdf. https://doi.org/10.1590/S0104-66322013000400009.

Chaikasem, S., Jacob, P., Visvanathan, C., 2015. Performance improvement in a two-stage thermophilic anaerobic membrane bioreactor using PVA-gel as biocarrier. Desalin. Water Treat. 53 (10), 2839–2849. Bellwether Publishing, Ltd., Thailand, 19443986. Available from: http://www.deswater.com/openaccess.php. https://doi.org/10.1080/19443994.2014.931531.

Chen, R., Cao, S., Zhang, L., Zhou, Y., 2022. NOB suppression strategies in a mainstream membrane aerated biofilm reactor under exceptionally low lumen pressure. Chemosphere 290, 133386. https://doi.org/10.1016/j.chemosphere.2021.133386. 00456535.

Dopheide, A., Lear, G., Stott, R., Lewis, G., 2011. Preferential feeding by the Ciliates Chilodonella and Tetrahymena spp. and effects of these protozoa on bacterial biofilm structure and composition. Appl. Environ. Microbiol. 77 (13), 4564–4572. https://doi.org/10.1128/aem.02421-10. 0099-2240.

Downing, L.S., Bibby, K.J., Esposito, K., Fascianella, T., Tsuchihashi, R., Nerenberg, R., 2010. Nitrogen removal from wastewater using a hybrid membrane-biofilm process: pilot-scale studies. Water Environ. Res. 82 (3), 195–201. Water Environment Federation, United States, 15547531. Available from: http://docserver.ingentaconnect.com/deliver/connect/wef/10614303/v82n3/s1.pdf?expires=1267502512%26id=55319098%26titleid=11548%26accname=Elsevier+Science%26checksum=1718D3FCC53E59978302636B74C4CF08. https://doi.org/10.2175/106143009X426103.

Elad, T., Hally, M.P., Domingo-Félez, C., Knoop, O., Drewes, J.E., Valverde-Pérez, B., Smets, B.F., 2023. Exploring the effects of intermittent aeration on the performance of nitrifying membrane-aerated biofilm reactors. Sci. Total Environ. 891. Elsevier B.V., Denmark, 18791026. Available from: www.elsevier.com/locate/scitotenv. https://doi.org/10.1016/j.scitotenv.2023.164329.

Elsamadony, M., Tawfik, A., Danial, A., Suzuki, M., 2015. Use of carica papaya enzymes for enhancement of H_2 production and degradation of glucose, protein, and lipids. Energy Procedia 75, 975–980. https://doi.org/10.1016/j.egypro.2015.07.308. 18766102.

Elsamadony, M., Elreedy, A., Mostafa, A., Fujii, M., Gescher, J., Shakeri Yekta, S., Schnürer, A., Gaillard, J.F., Pant, D., 2021. Perspectives on potential applications of nanometal derivatives in gaseous bioenergy pathways: mechanisms, life cycle, and toxicity. ACS Sustain. Chem. Eng. 9 (29), 9563–9589. American Chemical Society, Japan, 21680485. Available from: http://pubs.acs.org/journal/ascecg. https://doi.org/10.1021/acssuschemeng.1c02260.

Elsharkawy, K., Elsamadony, M., Afify, H., Guo, Y., 2019. Comparative analysis of common full scale reactors for dry anaerobic digestion process. E3S Web Conf. 83, 01011. https://doi.org/10.1051/e3sconf/20198301011. 2267-1242.

Elsharkawy, K., Alalm, M.G., Fujii, M., Afify, H., Tawfik, A., Elsamadony, M., 2020. Paperboard mill wastewater treatment via combined dark and LED-mediated fermentation in the absence of external chemical addition. Bioresour. Technol. 295, 122312. https://doi.org/10.1016/j.biortech.2019.122312. 09608524.

Elsheikh, M.A., Saleh, H.I., Rashwan, I.M., El-Samadoni, M.M., 2013. Hydraulic modelling of water supply distribution for improving its quantity and quality. Sustain. Environ. Res. 23 (6), 403–411. Chinese Institute of Environmental Engineering, Egypt, 24682039. Available from: https://link.springer.com/journal/42834.

Fenchel, T., 2014. Protozoa and oxygen. Acta Protozool. 53 (1), 3–12. Denmark, 16890027. Available from: http://www.eko.uj.edu.pl/ap/index.php?option=com_docman%26task=doc_view%26gid=269%26Itemid=27. https://doi.org/10.4467/16890027AP.13.0020.1117.

Flemming, H.C., Wingender, J., 2010. The biofilm matrix. Nat. Rev. Microbiol. 8 (9), 623–633. https://doi.org/10.1038/nrmicro2415. 17401526.

Fu, Z., Yang, F., Zhou, F., Xue, Y., 2009. Control of COD/N ratio for nutrient removal in a modified membrane bioreactor (MBR) treating high strength wastewater. Bioresour. Technol. 100 (1), 136–141. https://doi.org/10.1016/j.biortech.2008.06.006. 09608524.

Gong, Z., Liu, S., Yang, F., Bao, H., Furukawa, K., 2008. Characterization of functional microbial community in a membrane-aerated biofilm reactor operated for completely autotrophic nitrogen removal. Bioresour. Technol. 99 (8), 2749–2756. https://doi.org/10.1016/j.biortech.2007.06.040. 09608524.

He, H., Wagner, B.M., Carlson, A.L., Yang, C., Daigger, G.T., 2021. Recent progress using membrane aerated biofilm reactors for wastewater treatment. Water Sci. Technol. 84 (9), 2131–2157. IWA Publishing, United States, 19969732. Available from: https://watermark.silverchair.com/wst084092131.pdf. https://doi.org/10.2166/wst.2021.443.

Hou, F., Li, B., Xing, M., Wang, Q., Liang, H., Wang, S., 2013. Surface modification of PVDF hollow fiber membrane and its application in membrane aerated biofilm reactor (MABR). Bioresour. Technol. 140, 1–9. https://doi.org/10.1016/j.biortech.2013.04.056. 09608524.

Houweling, D., Peeters, J., Cote, P., Long, Z., Adams, N., 2017. Proving membrane aerated biofilm reactor (MABR) performance and reliability: results from four pilots and a full-scale plant. In: Water Environment Federation Technical Exhibition and Conference, WEFTEC 2017, Water Environment Federation Canada, 5, pp. 3420–3432. 9781510863682.

Hunt, S.M., Werner, E.M., Huang, B., Hamilton, M.A., Stewart, P.S., 2004. Hypothesis for the role of nutrient starvation in biofilm detachment. Appl. Environ. Microbiol. 70 (12), 7418–7425. https://doi.org/10.1128/AEM.70.12.7418-7425.2004.

Ismail, S., Tawfik, A., 2016. Performance of passive aerated immobilized biomass reactor coupled with Fenton process for treatment of landfill leachate. Int. Biodeterior. Biodegrad. 111, 22–30. https://doi.org/10.1016/j.ibiod.2016.04.010. 09648305.

Ismail, S., Elsamadony, M., Elreedy, A., Fujii, M., Tawfik, A., 2019. Physico-chemical and microbial characterization of compartment-wise profiles in an anammox baffled reactor. J. Environ. Manag. 232, 875–886. https://doi.org/10.1016/j.jenvman.2018.11.134. 03014797.

Ismail, S., Elreedy, A., Elsamadony, M., Abdelrazek, E., Fujii, M., Ni, S.-Q., Tawfik, A., 2021a. Response of anammox bacteria to short-term exposure of 1,4-dioxane: bacterial activity and community dynamics. Sep. Purif. Technol. 266, 118539. https://doi.org/10.1016/j.seppur.2021.118539. 13835866.

Ismail, S., Elreedy, A., Fujii, M., Ni, S.Q., Tawfik, A., Elsamadony, M., 2021b. Fatigue of anammox consortia under long-term 1,4-dioxane exposure and recovery potential: N-kinetics and microbial dynamics. J. Hazard. Mater. 414.

Elsevier B.V., China, 18733336. Available from: www.elsevier.com/locate/jhazmat. https://doi.org/10.1016/j.jhazmat.2021.125533.

Jiang, B., Zeng, Q., Liu, J., Hou, Y., Jin, X., Li, H., Shi, S., Ma, F., 2020. Enhanced treatment performance of phenol wastewater and membrane antifouling by biochar-assisted EMBR. Bioresour. Technol. 306, 123147. https://doi.org/10.1016/j.biortech.2020.123147. 09608524.

Kim, B., Perez-Calleja, P., Li, M., Nerenberg, R., 2020. Effect of predation on the mechanical properties and detachment of MABR biofilms. Water Res 186. https://doi.org/10.1016/j.watres.2020.116289.

Klein, T., Zihlmann, D., Derlon, N., Isaacson, C., Szivak, I., Weissbrodt, D.G., Pronk, W., 2016. Biological control of biofilms on membranes by metazoans. Water Res. 88, 20–29. Elsevier Ltd, Switzerland, 18792448. Available from: www.elsevier.com/locate/watres. https://doi.org/10.1016/j.watres.2015.09.050.

Kunlasubpreedee, P., Visvanathan, C., 2020. Performance evaluation of membrane-aerated biofilm reactor for acetonitrile wastewater treatment. J. Environ. Eng. 146 (7). https://doi.org/10.1061/(asce)ee.1943-7870.0001706. 0733-9372.

LaMotta, E.J., 1976. Internal diffusion and reaction in biological films. Environ. Sci. Technol. 10 (8), 765–769. https://doi.org/10.1021/es60119a003. 15205851.

Lan, M., Li, M., Liu, J., Quan, X., Li, Y., Li, B., 2018. Coal chemical reverse osmosis concentrate treatment by membrane-aerated biofilm reactor system. Bioresour. Technol. 270, 120–128. https://doi.org/10.1016/j.biortech.2018.09.011. 09608524.

Li, Y., Zhang, K., 2018. Pilot scale treatment of polluted surface waters using membrane-aerated biofilm reactor (MABR). Biotechnol. Biotechnol. Equip. 32 (2), 376–386. https://doi.org/10.1080/13102818.2017.1399826. 1310-2818.

Li, T., Bai, R., Liu, J., 2008a. Distribution and composition of extracellular polymeric substances in membrane-aerated biofilm. J. Biotechnol. 135 (1), 52–57. https://doi.org/10.1016/j.jbiotec.2008.02.011. 01681656.

Li, S.L., Kuo, S.C., Lin, J.S., Lee, Z.K., Wang, Y.H., Cheng, S.S., 2008b. Process performance evaluation of intermittent-continuous stirred tank reactor for anaerobic hydrogen fermentation with kitchen waste. Int. J. Hydrog. Energy 33 (5), 1522–1531. https://doi.org/10.1016/j.ijhydene.2007.09.049. 03603199.

Li, P., Zhao, D., Zhang, Y., Sun, L., Zhang, H., Lian, M., Li, B., 2015. Oil-field wastewater treatment by hybrid membrane-aerated biofilm reactor (MABR) system. Chem. Eng. J. 264, 595–602. https://doi.org/10.1016/j.cej.2014.11.131. 13858947.

Li, M., Li, P., Chunyu, D., Sun, L., Li, B., 2016a. Pilot-scale study of an integrated membrane-aerated biofilm reactor system on urban river remediation. Ind. Eng. Chem. Res. 55 (30), 8373–8382. https://doi.org/10.1021/acs.iecr.6b00143. 0888-5885.

Li, P., Li, M., Zhang, Y., Zhang, H., Sun, L., Li, B., 2016b. The treatment of surface water with enhanced membrane-aerated biofilm reactor (MABR). Chem. Eng. Sci. 144, 267–274. https://doi.org/10.1016/j.ces.2016.01.030. 00092509.

Lin, J., Zhang, P., Li, G., Yin, J., Li, J., Zhao, X., 2016. Effect of COD/N ratio on nitrogen removal in a membrane-aerated biofilm reactor. Int. Biodeterior. Biodegrad. 113, 74–79. https://doi.org/10.1016/j.ibiod.2016.01.009. 09648305.

Liu, S., Gong, Z., Yang, F., Zhang, H., Shi, L., Furukawa, K., 2008. Combined process of urea nitrogen removal in anaerobic anammox co-culture reactor. Bioresour. Technol. 99 (6), 1722–1728. https://doi.org/10.1016/j.biortech.2007.03.048. 09608524.

Lu, J.J., Shen, Q., Li, X.Y., Sun, F.Y., Yi, J.B., Dong, W.Y., Yan, W.J., Du, H., Mu, J.L., 2022. Surface-manipulated membranes to accelerate biofilm formation and to resist bacterial detachment in MBfR for aerobic methane oxidation coupled to denitrification. Chem. Eng. J. 430. Elsevier B.V., China, 13858947. Available from: www.elsevier.com/inca/publications/store/6/0/1/2/7/3/index.htt. https://doi.org/10.1016/j.cej.2021.132629.

Ma, Y., Piscedda, A., Veras, A.D.L.C., Domingo-Félez, C., Smets, B.F., 2022. Intermittent aeration to regulate microbial activities in membrane-aerated biofilm reactors: energy-efficient nitrogen removal and low nitrous oxide emission. Chem. Eng. J. 433. Elsevier B.V., Denmark, 13858947. Available from: www.elsevier.com/inca/publications/store/6/0/1/2/7/3/index.htt. https://doi.org/10.1016/j.cej.2021.133630.

Madoni, P., 2011. Protozoa in wastewater treatment processes: a minireview. Ital. J. Zool. 78 (1), 3–11. https://doi.org/10.1080/11250000903373797. 1125-0003.

Martin, K.J., Picioreanu, C., Nerenberg, R., 2013. Multidimensional modeling of biofilm development and fluid dynamics in a hydrogen-based, membrane biofilm reactor (MBfR). Water Res. 47 (13), 4739–4751. Elsevier Ltd, United States, 18792448. Available from: www.elsevier.com/locate/watres. https://doi.org/10.1016/j.watres.2013.04.031.

Matar, G., Gonzalez-Gil, G., Maab, H., Nunes, S., Le-Clech, P., Vrouwenvelder, J., Saikaly, P.E., 2016. Temporal changes in extracellular polymeric substances on hydrophobic and hydrophilic membrane surfaces in a submerged membrane bioreactor. Water Res. 95, 27–38. Elsevier Ltd, Saudi Arabia, 18792448. Available from: www.elsevier.com/locate/watres. https://doi.org/10.1016/j.watres.2016.02.064.

Mei, X., Guo, Z., Liu, J., Bi, S., Li, P., Wang, Y., Shen, W., Yang, Y., Wang, Y., Xiao, Y., Yang, X., Liu, Y., Zhao, L., Wang, Y., Hu, S., 2019. Treatment of formaldehyde wastewater by a membrane-aerated biofilm reactor (MABR): the degradation of formaldehyde in the presence of the cosubstrate methanol. Chem. Eng. J. 372, 673–683. Elsevier B.V., China, 13858947. Available from: www.elsevier.com/inca/publications/store/6/0/1/2/7/3/index.htt. https://doi.org/10.1016/j.cej.2019.04.184.

Mei, X., Ding, Y., Li, P., Lijie, X., Wang, Y., Guo, Z., Shen, W., Yang, Y., Wang, Y., Xiao, Y., Yang, X., Liu, Y., Shen, Y., Yuxin, W., Jiang, C., Xue, C., 2020. A novel system for zero-discharge treatment of high-salinity acetonitrile-containing wastewater: combination of pervaporation with a membrane-aerated bioreactor. Chem. Eng. J. 384, 123338. https://doi.org/10.1016/j.cej.2019.123338. 13858947.

Mostafa, A., Tolba, A., Alalm, M.G., Fujii, M., Afify, H., Elsamadony, M., 2020. Application of magnetic multi-wall carbon nanotube composite into fermentative treatment process of ultrasonicated waste activated sludge. Bioresour. Technol. 306, 123186. https://doi.org/10.1016/j.biortech.2020.123186. 09608524.

Mostafa, A., Elsamadony, M., Khalil, T.E., Elhusseiny, A.F., Tawfik, A., Fujii, M., El-dissouky, A., Pant, D., 2023. Bioelectrochemical system for enhancing anaerobic digestion of pharmaceutical-containing domestic wastewater. Chemosphere 339. Elsevier Ltd, Egypt, 18791298. Available from: www.elsevier.com/locate/chemosphere. https://doi.org/10.1016/j.chemosphere.2023.139766.

Nagy, E., 2004. Nonlinear, coupled mass transfer through a dense membrane. Desalination 163 (1–3), 345–354. https://doi.org/10.1016/s0011-9164(04)90207-x. 00119164.

Nerenberg, R., 2016. The membrane-biofilm reactor (MBfR) as a counter-diffusional biofilm process. Curr. Opin. Biotechnol. 38, 131–136. https://doi.org/10.1016/j.copbio.2016.01.015. 09581669.

Nygaard, K., Tobiesen, A., 1993. Bacterivory in algae: a survival strategy during nutrient limitation. Limnol. Oceanogr. 38 (2), 273–279. https://doi.org/10.4319/lo.1993.38.2.0273. 0024-3590.

Ohandja, D.G., Stuckey, D.C., 2010. Effect of perchloroethylene (PCE) and hydraulic shock loads on amembrane-aerated biofilm reactor (MABR) biodegrading PCE.

J. Chem. Technol. Biotechnol. 85 (2), 294–301. United Kingdom, 10974660. Available from: http://www3.interscience.wiley.com/cgi-bin/fulltext/123193657/PDFSTART. https://doi.org/10.1002/jctb.2307.

Oliveira, S.V.W.B., Moraes, E.M., Adorno, M.A.T., Varesche, M.B.A., Foresti, E., Zaiat, M., 2004. Formaldehyde degradation in an anaerobic packed-bed bioreactor. Water Res. 38 (7), 1685–1694. https://doi.org/10.1016/j.watres.2004.01.013. 00431354.

Parry, J.D., 2004. Protozoan grazing of freshwater biofilms. Adv. Appl. Microbiol. 54, 167–196. Academic Press Inc., United Kingdom, 00652164. Available from: http://www.sciencedirect.com/science/journal/00652164. https://doi.org/10.1016/S0065-2164(04)54007-8.

Peeters, J., Adams, N., Long, Z., Côté, P., Kunetz, T., 2017. Demonstration of innovative MABR low-energy nutrient removal technology at Chicago MWRD. Water Pract. Technol. 12 (4), 927–936. https://doi.org/10.2166/wpt.2017.096. 1751-231X.

Potvin, C.M., Long, Z., Zhou, H., 2012. Removal of tetrabromobisphenol A by conventional activated sludge, submerged membrane and membrane aerated biofilm reactors. Chemosphere 89 (10), 1183–1188. Elsevier Ltd, Canada, 18791298. Available from: www.elsevier.com/locate/chemosphere. https://doi.org/10.1016/j.chemosphere.2012.07.011.

Premarathna, N.H.S.M., Visvanathan, C., 2019. Enhancement of organic matter and total nitrogen removal in a membrane aerated biofilm reactor using PVA-Gel bio-carriers. Bioresour. Technol. Rep. 8. Elsevier Ltd, Thailand, 2589014X. Available from: https://www.journals.elsevier.com/bioresource-technology-reports. https://doi.org/10.1016/j.biteb.2019.100325.

Raghupathi, P.K., Liu, W., Sabbe, K., Houf, K., Burmølle, M., Sørensen, S.J., 2018. Synergistic interactions within a multispecies biofilm enhance individual species protection against grazing by a pelagic protozoan. Front. Microbiol. 8. Frontiers Media S.A., Belgium, 1664302X. Available from: https://www.frontiersin.org/articles/10.3389/fmicb.2017.02649/full. https://doi.org/10.3389/fmicb.2017.02649.

Renuka, N., Guldhe, A., Prasanna, R., Singh, P., Bux, F., 2018. Microalgae as multifunctional options in modern agriculture: current trends, prospects and challenges. Biotechnol. Adv. 36 (4), 1255–1273. https://doi.org/10.1016/j.biotechadv.2018.04.004. 07349750.

Sanchez-Huerta, C., Medina, J.S., Wang, C., Fortunato, L., Hong, P.Y., 2023. Understanding the role of sorption and biodegradation in the removal of organic micropollutants by membrane aerated biofilm reactor (MABR) with different biofilm thickness. Water Res. 236. Elsevier Ltd, Saudi Arabia, 18792448. Available from: www.elsevier.com/locate/watres. https://doi.org/10.1016/j.watres.2023.119935.

Semmens, M.J., Dahm, K., Shanahan, J., Christianson, A., 2003. COD and nitrogen removal by biofilms growing on gas permeable membranes. Water Res. 37 (18), 4343–4350. Elsevier Ltd, United States, 00431354. Available from: www.elsevier.com/locate/watres. https://doi.org/10.1016/S0043-1354(03)00416-0.

Siddiqui, M.A., Biswal, B.K., Siriweera, B., Chen, G., Wu, D., 2022. Integrated self-forming dynamic membrane (SFDM) and membrane-aerated biofilm reactor (MABR) system enhanced single-stage autotrophic nitrogen removal. Bioresour. Technol. 345, 126554. https://doi.org/10.1016/j.biortech.2021.126554.

Siriweera, W.B.S., Yun-Je, L., Masumi, K., Visvanathan, C., 2021. Organic matter and total nitrogen removal from wastewater using a pilot-scale membrane-aerated biofilm reactor. Bioresour. Technol. Rep. 15, 100817. https://doi.org/10.1016/j.biteb.2021.100817. 2589014X.

Soltan, M., Elsamadony, M., Mostafa, A., Awad, H., Tawfik, A., 2019. Harvesting zero waste from co-digested fruit and vegetable peels via integrated fermentation and pyrolysis processes. Environ. Sci. Pollut. Res. 26 (10), 10429–10438. https://doi.org/10.1007/s11356-019-04647-8. 0944-1344.

Song, F., Koo, H., Ren, D., 2015. Effects of material properties on bacterial adhesion and biofilm formation. J. Dent. Res. 94 (8), 1027–1034. https://doi.org/10.1177/0022034515587690. 0022-0345.

Sun, S.P., Nàcher, C.P.I., Merkey, B., Zhou, Q., Xia, S.Q., Yang, D.H., Sun, J.H., Smets, B.F., 2010. Effective biological nitrogen removal treatment processes for domestic wastewaters with low C/N ratios: a review. Environ. Eng. Sci. 27 (2), 111–126. https://doi.org/10.1089/ees.2009.0100. 10928758.

Sun, L., Ma, J., Lipin Li, Y., Tian, Z.Z., Liao, H., Li, J., Tang, W., He, D., 2020. Exploring the essential factors of performance improvement in sludge membrane bioreactor technology coupled with symbiotic algae. Water Res. 181, 115843. https://doi.org/10.1016/j.watres.2020.115843. 00431354.

Syron, E., Semmens, M.J., Casey, E., 2015. Performance analysis of a pilot-scale membrane aerated biofilm reactor for the treatment of landfill leachate. Chem. Eng. J. 273, 120–129. Elsevier, Ireland, 13858947. Available from: www.elsevier.com/inca/publications/store/6/0/1/2/7/3/index.htt. https://doi.org/10.1016/j.cej.2015.03.043.

Tawfik, A., Elsamadony, M., 2018. Development of dry anaerobic technologies of bio-waste and unlock the barriers for valorization. In: Optimization and Applicability of Bioprocesses. Springer Singapore, Egypt, pp. 267–282. Available from: http://www.springer.com/in/book/9789811068621. https://doi.org/10.1007/978-981-10-6863-8_13.

Tawfik, A., Salem, A., 2012. The effect of organic loading rate on bio-hydrogen production from pre-treated rice straw waste via mesophilic up-flow anaerobic reactor. Bioresour. Technol. 107, 186–190. https://doi.org/10.1016/j.biortech.2011.11.086. 09608524.

Tawfik, A., El-Gohary, F., Ohashi, A., Harada, H., 2006. The influence of physical–chemical and biological factors on the removal of faecal coliform through down-flow hanging sponge (DHS) system treating UASB reactor effluent. Water Res. 40 (9), 1877–1883. https://doi.org/10.1016/j.watres.2006.02.038. 00431354.

Tawfik, A., Hassan, G.K., Awad, H., Hassan, M., Rojas, P., Sanz, J.L., Elsamadony, M., Pant, D., Fujii, M., 2021. Strengthen "the sustainable farm" concept via efficacious conversion of farm wastes into methane. Bioresour. Technol. 341. Elsevier Ltd, Egypt, 18732976. Available from: www.elsevier.com/locate/biortech. https://doi.org/10.1016/j.biortech.2021.125838.

Tawfik, A., Eraky, M., Alhajeri, N.S., Osman, A.I., Rooney, D.W., 2022a. Cultivation of microalgae on liquid anaerobic digestate for depollution, biofuels and cosmetics: a review. Environ. Chem. Lett. 20 (6), 3631–3656. Springer Science and Business Media Deutschland GmbH, Egypt, 16103661. Available from: https://www.springer.com/journal/10311. https://doi.org/10.1007/s10311-022-01481-2.

Tawfik, A., Hasanan, K., Abdullah, M., Badr, O.A., Awad, H.M., Elsamadony, M., El-Dissouky, A., Qyyum, M.A., Nizami, A.S., 2022b. Graphene enhanced detoxification of wastewater rich 4-nitrophenol in multistage anaerobic reactor followed by baffled high-rate algal pond. J. Hazard. Mater. 424. Elsevier B.V., Egypt, 18733336. Available from: www.elsevier.com/locate/jhazmat. https://doi.org/10.1016/j.jhazmat.2021.127395.

Tawfik, A., Eraky, M., Khalil, M.N., Osman, A.I., Rooney, D.W., 2023. Sulfonated graphene nanomaterials for membrane antifouling, pollutant removal, and production of chemicals from biomass: a review. Environ. Chem. Lett. 21 (2),

1093–1116. Springer Science and Business Media Deutschland GmbH, Egypt, 16103661. Available from: https://www.springer.com/journal/10311. https://doi.org/10.1007/s10311-022-01538-2.

Terada, A., Hibiya, K., Nagai, J., Tsuneda, S., Hirata, A., 2003. Nitrogen removal characteristics and biofilm analysis of a membrane-aerated biofilm reactor applicable to high-strength nitrogenous wastewater treatment. J. Biosci. Bioeng. 95 (2), 170–178. https://doi.org/10.1263/jbb.95.170. 1347-4421.

Thant, K.J.W., Anh-Vu, N., Yun-Je, K., Masumi, K., Visvanathan, C., 2023. Performance of pilot-scale membrane aerated biofilm reactors integrated with anoxic nano-biotechnological reactor for domestic wastewater treatment. Chemosphere 319, 137927. https://doi.org/10.1016/j.chemosphere.2023.137927. 00456535.

Tian, H., Yanzhuo, H., Xingjian, X., Hui, M., Yuansen, H., Qi, W., Hongru, X., Li, B., 2019. Enhanced wastewater treatment with high o-aminophenol concentration by two-stage MABR and its biodegradation mechanism. Bioresour. Technol. 289, 121649. https://doi.org/10.1016/j.biortech.2019.121649. 09608524.

Uri-Carreño, N., Nielsen, P.H., Gernaey, K.V., Flores-Alsina, X., 2021. Long-term operation assessment of a full-scale membrane-aerated biofilm reactor under Nordic conditions. Sci. Total Environ. 779. Elsevier B.V., Denmark, 18791026. Available from: www.elsevier.com/locate/scitotenv. https://doi.org/10.1016/j.scitotenv.2021.146366.

Wang, J., Liu, G.F., Lu, H., Jin, R.F., Zhou, J.T., Lei, T.M., 2012. Biodegradation of acid orange 7 and its auto-oxidative decolorization product in membrane-aerated biofilm reactor. Int. Biodeterior. Biodegrad. 67, 73–77. https://doi.org/10.1016/j.ibiod.2011.12.003. 09648305.

Wang, Z., Chen, Y., Sun, X., Duddu, R., Lin, S., 2018. Mechanism of pore wetting in membrane distillation with alcohol vs. surfactant. J. Membr. Sci. 559, 183–195. https://doi.org/10.1016/j.memsci.2018.04.045. 03767388.

Wang, R., Zeng, X., Wang, Y., Tong, Y., Lewandowski, Z., 2019. Two-step startup improves pollutant removal in membrane-aerated biofilm reactors treating high-strength nitrogenous wastewater. Environ. Sci. Water Res. Technol. 5 (1), 39–50. https://doi.org/10.1039/c8ew00668g. 2053-1400.

Wang, L., Yun, W., Ren, Y., Wang, Y., Wang, Y., Zhang, H., 2022. Transition of fouling characteristics after development of membrane wetting in membrane-aerated biofilm reactors (MABRs). Chemosphere 299, 134355. https://doi.org/10.1016/j.chemosphere.2022.134355. 00456535.

Wazeri, A., Elsamadony, M., Tawfik, A., 2018. Carbon emissions reduction by catalyzing H_2 gas harvested from water hyacinth fermentation process using metallic salts. Energy Procedia 152, 1254–1259. https://doi.org/10.1016/j.egypro.2018.09.178. 18766102.

Wei, X., Li, B., Zhao, S., Qiang, C., Zhang, H., Wang, S., 2012a. COD and nitrogen removal in facilitated transfer membrane-aerated biofilm reactor (FT-MABR). J. Membr. Sci. 389, 257–264. https://doi.org/10.1016/j.memsci.2011.10.038. 03767388.

Wei, X., Li, B., Zhao, S., Wang, L., Zhang, H., Li, C., Wang, S., 2012b. Mixed pharmaceutical wastewater treatment by integrated membrane-aerated biofilm reactor (MABR) system—a pilot-scale study. Bioresour. Technol. 122, 189–195. https://doi.org/10.1016/j.biortech.2012.06.041. 09608524.

Weitere, M., Bergfeld, T., Rice, S.A., Matz, G., Kjelleberg, S., 2005. Grazing resistance of Pseudomonas aeruginosa biofilms depends on type of protective mechanism, developmental stage and protozoan feeding mode. Environ. Microbiol. 7 (10), 1593–1601. https://doi.org/10.1111/j.1462-2920.2005.00851.x.

Xia, Z., Ng, H.Y., Xu, D., Bae, S., 2024. Lumen air pressure regulated multifunctional microbiotas in membrane-aerated biofilm reactors for simultaneous nitrogen removal and antibiotic elimination from aquaculture wastewater. Water Res. 251, 121102. https://doi.org/10.1016/j.watres.2024.121102. 00431354.

Xiao, K., Wang, X., Huang, X., Waite, T.D., Wen, X., 2011. Combined effect of membrane and foulant hydrophobicity and surface charge on adsorptive fouling during microfiltration. J. Membr. Sci. 373 (1–2), 140–151. https://doi.org/10.1016/j.memsci.2011.02.041. 03767388.

Xin, Y., Bligh, M.W., Kinsela, A.S., Waite, T.D., 2016. Effect of iron on membrane fouling by alginate in the absence and presence of calcium. J. Membr. Sci. 497, 289–299. Elsevier, Australia, 18733123. Available from: www.elsevier.com/locate/memsci. https://doi.org/10.1016/j.memsci.2015.09.023.

Zhao, L., Zhang, M., He, Y., Chen, J., Hong, H., Liao, B.Q., Lin, H., 2016. A new method for modeling rough membrane surface and calculation of interfacial interactions. Bioresour. Technol. 200, 451–457. Elsevier Ltd, China, 18732976. Available from: www.elsevier.com/locate/biortech. https://doi.org/10.1016/j.biortech.2015.10.055.

9

Wastewater disinfection

Ye Du, Bao-Jun Xu, Chang-Jie Yuan, and Jie-Yu Cao
College of Architecture and Environment, Sichuan University, Chengdu, China

Abbreviations

AOM	algal organic matter
AOPs	advanced oxidation processes
ATP	adenosine triphosphate
BCAA	bromochloroacetic acid
BCAN	bromochloroacetonitrile
BDCAA	bromodichloroacetic acid
BDCAcAm	bromodichloroacetamide
BDCM	bromodichloromethane
DBAA	dibromoacetic acid
DBAcAm	dibromoacetamide
DBAN	dibromoacetonitrile
DBCAcAm	dibromochloroacetamide
DBCM	dibromochloromethane
DBPs	disinfection byproducts
DCAA	dichloroacetic acid
DCAcAm	dichloroacetamide
DCAN	dichloroacetonitrile
DCANFPs	dichloroacetonitrile generation potentials
DCP	1,1-dichloropropanone
DOM	dissolved organic matter
E. coli	Escherichia coli
EDCs	endocrine disrupting chemicals
EfOM	effluent organic matter
HAAs	haloacetic acids
HAN	haloacetonitrile
HKs	haloketones
I-THMs	iodinated trihalomethanes
LBT	luminescent bacteria test
MBAA	bromoacetic acid
MBAN	bromoacetonitrile
MCAA	chloroacetic acid
MIAA	iodoacetic acid
MTT	3-(4,5-dimethylthiazol-2-yl)-2,5-diphenyltetrazolium bromide
NDMA	N-nitrosodimethylamine
NOM	natural organic matter
NVOF	nonvolatile organic fraction
PMS	peroxymonosulfate
RCS	reactive chlorine species

Municipal Wastewater Treatment. https://doi.org/10.1016/B978-0-443-24826-9.00009-7

RNS	reactive nitrogen species
SMPs	soluble microbial products
SPE	solid-phase extraction
TBAcAm	tribromoacetamide
TBAN	tribromoacetonitrile
TBM	bromoform
tBM	tribromomethane
TCA	chlorohydrate
TCAA	trichloroacetic acid
TCAcAm	trichloroacetamide
TCM	chloroform
TCP	1,1,1-trichloropropanone
THMFPs	trihalomethanes generation potentials
THMs	trihalomethanes
TOBr	total organic bromide
TOCl	total organic chlorine
TOI	total organic iodine
TOX	total organic halogen
UV	ultraviolet
UVT	ultraviolet penetration rate
VOF	volatile organic fraction

9.1 Introduction

Disinfection serves as the ultimate barrier in municipal wastewater treatment, effectively deactivating pathogens to mitigate waterborne diseases. Pathogens encompassing bacteria, viruses, protozoa, and helminths are commonly found in untreated or inadequately treated wastewater (Richardson and Kimura, 2020; Cheng et al., 2024). In the absence of robust disinfection measures, these microorganisms can persist and proliferate, facilitating the transmission of waterborne illnesses such as cholera, dysentery, hepatitis, and gastroenteritis. The primary objective of disinfection is to render these pathogens inert or eliminate them entirely to ensure that treated wastewater meets safety standards for discharge into natural aquatic ecosystems or for reuse.

The public health implications of inadequate disinfection are significant. Waterborne diseases continue to be a major cause of illness and mortality, particularly in developing countries with limited access to clean water and proper sanitation (Li et al., 2008). Even in developed countries, outbreaks of waterborne diseases can occur as a result of failures in water treatment systems or contamination incidents (Stuart et al., 2012). Effective disinfection is crucial for preventing such outbreaks and safeguarding vulnerable populations, including children, the elderly, and individuals with compromised immune systems.

Various disinfection methods are employed in wastewater treatment, each with its own advantages, disadvantages, and implications for public health and environmental safety. The most commonly used disinfection processes include chlorination, chloramination, ozonation, and ultraviolet (UV) radiation. However, the use of chlorine, ozone, or other chemical disinfectants also raises a public health issue: the formation of disinfection byproducts (DBPs) (Sedlak and Von Gunten, 2011; Li and Mitch, 2018). In wastewater disinfection, DBPs are formed by the reaction of disinfectants with dissolved organic matter (DOM), anthropogenic contaminants and inorganic pollutants like bromides/iodides. It is reported that DBPs pose a risk to human health, with associated adverse health outcomes including cancer of the bladder, colon, and rectum (Villanueva et al., 2007; Cantor et al., 2010; Rahman et al., 2010; Bove et al., 2007), and a range of reproductive/developmental effects (Richardson et al., 2007; Wagner and Plewa, 2017).

Due to the diversity of DBPs and their inability to be identified individually, total organic halogens (TOX) are generally used to assess the overall organic halogenated DBPs (Xiao et al., 2023). TOX provides a comprehensive assessment of the total amount of halogenated organic compounds, offering insights into the potential health risks associated with DBP exposure. Besides, biotoxicity assays are important tools for evaluating the biological effects of disinfected wastewater. These assays evaluate the overall toxicity of water samples, considering the combined effects of all DBPs and other contaminants present. Common biotoxicity tests include assays using model organisms such as zebrafish, daphnia, algae and mammalian cells. These tests provide valuable information on the potential ecological and health risks of treated wastewater, guiding efforts to optimize disinfection processes and minimize adverse impacts.

The most widely employed disinfectants include chlorine, chloramines, chlorine dioxide and ozone, exhibiting disparities in the types of DBPs generated as well as their production characteristics (Sorlini and Collivignarelli, 2005; Hua and Reckhow 2007; Kimura and Ortega-Hernandez, 2019). Furthermore, estimating the potential ecological and health risks associated with disinfection cannot rely on a single or limited number of identified DBPs alone. Therefore, while exploring the extensive and intricate family of DBPs, it is imperative to evaluate and regulate the ecological risks and biological toxicity associated with different types of chemical disinfection processes.

Disinfection is a crucial element of wastewater treatment, playing a vital role in safeguarding public health and preventing the transmission of waterborne diseases. However, the occurrence of disinfection byproducts (DBPs) during chemical disinfection poses significant health hazards, necessitating meticulous consideration and management. By comprehending the mechanisms underlying DBP formation, evaluating their impact on human health, and implementing effective mitigation strategies, it becomes feasible to achieve safe and sustainable disinfection practices that protect both human well-being and the environment.

9.2 Wastewater disinfection process

To inactivate pathogenic microorganisms, reduce the incidence of waterborne diseases, and ensure water safety, various disinfection processes have emerged. The inactivation of pathogenic microorganisms by disinfectants primarily occurs through three mechanisms: (1) inducing structural damage to the cell by targeting its main tissue components such as the cell wall and membrane; (2) interfering with energy metabolism through enzymatic substrate chemosynthesis, resulting in a loss of enzyme function; (3) disrupting the synthesis of proteins, nucleic acids, coenzymes or cell walls to impede microbial growth.

9.2.1 Chlorine, chloramine, chlorine dioxide disinfection

To date, chlorine disinfection has been applied for more than 100 years as an important measure to safeguard the microbiological safety of drinking water. Chlorine (sodium hypochlorite or chlorine gas) and its compounds, including chloramine and chlorine dioxide, can effectively inactivate many pathogenic microorganisms (Ding et al., 2019; Huang et al., 1997; Shin and Sobsey, 2008; Wolfe et al., 1985). It is cost-effective, user-friendly, and allows for continuous disinfection. Even when alternative disinfection processes are utilized, residual chlorine is often added to effluent water to maintain a specific concentration within the distribution network and control microbial growth. Consequently, chlorine disinfection finds extensive application in sewage treatment due to its widespread use and well-established procedures.

9.2.1.1 Principles of microbial inactivation by chlorine disinfection

Chlorine and its compounds are a class of chemical disinfectants with multiple attack sites. At present, the mechanism of inactivation of microorganisms by chlorine disinfection is still unclear, and the effects and destruction of microorganisms by chlorine disinfection show a cross-cutting phenomenon. The specific mechanisms include oxidation of sulfhydryl enzymes and amino acids, chlorination of amino acid rings, oxidation of respiratory metabolites, leakage of cytoplasm, inhibition of nutrient and oxygen uptake, inhibition of protein and ATP synthesis, and DNA strand breaks or synthetic blockage.

The disinfection mechanism of chloramine is similar to that of chlorine disinfection, but the disinfection effect is weaker than that of chlorine disinfection, and it is often widely used as a secondary disinfectant. Chlorine dioxide, on the other hand, shows its disinfection ability by the oxidizing properties of positive tetravalent chlorine. Although chlorine dioxide disinfection exhibits rapid inactivation, ease of production, resistance to acidity and alkalinity, and minimal generation of halogenated disinfection by-products, it is accompanied by the formation of disinfection by-products such as chlorite and chlorate. Additionally, its high cost and challenging detectability limit its widespread application.

Although there are many different types of chlorine-containing disinfectants, chlorine is currently the most widely used of these chlorinated disinfectants. Chlorine disinfection technology will be discussed here as a representative.

9.2.1.2 The impact of chlorine disinfection

The influencing factors of chlorine disinfection primarily encompass chlorine dosage, water conditions (pH, temperature, turbidity, etc.), mixing efficiency, and reaction time (Winward et al., 2008). Generally, a higher chlorine dosage combined with lower pH levels and temperatures as well as longer reaction times result in enhanced disinfection efficacy. The pH level influences the disinfection process by affecting the various forms and proportions of active chlorine present in water. Additionally, organic matter within the water consumes chlorine thereby impacting its ability to deactivate pathogenic microorganisms; this influence varies depending on the functional group and structure of the organic matter.

9.2.1.3 Inactivation and dosage standards of chlorine disinfection for pathogenic microorganisms

Chlorine disinfection has a good inactivation effect on common pathogenic microorganisms. Chlorine disinfection is usually less effective in extinguishing pathogenic protozoa and spores than bacteria and viruses, and Gram-positive bacteria in bacteria are less effective than Gram-negative bacteria. Taking the inactivation rate of 2 log (99%) as an example, the chlorine disinfection dose required by intestinal viruses is usually one order of magnitude higher than that required by intestinal bacteria, and the chlorine disinfection dose required by pathogenic protozoa is two orders of magnitude higher than that required by intestinal viruses. In addition, Cryptosporidium and Giardia oocysts have strong chlorine disinfection resistance, which is mainly due to the special structure of oocysts.

Additionally, due to the complexity of wastewater, the determination of disinfection dosage in wastewater disinfection usually also needs to consider the water quality of wastewater. For example, for typical secondary effluent, to achieve a maximum of 3 coliforms per mL in the effluent, the chlorine addition concentration is usually between 7 and 25 mg/L, and the contact time is within 10–30 min.

9.2.2 Ozone disinfection

The potent oxidizing agent ozone can be generated on-site by subjecting oxygen to an electric discharge, effectively inactivating microorganisms. It undergoes rapid decomposition without leaving any residual disinfectant. However, when ozone reacts with organic matter present in wastewater, it can give rise to byproducts such as aldehydes. Additionally, the reaction between ozone and bromide in water leads to the formation of bromate, a regulated disinfection byproduct (DBP) that poses potential health risks.

9.2.2.1 Principles of microbial inactivation by ozone

As a strong oxidant, ozone has a good inactivation effect on pathogenic microorganisms in water, and the inactivation ability is mainly realized by strong oxidative destruction of cell composition. Ozone firstly acts on glycoprotein or glycolipid on the cell membrane of bacteria, and then ozone inhibits the activity of enzymes containing the thiol group, thereby inhibiting the physiological activity of bacteria. Once ozone breaks through the bacterial cell walls and membranes, it further acts by oxidizing the purine and pyrimidine bases in RNA. Relatively speaking, the mechanism of ozone inactivation of viruses is much simpler. Ozone firstly destroys the protein capsid of viruses, and high concentrations of ozone may even cause the separation of protein capsid from DNA or RNA, and then act on the nuclear region of viruses. Ozone inactivation of fungi and pathogenic protozoa mostly starts from the destruction of cell membranes and the enhancement of cell permeability (Wen et al., 2020).

9.2.2.2 The impact of ozone disinfection

Similar to chlorine disinfection, ozone is also a chemical disinfection method, and the factors affecting disinfection include the amount of ozone added, water quality conditions (pH, temperature, turbidity, etc.), mixing effects, and reaction time (Zuma et al., 2009). The ozone disinfection effect is enhanced by increasing the amount of added ozone, reducing the organic matter content and turbidity, as well as ensuring thorough mixing. The decomposition rate of ozone is influenced by pH and temperature, while the presence of organic matter consumes ozone and affects its inactivation on pathogenic microorganisms (Janex et al., 2000).

9.2.2.3 Inactivation and dosage standards of ozone disinfection for pathogenic microorganisms

Ozone has a broad spectrum of inactivation of microorganisms, and has a high inactivation ability to viruses, bacteria and pathogenic protozoa. The inactivation of ozone on pathogenic protozoa is poorer than that on bacteria and viruses, while the inactivation of ozone on viruses is between that on bacteria and spores. Protozoan oocysts and bacterial spores, being ozone-tolerant microorganisms, possess cell wall coatings (oocyst wall for the former and sporangium and blastocyst wall for the latter) that effectively impede ozone penetration into the cells, thereby preventing oxidative damage to cytoplasmic components, nucleus, or physiological metabolic processes Ozone-resistant (tolerant) microorganisms controlled by specific genes have not been reported.

Ozone can oxidize both reducing substances and pathogenic microorganisms simultaneously (von Gunten, 2003a). Due to the complexity of wastewater composition and high organic matter concentration, the dosage of wastewater disinfection needs to be determined based on the water quality as well. For example, in typical secondary effluent, to achieve a maximum of 3 coliforms per 100 mL of water, the ozone injection concentration should be between 10 and 30 mg/L, and the contact time should be within 10–30 min.

9.2.3 Emerging disinfectants

Conventional disinfection methods have problems such as the formation of disinfection by-products and microbial self-repair, which have spurred the development of new disinfection methods.

Ferrate (Fe(VI)) is a strong oxidant with redox potentials of +2.2 and +0.7 V (compared to NHE) in acidic and alkaline conditions, respectively (Wood 1958). Fe(VI) can kill microorganisms in two ways: (a) Due to its potent oxidizing capability, Fe(VI) can oxidize cell walls, protoplasm, DNA and other essential microbial organs, thus inactivating microorganisms immediately; (b) Ferric (Fe (III)), a decomposition product of Fe(VI), can remove microorganisms by coagulation. In addition, The formation of disinfection byproducts is of minor concern during Fe(VI) disinfection (Huang et al., 2016). Simultaneously, Fe(VI) exhibits the capability to effectively eliminate organic matters, toxic inorganic substances, emerging pollutants and toxicity from water (Song et al., 2016; Dong et al., 2019; Wu et al., 2021a). Therefore, these properties render Fe(VI) a potentially viable alternative to other traditional disinfectants.

The inactivation of bacteria by Fe(VI) has been demonstrated across various water conditions (phosphate buffer, tap water, surface water, ballast water, secondary effluent and wastewater), but variations in the disinfection efficacy have been observed (Cho et al., 2006; Jessen et al., 2008; Manoli et al., 2020; Zheng et al., 2021). The differences in disinfection efficacy were primarily attributed to variations in water conditions, pH values, Fe(VI) dosage, and disinfection duration. In a neutral environment, it has been observed that a concentration of 5 mg/L of Fe(VI) can result in an inactivation of *E. coli* exceeding 3 log. Furthermore, within a specific range, the efficacy of *E. coli* inactivation improves with lower pH levels and higher doses of Fe(VI). Comparative studies have also been conducted to assess the disinfection effects of Fe(VI) and chlorine, revealing that Fe(VI) exhibited greater efficacy than chlorine in deactivating *E. coli* at low doses (Mao et al., 2022; Jiang et al., 2007). Furthermore, Fe(VI) demonstrated control over chlorine-resistant bacteria (Sharma, 2007; Gombos et al., 2012; Mao et al., 2022). It has been demonstrated that Fe(VI) exhibits superior efficacy in the inactivation of Gram-negative bacteria (G−) compared to Gram-positive bacteria (G+), attributed to variations in cell wall thickness. Furthermore, Fe(VI) was observed to exhibit a higher reactivity towards proteins as opposed to other biomolecules such as phospholipids, peptidoglycans, and lipopolysaccharides (Mao et al., 2023).

Studies have demonstrated the ability of Fe(VI) to remove viruses (Talaiekhozani et al., 2017). The F2 virus is rapidly inactivated by Fe at low concentrations in both pure water and secondary effluent within the pH range of 6–8 (Schink and Waite, 1980). Fe(VI) was also able to successfully remove the SARS-CoV-2 virus

(Mališová et al., 2023). The high degradation efficiency of virus fragments by Fe(VI) was achieved in the natural pH range.

Except for Fe(VI), studies have affirmed the effectiveness of $SO_4^{-}\cdot$-based technology in inactivating pathogenic microorganisms reliably. Decades ago, Oxone and Virkon™ S (a trade name for potassium PMS) became widely used in the pool and spa industry as chlorine-free shock agents due to their ability to effectively inactivate over 500 strains of viruses, bacteria, and fungi (Li et al., 2022). It is widely accepted that the process of PMS inactivating pathogens occurs through the disruption of microbial cell membranes/walls, enzymes, and genetic material by $SO_4^{-}\cdot$ (Xiao et al., 2019).

9.3 Formation characteristics of disinfection byproduct

Chemical disinfectants can effectively kill pathogenic microorganisms, but they are also powerful oxidants that can react with organic matters and inorganic substances such as bromide and iodide ions present in wastewater, forming biotoxic disinfection byproducts (von Gunten, 2003b; Hua and Reckhow, 2007). The substances that react with the disinfectants are collectively known as disinfection byproduct precursors. When disinfected wastewater is used for municipal utilization, landscape and environmental water, or comes into contact with organisms in the natural environment, disinfection byproducts may pose a threat to their normal survival and reproduction, and may also affect human health (Chen et al., 2009; Krasner et al., 2009; Li and Mitch, 2018; Li et al., 2019). Therefore, the generation of disinfection byproducts and their biotoxicity has become an important indicator for evaluating the safety of disinfection.

9.3.1 Formation of disinfection byproducts

Over 700 disinfection byproducts (DBPs) have been identified. Approximately 100 DBPs have undergone cytotoxicity and genotoxicity testing in Chinese hamster ovary cells (Wagner and Plewa, 2017). Ultraviolet disinfection, a physical disinfection method, generates very few disinfection by-products. Chlorine, chloramines and ozone are the most commonly used disinfectants, each with its own unique characteristics for generating DBPs. Data on the types and concentrations of DBPs produced by chlorine and ozone disinfection are presented in Fig. 9.1A and B.

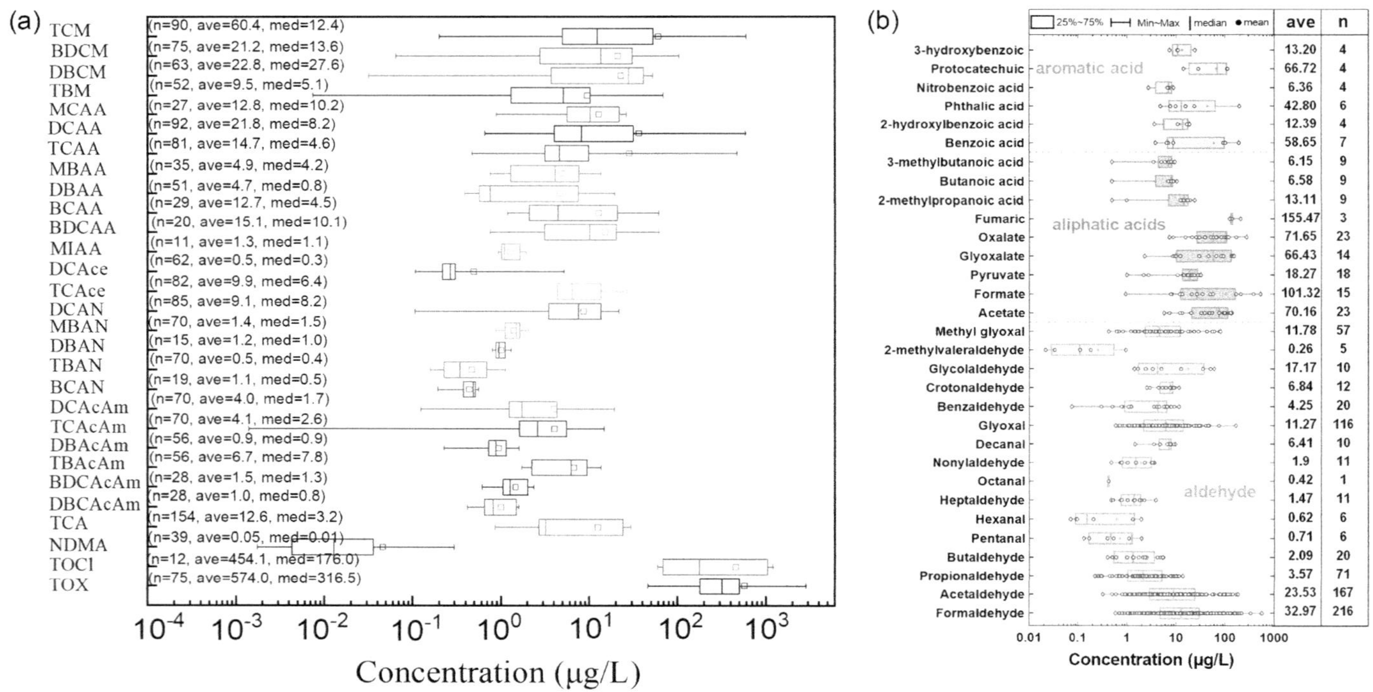

Fig. 9.1 Occurrence of (A) chlorination DBPs and (B) ozonation DBPs in wastewater disinfection processes. Copyright: Elsevier.

In chlorine disinfection, volatile trihalomethanes (THMs) and non-volatile haloacetic acids (HAAs) are the main types of disinfection byproducts, accounting for more than 80% of all chlorine disinfection byproducts. In addition, other types of DBPs such as haloacetonitrile (HANs) and haloketones (HKs) are also generated.

Ozone forms some non-halogenated disinfection byproducts, such as aldehydes, ketones and carboxylic acid DBPs, will be generated. In addition, it is worth paying special attention to that when wastewater contains a high dose of bromide, it will produce brominated disinfection byproducts such as bromate, tribromomethane and bromoacetonitrile.

The types of DBPs generated by chloramine and chlorine disinfection exhibit minimal disparity. however, under identical conditions, chloramine yields significantly fewer Cl-DBPs compared to chlorine under the same conditions due to its reduced reactivity with organic matter in water. In particular, the amount of DBPs such as THMs and HAAs is significantly lower than that of chlorine disinfection. Some studies have shown that chloramine can reduce HAAs by 90% and THMs by 70%. However, chloramine disinfection may be more likely to produce nitrogen-containing DBPs with greater potential harm. So far, the main N-DBPs detected include HANs, HAcAms, and HAMs.

While extensive research has been conducted on the formation of DBPs during wastewater disinfection, it is not feasible to detect every byproduct present. This limitation arises from challenges in sample preparation, analytical methods, and the associated human and material costs. Therefore, as a collective parameter, TOX can effectively characterize the overall halogenated DBPs formed during the disinfection process. The sum of known halogenated DBPs in chlorine disinfection is only about 50% of TOX, and about 20% of the TOX in chloraminated water (Hua and Reckhow, 2007). This indicates that a significant proportion of halogenated DBPs remains unknown.

Increased formation of TOX typically correlates with increased toxicity in disinfected water (Fig. 9.2). Therefore, TOX is often regarded as an indicator of toxicity (Du et al., 2017, 2023, Allen et al., 2022; Wu et al., 2020). During chlorine disinfection, hypochlorous acid (HOCl) can oxidize bromide to hypobromous acid (HOBr), and both HOCl and HOBr can produce organic halogens by electrophilic substitution or electrophilic addition to dissolved organic matter (DOM), thus both total organic chlorine (TOCl) and total organic bromine (TOBr) may lead to elevated toxicity. In the case of ozone and

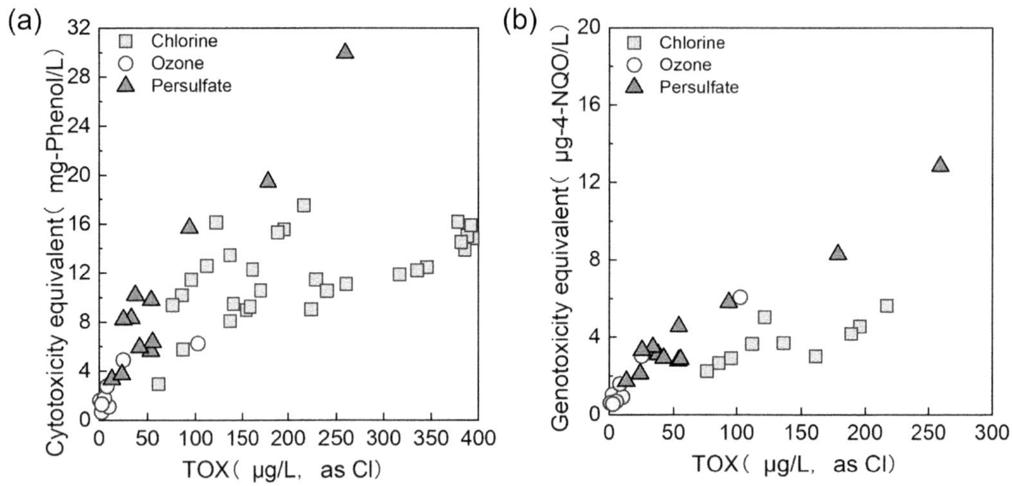

Fig. 9.2 Relationship of TOX to (A) cytotoxicity and (B) genotoxicity to mammalian cells.

persulfate disinfection, TOBr generation from Br^- is the main cause of elevated toxicity due to the weak reactivity of Cl^- with the oxidant (Wu et al., 2021a; Du et al., 2023).

9.3.2 Factors influencing the generation of disinfection byproducts

The amount and type of DBP formation in wastewater are influenced by many factors, including the properties and concentration of organic precursors, disinfection processes, temperature, pH, halide ions, and metal ions.

9.3.2.1 The properties and content of DBPs precursors

In drinking water, NOM is the main source of precursors. However, wastewater contains a high concentration and complexity of organic matter, the composition of DOM is significantly different from NOM of drinking water source water (Shi et al., 2021a,b). With a large number of synthetic organic compounds being discharged into wastewater through pathways such as industrial wastewater and domestic sewage, these organic pollutants can also serve as precursors for DBPs, such as surfactants, flame retardants, pesticides, pharmaceuticals, etc. (Wang and Chen, 2018; Shi et al., 2021b). In addition, wastewater contains a large proportion of microbial secretion products, i.e., soluble microbial products (SMPs), alongside recalcitrant organic compounds and biodegradation intermediates that have not been effectively eliminated. These substances are also likely to be precursors of toxic and harmful DBPs (Liu et al., 2014; Liu and Li, 2010).

The amount of DOM in wastewater is directly proportional to the generation of DBPs under specific conditions. In addition, the molecular weight, composition, and functional groups of DOM also significantly influence the formation of DBPs (Kitis et al., 2002; Zhang et al., 2018; Zhou et al., 2022). N-nitrosodimethylamine (NDMA) has received significant attention because ng/L levels in water are associated with 10^{-6} lifetime excess cancer risks (Li and Mitch, 2018). The presence of total nitrosamine precursors is closely associated with effluent organic matter (EfOM) and algal organic matter (AOM), whereas NDMA exhibits a strong correlation with EfOM (Zeng et al., 2016).

9.3.2.2 Disinfection process

Different types of disinfectants promote different DBPs classes. For instance, chlorine disinfection mainly generates chlorinated DBPs, and chlorine dioxide generates inorganic chlorite and chlorate (Rougé et al., 2018), while chloramine disinfection generates nitrosamines (such as N-nitrosodimethylamine received significant attention) and iodinated DBPs (Zhao et al., 2008; Ersan et al., 2019). Ozone mainly generates bromate, aldehydes, and halonitromethanes (Shi et al., 2021a,b; Li et al., 2022). In chlorinated water, TOCl and TOBr were highly significantly correlated with toxicity ($r=0.94$, $P<0.01$); in chloraminated water, TOI was highly significantly correlated with toxicity ($r=0.80$, $P<0.001$) (Allen et al., 2022). The formation of DBPs increases proportionally with the dosage and contact time of disinfectants within a specific range, given a fixed concentration of precursors.

9.3.2.3 Temperature, pH, halide ions, ammonia, other inorganic ions, etc.

The formation of DBPs generally increases with higher reaction temperatures. However, it has also been demonstrated that temperature does not significantly impact the production of THMs, while there is a notable decrease in the production of HAAs as temperature rises (Sun et al., 2009).

The influence of pH on DBPs is intricate. Firstly, pH impacts the distribution of hypochlorite and chloramine species. For instance, when the pH exceeds 7.5, hypochlorite predominantly exists as free chlorine, exhibiting lower reactivity compared to hypochlorous acid. Secondly, the pH level has a direct impact on the protonation extent of organic substances in water. As the pH increases, the oxidation-reduction potential of the system decreases, consequently altering the reactivity between precursor materials and chlorine. For chlorine disinfection, alkaline environments are more conducive to the formation of THMs and HAAs, while HANs and HAMs are produced to a greater extent in acidic or neutral conditions (Qian et al., 2021). In addition, pH also affects the stability of DBPs. In alkaline conditions, many DBPs, such as HAMs and HANs, will undergo hydrolysis and eventually convert to the corresponding HAAs.

In addition, the wastewater effluent generally contains higher concentrations of ammonia (N-NH_4^+) and halogen ions, especially bromide and iodide. These inorganic pollutants not only influence the disinfection efficacy but also contribute to the formation of more hazardous DBPs containing nitrogen, bromine, and iodine (Lu et al., 2021; Wu et al., 2020). Wastewater may exhibit elevated levels of bromide and iodide ions as a result of industrial discharges, waste leachate, groundwater infiltration, and intrusion of seawater. These water sources containing higher concentrations of bromide and iodide have the potential to alter the formation of DBPs into brominated and iodinated analogs (Hua and Reckhow, 2007; Dong et al., 2023). Many studies have shown that the presence of N-NH_4^+ significantly reduces the production of conventional DBPs such as trihalomethanes and haloacetic acids during chlorine disinfection. This is attributed to the reaction between ammonia and free chlorine, resulting in the formation of monochloramine, dichloroammonia, etc., which exhibit lower reactivity. However, a substantial amount of the more toxic N-DBP is generated. For example, the presence of N-NH_4^+ can potentially contribute to an elevated production of nitrogenous DBPs such as NDMA.

Due to the large number of copper pipes, cast iron pipes and other pipes used in the pipe network system, Fe^{2+}, Mn^{2+}, Mg^{2+}, Cu^{2+} are widely present in the wastewater. These metal ions will react with DOM or disinfectants, which not only consumes disinfectants but also affects the generation and types of DBPs. For example, copper ions may catalyze the formation of trichloromethane from humic acid, etc., while metal ions such as Mg^{2+} and Fe^{3+}, which can cooperate with some DOM, may affect the reaction between DOM and disinfectants, thus affecting the generation of DBPs.

9.4 Toxicity formation from different disinfection processes

The risks to ecosystems and human health resulting from disinfection cannot be adequately assessed based on just one or a few identified disinfection byproducts (DBPs). Biotoxicity tests offer a direct means of evaluating the diverse biological impacts of disinfected wastewater. Therefore, many researchers have assessed the full range of biological toxicity of disinfected wastewater (Rice et al., 2008; Watson et al., 2012; Yan et al., 2014; Bayo et al., 2009; Patterson et al., 1995). Over time, various bioassays with differing degrees of biological relevance have been developed for toxicity assessment (Jeong et al., 2012; Narotsky et al., 2008; Rice et al., 2008; Simmons et al., 2004; Patterson et al., 1995).

9.4.1 Chlorine disinfection

The cumulative distribution of different toxicities after wastewater chlorination was summarized and toxicity indicators included luminescent bacteria acute toxicity (Fig. 9.3A), *D. magna* acute toxicity measured by mortality (Fig. 9.3B), genotoxicity measured by the *umu*-test (Fig. 9.3C), estrogenic activity (Fig. 9.3D), anti-estrogenic activity (Fig. 9.3E) and cytotoxicity (Fig. 9.3F). These methods have been extensively used to evaluate the toxicity of wastewater, making them valuable indicators for ensuring water safety from multiple perspectives (Tang et al., 2014a,b; Wang et al., 2007a,b; Wu et al., 2009; Yang et al., 2015).

Effluent toxicity significantly increased after chlorine disinfection, with different bioassays revealing varying trends. In most cases, the formation and toxicity of DBPs were found to increase in effluents following chlorine disinfection.

Luminescent bacteria tests are commonly employed to assess acute toxicity in chlorinated effluents from wastewater treatment plants, as well as in batch studies (Wang et al., 2007b; Watson et al., 2012; Wei et al., 2006). As shown in Fig. 9.3A, the relative toxicity (defined as the ratio of post-chlorination toxicity to pre-chlorination toxicity) increases significantly after chlorination, with relative toxicity values ranging from 1 to 18 times, with 2–3 times being the most common. This suggests that luminescent bacteria are more sensitive to DBPs produced after chlorination, and that the magnitude of the brought toxicity is also influenced by factors such as the dose of chlorine and the duration of exposure.

Daphnia magna is a sensitive indicator of water quality, frequently used in ecotoxicological studies. Parameters such as mortality, reproduction rates, and motility inhibition of *D. magna* can be employed to evaluate the acute toxicity of effluent from wastewater treatment plants both before and after chlorination. The question of whether chlorinated water increases the mortality of *Daphnia magna* has not been conclusively answered. As shown in Fig. 9.3B, 78% of the values increased, while the remaining values exhibited a decreasing trend. Blatchley et al. (1997)

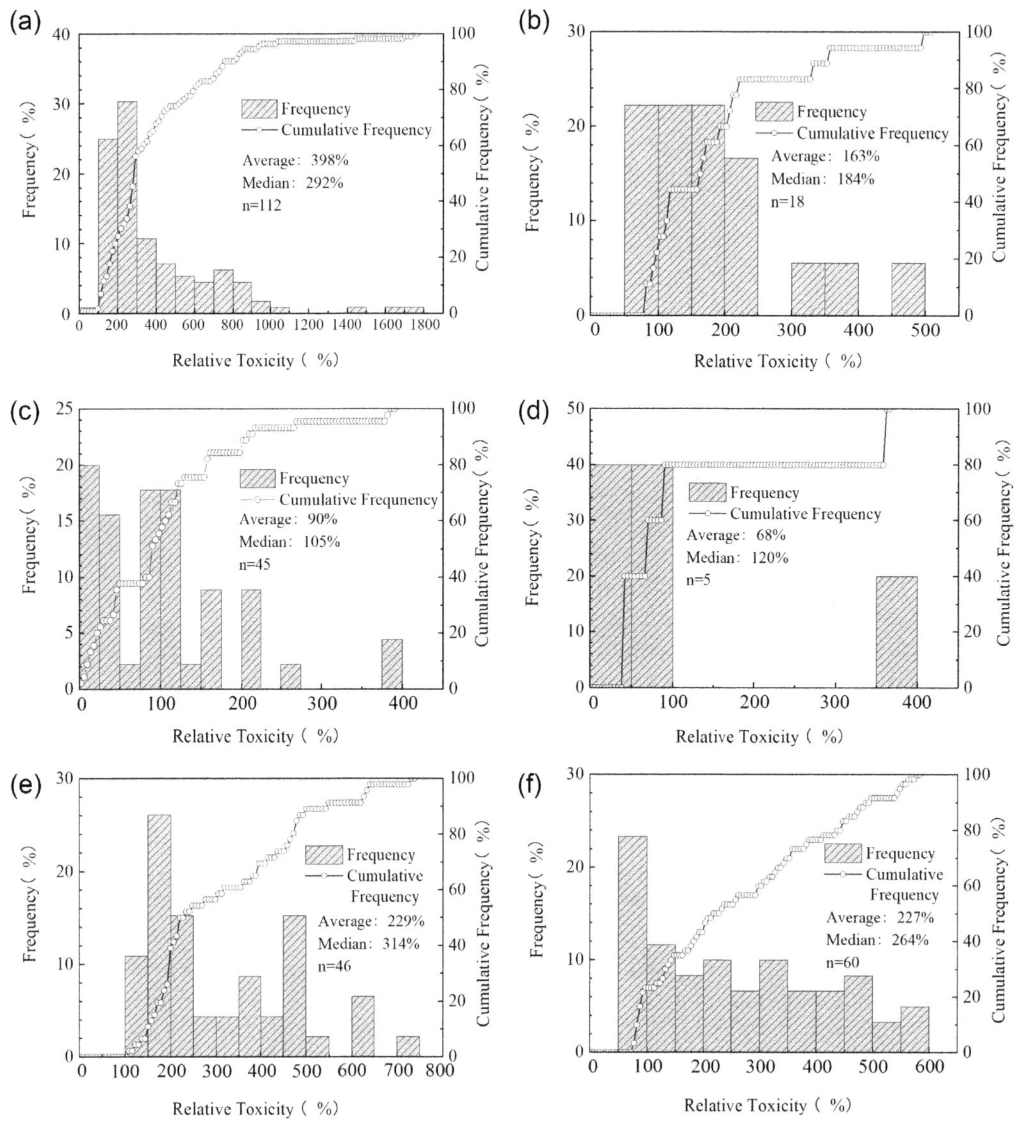

Fig. 9.3 Cumulative distribution of different toxicities after chlorination of wastewater. Relative toxicity is the ratio of the toxicity after chlorination to the toxicity before chlorination. The *dotted line* indicates the raw toxicity of unchlorinated wastewater. (A) acute toxicity of Luminescent bacteria; (B) acute toxicity of Daphnia magna as measured by mortality; (C) genotoxicity as measured by the *umu* test; (D) Estrogenic activity; (E) Antiestrogenic activity; (F) cytotoxicity (Du et al., 2017). Copyright: Elsevier.

suggested that wastewater does not elicit substantial toxicological responses either before or after chlorination. Conversely, increased toxicity of *D. magna* has been reported in other studies (Cao et al., 2009; Kontana et al., 2009). It is noteworthy that 100% mortality was frequently detected, despite variations in chlorine dose and duration of exposure. In addition, even with a fourfold dilution of chlorinated water, the 48-h mortality rate in neonates was still 70% (Cao et al., 2009).

Genotoxicity assesses DNA and/or chromosome damage caused by external chemicals. While assessment methods are standardized, the effects of chlorination on wastewater genotoxicity (*umu* assays) are more nuanced than with other bioassays. Statistical data show that 45% of cases indicate an increase in genotoxicity, while 55% show a decrease.

Endocrine disrupting chemicals (EDCs) have raised significant concern due to their capacity to disrupt the endocrine systems of both humans and animals (Schilirò et al., 2009). The estrogenic activity decreased in most cases, as depicted in Fig. 9.3D. Generally, the levels of estrogenic activity declined to 40–90% of the initial wastewater concentration (Schilirò et al., 2009; Wu et al., 2009; Zeng et al., 2016), suggesting that estrogenic chemicals were destroyed by chlorine. In addition, the presence of a mixture of industrial estrogen mimics, natural estrogens and anti-estrogens in the effluent samples may lead to mixed results (Watson et al., 2012).

The presence of anti-estrogenic chemicals in wastewater has been widely observed, as they exhibit a de-feminization effect and disrupt the estrous cycle of female animals, thereby adversely affecting their fertility (Wu et al., 2014). The increase in anti-estrogenic activity due to chlorine disinfection was evident, ranging from 1 to 7.5 times higher than that of the untreated wastewater (Fig. 9.3E), potentially attributed to the generation of anti-estrogenic DBPs. Additionally, significant precursors of these anti-estrogenic DBPs have been identified as aromatic amino acids and humic/fulvic acids (Tang et al., 2014a,b; Wu et al., 2009).

Cytotoxicity assays that employ mammalian cells to evaluate toxic effects offer a more relevant reference for assessing human health risks compared to bioassays involving bacteria. The data collected indicate that chlorine disinfection significantly increased the cytotoxicity of effluent by a factor ranging from 1 to 8, underscoring the imperative of considering the human health risks associated with chlorinated DBPs.

9.4.2 Ozone disinfection

Currently, there is a lack of clear and consistent understanding regarding the changes in toxicity during the ozone disinfection of wastewater, leading to controversial observations and interpretations. Some studies have reported that ozonation reduces the toxicity of wastewater (Du et al., 2020b; Reungoat et al., 2012), while others have reached an opposite conclusion suggesting that ozonation actually increases the toxicity in wastewater (Wang et al., 2019; Yan et al., 2014). Additionally, some studies have indicated that effluent toxicity remains relatively unchanged before and after ozone disinfection (Dong et al., 2017).

The variability in observed toxicity changes can be attributed to the complexity of disinfection byproducts (DBPs) formed during ozone disinfection. These include inorganic byproducts like bromate, volatile organic fractions (VOF) such as aldehydes and tribromomethane (tBM), as well as other byproducts present in the nonvolatile organic fractions (NVOF). While solid-phase extraction (SPE) is commonly used to pretreat water samples for toxicity assessments by concentrating them, it retains only non-volatile organic compounds (Du et al., 2020a; Hua et al., 2021; Le Roux et al., 2017). As a result, some studies using SPE to concentrate water samples may have overlooked the toxicity of bromate and VOF and therefore underestimated toxicity. Furthermore, the concentration of bromide in the effluent directly influences the formation of highly toxic Br-DBPs and subsequently affects observed levels of toxicity (Magazinovic et al., 2004; Wu et al., 2020; Yang et al., 2017).

Therefore, aldehydes, bromates, and organic bromines should be combined to assess the cytotoxicity (Fig. 9.4A and B) and genotoxicity (Fig. 9.4C and D) of ozonized wastewater (Wu et al., 2021b).

In a previous study, without bromide, the cytotoxicity equivalents of VOF (i.e., aldehydes) in sample 1 were quantified at 5.3 mg-phenol/L, which was 3.1 times higher than that of NVOF. Similarly, the cytotoxicity equivalents of VOF in sample 2 reached 4.0 mg-phenol/L, indicating a 3.9-fold higher than that of NVOF. After ozonation, the aldehyde byproducts formed in VOFs showed higher cytotoxic than those formed in NVOFs, indicating that VOFs are quite important for the overall cytotoxicity of ozonated wastewater. Furthermore, the cytotoxicity of TOF was further measured (considering both VOF and NVOF), and the cytotoxicity equivalents of 6.7 mg-phenol/L for sample 1 and 9.4 mg-phenol/L for sample 2, were higher than the cytotoxicity of NVOF or VOF alone, and even higher than the toxicity of the wastewater before ozonation. This suggests that the outcome in some studies that

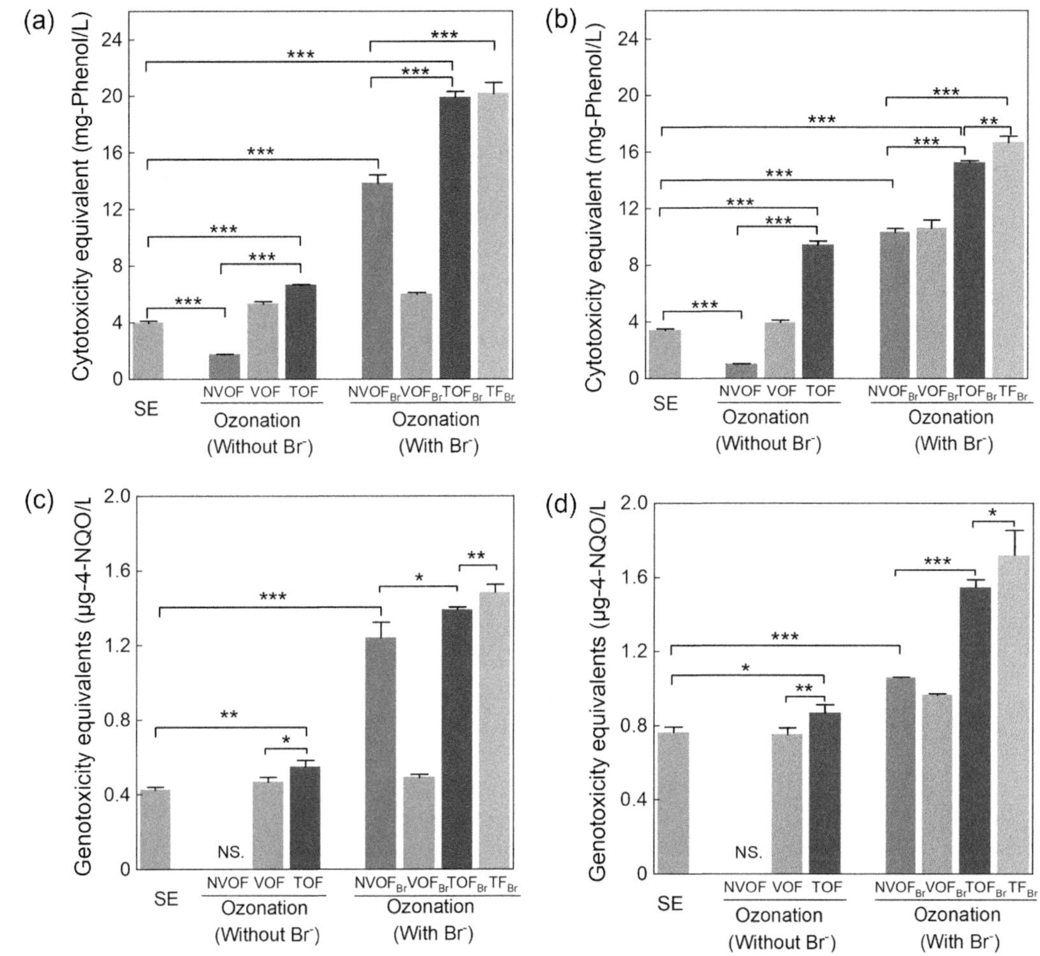

Fig. 9.4 Contribution of different organic fractions to cytotoxicity and genotoxicity after ozone disinfection: (A) cytotoxicity equivalents of sample 1; (B) cytotoxicity equivalents of sample 2; (C) genotoxicity equivalents of sample 1; (D) genotoxicity equivalents of sample 2 (Wu et al., 2021b). Copyright: ACS publication.

ozone disinfection removes the cytotoxicity of wastewater is most likely due to the overlooked toxicity associated with VOF.

The cytotoxicity of $NVOF_{Br}$ in the presence of bromide significantly increased from 4.0 to 13.9 mg-phenol/L for sample 1 and from 3.4 to 10.3 mg-phenol/L for sample 2, compared to the untreated wastewater. The increased cytotoxicity can be attributed to the formation of large amounts of TOBr (Plewa et al., 2010; Wagner and Plewa, 2017). Meanwhile, VOF_{Br} (including aldehydes and volatile brominated DBPs) also showed high cytotoxicity. The cytotoxicity of VOF_{Br} reached 6.0 mg-phenol/L for sample 1 and 10.6 mg-phenol/L for sample 2. The higher toxicity compared to VOF without bromide can be attributed to the presence of TBM and DBAN in addition to aldehydes within VOF_{Br}. Furthermore, compared to $NVOF_{Br}$, the cytotoxicity equivalents of TOF_{Br} ($NVOF_{Br}$ and VOF_{Br}) were further increased by 43–48%, emphasizing the importance of considering volatile ozonation DBPs when bromides are present in wastewater. In addition, bromate is by far the most worrisome DBPs of the ozonation process (von Gunten and Hoigne, 1994; Yang et al., 2017). Formed concentrations of bromate were spiked in TOF to obtain TF_{Br}. In sample 1, the addition of bromate did not increase the cytotoxicity of the sample (one-way ANOVA, $p > 0.05$). However, in sample 2, there was a statistically significant difference with only a 9% increase in cytotoxicity equivalents compared to TOF_{Br}. These findings suggest that while bromate is still a concern, its contribution to overall cytotoxicity appears limited (Han and Zhang, 2018; Wu et al., 2019).

Genotoxicity was similar to cytotoxicity. In the absence of bromide, ozone significantly reduced the genotoxicity of NVOF but increased the overall genotoxicity of the water samples ($p < 0.05$). This increase in genotoxicity was primarily caused by VOF. By adding bromide, the genotoxicity of the $NVOF_{Br}$ significantly increased compared with that of the secondary effluent (one-way ANOVA, $p < 0.001$), with genotoxicity equivalents of 1.2 μg-4-NQO/L in sample 1 and 1.1 μg-4-NQO/L in sample 2. This was also because of the formation of TOBr. Similar to cytotoxicity, the VOF_{Br} also showed considerable genotoxicity. In addition, compared with the genotoxicity of the $NVOF_{Br}$, co-exposure to the VOF_{Br} and $NVOF_{Br}$ (TOF_{Br}) further increased the genotoxicity (one-way ANOVA, $p < 0.05$). The genotoxicity of the TOF_{Br} in sample 1 increased by 12% compared with that of the $NVOF_{Br}$, up to 1.39 μg-4-NQO/L, while the genotoxicity of the TOF_{Br} in sample 2 increased by 46% compared with that of the $NVOF_{Br}$. Similar to cytotoxicity, bromate had a significant but limited impact on overall genotoxicity. The co-exposure of bromate and the TOF_{Br} (the TF_{Br}) increased the genotoxicity by 7% compared to that of the TOF_{Br} for sample 1 and increased it by 11% for sample 2. These results demonstrated that in the presence of bromide, the $NVOF_{Br}$ and VOF_{Br} both contributed greatly to the overall genotoxicity of the treated water.

9.5 Control of disinfection by-products and toxicity

The generation of TOX during disinfection is the primary factor contributing to increased water biotoxicity. Consequently, strategies that inhibit the formation of reactive halogens (HOCl, HOBr, etc.) have proven effective in controlling biotoxicity.

9.5.1 Pre-oxidation to remove precursors of DBPs

In chlorine disinfection, extensive research has been conducted on the elimination of precursors through pre-oxidation to achieve a reduction in the formation of DBPs (Jiang et al., 2019; Li et al., 2022). For instance, pre-ozonation can effectively destroy DBPs precursors, typically following the order of DHANs > THMs ≈ THAAs > DHAAs. A decrease of nearly 58% in haloacetic acids (HAAs) formation was observed at a dosage of 1.05 mg O_3/mg DOC (Naladala et al., 2018). Moreover, when applying an ozone concentration of 4.0 mg/L, there was a reduction ranging from 36% to 43% in trihalomethanes generation potentials (THMFPs), as well as over 30% decrease in dichloroacetonitrile generation potentials (DCANFPs) (Jiang et al., 2019). Additionally, the pre-ozonation of NOM resulted in a notable reduction of cytotoxicity formation by 30.7% after chlorination (Wang et al., 2021).

Some researchers use Fe(VI) for pre-oxidation to target electron-rich organic fractions, such as phenols and olefins. This process aims to reduce the aromaticity and unsaturation of DOM (Liu et al., 2020). Thus, Fe(VI) pretreatment has been shown to decrease the formation of THMs, HAAs, haloacetonitriles (HAN), and haloketones (HKs) during chlorination. It also reduces the formation of N-nitrosodimethylamine (NDMA) and iodinated trihalomethanes (I-THMs) during chloramination (Lee et al., 2008; Gan et al., 2015; Zhang et al., 2016).

However, the application of pre-oxidation is not universally applicable. A recent study found that pre-ozonation before chlorine disinfection can result in an increase in the generation of TOBr and toxicity when treating wastewater with high bromide concentrations (Yuan et al., 2023). The ozone-chlorine disinfection process leads to a "two-stage toxicity increase" in the cytotoxicity and genotoxicity of bromide-containing reclaimed water. The generation of TOBr was the main cause of the increase in toxicity, and DBPs such as TBM and DBAN were detected, while TOBr increased from 33.9 μg/L after ozonation (5 mg/L ozone and 1 mg/L bromide) to 100.5 μg/L after chlorine disinfection (10 mg/L as Cl).

9.5.2 Masking of reactive halogens with hydrogen peroxide

Hydrogen peroxide (H_2O_2) can rapidly reduce reactive halogens to halogen ions and block TOX generation.

For example, O_3/H_2O_2 co-oxidation processes can inhibit TOBr and reduce cytotoxicity and genotoxicity generation (Wu et al., 2020). It was shown that H_2O_2 could significantly inhibit the toxicity formation of organic byproducts (Fig. 9.5).

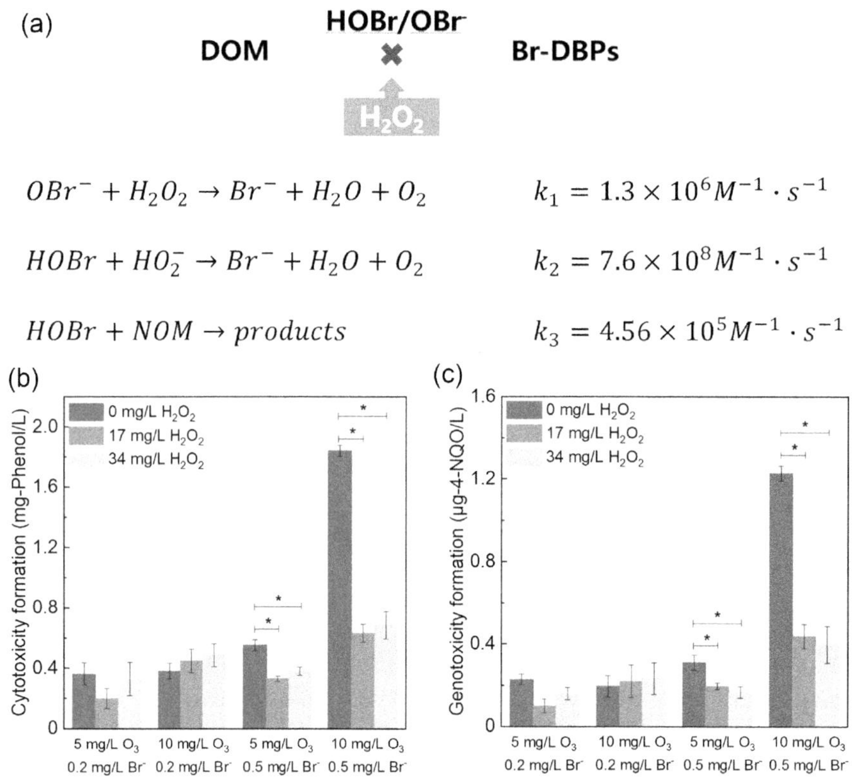

Fig. 9.5 (A) Scheme of the mechanism of H_2O_2 control of reactive halogens, control of (B) cytotoxicity and (C) genotoxicity of ozone-disinfected effluent by H_2O_2 (Wu et al., 2020). Copyright: ACS publication.

Fe(VI) has also been used to pre-oxidize wastewater, thereby reducing the generation of DBPs and toxicity after disinfection (Wu et al., 2021b). In addition to the removal of DBP precursors by Fe(VI), the use of Fe(VI) to control TOX and toxicity formation in wastewater is closely related to H_2O_2 from the self-decomposition of Fe(VI) (Fig. 9.6).

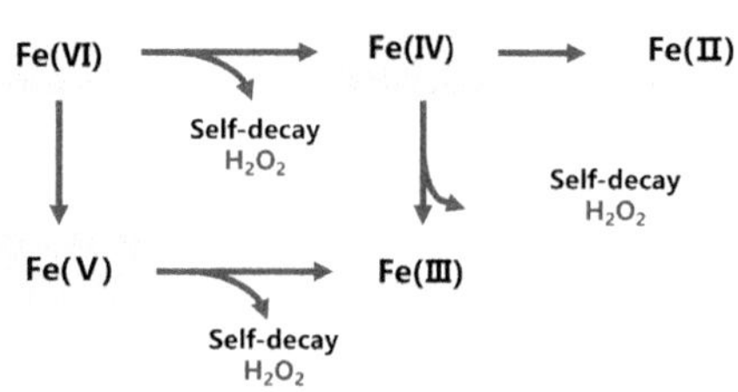

Fig. 9.6 Mechanism of hydrogen peroxide production by ferrate.

9.5.3 Activation of oxidants to generate free radicals to inhibit toxicity generation

Generation of free radicals by activated oxidants is another method of inhibiting byproduct and toxicity generation, the main mechanisms of which are (1) enhanced removal of byproduct precursors (electron-donating groups of DOM) (2) oxidation of hypohalous acids to orthohalogenates (ClO_3^-, BrO_3^-, IO_3^-), to reduce the generation of organic halogenated byproducts.

UV-based synergistic disinfection processes can effectively reduce the amount of oxidant and the formation of byproducts, improve the disinfection performance, to overcome the problem of photoactivation, which has received more and more attention. The disinfection mechanism of UV-based synergistic disinfection processes is shown in Fig. 9.7.

UV-based synergistic disinfection processes will not only enhance the disinfection effect, but also hold great promise for application in controlling DBPs and toxicity. A study reported that UV/chlorine treatment eliminated 90% and 62% of fluorescent components and UV_{254}, respectively, from municipal wastewater,

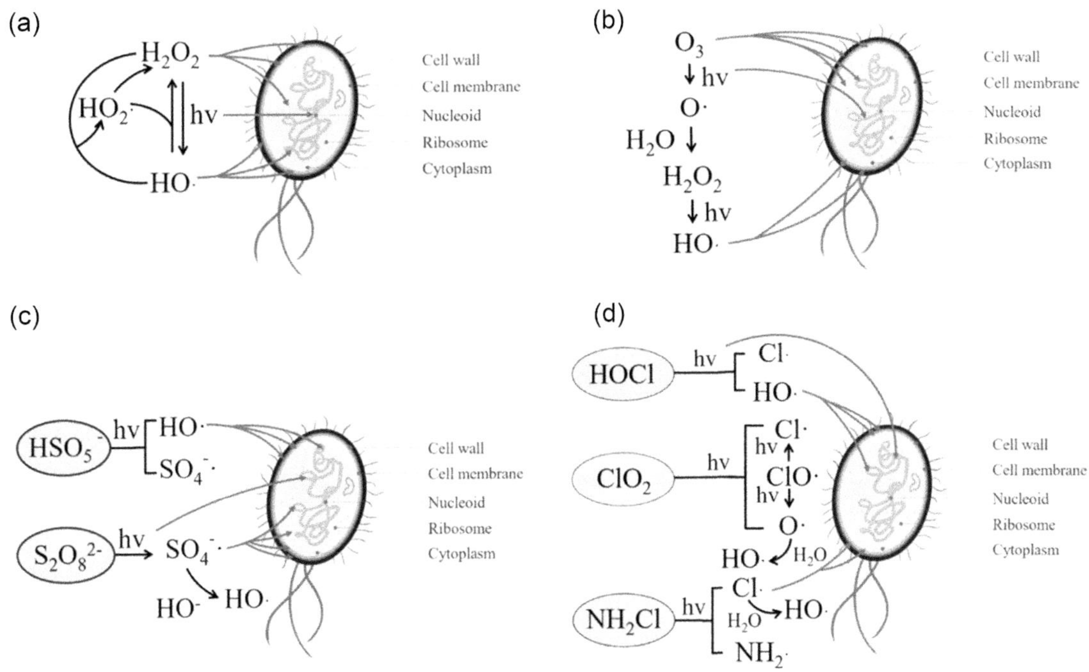

Fig. 9.7 Mechanism of UV-based synergistic disinfection processes. (A) UV/H_2O_2; (B) UV/O_3; (C) UV/PMS and UV/PDS; (D) UV/HOCl, UV/ClO_2, and UV/NH_2Cl (Zhang et al., 2023). Copyright: Elsevier.

the acute toxicity and genotoxicity were reduced by 19% and 76%, respectively (Hua et al., 2021). This is due to the generation of hydroxyl and other reactive chlorine radicals by UV/chlorine, which removes more precursors than chlorine disinfection alone and reduces effluent cytotoxicity and genotoxicity. UV/chlorine can also result in lower acute toxicity to luminescent bacteria than chlorine disinfection of effluent alone (Zhang et al., 2021).

Moreover, UV can be used in conjunction with ozone to reduce the TOBr and toxicity generation problems associated with ozone disinfection alone (Fig. 9.8). The UV/O_3 process (100 mJ/cm^2, 1 mg-O_3/mg-DOC) reduced organic bromine generation by 81% of TOBr, and the cytotoxicity and genotoxicity of ozonated wastewater (2 mg/L Br^-) by 72% and 91%, respectively. In addition, the UV/O_3 process removed about 60.0% of precursors, whereas the O_3 process alone removed only 39.5% of precursors.

In addition, UV/PMS is also effective in suppressing the elevated toxicity caused by PMS, a toxicity control phenomenon mainly attributed to the generation of sulfate and hydroxyl radicals and photolysis by UV. UV/PMS can mineralize typical DBPs precursors, such as humic acid and reduce the aromaticity of electron-rich DOM by dehydrogenation, dihydroxylation, and cyclo-cleavage to reduce the reactivity between DOM and HOBr reactivity. Additionally, as shown in Fig. 9.9, UV light may photolyze HOBr thereby reducing TOBr production, or may also destroy organic byproducts, especially TOBr formed during PMS treatment, thereby reducing toxicity (Du et al., 2023).

However, it does not mean that the generation of free radicals leads to more toxicity removal, and not all disinfection processes that enhance the generation of free radicals are effective in cutting

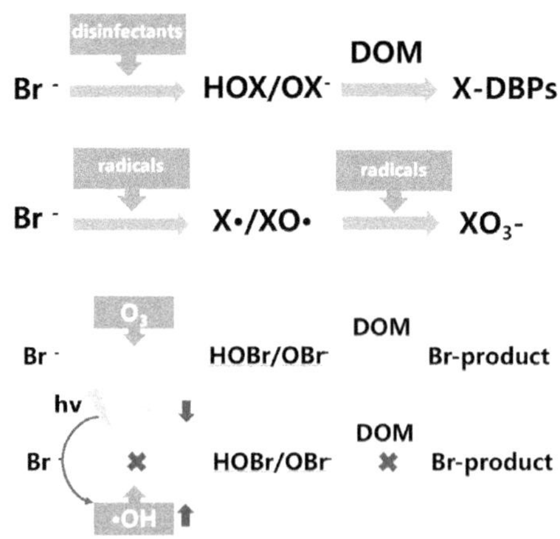

Fig. 9.8 Mechanisms of byproducts controlled by free radicals.

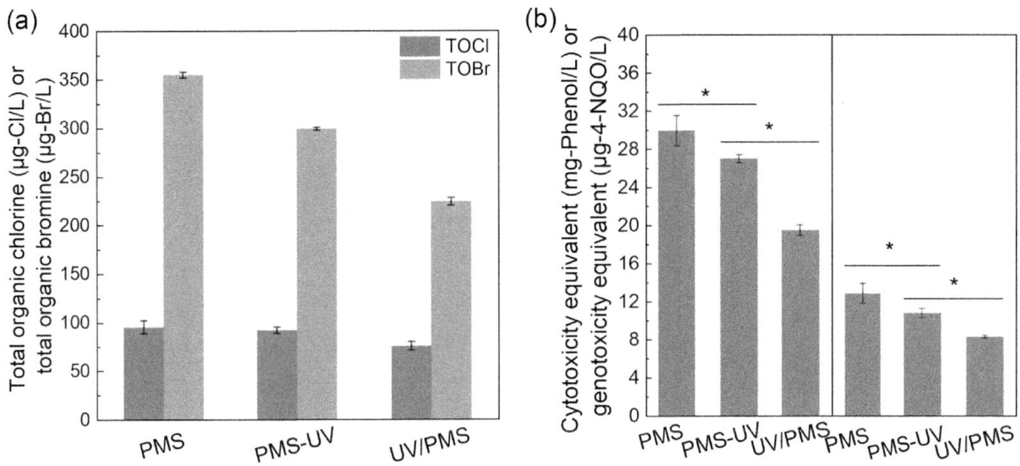

Fig. 9.9 Control of (A) TOX, (B) cytotoxicity and genotoxicity by UV/PMS process (Du et al., 2023). Copyright: ACS publication.

toxicity. For example, reactive chlorine species (RCS) and reactive nitrogen species (RNS) in the UV/chloramine process significantly enhance the formation of chloroform, chloral hydrate, chloropicrin and dichloroacetonitrile, which in turn increases toxicity (Chen et al., 2021).

9.5.4 Removal of DBPs and toxicity after disinfection

It has been observed that solar irradiation mitigates the cytotoxicity associated with the discharge of disinfected wastewater into surface water for ecological storage, as solar irradiation, particularly UV light in the solar spectrum, facilitates the degradation of formed DBPs (Fig. 9.10A) (Du et al., 2018).

Similarly, UV irradiation can significantly reduce the toxic components of chlorine-disinfected water. At a UV dose of $2400\,mJ/cm^2$, the reduction of TOCl ranged from 25.9% to 53.6%, while the reduction of TOBr was higher at about 60%. Both cytotoxicity and genotoxicity were significantly removed under UV irradiation, with the removal of cytotoxic equivalents ranging from 22.9% to 41.7%, and the removal of genotoxic equivalents was higher, ranging from 33.1% to 55.5% (Fig. 9.10B) (Huang et al., 2019). The similar phenomenon has been reported in many studies, for example, UV irradiation reduces CHO cytotoxicity and genotoxicity brought about by chlorine disinfection (Plewa et al., 2012).

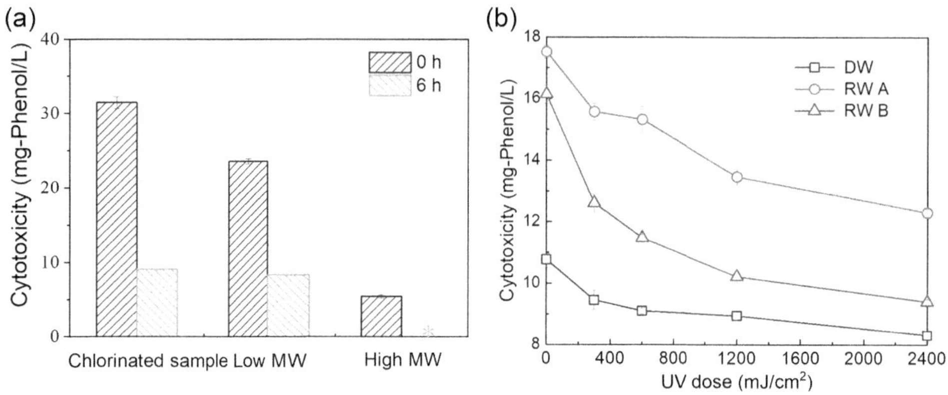

Fig. 9.10 Reduction of the toxicity of chlorine-disinfected water samples by exposure to light (Huang et al., 2019). Copyright: Elsevier.

9.6 Conclusions and future trends

The field of wastewater disinfection is continuously evolving with the advancement of technology and the growing understanding of public health and environmental impacts. This chapter explores the future trends and innovations that are shaping the next generation of wastewater disinfection technologies, specifically in the areas of disinfection byproducts detection, and toxicity assessment.

(1) Developing technology and methodology systems that objectively reflects the biological toxicity of wastewater

Biotoxicity evaluation usually requires pretreatment to concentrate and enrich byproducts in wastewater samples. Common pretreatment methods include solid-phase extraction and liquid-liquid extraction. However, both solid-phase extraction and liquid-liquid extraction make it difficult to retain volatile substances in water, which may lead to underestimation of toxicity. For example, the use of solid-phase extraction to enrich organic matter in oxygen-oxidized effluent led to the conclusion that ozone oxidation could reduce the toxicity of effluent to human hepatocytes (removal rate 56–70%). In contrast, after volatile byproducts such as aldehydes were collected, it was found that ozone oxidation actually elevated the toxicity. In the toxicity assessment work, to objectively reflect the toxicity of water quality, special attention should be paid to the enrichment of full-component pollutants, and the development of water sample pretreatment techniques and devices that can efficiently enrich full-component pollutants.

(2) **Identifying and scientifically controlling key toxic components of disinfected wastewater**

The presence of various pollutants in wastewater, particularly following the disinfection process, leads to the formation of byproducts and consequently results in a more intricate water quality. Existing water quality standards may not target the key toxic-causing components.

For example, in the case of ozonation, existing research has centered on the generation and control of bromate because water quality standards regulate the inorganic byproduct bromate. Recent studies have found that it is the organic bromine produced during ozonation that is the main contributor to toxicity. Similarly, for chlorine-disinfected water, China, the European Union and the United States all regulate the byproducts of trihalomethanes in water quality standards. However, recent studies have shown that volatile byproducts such as trihalomethanes make a very limited contribution to toxicity, and the non-volatile components of the byproducts that are the main contributors to elevated toxicity (Wu et al., 2020). A study of potable reuse of reclaimed water in the United States showed that existing controlled byproducts contributed only 0.2% of overall toxicity, while non-volatile and unidentified by-products contributed up to 83% of the toxicity (Lau et al., 2023).

These studies show that current water quality standards do not accurately control toxic components. Future work will require a scientific understanding of key toxigenic components of complex water quality and their targeted control.

(3) **Developing wastewater treatment processes based on toxicity assessment**

In the development of wastewater treatment processes, it is crucial to adopt a systematic approach that goes beyond the removal of individual pollutants and encompasses comprehensive biotoxicity control, given the diverse range of pollutants and complex byproducts present in wastewater.

For example, existing water quality standards regulate bromate. Some water treatment plants use the addition of ammonia to control bromate. Recent studies have shown that while ammonia can inhibit the formation of bromate, it also promotes the formation of a large number of organic brominated nitrogen byproducts, such as bromoacetamide and bromonitromethane, which can increase cytotoxicity and genotoxicity by more than 2 times (Wu et al., 2020). In addition, a number of advanced oxidation processes based on persulfate are used for the degradation of trace pollutants. Since water often contains chloride and bromine ions, persulfate leads to the formation of halogenated byproducts and increases toxicity (Du et al., 2023). Therefore, in order to ensure the safety of water quality, a comprehensive biotoxicity assessment should be formed as the forerunner in the development process of wastewater treatment processes, and new processes should be developed on the premise of toxicity control.

Unlike specific pollutant control, the systematic and holistic nature of biotoxicity evaluation leads to the discovery of new phenomena and the formulation of novel conclusions. In recent years, water toxicity assessment has gained increasing attention. Comprehensive biotoxicity evaluation will play an increasingly pivotal role in developing wastewater treatment processes and establishing wastewater pollutant discharge standards.

Acknowledgments

This work was funded by National Natural Science Foundation of China (52170044).

References

Allen, J.M., Plewa, M.J., Wagner, E.D., Wei, X., Bokenkamp, K., Hur, K., Jia, A., Liberatore, H.K., Lee, C.F.T., Shirkhani, R., Krasner, S.W., Richardson, S.D., 2022. Drivers of disinfection byproduct cytotoxicity in U.S. drinking water: should other DBPs be considered for regulation? Environ. Sci. Technol. 56 (1), 392–402. American Chemical Society, United States, 15205851. Available from: http://pubs.acs.org/journal/esthag. https://doi.org/10.1021/acs.est.1c07998.

Bayo, J., Angosto, J.M., Gómez-López, M.D., 2009. Ecotoxicological screening of reclaimed disinfected wastewater by *Vibrio fischeri* bioassay after a chlorination–dechlorination process. J. Hazard. Mater. 172, 166–171. https://doi.org/10.1016/j.jhazmat.2009.06.157.

Blatchley, E.R., Hunt, B.A., Duggirala, R., Thompson, J.E., Zhao, J., Halaby, T., Cowger, R.L., Straub, C.M., Alleman, J.E., 1997. Effects of disinfectants on wastewater effluent toxicity. Water Res 31 (7), 1581–1588. https://doi.org/10.1016/S0043-1354(96)00396-X. ISSN 0043-1354.

Bove, G.E., Rogerson, P.A., Vena, J.E., 2007. Case control study of the geographic variability of exposure to disinfectant byproducts and risk for rectal cancer. Int. J. Health Geogr. 6, 18. https://doi.org/10.1186/1476-072X-6-18.

Cantor, K.P., Villanueva, C.M., Silverman, D.T., Figueroa, J.D., Real, F.X., Garcia-Closas, M., Malats, N., Chanock, S., Yeager, M., Tardon, A., Garcia-Closas, R., Serra, C., Carrato, A., Castaño-Vinyals, G., Samanic, C., Rothman, N., Kogevinas, M., 2010. Polymorphisms in GSTT1, GSTZ1, AND CYP2E1, disinfection by-products, and risk of bladder cancer in Spain. Environ. Health Perspect. 118 (11), 1545–1550. United States, 15529924. Available from: http://ehp03.niehs.nih.gov/article/fetchObjectAttachment.action?uri=info%3Adoi%2F10.1289%2Fehp.1002206%26representation=PDF. https://doi.org/10.1289/ehp.1002206.

Cao, N., Miao, T., Li, K., Zhang, Y., Yang, M., 2009. Formation potentials of typical disinfection byproducts and changes of genotoxicity for chlorinated tertiary effluent pretreated by ozone. J. Environ. Sci. 21 (4), 409–413. 10010742 https://doi.org/10.1016/s1001-0742(08)62284-6.

Chen, B., Nam, S.N., Westerhoff, P.K., Krasner, S.W., Amy, G., 2009. Fate of effluent organic matter and DBP precursors in an effluent-dominated river: a case study of wastewater impact on downstream water quality. Water Res. 43 (6), 1755–1765. Elsevier Ltd, United States, 00431354. Available from: www.elsevier.com/locate/watres. https://doi.org/10.1016/j.watres.2009.01.020.

Chen, C., Ye, D., Zhou, Y., Wu, Q., Zheng, S., Fang, J., 2021. Formation of nitro(so) and chlorinated products and toxicity alteration during the UV/monochloramine treatment of phenol. Water Res. 194, 116914. 00431354 https://doi.org/10.1016/j.watres.2021.116914.

Cheng, L., Wei, X., Gao, A., Zhou, L., Shi, X., Zhou, X., Bi, X., Yang, T., Huang, S., 2024. Performance and mechanism of sequential UV-NaClO disinfection: inactivation and reactivation of antibiotic-resistant bacteria, disinfection byproduct formation and microbial community variation. J. Water Process Eng. 58, 104824. 22147144 https://doi.org/10.1016/j.jwpe.2024.104824.

Cho, M., Lee, Y., Choi, W., Chung, H., Yoon, J., 2006. Study on Fe(VI) species as a disinfectant: quantitative evaluation and modeling for inactivating Escherichia coli. Water Res. 40 (19), 3580–3586. 00431354 https://doi.org/10.1016/j.watres.2006.05.043.

Ding, W., Jin, W., Cao, S., Zhou, X., Wang, C., Jiang, Q., Huang, H., Tu, R., Han, S.F., Wang, Q., 2019. Ozone disinfection of chlorine-resistant bacteria in drinking water. Water Res. 160, 339–349. Elsevier Ltd, China. 18792448. Available from: www.elsevier.com/locate/watres. https://doi.org/10.1016/j.watres.2019.05.014.

Dong, S., Masalha, N., Plewa, M.J., Nguyen, T.H., 2017. Toxicity of wastewater with elevated bromide and iodide after chlorination, chloramination, or ozonation disinfection. Environ. Sci. Technol. 51 (16), 9297–9304. American Chemical Society, United States. 15205851. Available from: http://pubs.acs.org/journal/esthag. https://doi.org/10.1021/acs.est.7b02345.

Dong, S., Mu, Y., Sun, X., 2019. Removal of toxic metals using ferrate (VI): a review. Water Sci. Technol. 80, 1213–1225. https://doi.org/10.2166/wst.2019.376.

Dong, H., Cuthbertson, A.A., Plewa, M.J., Weisbrod, C.R., McKenna, A.M., Richardson, S.D., 2023. Unravelling high-molecular-weight DBP toxicity drivers in chlorinated and chloraminated drinking water: effect-directed analysis of molecular weight fractions. Environ. Sci. Technol. 57 (47), 18788–18800. American Chemical Society, United States, 15205851. Available from: http://pubs.acs.org/journal/esthag. https://doi.org/10.1021/acs.est.3c00771.

Du, Y., Lv, X.T., Wu, Q.Y., Zhang, D.Y., Zhou, Y.T., Peng, L., Hu, H.Y., 2017. Formation and control of disinfection byproducts and toxicity during reclaimed water chlorination: a review. J. Environ. Sci. (China) 58, 51–63. Chinese Academy of Sciences, China. 18787320. Available from: http://www.journals.elsevier.com/journal-of-environmental-sciences/. https://doi.org/10.1016/j.jes.2017.01.013.

Du, Y., Wu, Q.Y., Lv, X.T., Ye, B., Zhan, X.M., Lu, Y., Hu, H.Y., 2018. Electron donating capacity reduction of dissolved organic matter by solar irradiation reduces the cytotoxicity formation potential during wastewater chlorination. Water Res. 145, 94–102. Elsevier Ltd, China. 18792448. Available from: www.elsevier.com/locate/watres. https://doi.org/10.1016/j.watres.2018.08.023.

Du, Y., Wang, W.-L., He, T., Sun, Y.-X., Lv, X.-T., Wu, Q.-Y., Hu, H.-Y., 2020a. Chlorinated effluent organic matter causes higher toxicity than chlorinated natural organic matter by inducing more intracellular reactive oxygen species. Sci. Total Environ. 701, 134881. https://doi.org/10.1016/j.scitotenv.2019.134881.

Du, Y., Yang, Y., Wang, W.-L., Zhou, Y.-T., Wu, Q.-Y., 2020b. Surrogates for the removal by ozonation of the cytotoxicity and DNA double-strand break effects of wastewater on mammalian cells. Environ. Int. 135, 105369. https://doi.org/10.1016/j.envint.2019.105369.

Du, Y., Wang, W.L., Wang, Z.W., Yuan, C.J., Ye, M.Q., Wu, Q.Y., 2023. Overlooked cytotoxicity and genotoxicity to mammalian cells caused by the oxidant peroxymonosulfate during wastewater treatment compared with the sulfate radical-based ultraviolet/peroxymonosulfate process. Environ. Sci. Technol. 57 (8), 3311–3322. American Chemical Society, China. 15205851. Available from: http://pubs.acs.org/journal/esthag. https://doi.org/10.1021/acs.est.2c06965.

Ersan, M.S., Liu, C., Amy, G., Plewa, M.J., Wagner, E.D., Karanfil, T., 2019. Chloramination of iodide-containing waters: formation of iodinated disinfection byproducts and toxicity correlation with total organic halides of treated waters. Sci. Total Environ. 697. Elsevier B.V., United States, 18791026. Available from: www.elsevier.com/locate/scitotenv. https://doi.org/10.1016/j.scitotenv.2019.134142.

Gan, W., Sharma, V.K., Zhang, X., Yang, L., Yang, X., 2015. Investigation of disinfection byproducts formation in ferrate(VI) pre-oxidation of NOM and its model compounds followed by chlorination. J. Hazard. Mater. 292, 197–204. Elsevier B.V., China, 18733336. Available from: www.elsevier.com/locate/jhazmat. https://doi.org/10.1016/j.jhazmat.2015.02.037.

Gombos, E., Felföldi, T., Barkács, K., Vértes, C., Vajna, B., Záray, G., 2012. Ferrate treatment for inactivation of bacterial community in municipal secondary effluent. Bioresour. Technol. 107, 116–121. 09608524 https://doi.org/10.1016/j.biortech.2011.12.053.

Han, J., Zhang, X., 2018. Evaluating the comparative toxicity of DBP mixtures from different disinfection scenarios: a new approach by combining freeze-drying or rotoevaporation with a marine polychaete bioassay. Environ. Sci. Technol. 52 (18), 10552–10561. 0013-936X https://doi.org/10.1021/acs.est.8b02054.

Hua, G., Reckhow, D.A., 2007. Comparison of disinfection byproduct formation from chlorine and alternative disinfectants. Water Res. 41 (8), 1667–1678. Elsevier Ltd, United States, 00431354. Available from: www.elsevier.com/locate/watres. https://doi.org/10.1016/j.watres.2007.01.032.

Hua, Z., Li, D., Wu, Z., Wang, D., Cui, Y., Huang, X., Fang, J., An, T., 2021. DBP formation and toxicity alteration during UV/chlorine treatment of wastewater and the effects of ammonia and bromide. Water Res. 188, 116549. 00431354 https://doi.org/10.1016/j.watres.2020.116549.

Huang, J., Wang, L., Ren, N., Ma, F., Juli, 1997. Disinfection effect of chlorine dioxide on bacteria in water. Water Res. 31 (3), 607–613. 00431354 https://doi.org/10.1016/s0043-1354(96)00275-8.

Huang, X., Deng, Y., Liu, S., Song, Y., Li, N., Zhou, J., 2016. Formation of bromate during ferrate(VI) oxidation of bromide in water. Chemosphere 155, 528–533. 00456535 https://doi.org/10.1016/j.chemosphere.2016.04.093.

Huang, W.C., Du, Y., Liu, M., Hu, H.Y., Wu, Q.Y., Chen, Y., 2019. Influence of UV irradiation on the toxicity of chlorinated water to mammalian cells: toxicity drivers, toxicity changes and toxicity surrogates. Water Res. 165. Elsevier Ltd, China, 18792448. Available from: www.elsevier.com/locate/watres. https://doi.org/10.1016/j.watres.2019.115024.

Janex, M.L., Savoye, P., Roustan, M., Do-Quang, Z., Laîné, J.M., Lazarova, V., 2000. Wastewater disinfection by ozone: influence of water quality and kinetics modeling. Ozone: Sci. Eng. 22 (2), 113–121. 0191-9512 https://doi.org/10.1080/01919510008547215.

Jeong, C.H., Wagner, E.D., Siebert, V.R., Anduri, S., Richardson, S.D., Daiber, E.J., McKague, A.B., Kogevinas, M., Villanueva, C.M., Goslan, E.H., Luo, W., Isabelle, L.M., Pankow, J.F., Grazuleviciene, R., Cordier, S., Edwards, S.C.,

Righi, E., Nieuwenhuijsen, M.J., Plewa, M.J., 2012. Occurrence and toxicity of disinfection byproducts in European drinking waters in relation with the HIWATE epidemiology study. Environ. Sci. Technol. 46 (21), 12120–12128. 15205851 https://doi.org/10.1021/es3024226.

Jessen, A., Randall, A., Reinhart, D., Daly, L., 2008. Effectiveness and kinetics of ferrate as a disinfectant for ballast water. Water Environ. Res. 80 (6), 561–569. 1061-4303 https://doi.org/10.2175/193864708x267423.

Jiang, J.-Q., Wang, S., Panagoulopoulos, A., 2007. The role of potassium ferrate(VI) in the inactivation of *Escherichia coli* and in the reduction of COD for water remediation. Desalination 210, 266–273. https://doi.org/10.1016/j.desal.2006.05.051. Ninth Environmental Science and Technology Symposium.

Jiang, Y., Goodwill, J.E., Tobiason, J.E., Reckhow, D.A., 2019. Comparison of ferrate and ozone pre-oxidation on disinfection byproduct formation from chlorination and chloramination. Water Res. 156, 110–124. Elsevier Ltd, United States, 18792448. Available from: www.elsevier.com/locate/watres. https://doi.org/10.1016/j.watres.2019.02.051.

Kimura, S.Y., Ortega-Hernandez, A., 2019. Formation mechanisms of disinfection byproducts: recent developments. Curr. Opin. Environ. Sci. Health **7**, 61–68. Elsevier B.V., Canada, 24685844. Available from: www.journals.elsevier.com/current-opinion-in-environmental-science-and-health. https://doi.org/10.1016/j.coesh.2018.11.002.

Kitis, M., Karanfil, T., Wigton, A., Kilduff, J.E., 2002. Probing reactivity of dissolved organic matter for disinfection by-product formation using XAD-8 resin adsorption and ultrafiltration fractionation. Water Res. 36 (15), 3834–3848. Elsevier Ltd, United States, 00431354. Available from: https://www.sciencedirect.com/science/journal/00431354. https://doi.org/10.1016/S0043-1354(02)00094-5.

Kontana, A., Papadimitriou, C.A., Samaras, P., Zdragas, A., Yiangou, M., 2009. Effectiveness of ozonation and chlorination on municipal wastewater treatment evaluated by a battery of bioassays and biomarkers. Water Sci. Technol. 60 (6), 1497–1505. 0273-1223 https://doi.org/10.2166/wst.2009.478.

Krasner, S.W., Westerhoff, P., Chen, B., Rittmann, B.E., Nam, S.N., Amy, G., 2009. Impact of wastewater treatment processes on organic carbon, organic nitrogen, and DBP precursors in effluent organic matter. Environ. Sci. Technol. 43 (8), 2911–2918. United States. 15205851. Available from: http://pubs.acs.org/doi/pdfplus/10.1021/es802443t. https://doi.org/10.1021/es802443t.

Lau, S.S., Bokenkamp, K., Tecza, A., et al., 2023. Toxicological assessment of potable reuse and conventional drinking waters. Nat. Sustain. 6, 39–46. https://doi.org/10.1038/s41893-022-00985-7.

Le Roux, J., Plewa, M.J., Wagner, E.D., Nihemaiti, M., Dad, A., Croué, J.P., 2017. Chloramination of wastewater effluent: toxicity and formation of disinfection byproducts. J. Environ. Sci. (China) 58, 135–145. Chinese Academy of Sciences, France, 18787320. Available from: http://www.journals.elsevier.com/journal-of-environmental-sciences/. https://doi.org/10.1016/j.jes.2017.04.022.

Lee, C., Lee, Y., Schmidt, C., Yoon, J., Von Gunten, U., 2008. Oxidation of suspected N-nitrosodimethylamine (NDMA) precursors by ferrate (VI): kinetics and effect on the NDMA formation potential of natural waters. Water Res. 42 (1-2), 433–441. 00431354 https://doi.org/10.1016/j.watres.2007.07.035.

Li, X.F., Mitch, W.A., 2018. Drinking water disinfection byproducts (DBPs) and human health effects: multidisciplinary challenges and opportunities. Environ. Sci. Technol. 52 (4), 1681–1689. American Chemical Society, Canada, 15205851. Available from: http://pubs.acs.org/journal/esthag. https://doi.org/10.1021/acs.est.7b05440.

Li, Q., Mahendra, S., Lyon, D.Y., Brunet, L., Liga, M.V., Li, D., Alvarez, P.J.J., 2008. Antimicrobial nanomaterials for water disinfection and microbial control:

potential applications and implications. Water Res. 42 (18), 4591–4602. Elsevier Ltd, United States, 00431354. Available from: www.elsevier.com/locate/watres. https://doi.org/10.1016/j.watres.2008.08.015.

Li, Z., Liu, X., Huang, Z., Hu, S., Wang, J., Qian, Z., Feng, J., Xian, Q., Gong, T., 2019. Occurrence and ecological risk assessment of disinfection byproducts from chlorination of wastewater effluents in East China. Water Res. 157, 247–257. 00431354 https://doi.org/10.1016/j.watres.2019.03.072.

Li, T., Shang, C., Xiang, Y., Yin, R., Pan, Y., Fan, M., Yang, X., 2022. ClO_2 pre-oxidation changes dissolved organic matter at the molecular level and reduces chloro-organic byproducts and toxicity of water treated by the UV/chlorine process. Water Res. 216, 118341. 00431354 https://doi.org/10.1016/j.watres.2022.118341.

Liu, J.-L., Li, X.-Y., 2010. Biodegradation and biotransformation of wastewater organics as precursors of disinfection byproducts in water. Chemosphere 81 (9), 1075–1083. 00456535 https://doi.org/10.1016/j.chemosphere.2010.09.041.

Liu, J.-L., Li, X.-Y., Xie, Y.-F., Tang, H., 2014. Characterization of soluble microbial products as precursors of disinfection byproducts in drinking water supply. Sci. Total Environ. 472, 818–824. 00489697 https://doi.org/10.1016/j.scitotenv.2013.11.139.

Liu, J., Lujan, H., Dhungana, B., Hockaday, W.C., Sayes, C.M., Cobb, G.P., Sharma, V.-K., 2020. Ferrate(VI) pretreatment before disinfection: an effective approach to controlling unsaturated and aromatic halo-disinfection byproducts in chlorinated and chloraminated drinking waters. Environ. Int. 138. Elsevier Ltd, United States, 18736750. Available from: www.elsevier.com/locate/envint. https://doi.org/10.1016/j.envint.2020.105641.

Lu, Y., Wang, C., Zhang, X.Y., Wang, Z.W., Song, Z.M., Du, Y., Hu, Q., Wu, Q.Y., Hu, H.-Y., 2021. Tracing nitrogenous byproducts during ozonation in the presence of bromide and ammonia using stable isotope labeling and high resolution mass spectrometry. J. Hazard. Mater. 403. Elsevier B.V., China. 18733336. Available from: www.elsevier.com/locate/jhazmat. https://doi.org/10.1016/j.jhazmat.2020.123612.

Magazinovic, R.S., Nicholson, B.C., Mulcahy, D.E., Davey, D.E., 2004. Bromide levels in natural waters: Its relationship to levels of both chloride and total dissolved solids and the implications for water treatment. Chemosphere 57 (4), 329–335. Elsevier Ltd, Australia. 00456535. Available from: www.elsevier.com/locate/chemosphere. https://doi.org/10.1016/j.chemosphere.2004.04.056.

Mališová, E., Guštafík, A., Tamáš, M., Strečanský, T., Imreová, Z., Konečná, B., Červenková, A., Peciar, P., Mackuľak, T., Híveš, J., 2023. Effective stabilization of electrochemically prepared ecological oxidizing agent—ferrate(VI)—by encapsulation in zeolite and its application to water containing SARS-CoV-2 virus. Water Environ. Res. 95 (10), 1061–4303. https://doi.org/10.1002/wer.10940.

Manoli, K., Maffettone, R., Sharma, V.K., Santoro, D., Ray, A.K., Passalacqua, K.D., Carnahan, K.E., Wobus, C.E., Sarathy, S., 2020. Inactivation of murine norovirus and fecal coliforms by ferrate(VI) in secondary effluent wastewater. Environ. Sci. Technol. 54 (3), 1878–1888. American Chemical Society, Canada, 15205851. Available from: http://pubs.acs.org/journal/esthag. https://doi.org/10.1021/acs.est.9b05489.

Mao, Y., Chen, Z., Zhang, Z.W., Xue, S., Lu, Y., Shi, Q., Cao, K.F., Chen, X.W., Wu, Y.H., Hu, H.Y., 2022. Comparison of the disinfection efficacy between ferrate(VI) and chlorine in secondary effluent. Sci. Total Environ. 848. Elsevier B.V., China, 18791026. Available from: www.elsevier.com/locate/scitotenv. https://doi.org/10.1016/j.scitotenv.2022.157712.

Mao, Y., Chen, Z., Lu, Y., Cao, K.F., Wu, Y., Hu, H.Y., 2023. Inactivation of bacteria in water by ferrate(VI): efficiency and mechanisms. Environ. Sci. Technol. 57 (49), 20893–20904. American Chemical Society, China, 15205851. Available from: http://pubs.acs.org/journal/esthag. https://doi.org/10.1021/acs.est.3c05118.

Naladala, N.R., Singh, R., Katiyar, K.L.D., Bose, P., Dutta, V., 2018. Effect of pre-ozonation on haloacetic acids formation in Ganga River water at Kanpur, India. J. Inst. Eng. (India): Ser. A 99 (1), 37–44. Springer India, India. 22502157. Available from: http://www.springer.com/engineering/civil+engineering/journal/40030. https://doi.org/10.1007/s40030-017-0226-y.

Narotsky, M.G., Best, D.S., Rogers, E.H., McDonald, A., Sey, Y.M., Simmons, J.E., 2008. Integrated disinfection by-products mixtures research: assessment of developmental toxicity in sprague-dawley rats exposed to concentrates of water disinfected by chlorination and ozonation/postchlorination. J. Toxicol. Environ. Health A 71 (17), 1216–1221. 1528-7394 https://doi.org/10.1080/15287390802182623.

Patterson, K.S., Lykins, B.W., Richardson, S.D., 1995. Mutagenicity of drinking water following disinfection. J. Water Supply 44, 1–9. Aqua London.

Plewa, M.J., Simmons, J.E., Richardson, S.D., Wagner, E.D., 2010. Mammalian cell cytotoxicity and genotoxicity of the haloacetic acids, a major class of drinking water disinfection by-products. Environ. Mol. Mutagen. 51 (8-9), 871–878. 10982280 https://doi.org/10.1002/em.20585.

Plewa, M.J., Wagner, E.D., Metz, D.H., Kashinkunti, R., Jamriska, K.J., Meyer, M., 2012. Differential toxicity of drinking water disinfected with combinations of ultraviolet radiation and chlorine. Environ. Sci. Technol. 46 (14), 7811–7817. 15205851 https://doi.org/10.1021/es300859t.

Qian, Y., Chen, Y., Hu, Y., Hanigan, D., Westerhoff, P., Dong, A., 2021. Formation and control of C- and N-DBPs during disinfection of filter backwash and sedimentation sludge water in drinking water treatment. Water Res. 194, 116964. 00431354 https://doi.org/10.1016/j.watres.2021.116964.

Rahman, M.B., Driscoll, T., Cowie, C., Armstrong, B.K., 2010. Disinfection by-products in drinking water and colorectal cancer: a meta-analysis. Int. J. Epidemiol. 39 (3), 733–745. 0300-5771 https://doi.org/10.1093/ije/dyp371.

Reungoat, J., Escher, B.I., Macova, M., Argaud, F.X., Gernjak, W., Keller, J., 2012. Ozonation and biological activated carbon filtration of wastewater treatment plant effluents. Water Res. 46 (3), 863–872. 00431354 https://doi.org/10.1016/j.watres.2011.11.064.

Rice, G., Teuschler, L.K., Speth, T.F., Richardson, S.D., Miltner, R.J., Schenck, K.M., Chris Gennings, E., Hunter, S., Narotsky, M.G., Simmons, J.E., 2008. Integrated disinfection by-products research: assessing reproductive and developmental risks posed by complex disinfection by-product mixtures. J. Toxicol. Environ. Health A 71 (17), 1222–1234. 1528-7394 https://doi.org/10.1080/15287390802182649.

Richardson, S.D., Kimura, S.Y., 2020. Water analysis: emerging contaminants and current issues. Anal. Chem. 92 (1), 473–505. American Chemical Society, United States, 15206882. Available from: http://pubs.acs.org/journal/ancham. https://doi.org/10.1021/acs.analchem.9b05269.

Richardson, S.D., Plewa, M.J., Wagner, E.D., Schoeny, R., DeMarini, D.M., 2007. Occurrence, genotoxicity, and carcinogenicity of regulated and emerging disinfection by-products in drinking water: a review and roadmap for research. Mutat. Res. 636 (1-3), 178–242. 13835742 https://doi.org/10.1016/j.mrrev.2007.09.001.

Rougé, V., Allard, S., Croué, J.P., Von Gunten, U., 2018. In situ formation of free chlorine during ClO_2 treatment: implications on the formation of disinfection byproducts. Environ. Sci. Technol. 52 (22), 13421–13429. American Chemical Society, Australia. 15205851. Available from: http://pubs.acs.org/journal/esthag. https://doi.org/10.1021/acs.est.8b04415.

Schilirò, T., Pignata, C., Rovere, R., Fea, E., Gilli, G., 2009. The endocrine disrupting activity of surface waters and of wastewater treatment plant effluents in relation to chlorination. Chemosphere 75 (3), 335–340. 00456535 https://doi.org/10.1016/j.chemosphere.2008.12.028.

Schink, T., Waite, T.D., 1980. Inactivation of f2 virus with ferrate (VI). Water Res. 14 (12), 1705–1717. 00431354 https://doi.org/10.1016/0043-1354(80)90106-2.

Sedlak, D.L., Von Gunten, U., 2011. The chlorine dilemma. Science 331 (6013), 42–43. United States, 10959203. Available from: http://www.sciencemag.org/content/331/6013/42.full.pdf. https://doi.org/10.1126/science.1196397.

Sharma, V.K., 2007. Disinfection performance of Fe(VI) in water and wastewater: a review. Water Sci. Technol. 55 (1-2), 225–232. 0273-1223 https://doi.org/10.2166/wst.2007.019.

Shi, J.L., Plata, S.L., Kleimans, M., Childress, A.E., McCurry, D.L., 2021a. Formation and fate of nitromethane in ozone-based water reuse processes. Environ. Sci. Technol. 55 (9), 6281–6289. American Chemical Society, United States, 15205851. Available from: http://pubs.acs.org/journal/esthag. https://doi.org/10.1021/acs.est.0c07895.

Shi, W., Zhuang, W.-E., Hur, J., Yang, L., 2021b. Monitoring dissolved organic matter in wastewater and drinking water treatments using spectroscopic analysis and ultra-high resolution mass spectrometry. Water Res. 188, 116406. https://doi.org/10.1016/j.watres.2020.116406.

Shin, G.A., Sobsey, M.D., 2008. Inactivation of norovirus by chlorine disinfection of water. Water Res. 42 (17), 4562–4568. Elsevier Ltd, United States, 00431354. Available from: www.elsevier.com/locate/watres. https://doi.org/10.1016/j.watres.2008.08.001.

Simmons, J.E., Teuschler, L.K., Gennings, C., Speth, T.F., Richardson, S.D., Miltner, R.J., Narotsky, M.G., Schenck, K.D., Hunter, E.S., Hertzberg, R.C., Rice, G., 2004. Component-based and whole-mixture techniques for addressing the toxicity of drinking-water disinfection by-product mixtures. J. Toxicol. Environ. Health A 67, 741–754. https://doi.org/10.1080/15287390490428215.

Song, Y., Deng, Y., Jung, C., 2016. Mitigation and degradation of natural organic matters (NOMs) during ferrate(VI) application for drinking water treatment. Chemosphere 146, 145–153. https://doi.org/10.1016/j.chemosphere.2015.12.001.

Sorlini, S., Collivignarelli, C., 2005. Trihalomethane formation during chemical oxidation with chlorine, chlorine dioxide and ozone of ten Italian natural waters. Desalination. 176 (1-3), 103–111. 00119164 https://doi.org/10.1016/j.desal.2004.10.022.

Stuart, M., Lapworth, D., Crane, E., Hart, A., 2012. Review of risk from potential emerging contaminants in UK groundwater. Sci. Total Environ. 416, 1–21. 00489697 https://doi.org/10.1016/j.scitotenv.2011.11.072.

Sun, Y.X., Wu, Q.Y., Hu, H.Y., Tian, J., 2009. Effects of operating conditions on THMs and HAAs formation during wastewater chlorination. J. Hazard. Mater. 168 (2-3), 1290–1295. 03043894 https://doi.org/10.1016/j.jhazmat.2009.03.013.

Talaiekhozani, A., Talaei, M.R., Rezania, S., 2017. An overview on production and application of ferrate (VI) for chemical oxidation, coagulation and disinfection of water and wastewater. J. Environ. Chem. Eng. 5, 1828–1842. https://doi.org/10.1016/j.jece.2017.03.025.

Tang, X., Wu, Q.Y., Du, Y., Yang, Y., Hu, H.Y., 2014a. Anti-estrogenic activity formation potential assessment and precursor analysis in reclaimed water during chlorination. Water Res. 48, 490–497. https://doi.org/10.1016/j.watres.2013.10.009.

Tang, X., Wu, Q.-Y., Zhao, X., Du, Y., Huang, H., Shi, X.-L., Hu, H.-Y., 2014b. Transformation of anti-estrogenic-activity related dissolved organic matter in secondary effluents during ozonation. Water Res. 48, 605–612. https://doi.org/10.1016/j.watres.2013.10.016.

Villanueva, C.M., Cantor, K.P., Grimalt, J.O., Malats, N., Silverman, D., Tardon, A., Garcia-Closas, R., Serra, C., Carrato, A., Castaño-Vinyals, G., Marcos, R., Rothman, N., Real, F.X., Dosemeci, M., Kogevinas, M., 2007. Bladder cancer and exposure to water disinfection by-products through ingestion, bathing, showering, and swimming in pools. Am. J. Epidemiol. 165 (2), 148–156. 14766256 https://doi.org/10.1093/aje/kwj364.

von Gunten, U., 2003a. Ozonation of drinking water: part II. Disinfection and by-product formation in presence of bromide, iodide or chlorine. Water Res. 37, 1469–1487. https://doi.org/10.1016/S0043-1354(02)00458-X.

von Gunten, U., 2003b. Ozonation of drinking water: part I. Oxidation kinetics and product formation. Water Res. 37, 457–458. https://doi.org/10.1016/S0043-1354(02.

von Gunten, U., Hoigne, J., 1994. Bromate formation during ozonization of bromide-containing waters: interaction of ozone and hydroxyl radical reactions. Environ. Sci. Technol. 28 (7), 1234–1242. 0013-936X https://doi.org/10.1021/es00056a009.

Wagner, E.D., Plewa, M.J., 2017. CHO cell cytotoxicity and genotoxicity analyses of disinfection by-products: an updated review. J. Environ. Sci. (China) 58, 64–76. Chinese Academy of Sciences, United States, 18787320. Available from: http://www.journals.elsevier.com/journal-of-environmental-sciences/. https://doi.org/10.1016/j.jes.2017.04.021.

Wang, M., Chen, Y., 2018. Generation and characterization of DOM in wastewater treatment processes. Chemosphere 201, 96–109. 00456535 https://doi.org/10.1016/j.chemosphere.2018.02.124.

Wang, L.S., Wei, D.B., Wei, J., Hu, H.Y., 2007a. Screening and estimating of toxicity formation with photobacterium bioassay during chlorine disinfection of wastewater. J. Hazard. Mater. 141 (1), 289–294. 03043894 https://doi.org/10.1016/j.jhazmat.2006.07.001.

Wang, L.-S., Hu, H.-Y., Wang, C., 2007b. Effect of ammonia nitrogen and dissolved organic matter fractions on the genotoxicity of wastewater effluent during chlorine disinfection. Environ. Sci. Technol. 41, 160–165. https://doi.org/10.1021/es0616635.

Wang, W.L., Hu, H.Y., Liu, X., Shi, H.X., Zhou, T.H., Wang, C., Huo, Z.Y., Wu, Q.Y., 2019. Combination of catalytic ozonation by regenerated granular activated carbon (rGAC) and biological activated carbon in the advanced treatment of textile wastewater for reclamation. Chemosphere 231, 369–377. Elsevier Ltd, China, 18791298. Available from: www.elsevier.com/locate/chemosphere. https://doi.org/10.1016/j.chemosphere.2019.05.175.

Wang, W.L., Lee, M.Y., Du, Y., Zhou, T.H., Yang, Z.W., Wu, Q.Y., Hu, H.Y., 2021. Understanding the influence of pre-ozonation on the formation of disinfection byproducts and cytotoxicity during post-chlorination of natural organic matter: UV absorbance and electron-donating-moiety of molecular weight fractions. Environ. Int. 157. Elsevier Ltd, China, 18736750. Available from: www.elsevier.com/locate/envint. https://doi.org/10.1016/j.envint.2021.106793.

Watson, K., Shaw, G., Leusch, F.D.L., Knight, N.L., 2012. Chlorine disinfection by-products in wastewater effluent: bioassay-based assessment of toxicological impact. Water Res. 46 (18), 6069–6083. 00431354 https://doi.org/10.1016/j.watres.2012.08.026.

Wei, D., Wang, L., Wei, J., Hu, H.Y., 2006. Toxicity screening and evaluating in chlorination disinfection of wastewater reclamation processes. Water Sci. Technol. 53 (9), 239–246. 02731223 https://doi.org/10.2166/wst.2006.281.

Wen, G., Liang, Z., Xu, X., Cao, R., Wan, Q., Ji, G., Lin, W., Wang, J., Yang, J., Huang, T., 2020. Inactivation of fungal spores in water using ozone: Kinetics, influencing factors and mechanisms. Water Res. 185, 116218. 00431354 https://doi.org/10.1016/j.watres.2020.116218.

Winward, G.P., Avery, L.M., Stephenson, T., Jefferson, B., 2008. Chlorine disinfection of grey water for reuse: effect of organics and particles. Water Res. 42 (1-2),

483–491. Elsevier Ltd, United Kingdom, 00431354. Available from: https://www.sciencedirect.com/science/journal/00431354. https://doi.org/10.1016/j.watres.2007.07.042.

Wolfe, R.L., Ward, N.R., Olson, B.H., 1985. Inactivation of heterotrophic bacterial populations in finished drinking water by chlorine and chloramines. Water Res. 19, 1393–1403. https://doi.org/10.1016/0043-1354(85)90306-9.

Wood, R.H., 1958. The heat, free energy and entropy of the ferrate(VI) ion. J. Am. Chem. Soc. 80 (9), 2038–2041. 15205126 https://doi.org/10.1021/ja01542a002.

Wu, Q.Y., Hu, H.Y., Zhao, X., Sun, Y.X., 2009. Effect of chlorination on the estrogenic/antiestrogenic activities of biologically treated wastewater. Environ. Sci. Technol. 43 (13), 4940–4945. China, 15205851. Available from: http://pubs.acs.org/doi/pdfplus/10.1021/es8034329. https://doi.org/10.1021/es8034329.

Wu, Q.Y., Tang, X., Huang, H., Li, Y., Hu, H.Y., Ding, Y.N., Shao, Y.R., 2014. Antiestrogenic activity and related disinfection by-product formation induced by bromide during chlorine disinfection of sewage secondary effluent. J. Hazard. Mater. 273, 280–286. Elsevier, China, 18733336. Available from: www.elsevier.com/locate/jhazmat. https://doi.org/10.1016/j.jhazmat.2014.03.028.

Wu, Q.Y., Zhou, Y.T., Li, W., Zhang, X., Du, Y., Hu, H.Y., 2019. Underestimated risk from ozonation of wastewater containing bromide: Both organic byproducts and bromate contributed to the toxicity increase. Water Res. 162, 43–52. Elsevier Ltd, China, 18792448. Available from: www.elsevier.com/locate/watres. https://doi.org/10.1016/j.watres.2019.06.054.

Wu, Q.Y., Yang, L.L., Zhang, X.Y., Wang, W.L., Lu, Y., Du, Y., Lu, Y., Hu, H.Y., 2020. Ammonia-mediated bromate inhibition during ozonation promotes the toxicity due to organic byproduct transformation. Environ. Sci. Technol. 54 (14), 8926–8937. American Chemical Society, China, 15205851. Available from: http://pubs.acs.org/journal/esthag. https://doi.org/10.1021/acs.est.0c02984.

Wu, Q.-Y., Lu, X.-S., Feng, M.-B., Wang, W.-L., Du, Y., Yang, L.-L., Hu, H.-Y., 2021a. Reduction of cytotoxicity and DNA double-strand break effects of wastewater by ferrate(VI): roles of oxidation and coagulation. Water Res. 205, 117667. https://doi.org/10.1016/j.watres.2021.117667.

Wu, Q.-Y., Yang, L.-L., Du, Y., Liang, Z.-F., Wang, W.-L., Song, Z.-M., Wu, D.-X., 2021b. Toxicity of ozonated wastewater to HepG2 cells: taking full account of nonvolatile, volatile, and inorganic byproducts. Environ. Sci. Technol. 55, 10597–10607. https://doi.org/10.1021/acs.est.1c02171.

Xiao, R., Liu, K., Bai, L., Minakata, D., Seo, Y., Kaya Göktaş, R., Dionysiou, D.D., Tang, C.J., Wei, Z., Spinney, R., 2019. Inactivation of pathogenic microorganisms by sulfate radical: present and future. Chem. Eng. J. 371, 222–232. Elsevier B.V., China, 13858947. Available from: www.elsevier.com/inca/publications/store/6/0/1/2/7/3/index.htt. https://doi.org/10.1016/j.cej.2019.03.296.

Xiao, R., Xu, Y., Fang, C., Zhang, R., Chu, W., 2023. Total organic halogen (TOX) in drinking water: occurrence, correlation analysis, and precursor removal during drinking water treatment. Sci. Total Environ. 905, 167445. 00489697 https://doi.org/10.1016/j.scitotenv.2023.167445.

Yan, Z., Zhang, Y., Yuan, H., Tian, Z., Yang, M., 2014. Fish larval deformity caused by aldehydes and unknown byproducts in ozonated effluents from municipal wastewater treatment systems. Water Res. 66, 423–429. 00431354 https://doi.org/10.1016/j.watres.2014.08.019.

Yang, Y., Lu, Y., Wu, Q.Y., Hu, H.Y., Chen, Y.H., Liu, W.L., 2015. Evidence of ATP assay as an appropriate alternative of MTT assay for cytotoxicity of secondary effluents from WWTPs. Ecotoxicol. Environ. Saf. 122, 490–496. Academic Press, China, 10902414. Available from: http://www.elsevier.com/inca/publications/store/6/2/2/8/1/9/index.htt. https://doi.org/10.1016/j.ecoenv.2015.09.006.

Yang, J., Li, J., Dong, W., Ma, J., Yang, Y., Li, J., Yang, Z., Zhang, X., Jia, G., Xie, W., Cang, Y., 2017. Enhancement of bromate formation by pH depression during ozonation of bromide-containing water in the presence of hydroxylamine. Water Res. 109, 135–143. 00431354 https://doi.org/10.1016/j.watres.2016.11.037.

Yuan, C.J., Song, Z.M., Liu, M., Chen, Y., Ye, M.Q., Wang, W.L., Du, Y., Wu, Q.Y., 2023. Unraveling the evolution of organohalogens and their associated toxicity changes in the chlorination of preozonated reclaimed water containing bromide. ACS ES&T Water 3 (11), 3570–3580. American Chemical Society, China, 26900637. Available from: https://pubs.acs.org/page/aewcaa/about.html. https://doi.org/10.1021/acsestwater.3c00317.

Zeng, T., Glover, C.M., Marti, E.J., Woods-Chabane, G.C., Karanfil, T., Mitch, W.A., Dickenson, E.R.V., 2016. Relative importance of different water categories as sources of N-nitrosamine precursors. Environ. Sci. Technol. 50 (24), 13239–13248. American Chemical Society, United States, 15205851. Available from: http://pubs.acs.org/journal/esthag. https://doi.org/10.1021/acs.est.6b04650.

Zhang, M.S., Xu, B., Wang, Z., Zhang, T.Y., Gao, N.Y., 2016. Formation of iodinated trihalomethanes after ferrate pre-oxidation during chlorination and chloramination of iodide-containing water. J. Taiwan Inst. Chem. Eng. 60. 453–459, Taiwan Institute of Chemical Engineers, China, 18761070. Available from: http://www.elsevier.com/wps/find/journaldescription.cws_home/715607/description#description. https://doi.org/10.1016/j.jtice.2015.11.007.

Zhang, Y., Zhang, N., Zhao, P., Niu, Z., 2018. Characteristics of molecular weight distribution of dissolved organic matter in bromide-containing water and disinfection by-product formation properties during treatment processes. J. Environ. Sci. 65, 179–189. 10010742 https://doi.org/10.1016/j.jes.2017.03.013.

Zhang, B., Wang, X., Fang, Z., Wang, S., Shan, C., Wei, S., Pan, B., 2021. Unravelling molecular transformation of dissolved effluent organic matter in UV/H_2O_2, UV/persulfate, and UV/chlorine processes based on FT-ICR-MS analysis. Water Res. 199, 117158. 00431354 https://doi.org/10.1016/j.watres.2021.117158.

Zhang, Y.X., Xiang, J.L., Wang, J.J., Du, H.S., Wang, T.T., Huo, Z.Y., Wang, W.L., Liu, M., Du, Y., 2023. Ultraviolet-based synergistic processes for wastewater disinfection: a review. J. Hazard. Mater. 453. Elsevier B.V., China, 18733336. Available from: www.elsevier.com/locate/jhazmat. https://doi.org/10.1016/j.jhazmat.2023.131393.

Zhao, Y.Y., Boyd, J.M., Woodbeck, M., Andrews, R.C., Qin, F., Hrudey, S.E., Li, X.F., 2008. Formation of N-nitrosamines from eleven disinfection treatments of seven different surface waters. Environ. Sci. Technol. 42 (13), 4857–4862. 0013936X https://doi.org/10.1021/es7031423.

Zheng, L., Feng, H., Liu, Y., Gao, J., Sarkar, D., Deng, Y., 2021. Chemically enhanced primary treatment of municipal wastewater with ferrate(VI). Water Environ. Res. 93 (6), 817–825. 1061-4303 https://doi.org/10.1002/wer.1473.

Zhou, H., Tian, L., Ni, M., Zhu, S., Zhang, R., Wang, L., Wang, M., Wang, Z., 2022. Effect of dissolved organic matter and its fractions on disinfection by-products formation upon karst surface water. Chemosphere 308, 136324. 00456535 https://doi.org/10.1016/j.chemosphere.2022.136324.

Zuma, F., Lin, J., Jonnalagadda, S.B., 2009. Ozone-initiated disinfection kinetics of *Escherichia coli* in water. J. Environ. Sci. Health A 44, 48–56. https://doi.org/10.1080/10934520802515335.

10

Advances in anaerobic digestion for municipal wastewater treatment

Ahmed Alengebawy[a], Mohamed Eraky[a,b], Tarek Mahrous[a], Khaled Touil[c], Nourhan M. Elmahsoub[d], Abdelrahman Ibrahim[e], and Abdelbaky Hossam Elgarhy[c,f]

[a]College of Engineering, Huazhong Agricultural University, Hongshan District, Wuhan, China. [b]Water Pollution Research Department, National Research Centre, Dokki, Giza, Egypt. [c]State Environmental Protection Key Laboratory of Soil Health and Green Remediation, College of Resources and Environment, Huazhong Agricultural University, Wuhan, China. [d]Biotechnology Department, Faculty of Science, Mansoura University, Mansoura, Egypt. [e]Agricultural and Biosystems Engineering Department, Faculty of Agriculture, Alexandria University, Alexandria, Egypt. [f]Central Laboratory for Environmental Quality Monitoring (CLEQM), NWRC, Qalyobia, Egypt

Abbreviations

ABR	anaerobic baffled reactor
AD	anaerobic digestion
AnMBR	anaerobic membrane bioreactor
ASBR	anaerobic sequencing batch reactor
BOD	biochemical oxygen demand
C/N	carbon to nitrogen
COD	chemical oxygen demand
CSTR	continues stirred tank reactor
EBIT	earnings before interest and taxes
EBITDA	earnings before interest, tax, depreciation, and amortization
ECSB	external circulation sludge bed
FBR	fluidized bed reactor
FFB	fixed film bioreactor
GHG	greenhouse gas
HRT	hydraulic retention time
ICR	internal circulation reactor
IoT	internet of things
KPI	key performance indicator
MWW	municipal wastewater
OLR	organic loading rate
RCB	rotating biological contactor
SBR	sequencing batch reactor

Municipal Wastewater Treatment. https://doi.org/10.1016/B978-0-443-24826-9.00010-3

SRT	solid retention time
SS	suspended solids
TDS	total dissolved solids
TFS	total fixed solids
TKN	total Kjeldahl nitrogen
TOC	total organic carbon
TP	total phosphorus
TS	total solids
TSS	total suspended solids
TVS	total volatile solids
UASB	upflow anaerobic sludge blanket
VFAs	volatile fatty acids
VOCs	volatile organic compounds
WWDR	World Water Development Report
WWTPs	wastewater treatment plants

10.1 Introduction

Water scarcity and climate change are the most crucial issues facing human existence. However, climate variability and water are inseparably related, whereas fluctuating weather circumstances make water deficient, unanticipated, and contaminated. These effects on the water cycle pose a threat to biodiversity, sustainable development, and public access to water (The United Nations World Water Development Report 2024: water for prosperity and peace, 2024). Water scarcity is an urgent worldwide problem with broad implications for human health, agriculture, and the environment. As stated by the 2024 United Nations World Water Development Report (WWDR) (Boretti and Rosa, 2019), approximately 50% of the global population presently faces significant water scarcity for a portion of the year. Approximately 25% of the global population experiences "very high" levels of water stress, consuming more than 80% of their yearly renewable freshwater resources. According to Boretti and Rosa (2019), the number is now larger than four billion humans, or about 52% of the world's populace. Another article mentioned that by 2025, it is projected that half of the world's populace will be residing in water-stressed zones (Belhassan, 2023).

The impact of water scarcity is particularly severe for poorer communities, with 1.42 billion people, of which 450 million are children, residing in regions with highly high-water stress (Popoola, 2022). In addition, only 3% of the planet's water is available for use, and 2/3 of that is inaccessible for human use, leaving 1.1 billion people without access to clean water (Mishra, 2023). Water scarcity is further exacerbated by profligate use, pollution, climate change, and populace increment. The water request intensifies as the global population grows and economic development continues, leading to water resources and infrastructure failing to meet accelerating demand (Boretti and Rosa, 2019).

The implications of water scarcity are far-reaching, affecting not only access to drinking water but also agriculture, energy production, and overall economic development. As such, addressing water scarcity is crucial for ensuring a sustainable future. Efforts to mitigate water scarcity involve valuing and conserving water, preserving and rehabilitating freshwater ecosystems, and implementing sustainable water management practices (Belhassan, 2021). By understanding the scale of the problem and implementing practical solutions, it is possible to work towards a more sustainable and equitable water future for all.

MWW primarily consists of water containing low amounts of suspended and dissolved organic and inorganic particles. Organic substances encompass a range of compounds, such as carbohydrates, lignin, lipids, soaps, synthetic detergents, proteins, and their decomposition products. They also include a variety of natural and manufactured organic chemicals derived from industrial processes. This section provides an overview of sources and composition of MWW, in addition to some challenges and contaminants, coupled with treatment strategies. According to Preisner (2020), wastewater from municipalities poses a significant risk to water ecosystems because of its abundant nutrient and organic matter content. Researchers are therefore exploring ways to use this wastewater to power, water, and feed. In this way, we can ensure proper wastewater treatment and protect the aquatic environment while recovering valuable resources (energy, water, and nutrients) for growing crops (Pfafflin and MacInnis, 2010). The composition of wastewater varies depending on the sources of contamination, including residential areas, industries, businesses, and institutions. Table 10.1 provides a comparison of the composition of municipal, domestic, and industrial wastewater. Also, the most common components of MWW are presented in Fig. 10.1.

Anaerobic digestion (AD) plays a significant role in municipal water treatment, particularly in treating high-strength wastes and reducing organic matter. It is a well-established technology used for various purposes within wastewater management. The process involves decomposing organic matter under operational control in the absence of oxygen (Xiong et al., 2020). One of its critical applications is the treatment of municipal sewage sludge, both as an initial treatment for organic manufacturing effluent and as a technique for dealing with the sludge that exists in municipal wastewater treatment plants (WWTPs). AD offers

Table 10.1 Comparison of different characteristics of municipal, domestic, and industrial wastewaters.

Parameters and constituents	Municipal wastewater	Domestic wastewater	Industrial wastewater	References
pH	7.1–8.8	6.5–9.0	5.3–11.8	Kabera et al. (2013), Praneeth et al. (2014), Abou-Elela and El-Khateeb (2015), and Kashyap et al. (2015)
TSS (mg/L)	143–536	100–400	774–86,530	Nasr and Moustafa (2014) and Singh and Srivastava (2014)
TDS (mg/L)	–	270–1200	5975–16,000	Singh and Srivastava (2014) and Fahad et al. (2019)
COD (mg/L)	478 ± 132	470 ± 50	610–1168	Abou-Elela and El-Khateeb (2015) and Dolar et al. (2019)
BOD (mg/L)	98.49 ± 11.57	100–300	330–700	Praneeth et al. (2014) and Abou-Elela and El-Khateeb (2015)
Total alkalinity (mg/L $CaCO_3$)	232.22–340	50–200	–	Darwish et al. (2015) and Singh et al. (2017)
Zinc (mg/L)	0.91 ± 0.21	334.6 ± 30.0	–	Singh et al. (2017) and Gani et al. (2017)
Lead (mg/L)	0.22 ± 0.02	–	–	Singh et al. (2017)
Copper (mg/L)	0.12–0.18	–	–	Singh et al. (2017)
Cadmium (mg/L)	0.07–1.01	0.7	–	Singh et al. (2017) and Gani et al. (2017)
Chromium (mg/L)	0.02–0.04	0.17	7.2–26.2	Singh et al. (2017) and Mojiri et al. (2015)
Grease and oil (mg/L)	5.3–7.14	100–400	1.5–4.0	Kashyap et al. (2015), Fahad et al. (2019), and Singh et al. (2017)
Turbidity (NTU)	248 ± 11	270 ± 20	–	Praneeth et al. (2014) and Dolar et al. (2019)
Total nitrogen (mg/L)	85 ± 10	20–85	40–58.8	Abou-Elela and El-Khateeb (2015) and Nasr and Moustafa (2014)
Total phosphorus (mg/L)	6 ± 2	6–20	0.005–25	Abou-Elela and El-Khateeb (2015), Fahad et al. (2019), and Moradgholi et al. (2019)
Gentamicin (μL/L)	8 ± 0.42	–	–	Singh et al. (2017)

Abbreviations: *BOD*, biochemical oxygen demand; *COD*, chemical oxygen demand; *TDS*, total dissolved solids; *TSS*, total suspended solids.

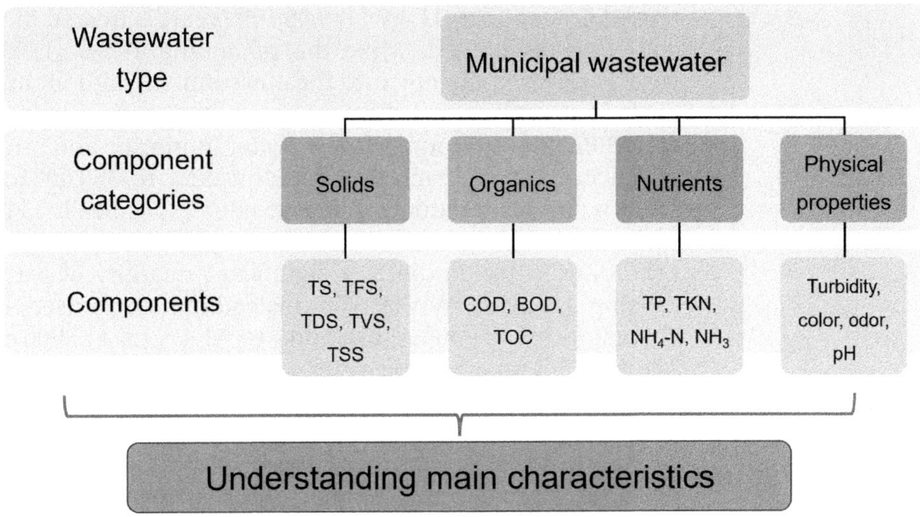

Fig. 10.1 The main characteristics and analysis parameters of municipal wastewater. Abbreviations: *BOD*, biochemical oxygen demand; *COD*, chemical oxygen demand; *NH_3*, ammonia; *NH_4-N*, ammoniacal nitrogen; *TDS*, total dissolved solids; *TFS*, total fixed solids; *TKN*, total Kjeldahl nitrogen; *TOC*, total organic carbon; *TP*, total phosphorus; *TS*, total solids; *TSS*, total suspended solids; *TVS*, total volatile solids.

several advantages in municipal water treatment. It requires low energy inputs and yields relatively tiny amounts of stabilized sludge, making it an efficient and cost-effective treatment technology. The process also facilitates energy recovery from wastewater, producing biogas that could be captured and utilized as an onsite energy source (Di Giacomo and Romano, 2022).

This chapter aims to offer technical expertise and relevant knowledge on waste management options specifically for the organic component of municipal wastewater (MWW). The chapter discusses the promise of AD as a sustainable method for managing organic waste and producing renewable energy. Moreover, it explores the prospects and challenges associated with adopting AD systems. Furthermore, the chapter presents a systematic framework for decision support that may be used to guide investors and government officials in promoting the development of a secure and adaptable wastewater industry.

10.2 Municipal wastewater treatment

10.2.1 Challenges and contaminants

According to Ehalt MacEdo et al. (2022) and Jones et al. (2021), 360,000 m^3 of domestic and MWW is globally produced annually. Out of this, only 11.4% (41,000 m^3/year) is treated in WWTPs and reused. In contrast, 41.4% (149,000 m^3/year) is treated in WWTPs and then discharged, while the remaining 47.2% (170,000 m^3/year) is released directly into the environment without any treatment in WWTPs. According to recent studies by (Ehalt MacEdo et al., 2022), the amount of wastewater (domestic and municipal) produced globally each year is equivalent to 360,000 m^3. Each year, a small fraction of wastewater, specifically 11.4% or 41,000 m^3, undergoes treatment and is subsequently reused. According to the findings, a significant quantity of wastewater is being treated by WWTPs, but instead of being reused, it is discharged. This discharge amounts to 41.4% or 149,000 m^3/year. Additionally, an alarming 47.2% or 170,000 m^3/year of untreated wastewater is being directly released into the environment. In areas that have low water availability, MWW can be a significant solution due to its potential to provide water, valuable metals, essential nutrients, and energy. Still, its treatment faces many challenges: regulations, social, economic, environmental or technologies, capacities, and knowledge. These challenges are the main reason for the delay and limitation of MWW treatment worldwide (UN-Water, 2020). To sum up, Table 10.2 lists the main challenges of MWW treatment.

10.2.2 Treatment strategies

MWW functions as a sink and a source of chemical, physical, and biological components. It is mainly composed of water, which makes up almost 99.9% of its weight (Yang et al., 2020). At the same time, the remaining 0.1% of wastewater may contain (microorganisms, micropollutants, suspended solids, biodegradable organic matter, nutrients, and metals). MWW is contaminated with various chemicals, ranging from surfactants to heavy metals. These include medical waste, household chemicals, and agricultural and biological products. Table 10.3 shows the different types of contaminants in MWW and their potential impacts. Its ecological fate and potential to harm aquatic organisms depend largely on how it is managed at WWTPs. Analyzing the types of contaminants present in MWW is essential to developing and implementing effective treatment strategies (Margot et al., 2015).

The amount of MWW being produced is increasing as the population grows. It is essential to expand the coverage of wastewater treatment to address this issue. Currently, high-income countries treat 70% of MWW, while middle-income countries treat only 38%, lower-middle-income countries treat 28%, and low-income countries treat just 8% of MWW (Cárdenas-Talero et al., 2022).

Table 10.2 Main challenges of municipal wastewater treatment.

Type of challenges	Description	References
Technological	• High energy demands • Greenhouse gas emissions • Inefficient treatment processes • Distance between treatment and reuse sites • Efficiency limitations in contaminants removal • Reliability of water quality monitoring systems • Development of new technologies for sustainable and secure water management	UN-Water (2020), Capodaglio (2020), and Drechsel et al. (2022a)
Regulatory	• Low-income urban farmers • Land availability and accessibility • Guidelines set by authorities to maintain water quality • Poor systems and for management and governance • Legal regulations for safeguarding the environment	UN-Water (2020) and Kryston et al. (2024)
Economic	• Cost recovery • Available subsidies • Investment policies • Valuation of water reuse benefits • Limited alternative water supply options	UN-Water (2020) and Drechsel et al. (2022a)
Social or cultural	• Safety concerns • Insufficient communication • Lack of adequate legislation • Trust in providers and reuse programs • Acceptance by people and policy makers	UN-Water (2020), Capodaglio (2021), and Mateo-Sagasta and Drechsel (2022)
Health and environmental	• Nutrients and energy recovery • Potentially harmful pathogens • Wastewater quality monitoring • Environmental risk assessment • Spread of antimicrobial resistance • Presence of emerging contaminants • Health risks incurred by users and consumers	Mateo-Sagasta and Drechsel (2022) and Drechsel et al. (2022b)

Table 10.3 Classification of contaminants in municipal wastewater and their potential impacts.

Categories	Contaminant forms	Impacts	References
Physical	Suspended solids, sediments, and organic matter	Accumulation in aquatic organisms and negative impacts on the ecological system	Kataki et al. (2021)
Chemical	Nutrients, heavy metals, organic compounds	Harmful to human health and the environment	Espinoza-Guillen et al. (2024)
Biological	Bacteria (*Fecal coliforms, E. coli, Salmonella, Shigella,* and *Vibrio cholera*) Viruses (Adenovirus, hepatitis A virus, and rotavirus) Fungi (*Sphaerotilus natans, Beggiatoa alba, Zoogloea* spp., and *Rhodoferax* spp.) Parasite cysts and eggs	Public health hazards, especially for humans	Aghalari et al. (2020)

Modern technologies have revolutionized resource recovery and treatment of MWW. By implementing various treatment strategies, multiple benefits can be obtained. Sewage treatment plants consume a large amount of energy, but urine and dried sludge contain valuable resources (Patel et al., 2021a). Several technologies, including membrane bioreactors and sequencing batch reactors, are designed for MWW treatment (Sikosana et al., 2019; Kamble et al., 2019). However, aside from the other physical, chemical, and biological methods, AD is recognized as a well-matured technology for MMW. In this context, Fig. 10.2 shows the application of AD for MWW treatment.

10.3 Anaerobic digestion

AD process is a series of processes in which microorganisms decompose biodegradable substances without the presence of oxygen. The process is characterized by low energy consumption and the production of small amounts of sludge, resulting in lower costs and simplified operation compared to aerobic processes. This section provides an overview of using AD technology for MWW treatment, emphasizing the fundamentals and principles, mechanisms and microbial processes, and key factors affecting digestion performance.

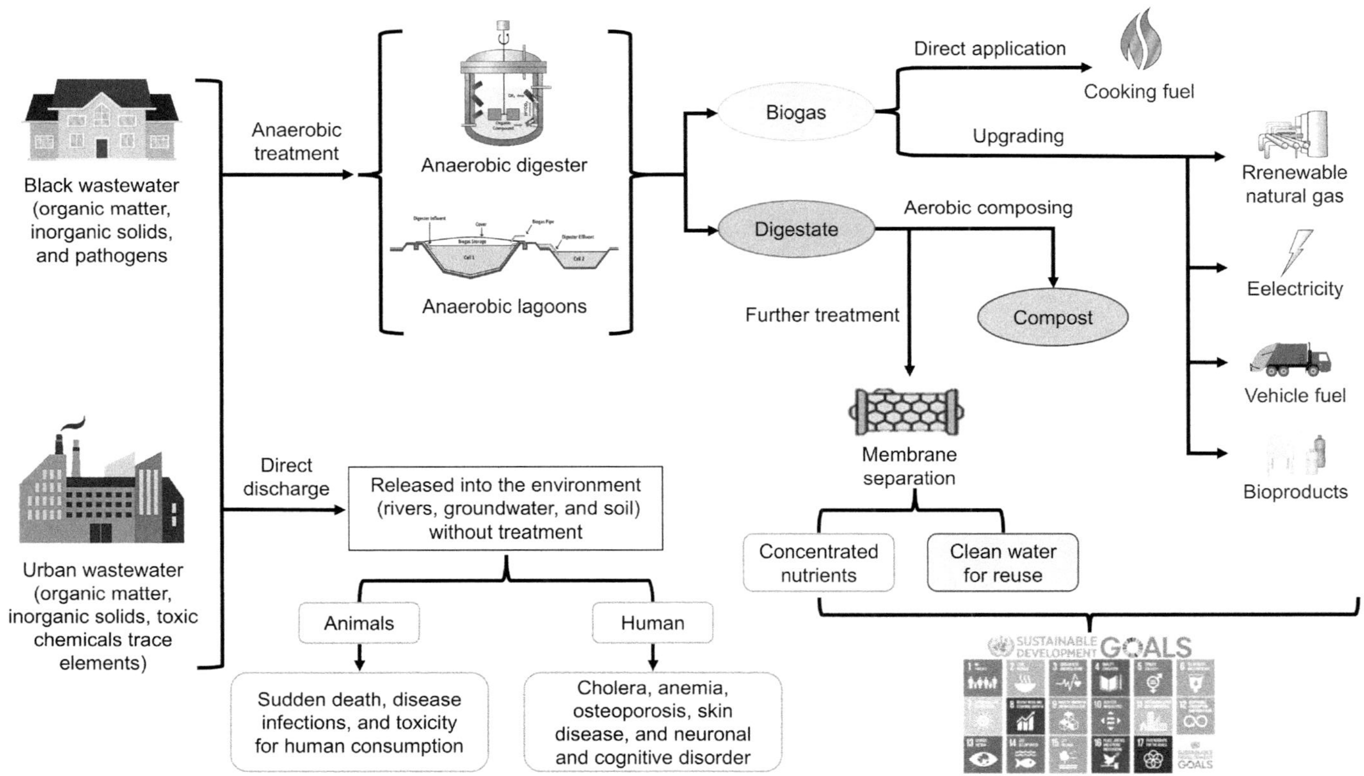

Fig. 10.2 Application of AD for sustainable energy production, water reuse, and nutrient recovery from municipal wastewater.

10.3.1 Process fundamentals and principles

AD is the biochemical conversion of the organic waste into biogas, mainly 60–70% methane (CH_4) and 30–40% carbon dioxide (CO_2) via the bacterial action in oxygen-free conditions. AD consists of four stages: hydrolysis, acidogenesis, acetogenesis, and methanogenesis (Eraky et al., 2021). AD technology has a dual benefit, not only in bioenergy production but also in the treatment of organic waste. Organic wastes are produced continuously due to human daily practices and are considered a serious environmental challenge. The accumulation of organic wastes may lead to greenhouse gas (GHG) emissions and the global warming crisis. Therefore, AD is a cost-effective and eco-friendly technology for organic waste valorization compared to other traditional methods, including landfilling and incineration (Alengebawy et al., 2022b). The hydrolysis stage is the key stage that affects AD efficiency (Eraky et al., 2023).

In the hydrolysis stage, the complex components of the organic waste, including proteins, lipids, and carbohydrates, are degraded and hydrolyzed by the action of the hydrolyzing bacteria into simple soluble components (amino acids, fatty acids, and sugars). Carbohydrates are the easiest and the earliest complex compounds to be hydrolyzed into simple sugars through the α-amylase enzyme. Protein and lipids are hydrolyzed by proteases and lipase enzymes into amino acids and long-chain fatty acids. The simple components (sugars, amino acids, and fatty acids) are converted into pyruvate, which is the fundamental intermediate to be converted into volatile fatty acids (VFAs) at the acidogenesis stage with the generation of hydrogen (H_2). The main VFAs are acetate, propionate, and butyrate. At this stage, the pH value is dropped. In the acetogenesis stage, all VFAs are converted into acetate to be the dominant fatty acid, with the feasibility of ethanol production. The final stage is the consumption of acetate in the presence of H_2 by acetotrophic or acetoclastic methanogenesis to generate CH_4 and CO_2 (Alhajeri et al., 2023; Parvez and Ahammed, 2024).

10.3.2 Mechanisms and microbial processes

Biochemically, AD is the microbial degradation of organic substances into simple components and biogas in oxygen-free conditions. As mentioned before, AD consists of a cascade of biochemical reactions, including hydrolysis, acidogenesis, acetogenesis, and methanogenesis. Bacteria-Archaea interaction was important in the complete cascade reactions to generate the final CH_4. Most of the bacterial action is hydrolysis and fermentation of the complex organic substances to produce H_2 and short-chain fatty acids (mainly acetate). Archaea in the role can utilize acetate and CO_2 to generate CH_4. *Methanobacteriales* (hydrogenotrophic methanogens) can use H_2 and CO_2 (Lee et al., 2023). *Methanosarcinales* can use acetate (acetoclastic methanogenesis) or methanol and methylamines (methylotrophic methanogenesis) (Lee et al., 2023). *Methanomicrobiales* can use H_2 and formate (Lv et al., 2020). Hyperthermophilic conditions of AD are suitable for *Methanococcales*, which can utilize H_2 and CO_2 to generate CH_4 (Lü et al., 2022).

The most important bacterial phyla and their function in AD are summarized in Table 10.4. *Firmicutes* are known for their capacity to degrade proteins, lipids, and carbohydrates with hydrolytic enzymes. They also contribute to acidogenesis and acetogenesis, which are processes that produce VFAs and intermediates. *Bacteroidetes*, on the other hand, are experts in the breakdown of proteins and carbohydrates. They also play a part in acetogenesis, which produces VFAs and H_2 gas (Alhajeri et al., 2023). *Proteobacteria* perform a variety of tasks, including organic matter hydrolysis, VFAs usage and digestion, syntrophic interaction facilitation, and H_2 generation. Some *Proteobacteria* species have symbiotic interactions with methanogens, helping to remove H_2 and provide an environment conducive to methanogenesis (Eraky et al., 2023). *Spirochaetes* are remarkable glucose fermenters that may form a syntrophic connection with hydrogenotrophic methanogens, converting acetate to H_2 and then to CH_4 at high concentrations (Zhen et al., 2023; Lee et al., 2013). *Synergistetes* are engaged in protein degradation and have a synergistic connection with *Methanosaeta* for acetate utilization, which results in CH_4 generation (Sun et al., 2023). *Chloroflexi* contributes to hydrolysis and acetogenesis, with some species capable of breaking down complex organic molecules and producing acetate, a precursor to methanogenesis (Fan et al., 2024). Finally, *Actinobacteria* are involved in the hydrolysis and degradation of complex substances, actively breaking down organic substrates to produce fatty acids (Ahmed et al., 2023).

10.3.3 Key factors affecting anaerobic digestion performance

The AD process needs a combination of several microbial organisms in the subsequent stages; however, numerous factors affect biogas productivity and overall AD performance.

- **Substrate composition and nutrient content**

  The composition of the substrate and its biodegradability are significant factors affecting microbial hydrolysis and fermentation. For example, the lignocellulosic materials of the

Table 10.4 Roles of bacterial phyla and archaea in anaerobic digestion.

Types	Groups	Functions	References
Bacteria	*Firmicutes*	• Hydrolysis of proteins, lipids, and carbohydrates due to its hydrolytic enzymes • Acidogenesis and Acetogenesis for VFAs production and intermediate products	Eraky et al. (2023), Alhajeri et al. (2023), and Ibrahim et al. (2022)
	Bacteroidetes	• Hydrolysis of proteins and carbohydrates • Acetogenesis to produce VFAs and hydrogen	Alhajeri et al. (2023), Ibrahim et al. (2022), and Le et al. (2024)
	Proteobacteria	• Hydrolysis of organic matters • VFAs utilization and digestion • Syntrophic interactions and hydrogen production • Some *Proteobacteria* species form symbiotic relationships with methanogens, helping to remove hydrogen and maintain a favorable environment for methanogenesis	Eraky et al. (2023), Öztep et al. (2023), and Kumar et al. (2021)
	Spirochaetes	• Glucose fermenters • Syntrophic relations with hydrogenotrophic methanogens (*Methanolinea*) might convert acetate to hydrogen first and then to methane under high acetate concentrations	Zhen et al. (2023) and Lee et al. (2013)
	Synergistetes	• Protein degradation • Syntrophic relation with *Methanosaeta* in the utilization of acetate to produce methane	Sun et al. (2023)
	Chloroflexi	• Hydrolysis and acetogenesis • Certain *Chloroflexi* species are capable of degrading complex organic compounds and producing acetate, which serves as a precursor for methanogenesis	Fan et al. (2024) and Li et al. (2023)
	Actinobacteria	• Hydrolysis and complex compound degradation *Actinobacteria* contribute to the breakdown of organic substrates into fatty acids	Ahmed et al. (2023) and Öztep et al. (2023)
Archaea	*Methanobacteriales*	• Utilize hydrogen and carbon dioxide to produce methane	Lee et al. (2023) and Sha et al. (2024)
	Methanosarcinales	• Can utilize a wide range of substrates for methane production	Lee et al. (2023) and Ni et al. (2022)
	Methanomicrobiales	• Can utilize a wide range of substrates for methane production	Lv et al. (2020)
	Methanococcales	• Thrive in high-temperature environments and produce methane	Lü et al. (2022)

Abbreviation: *VFAs*, volatile fatty acids.

plant-based substrates are challenging to hydrolyze easily. Therefore, pretreatment processes may be necessary to facilitate and maximize hydrolysis and AD performance. Chemical, physical, thermal, and biological pretreatment are common pretreatment methods. Chemical pretreatment is the dissolving of lignocellulosic materials by adding alkali, such as sodium hydroxide, or acidic, such as sulfuric acid (Cai et al., 2021).

Physical pretreatment methods include mechanical and thermal pretreatment methods. The mechanical crushing of the lignocellulosic materials into small particle sizes resulted in accelerating the AD process (Kang et al., 2019). Thermal pretreatment by heating the substrate using autoclaves, microwaves, or hot water baths can increase the solubility of the lignocellulosic materials (Kainthola et al., 2019). Biological pretreatment is the utilization of microorganisms, such as bacteria, fungi, and protozoa to degrade and hydrolyze the substrate to more biodegradable components. Biological pretreatment is an eco-friendly and cost-effective method compared to chemical and physical pretreatment methods (Cai et al., 2021; Tamilselvan and Selwynraj, 2024). Furthermore, the carbon/nitrogen (C/N) ratio of the substrate is a significant factor affecting AD performance. The most suitable C/N ratio is 20–35 (Phan et al., 2023). The high C/N ratio could minimize the negative impacts of the inhibitor substances, such as ammonia (Rocamora et al., 2023).

Microbial development and activity rely heavily on adequate nutrient availability, particularly nitrogen, phosphorus, and trace elements. Inadequate nutrition levels can reduce microbial population and CH_4 generation. Supplementation with sufficient nutrients may be required, particularly when handling waste with an uneven nutrient composition (Acién Fernández et al., 2018; Fricke et al., 2007).

- **Temperature**

  Temperature plays a crucial role in various aspects of AD. It not only affects the activities of microorganisms but also influences the physicochemical properties of the substrate and the partial pressure of the biogas produced. AD can occur under two temperature ranges: mesophilic, which typically ranges from 35°C to 40°C, and thermophilic, which ranges from 50°C to 70°C (Phan et al., 2023). The temperature directly influences the microbial community and the rate at which digestion occurs. Generally, thermophilic conditions lead to faster reaction rates and higher biogas production, but they require more energy for heating (Ran et al., 2024). Mesophilic conditions are more commonly employed due to their lower energy requirements and broader substrate adaptability. Among the microbial groups involved, acetotrophic methanogens are particularly sensitive to increasing temperatures. Additionally, the degradation of propionate and butyrate is also sensitive to temperatures exceeding 70 degrees Celsius. Moreover, the temperature significantly impacts the partial pressure of H_2 in digesters, thereby influencing the kinetics of the syntrophic metabolism. According to thermodynamics, at higher temperatures, endergonic reactions (such as the breakdown of propionate into acetate, CO_2, and H_2) become energetically more favorable, whereas exergonic reactions (e.g., hydrogenotrophic methanogenesis) are less favored at higher temperatures (Phan et al., 2023).

- **pH**

  The pH level within the anaerobic system plays a critical role in influencing microbial activity and ensuring the stability of the AD process. Studies have demonstrated that a pH of 7 maximizes biogas yield by facilitating the growth of methanogenic bacteria. Hydrolysis occurs optimally at a pH of 5.5, while acidogenesis takes place around a pH of 6.5. Even small pH variations of approximately 0.5 can have a significant impact on microbial metabolism and reaction kinetics, ultimately affecting gas production. Although AD can occur within a broad pH range of 5.50–8.50, methanogens are particularly sensitive to changes in pH and perform best within a pH range close to 7 (Prasanna Kumar et al., 2024). If the pH falls below 6.30 or exceeds 7.80, the survival of methanogenic bacteria is severely compromised, leading to process failure (Meegoda et al., 2018). However, unlike methanogenesis, hydrolysis and acidogenesis can function optimally within a pH range of 5.50–6.50. Non-metabolize VFAs produced during acidogenesis can cause the pH to drop significantly below 3, resulting in process failure. Moreover, high alkalinity during hydrolysis promotes the formation of total ammonium nitrogen and can disrupt the process. pH extremes of around 3 or lower and 12 or higher can negatively impact acidogenesis and inhibit the hydrolysis rate (Tawfik et al., 2023).

- **Organic loading rate**

  Organic loading rate (OLR) refers to the quantity of the added organic components of substrate per day per unit volume of the anaerobic reactor. It can be measured based on the addition of volatile solids or chemical oxygen demand (COD). High OLRs can promote the enrichment of microbial organisms. However, the excessively high OLRs may be due to the overloaded anaerobic system leading to decreased performance (Chuenchart et al., 2020). It is crucial to consider the hydraulic retention time (HRT) when operating an anaerobic digester to avoid potential issues. Decreasing the HRT can lead to microbial washout and eventual digester failure. Additionally, operating the digester at a higher OLR or a lower HRT can result in the accumulation of VFAs. To maintain stability, methanogenic reactors are typically operated at a slightly higher HRT, typically between 10 and 25 days for continuous stirred tank reactor (CSTR) configurations (Prasanna Kumar et al., 2024).

- **Hydraulic retention time**

  The HRT is a critical parameter that affects the performance of the AD process. Longer HRTs allow for better degradation of complex compounds and higher CH_4 yields, but excessively long HRTs can lead to inefficiency and increased costs. Determining the optimal HRT depends on substrate characteristics, desired biogas production, and digester size (Shi et al., 2017). Shortening the HRT can reduce capital expenditures and improve process efficiency, but it should not fall below 10 days to prevent bacterial washout. The design of AD systems is influenced by both HRT and solid retention time (SRT), with HRT reflecting the duration of the liquid phase and SRT referring to the retention of microbial culture. However, for certain substrates, such as primary sludge and waste activated sludge, the interaction between solids and microbial cultures occurs in two phases, resulting in distinct HRT and SRT. Feedstocks like food waste, kitchen trash, and municipal solid waste generally exhibit similar SRT and HRT characteristics, making them suitable for the AD process (Magdalena et al., 2019).

10.4 Advances in anaerobic digestion

AD is a proven and widely used treatment method that is particularly effective for digesting waste with high amounts of organic matter or solid content. This section provides an overview of high-rate anaerobic reactors, two-stage processes, and some advances in AD technology.

10.4.1 Two-stage anaerobic digestion

Two-stage AD is a process that involves two distinct stages, acidogenesis, and methanogenesis, to enhance the efficiency and performance of the overall AD system. This two-stage approach offers several advantages over single-stage systems, but it also comes with some disadvantages that need to be considered. One of the key advantages of two-stage AD is the separate optimization of each stage. Acidogenesis and methanogenesis have different requirements in terms of pH, temperature, and environmental conditions. By optimizing each stage independently, it is possible to have better control over microbial populations and metabolic pathways, leading to improved overall system performance and biogas production. This separate optimization allows for adjustments and fine-tuning of each stage, maximizing the degradation of complex organic materials and increasing biogas yields (Cremonez et al., 2021).

The two-stage process also enhances the degradation of organic matter. During acidogenesis, complex organic materials, such as carbohydrates, fats, and proteins are broken down into simpler compounds, which makes it easier for the methanogenic bacteria in the second stage to convert the VFAs into CH_4 and CO_2. The sequential breakdown of complex materials increases the overall efficiency of biogas production and improves the utilization of organic waste (Alhajeri et al., 2023). Furthermore, the separation of acidogenesis and methanogenesis reduces the risk of process inhibition. Fluctuations in feedstock composition and loading rates can negatively impact the performance of AD systems. However, by maintaining different pH levels and environmental conditions in each stage, the two-stage AD system becomes more resistant to such fluctuations. This adaptability allows for better stability and ensures efficient performance even when faced with changes in organic load and composition.

Two-stage AD systems also offer the advantage of flexibility in reactor configuration. The system can be implemented using different types of reactors, such as two-stage CSTR or anaerobic sequencing batch reactor (ASBR), or a combination of different reactor types. The selection of the appropriate configuration depends on factors, such as feedstock characteristics, process requirements, and available infrastructure. This flexibility enhances the versatility and applicability of two-stage AD systems (Holl et al., 2022). However, there are also several disadvantages associated with two-stage AD. One significant drawback is the higher implementation costs compared to single-stage systems. The need for separate reactors and infrastructure increases the initial investment required for setting up a two-stage AD system. This cost factor needs to be carefully evaluated and considered when deciding on the feasibility of implementing such a system (Rajendran et al., 2020). Another potential disadvantage is the possibility of accumulation during the acidogenesis stage. If not properly managed, VFAs can accumulate and impact the overall process efficiency. Proper sizing and management of the system are crucial to avoid this issue and ensure smooth operation. Additionally, there is a time lag between the acidogenesis and methanogenesis stages. The time required for the breakdown of complex organic materials and the subsequent conversion of VFAs into CH_4 needs to be accounted for in system design and operation. Proper synchronization between the stages is essential to ensure optimal performance and efficiency (Chakraborty et al., 2022).

10.4.2 High-rate anaerobic reactors

High-rate anaerobic reactors are systems designed to maintain better contact between the inoculum and the substrate. The most common feature of these reactors is the high retention time with high OLRs. Small-sized reactors are preferred to keep the high inoculum-substrate contact with minimal energy requirements (de Mello et al., 2021).

- **Upflow anaerobic sludge blanket**

  Upflow anaerobic sludge blanket (UASB) is the most operated AD system. The organic substrate or the movement of wastewater down and upward through sludge blankets is the main characteristic of the UASB design. The flow direction maximizes the contact between the substrate and the anaerobic inoculum and retains the active settling biomass with a reduced sludge volume index (Mainardis et al., 2020).
- **Anaerobic baffled reactor**

  Anaerobic baffled reactor (ABR) is a reactor design that employs a sequence of baffles to induce the flow of wastewater containing organic contaminants beneath and above (or through) the baffles as it moves from the input to the output. The bacteria in the reactor exhibit buoyancy and sedimentation because of flow properties and gas generation, yet they migrate downwards through the reactor at a sluggish pace. The baffles divide the reactor into chambers, which separate the acidogenesis and methanogenesis stages of the AD process for more stability and effectiveness. Therefore, ABR can work as a two-phase anaerobic rector (de Mello et al., 2021). Its capacity to longitudinally separate acidogenesis and methanogenesis—thereby enabling the reactor to function as a two-phase system without the related control issues and expensive costs—is likely its greatest benefit. Because of its straightforward design and lack of mechanical mixing or moving parts, the reactor is reasonably affordable to build. There is little sludge generation and a high SRT. It is also feasible to operate intermittently, which makes treating seasonal wastewater easier (Moradgholi et al., 2019).

- **Expanded granular sludge bed**

  Expanded granular sludge bed (EGSB) reactors have gained widespread recognition for their exceptional performance in treating waste with high organic loads, particularly in the removal of phenolic pollutants like linear alkylbenzene sulfonate. These reactors employ a high upflow velocity ranging from 2 to 10 m/h, which serves to intensify hydraulic mixing within the reactor and expand the sludge bed. Consequently, this configuration promotes extensive contact between the biomass and the substrate, resulting in enhanced treatment efficiency. The impressive capabilities of EGSB reactors have led to their widespread adoption for the treatment of various domestic and industrial effluents on a global scale. These reactors, characterized by their tall columnar structures filled with granular sludge, excel at accommodating high OLRs, operating in an upflow mode, and facilitating efficient biogas production. As a result, EGSB reactors find versatile applications in diverse wastewater treatment scenarios (Gutiérrez et al., 2022).
- **Fluidized bed reactor**

  Fluidized bed reactor (FBR) is a type of reactor where solid particles are suspended and display fluid-like behavior due to the upward flow of a gas or liquid. These reactors provide several advantages, including improved mass transfer, efficient heat transfer, and uniform distribution of reactants. They have proven to be effective in both advanced oxidation processes and biological processes. FBRs are known for their low operating costs, high resilience to system upsets, and high rates of mass transfer (Bello et al., 2017). They have gained extensive use in wastewater treatment, particularly in biological processes, with the operation of large-scale FBRs. The functioning of FBRs is based on the principle of fluidization, where the packing material expands because of fluid movement. The extent of bed expansion depends on various factors. FBRs involve multiphase flow systems and find wide applications in chemical and biotechnology fields. While initially employed primarily in chemical processes, FBRs have also found significant utility in wastewater treatment (Dong et al., 2023).
- **Anaerobic membrane bioreactor**

  The combination of AD technology and the membrane filtration technique has been reported to enhance AD efficiency. The anaerobic membrane bioreactor (AnMBR) design is characterized by the presence of a micro- or ultra-filtration membrane that retains the microbial biomass and prevents its elution with the digested substrate. The most important advantages of AnMBRs are retaining the active biomass and generation of high-quality effluent with lower turbidity to be useful in the wastewater treatment process. However, membrane fouling is a serious challenge that leads to the use of some chemicals, such as sodium hypochlorite (NaOCl), to remove the accumulated filtrating cake (Osman and Hodaifa, 2023).

10.4.3 Integration of bioreactors

High-rate anaerobic reactors are often used in WWTPs to reduce costs and raise the financial income from the production of fuel and manure. UASB is a common configuration for AD and is a natural reactor that feeds wastewater from the bottom and recovers treated water from the top. Microorganisms digest sludge pollutants, producing biogas and biomass. This biogas is collected from reactor domes, purified, and used for energy (Patel et al., 2021a). AD systems are used to enhance enzymatic activity and CH_4 production by employing a combination of physical, chemical, and biological pre-treatment methods. Microorganisms attach to a non-reactive support material, producing a biofilm, while the support rotates slowly horizontally within a tank through which the wastewater passes. Bioreactors like EGSB and internal circulation reactors are commonly utilized to improve the breakdown of pollutants in wastewater. AD processes in WWTPs generate VFAs, which have useful properties in various industries (Patel et al., 2021a).

10.5 Microbial ecology in anaerobic digestion

The microbial populations in anaerobic digesters are very responsive and adaptable to environmental fluctuations within the reactor. The presence of highly specialized genera, such as *Methanothrix* (previously *Methanosaeta*) or *Methanosarcina* that can break down acetate or use both acetate and hydrogen as energy sources, is a trustworthy indicator of the methane generation process and stability. Hence, comprehending and researching the microbial ecology of AD is crucial and essential for assessing the efficiency and consistency of the process. This section provides an overview of the role of microorganisms in AD process performance, highlighting the microbial diversity and structure.

10.5.1 Microbial community in anaerobic digestion

AD is a complex process and involves four different phases and complex relationships between microorganisms. These microorganisms are classified into four categories based on their functions: hydrolyzers, acidogens, acetogens, and methanogens (Amin et al., 2021), which are further categorized into bacteria and archaea domains and arranged in a taxonomic hierarchy. Each group contributes in unique ways to the AD process.

The hydrolyzers, which are involved in the initial stage of the AD process, play an essential role in the breakdown of complex organic materials, like proteins, carbohydrates, and lipids, into small molecules like amino acids, peptides, and sugar. The AD process cannot happen naturally in the absence of hydrolyzers (Bhushan et al., 2023). Generally, healthy AD systems must contain an efficient group of hydrolytic bacteria. These bacteria are present in a variety of phyla, e.g., *Proteobacteria, Firmicutes, Thermogae, Chloroflexi,* and *Spirochaete* organisms. The most dominant phyla of AD are *Bacteroides* and *Firmicutes* (Achinas et al., 2020). Hydrolytic bacteria can produce *cellulosomes* (Alves et al., 2021), which are a complex of many enzymes that allow them to release different hydrolases, such as *chitinases, hemicellulases, glucanases,* and *lihanases* (Azman et al., 2015). These enzymes act as a connection between hydrolytic bacteria, enzymes, and substrates, allowing the former to decompose many difficult organic waste types.

In the second phase, acidogens use the output of hydrolysis from phase 1 as acceptors of electrons to produce substances from fermentation, such as pentatonic acid, butyric acid, acetic acid, formic acid, alcohols, CO_2, and H_2. Phase 4 methanogens can directly consume acetate, CO_2, and H_2, while Phase 3 acetogenic bacteria break down other organic acids to acetates and H_2 (Mutungwazi et al., 2023). Acidogen species exist in phyla *Proteobacteria, Actinobacteria, Bacteroidetes,* and *Firmicutes.* Some species, including *Bifidobacterium, Corynebacterium, Bacillus, Pseudomonas, Clostridium,* and *Desulfobacter,* have been isolated from AD (Venkata Mohan et al., 2019).

In the third stage, acetic microorganisms break down organic acids like butyric, pentatonic, and propionic acids into acetates and H_2 (Lee et al., 2015). Certain acetogens are part of the genus called *Syntrophomonas* (*Syntrophomonas wolfei* and *Syntrophobacter wolinii*), which is an oxidation microbe of fatty acid syntrophic. In the AD process, it is essential to maintain favorable conditions for converting VFAs into acetate for efficient biogas production. Acetogenesis is crucial to the production of biogas, as 70% of CH_4 production is produced byacetate reduction. Syntrophic acetate oxidizing bacteria stabilize AD processes and maintain stability during environmental changes. Under harsh conditions, syntrophic acetate oxidizing bacteria can convert acetate to H_2 and CO_2, which hydrogenotrophic methanogens use for

CH_4 production in the fourth stage. There are also Synchronous acetogens, but they often dominate hydrogenotrophic methanogens. *Syntrophaceticus schinkii, Pseudothermotoga lettingae, Thermacetogenium phaeum*, and a few bacteria belonging to the phylum *Spirochaetes* are among the syntrophic acetate oxidizing microbes found in mesophile and thermophile AD conditions (Lee et al., 2015). Homoacetogenic bacteria, like *Clostridium aceticum*, are present but not as favorable in the AD process.

Methanogens are archaea that can produce CH_4 gas. *Methanosarcinales, Methanopyrales, Methanomicrobiales, Methanobacteriales*, and *Methanococcales* are the five categories into which researchers have so far classified the 65 methanogenic species that they have discovered. These species are very sensitive to oxygen and have slow rates of growth as they are limited to restricted organic substances of energy and carbon sources (Ye and Li, 2023). Methanogens can be divided into three categories: aceticlastic, hydrogenotrophic, and methylotrophic, according to how they use substrates. Acetoclastic methanogens produce most of the CH_4 produced during AD, although hydrogenotrophic methanogens are present in AD at a lower relative abundance (Conrad, 2020). Fortunately, the presence of hydrogenotrophic methanogens is crucial to maintaining the low partial H_2 pressure and enhancing the acetogenesis of VFAs.

10.5.2 Implications for treatment efficiency

Although AD is a mature technology for waste management and renewable energy production, its treatment efficiency is paramount for its widespread adoption and success. Understanding the implications of treatment efficiency in AD processes is essential for optimizing resource recovery, minimizing environmental impacts, and ensuring economic viability.

There are several reasons why it is essential to improve the efficiency of the production of biogas. Biogas is an energy source that is renewable and can be produced from many kinds of organic waste, including municipal solid waste, agricultural residues, as well as industrial waste, which can help mitigate climate change by lowering GHG emissions and serving as a fossil fuel substitute (Afridi et al., 2023). Biogas production provides an effective way to handle organic wastes, eliminating the need for landfills and lowering pollution levels in the environment (Chen et al., 2022). By redirecting organic waste from landfills, biogas production additionally minimizes the emission of CH_4, a potent GHG. Biogas production provides benefits to farmers by providing an alternative source of income, promoting sustainable agriculture practices, and reducing the need for synthetic fertilizers (Alengebawy et al., 2022a). High-quality organic fertilizer is another output of biogas generation that can increase soil fertility and lower nutrient losses. Biogas production can generate income for rural communities and contribute to rural development (Meng et al., 2020). Maximizing the efficiency of biogas production may be beneficial to the security of energy by diversifying the energy mix, decreasing the use of fossil fuels, and increasing local production of energy.

The efficient degradation of organic waste is central to treatment efficiency in AD. Proper substrate selection, pretreatment methods, and process optimization are crucial for enhancing the breakdown of complex organic compounds into simpler molecules. This ensures a robust microbial metabolism and promotes stable biogas production throughout the AD process (Van Hung et al., 2019). Moreover, AD treatment efficiency extends beyond biogas production to encompass nutrient recycling and resource recovery. Efficient digestion processes facilitate the conversion of organic nutrients into valuable products, such as biofertilizer and nutrient-rich digestate (Alengebawy et al., 2023). By recovering and recycling nutrients, AD contributes to circular economy principles, reducing reliance on synthetic fertilizers and closing nutrient loops. Also, efficient AD has a significant impact on minimizing the environmental impact of organic waste treatment and has a direct impact on the economic viability and operational sustainability of AD facilities (Alengebawy et al., 2022a).

10.6 Performance evaluation and monitoring

The advancement of the anaerobic sector depends on a specific regulatory framework at regional and national levels that differentiates each type. This sector is primarily a service-oriented business, incorporating multiple perspectives by supplying energy and often dealing with substantial amounts of biowaste related to, compelling point of view, operation expenditure. Therefore, evaluation of AD performance is important to optimize the overall efficiency of the process. This section provides an overview of key performance indicators (KPIs), real-time monitoring technologies, and smart monitoring equipment used in AD.

10.6.1 Key performance indicators

According to Feiz et al. (2020), the economic aspect of AD processes, KPI has been revealed in every revenue and sacrifice, commonly presented as the income statement, earnings before interest and taxes (EBIT)-module and earnings before interest, tax, depreciation, and amortization (EBITDA), expressed as a percentage. The EBITDA module specifically defines operational cost, whereas the EBIT margin factors in the impact of consumption, regardless of the tax ratio, familiar desires, and trends (Balcioglu et al., 2022), allowing for a direct comparison of financial outcomes. They also include the estimation of the ultimate revenues while using various energy sources during the wastewater treatment processes, then roughly estimate the feasibility of the operation.

Waste to the economy is considered a convincing way to evaluate the overall business performance of the sector, which benefits managers, investors, and lenders in making loan financing decisions (Bortot Coelho et al., 2023). Operating WWTPs reveals significant challenges concerning both economic and environmental sustainability. These challenges encompass heightened resource utilization, GHG impacts, and escalating energy expenditures.

Consequently, it is compelling to ensure that the downside impacts of these advantages do not outweigh their benefits to promote sustainable development. Novel approaches aimed at increasing energy and resource recovery have been initiated to assess and enhance the energy aptitude of existing facilities. These approaches aim at transitioning to energy-neutral plants and are crucial for ensuring the long-term sustainability of the sector (Krueger et al., 2015). Stakeholders must reduce costs inherent to operating and increase environmental security; sludge production is also considered one of the most critical points of view that must be well integrated with the KPIs of WWTPs, which are crucial for monitoring and enhancing the quality of services delivered. An analysis of European WWTPs highlighted a substantial potential for enhancing energy aptitude (Srisamran et al., 2022). The analysis also indicated that cost inherent to energy savings of up to 25% could be achieved without compromising effluent quality.

For the sake of a feasible and eco-friendly system, relying on progressive KPIs assessments (Ma et al., 2023), analysis of key parameters is crucial for assessing water quality and ensuring compliance with standards. These parameters include the 5-day amount of oxygen required for microbial metabolism of organic compounds in water, chemical oxygen required, total fine particles of sediment in the water, total nitrogen, and total phosphorus. The compliance criteria for these parameters encompass parametric values, the maximum allowable number of samples that can deficient the requirement, and the optimum allowable digression from the specified parametric value (Silva and Rosa, 2020).

10.6.2 Process monitoring methods

Powerful instrumentation and control tools play a crucial role in boosting CH_4 energy production and resource retrieval from digestate during AD processes. Over the years, research in these domains has progressed in parallel with the primary uses of AD. Initially concentrated on municipal sewage sludge, investigations have expanded to cover liquid waste, mainly from domestic sources, along with the organic proportion of solid waste and agricultural residues prevalent in urban areas. Moreover, significant research efforts have been dedicated to analysis the inherent time constants within these processes. (Valdés et al., 2020).

AD has emerged as a highly appealing alternative for managing organic waste (Silva and Rosa, 2020). For instance, in France, the energy consumption of a WWTP amounts to approximately 20 kWh per person equivalent annually, considering a plant with a capacity of 100,000 person equivalents. Given these insights and the substantial energy potential present in wastewater sludge, it is well observed that future WWTPs (water resource reclamation facilities) are expected to continue performance improvement.

AD is a process that is highly responsive to changes in operational conditions, and even minor variations can lead to significant fluctuations in both the composition and activity of the microbial community involved, so compelling monitoring and management are required to use the produced sludge from MWW (Patel et al., 2021b). Consequently, online instrumentation shall be considered, including specific H_2, temperature, pH, flow, and oxidation-reduction potential (Longis et al., 2023). According to Rawoof et al. (2021) and Longis et al. (2023), temperature plays an important role in AD reactors, and controlling temperature is frequently implemented. There are three main types commonly used in process-measured data: resistance thermometer, thermocouple, and thermistor. Secondly, pH electrodes should be continuously cleaned. In addition, flow meters or ultrasound sensors are frequently used in calculating a given flow rate, and oxidation-reduction potential sensors are frequently used to indicate both the oxygen percentage and sulfate amongst the surrounding ambient (Saiyad et al., 2024). In this context, alkalinity is a major factor in AD, as methanogens literally generate alkalinity through CO_2, ammonia, and bicarbonate production. This contributes to neutralizing VFAs produced during the process. Therefore, real-time monitoring of alkalinity can enhance the stability of AD, especially when CO_2 concentration remains consistent (Yan et al., 2021).

10.6.3 Real-time monitoring technologies

To increase the proportion of revenues from WWTPs, improved technologies have been created to raise the possibility of immediately collected data; from then on, real-time monitoring is the clue for optimizing the attained outcomes. For instance, the Internet of Things (IoT) illustrates a networked environment formed by interacted devices that are capable of communicating with each other over the internet (Cartlidge, 2017). It depends almost entirely on three main functions: data transmission, data reception, and the processing of received data. The main goal of IoT is to outperform our daily progress and streamline task execution, offering significant benefits, such as optimized tracking methods, adjustability, and real-time monitoring facilities. Secondly, integrated spectroscopic techniques with optical fibers have become incrementally crucial in the biotechnology sector. These techniques increase the ability to measure a diverse array of parameters, including VFAs, volatile solids, CH_4 proportion, and biodegradability of wastewater, COD, and others. This versatility stands out as the most booming utility advantage of utilizing spectroscopic techniques in biotechnology (Pourbavarsad et al., 2021; Lamb et al., 2019). Electrochemical sensors utilize potentiometric and voltammetric measurements, as outlined in (Awhangbo et al., 2020). This involves capturing the signal from a specific sample using an electrode, which is used to establish a relationship for determining the concentration of the samples while processing (Zhu et al., 2023). In this regard, optimizing electrochemical sensors offers several added-values for detecting various organic structures, including simplicity, robustness, low cost, rapid response time, and reduced sample implementation. These sensors have been spread out for both offline and online monitoring in AD processes (Akter et al., 2023).

10.7 Case studies

The success of any technology is proven by cases that have been implemented on the ground. Like any technology, AD also has many cases that have been implemented in many countries of the world in the field of water treatment. This section gives an overview of these successful cases, indicating successful applications in urban areas, challenges and countermeasures, and lessons for future applications.

10.7.1 Successful applications in urban settings

The treated MWW reclamation and reuse have climbed from 10% to 29% annually in the United States, Europe, and China, and as much as 41% in Australia (Dolgen and Alpaslan, 2023). China is distinguished as the top nation in Asia for wastewater treatment volume, where China's metropolitan regions release a significant amount of wastewater, with a 65.7 billion m^3 yearly volume, of which 63.3 billion m^3 were treated, or roughly 96.3%. However, the reuse of wastewater is just 12.62 billion m^3, which is around 19.9% (Hong-Ying et al., 2021). According to estimates, just 37.6% of India's local waste water is currently being handled (Singh et al., 2019). In addition, Israel is the country that uses the most handled or treated wastewater, especially for agricultural irrigation, where 90% of the improved water is used (Angelakis and Snyder, 2015).

AD technology offers a chance to advance sustainable agriculture and raise rural residents' standards of living. It has been successfully utilized in developing nations to address energy needs and manage waste. For instance, Colombia uses the AD process to produce about 15–28% of the country's energy from agricultural activities and agro-industry residual biomass. A fruitful case of AD application on a large scale in Bogota city, the project is planned in three plants to process residues from leached urban solid waste and some neighboring municipalities for obtaining biogas and generate electrical energy; with this project, Colombia achieves 35% of goals of agreed Colombia targets at the United Nations (Tavera-Ruiz et al., 2023).

In addition, Bautista Angeli et al. (2018) listed 15 case studies involving the application of AD to treat blackwater and other household wastes. The study thoroughly analyzes waste valorization and AD in urban contexts. It covers the essential components and procedures of AD-based urban waste valorization, such as waste selection, physical-chemical characterization, evaluation of methanogenic potential, collection, storage, pretreatment, reactor conditions, production of biogas, digestate management, and biogas valorization. The study also considers several other criteria, including social, financial, and ecological issues. It emphasizes how crucial the success of the AD project is to the local community's involvement, economic viability, and environmental impact.

The successful case studies include the AD processes designed for the Western Harbor of Malmö city in Sweden project, where a treatment plant was established to serve 1303 dwellings on 22 ha with a density of buildings 600 units or parts for 9 ha and a density of population of 122 people/ha. The plant treated the blackwater and household wastes with AD, both in-situ and out-of-situ. Biogas is produced from waste and utilized as vehicle fuel to power the city's public transportation system. Agronomic use is made of the digestate that is produced.

In addition, AD has been successfully applied in urban settings for MWW treatment across the globe. This technology offers a sustainable way to manage sewage sludge, reducing its volume, producing biogas that can be used for energy, and producing digestate that can be used as a soil conditioner. Table 10.5 summarizes some notable case studies that highlight the successful application of AD in urban wastewater treatment.

10.7.2 Challenges and countermeasures

The application of AD for organic portion treatment of urban wastes provides numerous ecological and financial benefits; however, the large-scale application of these techniques rises crucial political, financial, and societal challenges that need to be considered to create enduring and lucrative AD plants (Ibarra-Esparza et al., 2023). This is especially crucial for developing nations, where investments in energy from waste technologies are constrained by restrictive waste management budgets, and large-scale deployment is still in its early stages (Atelge et al., 2020).

The biggest issue from a sociological standpoint is how acceptable AD plants are in various societies. Most nations, however, acknowledge and even call for a shift to clean energy; locally, residents, on the other hand, are frequently against the establishment of AD plants next to their homes (Bourdin and Chassy, 2023). Furthermore, the public's opposition to AD plants is mostly caused by worries about releases of smells, noise and visual contamination, high traffic, explosion hazards, and the potential decline in property value as a result of these adverse effects (Bourdin et al., 2020). However, the distribution of expenses and benefits is the reason why AD plants are close to communities (Bourdin et al., 2020). Establishing small-scale AD facilities could aid in the spread of this technology, which has been installed as a result of initiatives to help cover the expenses of building AD reactors in nations like China and India (Jaeger and Blanchard, 2022).

Table 10.5 Typical anaerobic digesters used in anaerobic-aerobic municipal wastewater treatment systems.

Facility and location	Overview	Success factors	Outcomes	References
Gaobeidian Wastewater Treatment Plant, Beijing, China	Largest wastewater treatment facility in Asia, incorporating AD	Integration with advanced wastewater treatment technologies in between them AD, focus on how to reach energy sufficiency and emissions reduction neutralization	In large-scale biogas production, the energy neutralization ratio reaches 95.8%, and only 56.7% of carbon emissions reduced	Pang et al. (2023)
Bureau of Sewerage, Tokyo, Japan	Implements AD in wastewater treatment to enhance sustainability also energy and nutrient recovery	Advanced processes tailored to densely populated areas	Effective sludge reduction, biogas production for energy, green energy initiatives contribution	Guo et al. (2022)
Changi Water Reclamation Plant, Changi, Singapore	Part of the NEWater project uses AD for efficient sludge management and energy recovery. Treatment for potable and non-potable use	State-of-the-art AD technology, advanced water reclamation techniques	Energy self-sufficiency, support for water independence strategy, sustainable management. 59% of wastewater treated in this plant	Johnson et al. (2021) and Kog (2020)
Barueri Wastewater Treatment Plant, Sao Paulo, Brazil	One of the largest facilities in Brazil, incorporating AD for sludge management and recovery of energy, water, cellulose, polyhydroxyalkanoates, and nutrients	Adaptation to treat high volumes of urban wastewater and reduce municipal river pollution	Reduction in sludge volume, biogas for energy, recovery of 5.3 tons of phosphorus per day, and produce 38 tons of struvite particles per day	Sánchez (2020)
Humber Wastewater Treatment Plant, Toronto, Canada	Implements AD for sludge treatment, crucial for Canada's largest city	Advanced AD technologies for enhanced biogas production apply the tertiary treatment for water recovery	Increased energy efficiency, greenhouse gas emission reduction, cost savings on energy	Zhou et al. (2019)
Lynetten Wastewater Treatment Plant, Copenhagen, Denmark	Advanced wastewater treatment plant serving Copenhagen utilizes AD	Innovative digestion technologies, integration with the city's energy systems	Biogas production for energy, contribution to renewable energy targets, reductions of greenhouse gas emissions	Delre et al. (2019)
Valdebebas Wastewater Treatment Plant, Madrid, Spain	Advanced treatment plant using AD to manage urban wastewater effectively. Potential integration for using the biogas in the district heating system energy supply	Cutting-edge technology for AD-enhanced energy recovery	High levels of biogas production for electricity energy self-sufficiency. Supply at least 50,000 people with the energy required	Picardo et al. (2019)

			for water heating for municipal using. Reduction in CO_2 emissions	
Henriksdal Wastewater Treatment Plant, Stockholm, Sweden	One of the largest sewage treatment plants in Europe using AD	Innovative AD technologies like temperature-phased AD	Significant energy production, reduced environmental impact, efficient sludge management	Andersson et al. (2016)
Durban Water Recycling Plant, Durban, South Africa	A comprehensive approach to wastewater management with AD	Integration of AD with water reuse technologies, focus on wastewater treatment for non-potable uses, i.e., irrigation	Reduced environmental impact and potable water usage, provision of recycled water for industrial use, 50% of non-potable uses could be covered in the investigated area	Kumarasamy and Dube (2016)
East Bay Municipal Utility District, Oakland, California, United States	First plant in America North to extract renewable energy than it needs, becoming a net energy producer	High-efficiency anaerobic digesters, co-digestion of high-strength organic wastes, advanced biogas utilization systems	Gas emissions mitigation, replenishable energy generation, and biogas production by almost 70%. Energy self-sufficiency where produce 126% of plant demand	Shen et al. (2015)
Beckton Sewage Treatment Works, London, United Kingdom	Part of the Thames Tideway Tunnel project, one of Europe's largest sewage treatment works, serves more than 3.5 million people, enhancing capacity for AD	Upgrading AD facilities, innovations in biogas utilization. Using the thermal hydrolysis process	Improved energy self-sufficiency, higher electricity generation, and reduced natural gas consumption. Reduced carbon footprint, enhanced urban wastewater management capacity	Stocks (2016)
Sha Tin Sewage Treatment Works, Sha Tin, Hong Kong	Serves a large urban population with integrated AD for sludge treatment	Thermophilic AD technology for increased biogas production and pathogen reduction	Providing the sewage treatment reactor with heated water and power. Enhanced biogas production, reduced sludge volume, improved environmental protection	Fung and Yeung (2012)

Abbreviation: *AD*, anaerobic digestion.

By producing biogas for self-sufficiency, one benefit of these tiny reactors is that they allow people to become energy-independent (Sime and Jin, 2020).

Small-scale AD is more accepted than large-scale reactors because of the resident benefits they offer to societies. Small-scale AD plants are more able to adapt to the unique requirements and features of their immediate geographic setting, leading to benefits, such as enhanced waste management, decreased pollution, and the production of locally produced renewable energy. Additionally, the smaller AD plants assist in reducing possible odor, noise, and aesthetic difficulties, which are highly noticeable in large-scale systems (Chodkowska-Miszczuk et al., 2019).

Another consideration is that people in lower socioeconomic strata tend to make less effort when it comes to ecological issues, such as trash segregation and supporting sustainable energy programs. This can be caused by two items: a lack of knowledge about the severity of climate change and alternative energies and the general perception that using alternative clean energy sources like biogas will raise energy prices (Bourdin and Chassy, 2023).

10.7.3 Lessons for future implementations

Due to the vast population and water scarcity, which will be more complicated, there is an urgent need to improve MWW treatment processes for renewable energy, nutrients, and potable and non-potable water recovery. Furthermore, wastewater treatment significantly reduces the environmental effects of releasing pollutants without treatment. Therefore, there is a need for deep and advanced research in methods and technologies to improve and optimize the treatment process, especially AD, by applying outcomes to full-scale systems. Furthermore, some thoughts must be considered for better future outcomes, such as the need for more research on psychrophilic AD (unheated, low-cost bioreactors) that developing countries must deal with, where most bioreactors applied in underdeveloped nations operate in lower temperatures conditions (psychrophilic). The methods above, such as biochemical methane potential tests, concentrate on higher temperature conditions (mesophilic), especially those that are applied in developed countries with enough energy to heat the bioreactors; thus, a broader range of temperatures needs to be investigated (Tavera-Ruiz et al., 2023).

Moreover, exploring the potential of co-digestion with sludge-rich feedstocks enhances biogas production by tackling nutrient imbalances (Memon and Memon, 2020). It also draws more attention to evaluating the potential of advanced technology, such as biogas upgrading solutions, monitoring systems, microplastic elimination, and pre-treatment technologies. Where effective pre-treatment of MWW is vital to remove inorganic and non-biodegradable materials that can inhibit AD. Moreover, robust pre-treatment techniques can improve the digestive process's dependability and efficiency (Sadia et al., 2022). Additionally, thorough feasibility studies should be conducted before implementation, and the technical, financial, and social viability of the local area should be carefully evaluated while taking the needs and circumstances of the local area into account, where it is vital for the design and implementation of digesters. Besides, the study and analysis of the local area wastewater characteristics are needed to optimize the configuration and operational parameters needed for digesters to be more effective (Nsair et al., 2020). By implementing these lessons, the challenges of AD can be minimized, and its benefits in treating MWW can be optimized. This results in waste reduction, energy recovery, and environmental protection.

10.8 Conclusions and future trends

Anaerobic digestion has considerable potential for treating municipal wastewater by efficiently removing organic matter and producing biogas. Recent advancements in AD technology, such as the implementation of two-stage digestion and high-rate anaerobic reactors, have shown enhanced performance and increased efficiency in the process. Nevertheless, there are still obstacles that need to be addressed, such as the requirement for improved comprehension of microorganisms, the optimization of performance metrics, and the creation of intelligent monitoring devices. In the future, developments in advanced treatment for municipal wastewater treatment are anticipated to prioritize the integration of cutting-edge technology, improving treatment efficiency, and addressing sustainability issues. Case studies highlight the significance of ongoing learning and adjustment to specific local circumstances, providing useful knowledge for future deployments. In summary, the ongoing progress of anaerobic digestion technology shows potential for attaining environmentally friendly and economically viable methods of treating wastewater in urban areas.

References

Abou-Elela, S.I., El-Khateeb, M.A., 2015. Performance evaluation of activated sludge process for treating pharmaceutical wastewater contaminated with β-lactam antibiotics. J. Ind. Pollut. Control 31 (1), 1–5. Enviro Media, Egypt, 09702083. Available from: http://www.indiaenvironmentportal.org.in/taxonomy/term/8906.

Achinas, S., Achinas, V., Euverink, G.J.W., 2020. Microbiology and biochemistry of anaerobic digesters: an overview. In: Bioreactors: Sustainable Design and Industrial Applications in Mitigation of GHG Emissions. Elsevier, Netherlands, pp. 17–26. Available from: https://www.elsevier.com/books/bioreactors/singh/978-0-12-821264-6. https://doi.org/10.1016/B978-0-12-821264-6.00002-4.

Acién Fernández, F.G., Gómez-Serrano, C., Fernández-Sevilla, J.M., 2018. Recovery of nutrients from wastewaters using microalgae. Front. Sustain. Food Syst. 2. Frontiers Media S.A., Spain, 2571581X. Available from: www.frontiersin.org/journals/sustainable-food-systems. https://doi.org/10.3389/fsufs.2018.00059.

Afridi, Z.U.R., Ullah, K., Mustafa, M.F., Saleem, H., Shaker, B., Ashraf, N., Aslam, S., 2023. Biogas as sustainable approach for social uplift in South East Asian region. Energy Rep. 10, 4808–4818. Elsevier Ltd, Pakistan, 23524847. Available from: https://www.sciencedirect.com/science/journal/23524847. https://doi.org/10.1016/j.egyr.2023.11.037.

Aghalari, Z., Dahms, H.U., Sillanpää, M., Sosa-Hernandez, J.E., Parra-Saldívar, R., 2020. Effectiveness of wastewater treatment systems in removing microbial agents: a systematic review. Glob. Health 16 (1). BioMed Central Ltd., Iran, 17448603. Available from: http://www.globalizationandhealth.com/. https://doi.org/10.1186/s12992-020-0546-y.

Ahmed, B., Gahlot, P., Balasundaram, G., Tyagi, V.K., Banu, R., Vivekanand, J.V., Kazmi, A.A., 2023. Semi-continuous anaerobic co-digestion of thermal and thermal-alkali processed organic fraction of municipal solid waste: methane yield, energy analysis, anaerobic microbiome. J. Environ. Manag. 345. Academic Press, India, 10958630. Available from: https://www.sciencedirect.com/journal/journal-of-environmental-management. https://doi.org/10.1016/j.jenvman.2023.118907.

Akter, J., Kim, W., Kim, I., Lee, J., 2023. Real-time organic compound monitoring in advanced oxidation process (AOP) of anaerobic digestion effluent: introduction of UV-VIS spectrum representative index by discrete Fourier transform. Environ. Monit. Assess. 195 (11). Springer Science and Business Media Deutschland GmbH, South Korea, 15732959. Available from: https://www.springer.com/journal/10661. https://doi.org/10.1007/s10661-023-11984-2.

Alengebawy, A., Mohamed, B.A., Jin, K., Liu, T., Ghimire, N., Samer, M., Ai, P., 2022a. A comparative life cycle assessment of biofertilizer production towards sustainable utilization of anaerobic digestate. Sustain. Prod. Consum. 33, 875–889. Elsevier B.V., China, 23525509. Available from: http://www.journals.elsevier.com/sustainable-production-and-consumption/. https://doi.org/10.1016/j.spc.2022.08.016.

Alengebawy, A., Mohamed, B.A., Ghimire, N., Jin, K., Liu, T., Samer, M., Ai, P., 2022b. Understanding the environmental impacts of biogas utilization for energy production through life cycle assessment: an action towards reducing emissions. Environ. Res. 213, 113632. https://doi.org/10.1016/j.envres.2022.113632. 00139351.

Alengebawy, A., Ran, Y., Ghimire, N., Osman, A.I., Ai, P., 2023. Rice straw for energy and value-added products in China: a review. Environ. Chem. Lett. 21 (5), 2729–2760. Springer Science and Business Media Deutschland GmbH, China,

16103661. Available from: https://www.springer.com/journal/10311. https://doi.org/10.1007/s10311-023-01612-3.

Alhajeri, N.S., Eraky, M., Qyyum, M.A., Tawfik, A., 2023. Eco-friendly fermentation module for maximization of hydrogen harvesting from fatty restaurant waste diluted with grey water. Int. J. Hydrogen Energy 48 (68), 26461–26474. Elsevier Ltd, Kuwait, 03603199. Available from: http://www.journals.elsevier.com/international-journal-of-hydrogen-energy/. https://doi.org/10.1016/j.ijhydene.2022.11.015.

Alves, V.D., Fontes, C.M.G.A., Bule, P., 2021. Cellulosomes: highly efficient cellulolytic complexes. Subcell. Biochem. 96, 323–354. Springer Science and Business Media B.V., Portugal, Available from: http://www.springer.com/series/6515. https://doi.org/10.1007/978-3-030-58971-4_9.

Amin, F.R., Khalid, H., El-Mashad, H.M., Chen, C., Liu, G., Zhang, R., 2021. Functions of bacteria and archaea participating in the bioconversion of organic waste for methane production. Sci. Total Environ. 763. Elsevier B.V., China, 18791026. Available from: www.elsevier.com/locate/scitotenv. https://doi.org/10.1016/j.scitotenv.2020.143007.

Andersson, S., Ek, P., Berg, M., Grundestam, J., Lindblom, E., 2016. Extension of two large wastewater treatment plants in Stockholm using membrane technology. Water Pract. Technol. 11 (4), 744–753. https://doi.org/10.2166/wpt.2016.034. 1751-231X.

Angelakis, A., Snyder, S., 2015. Wastewater treatment and reuse: past, present, and future. Water 7 (9), 4887–4895. https://doi.org/10.3390/w7094887. 2073-4441.

Atelge, M.R., Atabani, A.E., Rajesh Banu, J., Krisa, D., Kaya, M., Eskicioglu, C., Kumar, G., Lee, C., Yildiz, Y.Ş., Unalan, S., Mohanasundaram, R., Duman, F., 2020. A critical review of pretreatment technologies to enhance anaerobic digestion and energy recovery. Fuel 270, 117494. https://doi.org/10.1016/j.fuel.2020.117494. 00162361.

Awhangbo, L., Bendoula, R., Roger, J.M., Béline, F., 2020. Multi-block data analysis for online monitoring of anaerobic co-digestion process. Chemom. Intell. Lab. Syst. 205, 104120. https://doi.org/10.1016/j.chemolab.2020.104120. 01697439.

Azman, S., Khadem, A.F., Van Lier, J.B., Zeeman, G., Plugge, C.M., 2015. Presence and role of anaerobic hydrolytic microbes in conversion of lignocellulosic biomass for biogas production. Crit. Rev. Environ. Sci. Technol. 45 (23), 2523–2564. Taylor and Francis Inc., Netherlands, 15476537. Available from: www.tandf.co.uk/journals/titles/10643389.asp. https://doi.org/10.1080/10643389.2015.1053727.

Balcioglu, G., Jeswani, H.K., Azapagic, A., 2022. Evaluating the environmental and economic sustainability of energy from anaerobic digestion of different feedstocks in Turkey. Sustain. Prod. Consum. 32, 924–941. Elsevier B.V., United Kingdom, 23525509. Available from: http://www.journals.elsevier.com/sustainable-production-and-consumption/. https://doi.org/10.1016/j.spc.2022.06.011.

Bautista Angeli, J.R., Morales, A., LeFloc'h, T., Lakel, A., Andres, Y., 2018. Anaerobic digestion and integration at urban scale: feedback and comparative case study. Energy Sustain. Soc. 8 (1). https://doi.org/10.1186/s13705-018-0170-3. 2192-0567.

Belhassan, K., 2021. Water Scarcity Management Advanced Sciences and Technologies for Security Applications. Springer, United Kingdom, pp. 443–462. Available from: https://link.springer.com/bookseries/5540. https://doi.org/10.1007/978-3-030-76008-3_19.

Belhassan, K., 2023. Managing Drought and Water Stress in Northern Africa. IntechOpen, https://doi.org/10.5772/intechopen.107391.

Bello, M.M., Abdul Raman, A.A., Purushothaman, M., 2017. Applications of fluidized bed reactors in wastewater treatment—a review of the major design and operational parameters. J. Clean. Prod. 141, 1492–1514. https://doi.org/10.1016/j.jclepro.2016.09.148. 09596526.

Bhushan, S., Jayakrishnan, U., Shree, B., Bhatt, P., Eshkabilov, S., Simsek, H., 2023. Biological pretreatment for algal biomass feedstock for biofuel production. J. Environ. Chem. Eng. 11 (3), 109870. https://doi.org/10.1016/j.jece.2023.109870. 22133437.

Boretti, A., Rosa, L., 2019. Reassessing the projections of the World Water Development Report. npj Clean Water 2 (1). Nature Research, Vietnam, 20597037. Available from: https://www.nature.com/npjcleanwater/. https://doi.org/10.1038/s41545-019-0039-9.

Bortot Coelho, F.E., Magnacca, G., Boffa, V., Candelario, V.M., Luiten-Olieman, M., Zhang, W., 2023. From ultra to nanofiltration: a review on the fabrication of ZrO_2 membranes. Ceram. Int. 49 (6), 8683–8708. Elsevier Ltd, Denmark, 02728842. Available from: https://www.journals.elsevier.com/ceramics-international. https://doi.org/10.1016/j.ceramint.2023.01.076.

Bourdin, S., Chassy, A., 2023. Are citizens ready to make an environmental effort? A study of the social acceptability of biogas in France. Environ. Manage. 71 (6), 1228–1239. Springer, France, 14321009. Available from: https://www.springer.com/journal/267. https://doi.org/10.1007/s00267-022-01779-5.

Bourdin, S., Raulin, F., Josset, C., 2020. On the (un)successful deployment of renewable energies: territorial context matters. A conceptual framework and an empirical analysis of biogas projects. Energy Stud. Rev. 24 (1). McMaster University, France, 08434379. Available from: https://energystudiesreview.ca/esr/article/view/4088. 10.15173/ESR.V24I1.4088.

Cai, Y., Zheng, Z., Schäfer, F., Stinner, W., Yuan, X., Wang, H., Cui, Z., Wang, X., 2021. A review about pretreatment of lignocellulosic biomass in anaerobic digestion: achievement and challenge in Germany and China. J. Clean. Prod. 299, 126885. https://doi.org/10.1016/j.jclepro.2021.126885. 09596526.

Capodaglio, A.G., 2020. Taking the water out of "wastewater": an ineluctable oxymoron for urban water cycle sustainability. Water Environ. Res. 92 (12), 2030–2040. John Wiley and Sons Inc, Italy, 15547531. Available from: https://onlinelibrary.wiley.com/loi/15547531. https://doi.org/10.1002/wer.1373.

Capodaglio, A.G., 2021. Fit-for-purpose urban wastewater reuse: Analysis of issues and available technologies for sustainable multiple barrier approaches. Crit. Rev. Environ. Sci. Technol. 51 (15), 1619–1666. https://doi.org/10.1080/10643389.2020.1763231. 1064-3389.

Cárdenas-Talero, J.L., Silva-Leal, J.A., Pérez-Vidal, A., Torres-Lozada, P., 2022. The influence of municipal wastewater treatment technologies on the biological stabilization of sewage sludge: a systematic review. Sustainability 14 (10), 5910. https://doi.org/10.3390/su14105910. 2071-1050.

Cartlidge, E., 2017. The internet of things: from hype to reality. Opt. Photonics News 28 (9), 26. https://doi.org/10.1364/OPN.28.9.000026. 1047-6938.

Chakraborty, D., Karthikeyan, O.P., Selvam, A., Palani, S.G., Ghangrekar, M.M., Wong, J.W.C., 2022. Two-phase anaerobic digestion of food waste: effect of semi-continuous feeding on acidogenesis and methane production. Bioresour. Technol. 346. Elsevier Ltd, Hong Kong, 18732976. Available from: www.elsevier.com/locate/biortech. https://doi.org/10.1016/j.biortech.2021.126396.

Chen, L., Msigwa, G., Yang, M., Osman, A.I., Fawzy, S., Rooney, D.W., Yap, P.S., 2022. Strategies to achieve a carbon neutral society: a review. Environ. Chem. Lett. 20 (4), 2277–2310. Springer Science and Business Media Deutschland GmbH, China, 16103661. Available from: https://www.springer.com/journal/10311. https://doi.org/10.1007/s10311-022-01435-8.

Chodkowska-Miszczuk, J., Kulla, M., Novotný, L., 2019. Biogas energy—a chance for agriculture and rural development? Insight from the post-communist Central Europe. DETUROPE—Central Eur. J. Tour. Reg. Dev. 11 (2), 30–53. https://doi.org/10.32725/det.2019.014. 18212506.

Chuenchart, W., Logan, M., Leelayouthayotin, C., Visvanathan, C., 2020. Enhancement of food waste thermophilic anaerobic digestion through synergistic effect with chicken manure. Biomass Bioenergy 136, 105541. https://doi.org/10.1016/j.biombioe.2020.105541. 09619534.

Conrad, R., 2020. Importance of hydrogenotrophic, aceticlastic and methylotrophic methanogenesis for methane production in terrestrial, aquatic and other anoxic environments: A mini review. Pedosphere 30 (1), 25–39. Soil Science Society of China, Germany, 10020160. Available from: http://pedosphere.issas.ac.cn. https://doi.org/10.1016/S1002-0160(18)60052-9.

Cremonez, P.A., Teleken, J.G., Weiser Meier, T.R., Alves, H.J., 2021. Two-stage anaerobic digestion in agroindustrial waste treatment: A review. J. Environ. Manag. 281. Academic Press, Brazil, 10958630. Available from: https://www.sciencedirect.com/journal/journal-of-environmental-management. https://doi.org/10.1016/j.jenvman.2020.111854.

Darwish, M., Abdulrahim, H.K., Mabrouk, A.N., Hassan, A., 2015. Reclaimed wastewater for agriculture irrigation in Qatar. Glob. Sci. Res. J. 3, 106–120.

de Mello, B.S., Rodrigues, B.C.G., Sarti, A., 2021. Comparative performance of high-rate anaerobic reactors for biodegradation of soybean molasses. Environ. Technol. Innov. 24. Elsevier B.V., Brazil, 23521864. Available from: http://www.journals.elsevier.com/environmental-technology-and-innovation/. https://doi.org/10.1016/j.eti.2021.101937.

Delre, A., ten Hoeve, M., Scheutz, C., 2019. Site-specific carbon footprints of Scandinavian wastewater treatment plants, using the life cycle assessment approach. J. Clean. Prod. 211, 1001–1014. Elsevier Ltd, Denmark, 09596526. Available from: https://www.journals.elsevier.com/journal-of-cleaner-production. https://doi.org/10.1016/j.jclepro.2018.11.200.

Di Giacomo, G., Romano, P., 2022. Evolution and prospects in managing sewage sludge resulting from municipal wastewater purification. Energies 15 (15), 5633. https://doi.org/10.3390/en15155633. 1996-1073.

Dolar, D., Racar, M., Košutić, K., 2019. Municipal wastewater reclamation and water reuse for irrigation by membrane processes. Chem. Biochem. Eng. Q. 33 (3), 417–425. Assoc. of Chemists and Chemical Engineers of Croatia, Croatia, 18465153. Available from: https://hrcak.srce.hr/file/329236. 10.15255/cabeq.2018.1571.

Dolgen, D., Alpaslan, M.N., 2023. Reuse of Wastewater from the Circular Economy (CE) Perspective A Sustainable Green Future: Perspectives on Energy, Economy, Industry, Cities and Environment. Springer International Publishing, Turkey, pp. 385–408. Available from: https://link.springer.com/book/10.1007/978-3-031-24942-6. https://doi.org/10.1007/978-3-031-24942-6_18.

Dong, Y., Zhang, J., Wang, Q., Dongyu, X., Pang, S., Campos, L.C., Ren, Z., Wang, P., 2023. Dual function of magnetic field in enhancing antibiotic wastewater treatment by an integrated photocatalysis and fluidized bed biofilm reactor (FBBR). J. Environ. Manag. 347, 119249. https://doi.org/10.1016/j.jenvman.2023.119249. 03014797.

Drechsel, P., Otoo, M., Hanjra, M.A., 2022a. Resource recovery from wastewater and the consumer point of view: social, cultural and economic aspects. IWA Publishing, pp. 383–414, https://doi.org/10.2166/9781780409566_0383.

Drechsel, P., Qadir, M., Galibourg, D., 2022b. The WHO guidelines for safe wastewater use in agriculture: a review of implementation challenges and possible solutions in the global south. Water 14 (6), 864. https://doi.org/10.3390/w14060864, 2073-4441.

Ehalt MacEdo, H., Lehner, B., Nicell, J., Grill, G., Li, J., Limtong, A., Shakya, R., 2022. Distribution and characteristics of wastewater treatment plants within the global river network. Earth Syst. Sci. Data 14 (2), 559–577. Copernicus Publications, Canada, 18663516. Available from: www.earth-system-science-data.net. https://doi.org/10.5194/essd-14-559-2022.

Eraky, M., Jin, K., Zhang, Q., Zhang, Z., Ai, P., Elsayed, M., 2021. Acidogenic biorefinery of rice straw for volatile fatty acids production via sequential two-stage fermentation: effects of pre-treatments. Environ. Technol. Innov. 23, 101686. https://doi.org/10.1016/j.eti.2021.101686. 23521864.

Eraky, M., Nasr, M., Elsayed, M., Ai, P., Tawfik, A., 2023. Synergistic interaction of tween 20 and magnesium@ functionalized graphene oxide nano-composite for dual productivity of biohydrogen and biochar from onion peel waste. Renew. Energy 216, 119082. https://doi.org/10.1016/j.renene.2023.119082. 09601481.

Espinoza-Guillen, J.A., Alderete-Malpartida, M.B., Gallegos-Huamán, R.L., Paz-Rosales, Y.M., Domínguez-Vivar, R.M., Bujaico-León, C., 2024. Ecological risk assessment and identification of sources of heavy metals contamination in sewage sludge from municipal wastewater treatment plants in the Metropolitan Area of Lima-Callao, Peru. Environ. Dev. Sustain. 26 (1), 1559–1590. Springer Science and Business Media B.V., Peru, 15732975. Available from: https://www.springer.com/journal/10668. https://doi.org/10.1007/s10668-022-02774-w.

Fahad, A., Mohamed, R.M.S., Radhi, B., Al-Sahari, M., 2019. Wastewater and its treatment techniques: an ample review. Indian J. Sci. Technol. 12 (25), 1–13. https://doi.org/10.17485/ijst/2019/v12i25/146059. 0974-5645.

Fan, Q., Shao, Z., Guo, X., Qiang, Q., Yao, Y., Zhang, Z., Qiu, L., 2024. Effects of Fe–N co-modified biochar on methanogenesis performance, microbial community, and metabolic pathway during anaerobic co-digestion of alternanthera philoxeroides and cow manure. J. Environ. Manag. 351, 120006. https://doi.org/10.1016/j.jenvman.2023.120006. 03014797.

Feiz, R., Johansson, M., Lindkvist, E., Moestedt, J., Påledal, S.N., Svensson, N., 2020. Key performance indicators for biogas production—methodological insights on the life-cycle analysis of biogas production from source-separated food waste. Energy 200, 117462. https://doi.org/10.1016/j.energy.2020.117462. 03605442.

Fricke, K., Santen, H., Wallmann, R., Hüttner, A., Dichtl, N., 2007. Operating problems in anaerobic digestion plants resulting from nitrogen in MSW. Waste Manag. 27 (1), 30–43. https://doi.org/10.1016/j.wasman.2006.03.003. 0956053X.

Fung, W.-C., Yeung, T.-K., 2012. Utilisation of biogas at Shatin sewage treatment works. In: Conf. 2012- Effic. Energy Manag. Water Waste Ind, pp. 1–12.

Gani, P., Mohamed Sunar, N., Matias-Peralta, H., Abdul Latiff, A.A., Parjo, U.K., Oyekanmi, A.A., 2017. Green approach in the bio-removal of heavy metals from wastewaters. MATEC Web Conf. 103. https://doi.org/10.1051/matecconf/201710306007. http://www.matec-conferences.org/. 2261236X EDP Sciences Malaysia.

Guo, Z., Hatakeyama, T., Yoshimura, C., Wang, T., Hatano, Y., 2022. Basic influent sewage quality reflects sewershed characteristics in Tokyo city. J. Water Health 20 (6), 972–984. IWA Publishing, Japan, 19967829. Available from: https://watermark.silverchair.com/jwh0200972.pdf. https://doi.org/10.2166/wh.2022.022.

Gutiérrez, J.E.V., Camargo, F.P., Sakamoto, I.K., Varesche, M.B.A., 2022. Expanded granular sludge bed reactor technology feasibility for removal of nonylphenol ethoxylate in co-digestion of domestic sewage and commercial laundry wastewater: taxonomic characterization and biogas production. Process Saf. Environ.

Prot. 161, 556–570. Institution of Chemical Engineers, Brazil, 09575820. Available from: http://www.elsevier.com/wps/find/journaldescription.cws_home/713889/description#description. https://doi.org/10.1016/j.psep.2022.03.055.

Holl, E., Steinbrenner, J., Merkle, W., Krümpel, J., Lansing, S., Baier, U., Oechsner, H., Lemmer, A., 2022. Two-stage anaerobic digestion: state of technology and perspective roles in future energy systems. Bioresour. Technol. 360, 127633. https://doi.org/10.1016/j.biortech.2022.127633. 09608524.

Hong-Ying, H., Zhuo, C., Shuran, H., Yinhu, W., 2021. Towards the new era of water reuse in China. Int. Water Assoc.

Ibarra-Esparza, F.E., González-López, M.E., Ibarra-Esparza, J., Lara-Topete, G.O., Senés-Guerrero, C., Cansdale, A., Forrester, S., Chong, J.P.J., Gradilla-Hernández, M.S., 2023. Implementation of anaerobic digestion for valorizing the organic fraction of municipal solid waste in developing countries: technical insights from a systematic review. J. Environ. Manag. 347. Academic Press, Mexico, 10958630. Available from: https://www.sciencedirect.com/journal/journal-of-environmental-management. https://doi.org/10.1016/j.jenvman.2023.118993.

Ibrahim, A.Y., Tawfik, A., El-Dissouky, A., Kassem, T.S., Alhajeri, N.S., Pant, D., Khalil, T.E., 2022. Sulphonated graphene catalyst incorporation with sludge enhanced the microbial activities for biomethanization of crude rice straw. Bioresour. Technol. 361. Elsevier Ltd, Egypt, 18732976. Available from: www.elsevier.com/locate/biortech. https://doi.org/10.1016/j.biortech.2022.127614.

Jaeger, A., Blanchard, R.E., 2022. Techno-economic analysis of an enhanced anaerobic digester in the Andean area of Peru. Int. J. Energy Environ. Eng. 13 (2), 805–819. Springer Science and Business Media Deutschland GmbH, Peru, 22516832. Available from: http://www.journal-ijeee.com/. https://doi.org/10.1007/s40095-022-00473-w.

Johnson, B.R., Kadiyala, R., Owens, G., Ping, M.Y., Grace, P., Newbery, C., Sing, S., Saxena, A., Green, J., 2021. Water reuse and recovery facility connected digital twin case study: Singapore PUB's Changi WRP process, control, and hydraulics digital twin. In: 94th Annual Water Environment Federation Technical Exhibition and Conference, WEFTEC. Water Environment Federation, United States, pp. 441–455.

Jones, E.R., Van Vliet, M.T.H., Qadir, M., Bierkens, M.F.P., 2021. Country-level and gridded estimates of wastewater production, collection, treatment and reuse. Earth Syst. Sci. Data 13 (2), 237–254. Copernicus Publications, Netherlands, 18663516. Available from: www.earth-system-science-data.net. https://doi.org/10.5194/essd-13-237-2021.

Kabera, T., Nishimwe, H., Rutayisire, B., Twagirayezu, E., 2013. Performance evaluation of a water treatment plant. vol. 2 Informa UK Limited, pp. 1103–1107, https://doi.org/10.1201/b16320-227.

Kainthola, J., Shariq, M., Kalamdhad, A.S., Goud, V.V., 2019. Enhanced methane potential of rice straw with microwave assisted pretreatment and its kinetic analysis. J. Environ. Manag. 232, 188–196. Academic Press, India, 10958630. Available from: https://www.sciencedirect.com/science/journal/03014797. https://doi.org/10.1016/j.jenvman.2018.11.052.

Kamble, S., Singh, A., Kazmi, A., Starkl, M., 2019. Environmental and economic performance evaluation of municipal wastewater treatment plants in India: A life cycle approach. Water Sci. Technol. 79 (6), 1102–1112. IWA Publishing, India, 19969732. Available from: https://watermark.silverchair.com/wst079061102.pdf. https://doi.org/10.2166/wst.2019.110.

Kang, X., Zhang, Y., Song, B., Sun, Y., Li, L., He, Y., Kong, X., Luo, X., Yuan, Z., 2019. The effect of mechanical pretreatment on the anaerobic digestion of hybrid Pennisetum. Fuel 252, 469–474. Elsevier Ltd, China, 00162361. Available from: http://www.journals.elsevier.com/fuel/. https://doi.org/10.1016/j.fuel.2019.04.134.

Kashyap, R., Singh, F., Sharma, A., 2015. Utilization of waste water in agriculture. Indian Farmer 2, 884–887.

Kataki, S., Chatterjee, S., Vairale, M.G., Dwivedi, S.K., Gupta, D.K., 2021. Constructed wetland, an eco-technology for wastewater treatment: a review on types of wastewater treated and components of the technology (macrophyte, biolfilm and substrate). J. Environ. Manag. 283. Academic Press, India, 10958630. Available from: https://www.sciencedirect.com/journal/journal-of-environmental-management. https://doi.org/10.1016/j.jenvman.2021.111986.

Kog, Y.C., 2020. Water reclamation and reuse in Singapore. J. Environ. Eng. (United States) 146 (4). American Society of Civil Engineers (ASCE), Singapore, 19437870. Available from: http://ascelibrary.org/toc/joeedu/current. https://doi.org/10.1061/(ASCE)EE.1943-7870.0001675.

Krueger, M., Luo, H., Ding, S.X., Dominic, S., Yin, S., 2015. Data-driven approach of KPI monitoring and prediction with application to wastewater treatment process. IFAC-PapersOnLine 48 (21), 627–632. https://doi.org/10.1016/j.ifacol.2015.09.596. 24058963.

Kryston, A., Woods, C.G., Manga, M., 2024. Social barriers to safe sanitation access among housed populations in the United States: a systematic review. Int. J. Hyg. Environ. Health 257, 114326. https://doi.org/10.1016/j.ijheh.2024.114326. 14384639.

Kumar, M., Yadav, A.N., Saxena, R., Rai, P.K., Paul, D., Tomar, R.S., 2021. Novel methanotrophic and methanogenic bacterial communities from diverse ecosystems and their impact on environment. Biocatal. Agric. Biotechnol. 33. Elsevier Ltd, India, 18788181. Available from: http://www.journals.elsevier.com/biocatalysis-and-agricultural-biotechnology/. https://doi.org/10.1016/j.bcab.2021.102005.

Kumarasamy, M.V., Dube, V.N., 2016. Study on recycling urban wastewater for non-potable uses for water conservation. Polish J. Environ. Stud. 25 (1), 167–171. HARD Publishing Company, South Africa, 12301485. Available from: http://www.pjoes.com/abstracts.html. 10.15244/pjoes/59273.

Lamb, J.J., Bernard, O., Sarker, S., Lien, K.M., Hjelme, D.R., 2019. Perspectives of optical colourimetric sensors for anaerobic digestion. Renew. Sustain. Energy Rev. 111, 87–96. Elsevier Ltd, Norway, 18790690. Available from: https://www.journals.elsevier.com/renewable-and-sustainable-energy-reviews. https://doi.org/10.1016/j.rser.2019.04.083.

Le, T.S., Bui, X.T., Nguyen, P.D., Hao Ngo, H., Dang, B.T., Le Quang, D.T., Thi Pham, T., Visvanathan, C., Diels, L., 2024. Bacterial community composition in a two-stage anaerobic membrane bioreactor for co-digestion of food waste and food court wastewater. Bioresour. Technol. 391. Elsevier Ltd, Vietnam, 18732976. Available from: www.elsevier.com/locate/biortech. https://doi.org/10.1016/j.biortech.2023.129925.

Lee, S.H., Park, J.H., Kang, H.J., Lee, Y.H., Lee, T.J., Park, H.D., 2013. Distribution and abundance of Spirochaetes in full-scale anaerobic digesters. Bioresour. Technol. 145, 25–32. Elsevier Ltd, South Korea, 18732976. Available from: www.elsevier.com/locate/biortech. https://doi.org/10.1016/j.biortech.2013.02.070.

Lee, S.H., Park, J.H., Kim, S.H., Yu, B.J., Yoon, J.J., Park, H.D., 2015. Evidence of syntrophic acetate oxidation by Spirochaetes during anaerobic methane production. Bioresour. Technol. 190, 543–549. Elsevier Ltd, South Korea, 18732976. Available from: www.elsevier.com/locate/biortech. https://doi.org/10.1016/j.biortech.2015.02.066.

Lee, E.S., Park, S.Y., Kim, C.G., 2023. Feasibility test anaerobically enhancing methane yield under the injection of hydrogen and carbon dioxide. Renew. Energy 212, 761–768. Elsevier Ltd, South Korea, 18790682. Available from: http://www.journals.elsevier.com/renewable-and-sustainable-energy-reviews/. https://doi.org/10.1016/j.renene.2023.05.104.

Li, M., Zhang, Q., Sun, F., Cui, M.-H., Liu, H., Chen, C., 2023. The effect of microbial electrolysis cell coupled with anaerobic digestion for landfill leachate treatment. J. Water Process Eng. 55, 104260. https://doi.org/10.1016/j.jwpe.2023.104260. 22147144.

Longis, M., Pereira, J.C., Högl, T.H., Neubauer, P., Junne, S., 2023. Plug-flow anaerobic digestion with multi-position sensors: the value of gradient measurement for process monitoring. Biomass Bioenergy 173. Elsevier Ltd, Germany, 18732909. Available from: http://www.journals.elsevier.com/biomass-and-bioenergy/. https://doi.org/10.1016/j.biombioe.2023.106803.

Lü, W., Ren, H., Ding, W., Li, H., Yao, X., Jiang, X., Qadeer, A., 2022. Biotic and abiotic controls on sediment carbon dioxide and methane fluxes under short-term experimental warming. Water Res. 226, 119312. https://doi.org/10.1016/j.watres.2022.119312. 00431354.

Lv, N., Zhao, L., Wang, R., Ning, J., Pan, X., Li, C., Cai, G., Zhu, G., 2020. Novel strategy for relieving acid accumulation by enriching syntrophic associations of syntrophic fatty acid-oxidation bacteria and H_2/formate-scavenging methanogens in anaerobic digestion. Bioresour. Technol. 313, 123702. https://doi.org/10.1016/j.biortech.2020.123702. 09608524.

Ma, J., Ni, X., Cai, C., Da, Y., Kan, H., Guo, R., Liu, J., Peng, K., Liu, E., Huang, X., 2023. Insights into greenhouse gas emissions from a wastewater treatment plant in vulnerable water areas of China. Sci. Total Environ. 902, 166017. https://doi.org/10.1016/j.scitotenv.2023.166017. 00489697.

Magdalena, J.A., Greses, S., González-Fernández, C., 2019. Impact of organic loading rate in volatile fatty acids production and population dynamics using microalgae biomass as substrate. Sci. Rep. 9 (1). Nature Research, Spain, 20452322. Available from: www.nature.com/srep/index.html. https://doi.org/10.1038/s41598-019-54914-4.

Mainardis, M., Buttazzoni, M., Goi, D., 2020. Up-flow anaerobic sludge blanket (UASB) technology for energy recovery: a review on state-of-the-art and recent technological advances. Bioengineering 7 (2), 43. https://doi.org/10.3390/bioengineering7020043. 2306-5354.

Margot, J., Rossi, L., Barry, D.A., Holliger, C., 2015. A review of the fate of micropollutants in wastewater treatment plants. WIREs Water 2 (5), 457–487. https://doi.org/10.1002/wat2.1090. 2049-1948.

Mateo-Sagasta, J., Drechsel, P., 2022. Guidelines to improve acceptance of water reuse. In: Mateo-Sagasta, J., Al-Hamdi, M., AbuZeid, K. (Eds.), Water Reuse in the Middle East and North Africa. IWMI, Colombo, Sri Lanka, pp. 142–155.

Meegoda, J.N., Li, B., Patel, K., Wang, L.B., 2018. A review of the processes, parameters, and optimization of anaerobic digestion. Int. J. Environ. Res. Public Health 15 (10), 2224. https://doi.org/10.3390/ijerph15102224. 1660-4601.

Memon, M.J., Memon, A.R., 2020. Wheat straw optimization via its efficient pretreatment for improved biogas production. Civil Eng. J. (Iran) 6 (6), 1056–1063. Salehan Institute of Higher Education, Pakistan, 24763055. Available from: https://www.civilejournal.org/index.php/cej/article/view/2144/pdf. 10.28991/cej-2020-03091528.

Meng, L., Alengebawy, A., Ai, P., Jin, K., Chen, M., Pan, Y., 2020. Techno-economic assessment of three modes of large-scale crop residue utilization projects in China. Energies 13 (14), 3729. https://doi.org/10.3390/en13143729. 1996-1073.

Mishra, R.K., 2023. Fresh water availability and it's global challenge. J. Mar. Sci. Res. 2 (1), 01–03. https://doi.org/10.58489/2836-5933/004. 28365933.

Mojiri, A., Aziz, H.A., Tajud, R.M., 2015. Sulfide, phenols and chromium (VI) removal from landfill leachate and domestic wastewater by ZELIAC, zeolite and activated carbon augmented sequencing batch reactor (SBR). Res. J. Environ. Toxicol. 9 (4), 179–187. https://doi.org/10.3923/rjet.2015.179.187. 18193420.

Moradgholi, M., Massoudinejad, M., Aghayani, E., Yazdanbakhsh, A., 2019. Performance of electrical stimulated anaerobic baffled reactor for removal of typical pollutants from low-strength municipal wastewater at low temperatures. Environ. Health Eng. Manag. 6 (2), 121–128. Kerman University of Medical Sciences, Iran, 24234311. Available from: http://ehemj.com/browse.php?a_id=501%26sid=1%26slc_lang=en%26ftxt=1. 10.15171/EHEM.2019.14.

Mutungwazi, A., Awosusi, A., Matambo, T.S., 2023. Comparative functional microbiome profiling of various animal manures during their anaerobic digestion in biogas production processes. Biomass Bioenergy 170, 106728. https://doi.org/10.1016/j.biombioe.2023.106728. 09619534.

Nasr, M., Moustafa, M., 2014. Performance evaluation of El-Agamy wastewater treatment plant-Egypt. In: Int. Conf. Role Eng. Towards A Better Environ, pp. 1–9.

Ni, R., Chenming, X., Shi, X., Yang, S., Li, L., Peng, X., Song, L., 2022. Acetoclastic methanogenesis pathway stability despite the high microbial taxonomic variability in the transition from acidogenesis to methanogenesis during food waste anaerobic digestion. J. Clean. Prod. 372, 133758. https://doi.org/10.1016/j.jclepro.2022.133758. 09596526.

Nsair, A., Cinar, S.O., Alassali, A., Qdais, H.A., Kuchta, K., 2020. Operational parameters of biogas plants: a review and evaluation study. Energies 13 (15), 3761. https://doi.org/10.3390/en13153761. 1996-1073.

Osman, R.M., Hodaifa, G., 2023. An overview of anaerobic membrane bioreactors: current developments, fouling problems, and future prospects. J. Environ. Chem. Eng. 11 (6), 111482. https://doi.org/10.1016/j.jece.2023.111482. 22133437.

Öztep, G., Güngören-Madenoğlu, T., Özdemir, G., Işık, E., Serez, H., Kabay, N., Yüksel, M., 2023. Optimization and microbial community analysis for anaerobic digestion of water hyacinth (Eichhornia crassipes) with waste sludge at different solid contents and temperatures. Environ. Technol. Innov. 32, 103395. https://doi.org/10.1016/j.eti.2023.103395. 23521864.

Pang, C., Luo, X., Rong, B., Nie, X., Jin, Z., Xia, X., 2023. Carbon emission accounting and the carbon neutralization model for a typical wastewater treatment plant in China. Int. J. Environ. Res. Public Health 20 (1), 140. https://doi.org/10.3390/ijerph20010140. 1660-4601.

Parvez, K., Ahammed, M.M., 2024. Effect of composition on anaerobic digestion of organic fraction of municipal solid wastes: a review. Bioresour. Technol. Rep. 25, 101777. https://doi.org/10.1016/j.biteb.2024.101777. 2589014X.

Patel, A., Arkatkar, A., Singh, S., Rabbani, A., Solorza Medina, J.D., Ong, E.S., Habashy, M.M., Jadhav, D.A., Rene, E.R., Mungray, A.A., Mungray, A.K., 2021a. Physico-chemical and biological treatment strategies for converting municipal wastewater and its residue to resources. Chemosphere 282. Elsevier Ltd, India, 18791298. Available from: www.elsevier.com/locate/chemosphere. https://doi.org/10.1016/j.chemosphere.2021.130881.

Patel, A., Sarkar, O., Rova, U., Christakopoulos, P., Matsakas, L., 2021b. Valorization of volatile fatty acids derived from low-cost organic waste for lipogenesis in oleaginous microorganisms—a review. Bioresour. Technol. 321, 124457. https://doi.org/10.1016/j.biortech.2020.124457. 09608524.

Pfafflin, J., MacInnis, C., 2010. Municipal Wastewater. Informa UK Limited, pp. 727–736, https://doi.org/10.1201/NOE0849398438.ch49.

Phan, K.H., Nguyen, A.T., Le, S.L., Ngo, T.M., Cao, T.N.D., Nguyen, T.T., Le, L.T., Mai, T.P., Bui, X.T., 2023. Thermophilic anaerobic digestion and emerging methods for organic waste treatment: a review. Bioresour. Technol. Rep. 22. Elsevier Ltd, Vietnam, 2589014X. Available from: https://www.journals.

elsevier.com/bioresource-technology-reports. https://doi.org/10.1016/j.biteb.2023.101402.

Picardo, A., Soltero, V.M., Peralta, M.E., Chacartegui, R., 2019. District heating based on biogas from wastewater treatment plant. Energy 180, 649–664. Elsevier Ltd, Spain, 03605442. Available from: www.elsevier.com/inca/publications/store/4/8/3/. https://doi.org/10.1016/j.energy.2019.05.123.

Popoola, K.O., 2022. Water accessibility challenges, and implications on children in rural border communities of Oyo-Kwara State, Nigeria. Econ. Reg. Stud./Studia Ekonomiczne i Regionalne 15 (4), 581–600. https://doi.org/10.2478/ers-2022-0039. 2451-182X.

Pourbavarsad, M.S., Jalalieh, B.J., Harkins, C., Sevanthi, R., Jackson, W.A., 2021. Nitrogen oxidation and carbon removal from high strength nitrogen habitation wastewater with nitrification in membrane aerated biological reactors. J. Environ. Chem. Eng. 9 (5). Elsevier Ltd, United States, 22133437. Available from: http://www.journals.elsevier.com/journal-of-environmental-chemical-engineering/. https://doi.org/10.1016/j.jece.2021.106271.

Praneeth, K., Bhargava, S.K., Tardio, J., Sridhar, S., 2014. Water recovery from domestic kitchen effluent through membrane separation processes. In: Process. Excell. Powering Our Future, pp. 1–19, https://doi.org/10.3316/informit.712944562887114.

Prasanna Kumar, D.J., Mishra, R.K., Chinnam, S., Binnal, P., Dwivedi, N., 2024. A comprehensive study on anaerobic digestion of organic solid waste: a review on configurations, operating parameters, techno-economic analysis and current trends. Biotechnol. Notes 5, 33–49. KeAi Communications Co., India, 26659069. Available from: https://www.sciencedirect.com/science/journal/26659069. https://doi.org/10.1016/j.biotno.2024.02.001.

Preisner, M., 2020. Surface water pollution by untreated municipal wastewater discharge due to a sewer failure. Environ. Process. 7 (3), 767–780. Springer, Poland, 21987505. Available from: www.springer.com/earth+sciences+and+geography/environmental+science+%26+engineering/journal/40710. https://doi.org/10.1007/s40710-020-00452-5.

Rajendran, K., Mahapatra, D., Venkatraman, A.V., Muthuswamy, S., Pugazhendhi, A., 2020. Advancing anaerobic digestion through two-stage processes: current developments and future trends. Renew. Sust. Energ. Rev. 123, 109746. https://doi.org/10.1016/j.rser.2020.109746. 13640321.

Ran, Y., Elsayed, M., Eraky, M., Dianlong, W., Ai, P., 2024. Sequential production of biomethane and bioethanol through the whole biorefining of rice straw: analysis of structural properties and mass balance. Biomass Convers. Biorefin. 14 (2), 2693–2705. Springer Science and Business Media Deutschland GmbH, China, 21906823. Available from: https://www.springer.com/journal/13399. https://doi.org/10.1007/s13399-022-02548-4.

Rawoof, S.A.A., Kumar, P.S., Vo, D.V.N., Subramanian, S., 2021. Sequential production of hydrogen and methane by anaerobic digestion of organic wastes: a review. Environ. Chem. Lett. 19 (2), 1043–1063. Springer Science and Business Media Deutschland GmbH, India, 16103661. Available from: https://www.springer.com/journal/10311. https://doi.org/10.1007/s10311-020-01122-6.

Rocamora, I., Wagland, S.T., Hassard, F., Villa, R., Peces, M., Simpson, E.W., Fernández, O., Bajón-Fernández, Y., 2023. Inhibitory mechanisms on dry anaerobic digestion: ammonia, hydrogen and propionic acid relationship. Waste Manag. 161, 29–42. Elsevier Ltd, United Kingdom, 18792456. Available from: www.elsevier.com/locate/wasman. https://doi.org/10.1016/j.wasman.2023.02.009.

Sadia, M., Mahmood, A., Ibrahim, M., Irshad, M.K., Quddusi, A.H.A., Bokhari, A., Mubashir, M., Chuah, L.F., Show, P.L., 2022. Microplastics pollution from wastewater treatment plants: a critical review on challenges, detection, sustainable removal techniques and circular economy. Environ. Technol. Innov. 28. Elsevier B.V., Pakistan, 23521864. Available from: http://www.journals.elsevier.com/environmental-technology-and-innovation/. https://doi.org/10.1016/j.eti.2022.102946.

Saiyad, M., Shah, N., Joshipura, M., Dwivedi, A., Pillai, S., 2024. Modified biopolymers in wastewater treatment: a review. Mater. Today Proc. https://doi.org/10.1016/j.matpr.2024.01.031. 22147853.

Sánchez, A.S., 2020. Technical and economic feasibility of phosphorus recovery from wastewater in São Paulo's Metropolitan Region. J. Water Process Eng. 38, 101537. https://doi.org/10.1016/j.jwpe.2020.101537. 22147144.

Sha, H., Wang, Q., Dong, Z., Cao, S., Zhao, B., Wang, G., Duan, J., 2024. NaOH-urea pretreatment enhanced H_2 and CH_4 yields via optimizing mixed alkali ratio, pretreatment time, and organic loading rate during anaerobic digestion of corn stover. Energy 288, 129595. https://doi.org/10.1016/j.energy.2023.129595. 03605442.

Shen, Y., Linville, J.L., Urgun-Demirtas, M., Mintz, M.M., Snyder, S.W., 2015. An overview of biogas production and utilization at full-scale wastewater treatment plants (WWTPs) in the United States: challenges and opportunities towards energy-neutral WWTPs. Renew. Sust. Energ. Rev. 50, 346–362. Elsevier Ltd, United States, 18790690. Available from: https://www.journals.elsevier.com/renewable-and-sustainable-energy-reviews. https://doi.org/10.1016/j.rser.2015.04.129.

Shi, X.S., Dong, J.J., Yu, J.H., Yin, H., Hu, S.M., Huang, S.X., Yuan, X.Z., 2017. Effect of hydraulic retention time on anaerobic digestion of wheat straw in the semicontinuous continuous stirred-tank reactors. Biomed. Res. Int. 2017. Hindawi Limited, China, 23146141. Available from: http://www.hindawi.com/journals/biomed/. https://doi.org/10.1155/2017/2457805.

Sikosana, M.L., Sikhwivhilu, K., Moutloali, R., Madyira, D.M., 2019. Municipal wastewater treatment technologies: a review. Procedia Manuf. 35, 1018–1024. 23519789 Elsevier B.V. South Africa http://www.journals.elsevier.com/procedia-manufacturing. https://doi.org/10.1016/j.promfg.2019.06.051.

Silva, C., Rosa, M.J., 2020. Performance assessment of 23 wastewater treatment plants—a case study. Urban Water J. 17 (1), 78–85. Taylor and Francis Ltd., Portugal, 17449006. Available from: http://www.tandf.co.uk/journals/titles/1573062X.asp. https://doi.org/10.1080/1573062X.2020.1734634.

Sime, G., Jin, Z., 2020. Technical and socioeconomic constraints to the domestication and functionality of biogas technology in rural areas of southern Ethiopia. Cogent Eng. 7 (1), 1765686. https://doi.org/10.1080/23311916.2020.1765686. 2331-1916.

Singh, M., Srivastava, R.K., 2014. Treatment efficacy of short rotation woody species and their economic appraisal in vegetation filter system. Ann. Plant Soil Res. 16, 289–293.

Singh, K.K., Vaishya, R.C., Gupta, A., 2017. Evaluation of consortia performance under continuous process treating municipal wastewater with low concentration of heavy metals, antibiotic (gentamicin) and diesel oil. Int. J. Eng. Technol. 9 (5), 3448–3457. https://doi.org/10.21817/ijet/2017/v9i5/170905023. 23198613.

Singh, A., Sawant, M., Kamble, S.J., Herlekar, M., Starkl, M., Aymerich, E., Kazmi, A., 2019. Performance evaluation of a decentralized wastewater treatment system in India. Environ. Sci. Pollut. Res. 26 (21), 21172–21188. Springer Verlag, India, 16147499. Available from: http://www.springerlink.com/content/0944-1344. https://doi.org/10.1007/s11356-019-05444-z.

Srisamran, J., Atwill, E.R., Chuanchuen, R., Jeamsripong, S., 2022. Detection and analysis of indicator and pathogenic bacteria in conventional and organic fruits and vegetables sold in retail markets. Food Qual. Saf. 6. Oxford University Press, Thailand, 23991402. Available from: www.academic.oup.com/fqs. https://doi.org/10.1093/fqsafe/fyac013.

Stocks, N., 2016. Proc. Aquaenviro's 19th Eur. Biosolids Org. Resour. Conf. Exhib. Manchester 1–23.

Sun, H., Yang, X., Zhang, L., Liu, Y., 2023. The role of protein contents in promoting wastewater phosphorus and bioenergy recovery during anaerobic digestion. Biomass Bioenergy 169, 106694. https://doi.org/10.1016/j.biombioe.2022.106694. 09619534.

Tamilselvan, R., Selwynraj, A.I., 2024. Enhancing biogas generation from lignocellulosic biomass through biological pretreatment: exploring the role of ruminant microbes and anaerobic fungi. Anaerobe 85, 102815. https://doi.org/10.1016/j.anaerobe.2023.102815. 10759964.

Tavera-Ruiz, C., Martí-Herrero, J., Mendieta, O., Jaimes-Estévez, J., Gauthier-Maradei, P., Azimov, U., Escalante, H., Castro, L., 2023. Current understanding and perspectives on anaerobic digestion in developing countries: Colombia case study. Renew. Sust. Energ. Rev. 173, 113097. https://doi.org/10.1016/j.rser.2022.113097. 13640321.

Tawfik, A., Eraky, M., Osman, A.I., Ai, P., Zhou, Z., Meng, F., Rooney, D.W., 2023. Bioenergy production from chicken manure: a review. Environ. Chem. Lett. 21 (5), 2707–2727. Springer Science and Business Media Deutschland GmbH, Egypt, 16103661. Available from: https://www.springer.com/journal/10311. https://doi.org/10.1007/s10311-023-01618-x.

United Nations, 2024. The United Nations World Water Development Report 2024: Water for Prosperity and Peace. United Nations Educational, Scientific and Cultural Organization.

UN-Water, 2020. UN-Water Analytical Brief on Unconventional Water Resources.

Valdés, J.P., Becerra, D., Rozo, D., Cediel, A., Torres, F., Asuaje, M., Ratkovich, N., 2020. Comparative analysis of an electrical submersible pump's performance handling viscous Newtonian and non-Newtonian fluids through experimental and CFD approaches. J. Pet. Sci. Eng. 187, 106749. https://doi.org/10.1016/j.petrol.2019.106749. 09204105.

Van Hung, N., Maguyon-Detras, M.C., Migo, M.V., Quilloy, R., Balingbing, C., Chivenge, P., Gummert, M., 2019. Rice Straw Overview: Availability, Properties, and Management Practices Sustainable Rice Straw Management. Springer International Publishing, Philippines, pp. 1–13. Available from: https://link.springer.com/book/10.1007/978-3-030-32373-8. https://doi.org/10.1007/978-3-030-32373-8_1.

Venkata Mohan, S., Chiranjeevi, P., Chandrasekhar, K., Babu, P.S., Sarkar, O., 2019. Acidogenic Biohydrogen Production From Wastewater Biomass, Biofuels, Biochemicals: Biohydrogen, second ed. Elsevier, India, pp. 279–320. Available from: https://www.sciencedirect.com/book/9780444642035. https://doi.org/10.1016/B978-0-444-64203-5.00011-3.

Xiong, W., Wang, L., Zhou, N., Fan, A., Wang, S., Haijia, S., 2020. High-strength anaerobic digestion wastewater treatment by aerobic granular sludge in a step-by-step strategy. J. Environ. Manag. 262, 110245. https://doi.org/10.1016/j.jenvman.2020.110245. 03014797.

Yan, P., Gai, M., Wang, Y., Gao, X., 2021. Review of soft sensors in anaerobic digestion process. Processes 9 (8), 1434. https://doi.org/10.3390/pr9081434. 2227-9717.

Yang, J., Monnot, M., Ercolei, L., Moulin, P., 2020. Membrane-based processes used in municipal wastewater treatment for water reuse: state-of-the-art and performance analysis. Membranes 10 (6), 131. https://doi.org/10.3390/membranes10060131, 2077-0375.

Ye, M., Li, Y.-Y., 2023. Methanogenic treatment of dairy wastewater: a review of current obstacles and new technological perspectives. Sci. Total Environ. 866, 161447. https://doi.org/10.1016/j.scitotenv.2023.161447. 00489697.

Zhen, G., Pan, Y., Han, Y., Gao, Y., Gadow, S.I., Zhu, X., Yang, L., Lu, X., 2023. Enhanced co-digestion of sewage sludge and food waste using novel electrochemical anaerobic membrane bioreactor (EC-AnMBR). Bioresour. Technol. 377, 128939. https://doi.org/10.1016/j.biortech.2023.128939. 09608524.

Zhou, P., Li, Z., Snowling, S., Baetz, B.W., Na, D., Boyd, G., 2019. A random forest model for inflow prediction at wastewater treatment plants. Stoch. Environ. Res. Risk Assess. 33 (10), 1781–1792. Springer New York LLC, Canada, 14363259. Available from: http://link.springer-ny.com/link/service/journals/00477/index.htm. https://doi.org/10.1007/s00477-019-01732-9.

Zhu, T.J., Lin, C.W., Liu, S.H., 2023. Sensitivity and reusability of a simple microbial fuel cell-based sensor for detecting bisphenol A in wastewater. Chemosphere 320. Elsevier Ltd, Taiwan, 18791298. Available from: www.elsevier.com/locate/chemosphere. https://doi.org/10.1016/j.chemosphere.2023.138082.

11

Emerging technologies for sludge management

Rodrigo Custodio Urban[a] and Liane Yuri Kondo Nakada[b]

[a]Department of Environmental Engineering, Institute of Science and Technology, São Paulo State University (UNESP), São José dos Campos, São Paulo, Brazil.

[b]Department of Environmental Engineering, Institute of Geosciences and Exact Sciences, São Paulo State University (UNESP), Rio Claro, São Paulo, Brazil

Abbreviations

APC	advanced process control
DAF	dissolved air flotation
DOF	dissolved ozone flotation
EAOP	electrochemical advanced oxidation process
ECHO	enforcement and compliance history online
EDW	electro-dewatering
EPS	extracellular polymeric substances
GIS	geographic information system
G-PEDW	pressurized electro-dewatering with constant voltage gradient mode
ICP	internal concentration polarization
IER	ion exchange resin
MAFB	modified areca husk fiber biochar
MNP	most probable number
PA	polyamide
PAC	polyaluminium chloride
PE	polyethylene
PET	polyethylene terephthalate
PFAS	per- and polyfluoroalkyl substances
PFU	plaque-forming units
PHMB	polyhexamethylene biguanide
PP	polypropylene
PPCP	pharmaceuticals and personal care products
PSRP	processes to significantly reduce pathogens
SDRB	sludge drying reed bed
SSA	sludge incinerator ash
STP	sewage treatment plant
TS	total solids
UASB	upflow anaerobic sludge blanket
USEPA	United States Environmental Protection Agency
VS	volatile solids
WAS	waste activated sludge
WWTP	wastewater treatment plant

Municipal Wastewater Treatment. https://doi.org/10.1016/B978-0-443-24826-9.00011-5

11.1 Introduction

Municipal sewage treatment involves a range of processes and unit operations that can generate by-products. It is important to note that there is no specific standard for this system as it can be a combination of different processes and operations which can have an impact on the type, quantity, and characteristics of the by-products.

In general terms, domestic sewage treatment can be composed of preliminary, primary, secondary, and tertiary treatments. The preliminary treatment aims to remove coarser materials and solids that may overload the treatment sequence. The primary treatment is predominantly physicochemical, with the goal of destabilizing and removing colloidal particles. The secondary treatment is essentially biological, aiming to remove and stabilize the organic matter found in sewage. The tertiary treatment aims to polish the treatment process, using physicochemical, biological, or alternative methods to optimize the removal of components that may have resisted previous stages.

Although effluent treatment methods are diverse, municipal sewage is predominantly organic due to its domestic origin. It consists of liquid waste from toilets, sinks, and other household uses. Therefore, the treatment of municipal sewage usually follows the sequence shown in Fig. 11.1. There are no strict requirements for the sequence of Wastewater Treatment Plants (WWTPs), as long as they comply with current regulations for discharging treated effluents into water bodies or for reuse purposes.

Fig. 11.1 shows the main sources of by-products in municipal sewage treatment. Gaseous and semi-solid residues are

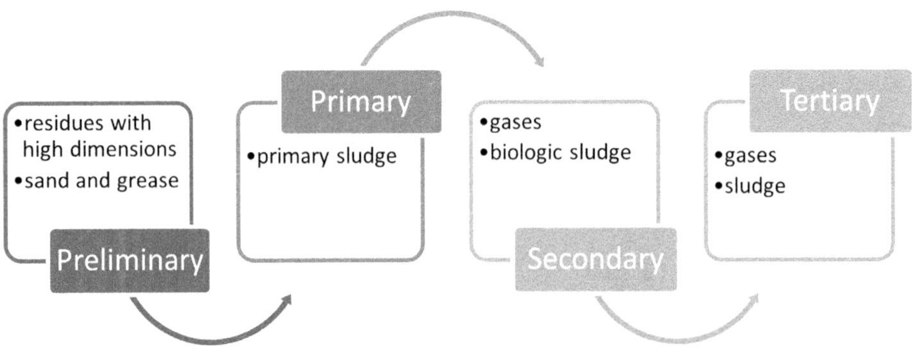

Fig. 11.1 Schematic flow of municipal wastewater treatment and generated byproducts.

predominant, with emphasis on the production of sludge from biological treatment. This sludge is typically an organic material that has not yet been fully stabilized.

The available biological treatment technologies for municipal sewage are diverse. They can be broadly categorized as aerobic and anaerobic, with significant implications for the quality and quantity of by-products generated. To optimize the treatment process, Wastewater Treatment Plants (WWTPs) often use combined aerobic-anaerobic treatments in various configurations.

In general, aerobic treatment technologies produce more sludge than anaerobic treatment, which also generates methane gas. However, the specific estimates depend on the technology used. This chapter does not aim to discuss the characteristics of each treatment technology in detail, but such information can be found in other chapters of this book or in Andreoli et al. (2007b).

The gases produced in anaerobic treatment systems can be combusted to decrease their potential as greenhouse gases. Alternatively, they can be collected, purified, and utilized as a source of thermal and/or electrical energy. This is an important approach for optimizing municipal sewage treatment processes, particularly considering the principles of Circular Economy and sustainable development. However, the focus of this chapter will be on the semi-solid by-product of treatment: sludge.

The sludge from WWTPs is a material of highly diverse composition, depending on the characteristics of municipal sewage and the type of treatment employed. Generally, it consists of a large proportion of organic matter, nutrients, pathogenic organisms, minerals, and possibly metals and toxic organic compounds (von Sperling and Gonçalves, 2007; Gonçalves et al., 2007). Additionally, the sludge may contain emerging contaminants such as pharmaceuticals, hormones, endocrine disruptors, microplastics, and other compounds resulting from illegal disposal into the sewage network.

Managing this waste is a complex, costly, and challenging task. Coupled with population growth, increasing water consumption and rising treatment rates, sludge management is becoming one of the most significant challenges facing the wastewater sector. Overall, although sludge represents only 1–2% of the total volume of treated wastewater, its management can account for 20–60% of the operating costs of a wastewater treatment plant (von Sperling and Andreoli, 2007; Andreoli et al., 2007b).

Although some treatment technologies have the ability to produce stabilized sludge, such as the Upflow Anaerobic Sludge Blanket (UASB), here we provide a more comprehensive overview of the steps required to manage this waste (Fig. 11.2).

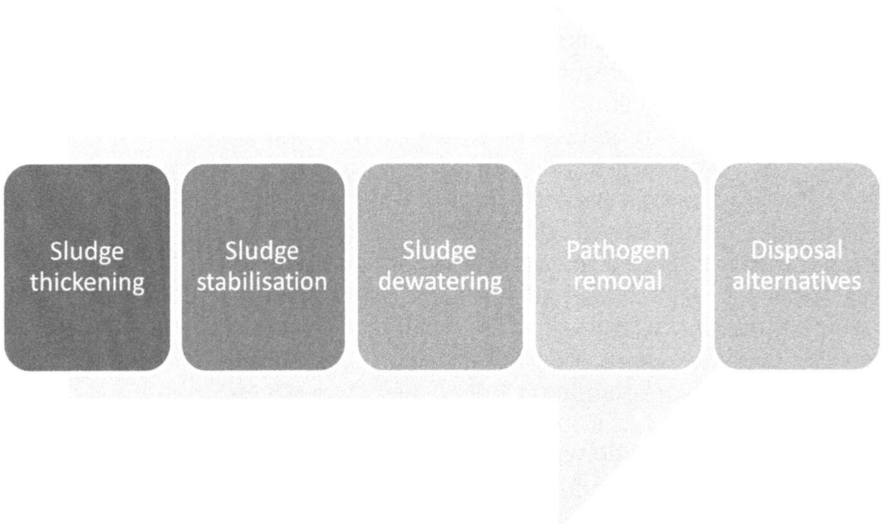

Fig. 11.2 General flow of WWTP sludge management.

The flow chart in Fig. 11.2 shows that the management of sewage sludge is a complex and challenging task. The responsibility of sanitation companies extends beyond just treating this waste to ensuring its proper disposal. In the latter case, it becomes not only a technical problem but also a logistical one, as it is necessary to determine how this treated material will be stored and transported, for instance. Additionally, the disposal of this material into water bodies, even when treated, is inconvenient due to its potential for pollution. Sending it to landfills, while appropriate, does not adhere to the principles of Circular Economy and resource utilization. Therefore, the study of suitable alternatives for the beneficial use of sludge should be part of its management.

The following sections will address each of the stages of WWTP sludge management, with a focus on emerging technologies that are being used or studied to optimize each process and operation.

11.2 Sludge thickening

The purpose of sludge thickening in wastewater treatment plants is to reduce the volume of water contained in the sludge. This step is essential for reducing the volume of material treated in the digestion stage, thus optimizing the subsequent biological process (Gonçalves et al., 2007).

Thickening techniques are effective in removing free water from sludge, while interstitial, capillary and bound water are not removed at this stage. Traditional thickening processes are capable of increasing the solids content of WWTP sludge from 2% to around 5%, which represents a volume reduction of 60% or more (Gonçalves et al., 2007).

In general, sludge thickening can be carried out in two ways: (i) gravity thickening and (ii) dissolved air flotation thickening (Gonçalves et al., 2007). Research in this area aims to evaluate ways to make this stage more effective, either by improving existing techniques, developing new techniques, or combining thickening with other stages of wastewater treatment, such as tertiary treatment, to optimize space and construction costs of structures.

Some emerging technologies for sludge thickening, grouped according to classical techniques are presented below.

11.2.1 Gravity thickening

The basic premise of gravity thickening is the process of sedimentation of solid particles. For this purpose, the tanks are sized to provide the minimum retention time required for particle sedimentation without allowing the sludge to begin digestion to prevent the propagation of odors (Gonçalves et al., 2007). The supernatant can be returned to wastewater treatment, while the sedimented material is directed to the sludge digestion stage.

Primary and biological (secondary) sludges are typically treated together in WWTPs configured in this way. However, separate thickening of the sludge can have advantages for the sludge line. This allows the digestion stage to be carried out only on sludge with biological characteristics. Modifications can be made to the treatment plant and gravity thickening is recommended for primary sludge, while flotation or another method is suggested for biological sludge (Tomei et al., 2016).

Although gravity thickening has a simple premise, innovations can be made to optimize the process. Romero-Güiza et al. (2022) applied various Advanced Process Control (APC) strategies to optimize processes in a wastewater treatment plant to increase its energy efficiency and sustainability. The WWTP in question has primary sludge thickening tanks where a controller was applied to the thickener purge valves to maintain a constant sludge level. Application of the APC tool resulted in a 9.5% increase in sludge thickening and a 10% increase in methane production in the sludge digesters. The authors suggest that the use of

APC tools is a simple and cost-effective solution to improve the sludge treatment process.

11.2.2 Dissolved air flotation thickening

Dissolved Air Flotation (DAF) thickening technology involves pressurizing air into a solution, causing it to dissolve. Upon depressurisation, the air is released, forming small bubbles that carry sludge particles to the surface. DAF requires larger tanks than gravity thickening and air dissolution equipment. However, it is highly effective for low density, poorly settling sludges such as those produced in activated sludge processes. Process optimisation can be achieved by the addition of auxiliary polymers to improve sludge particle capture (Gonçalves et al., 2007).

The application of Dissolved Ozone Flotation (DOF) for sludge thickening was studied by Xiao et al. (2022). The authors used DOF in conjunction with polyacrylamide as a flocculant and obtained promising results, such as an increase in total solids from 0.33% to 8.03%, and a reduction in sludge volume of approximately 38.61% compared to dissolved air flotation. However, the study was conducted at the bench scale and needs to have its feasibility evaluated for scale-up.

11.2.3 Other sludge thickening technologies

Sludge thickening can be conducted in conjunction with other stages of wastewater treatment. Takabe and Ida (2023) evaluated the process of sludge thickening together with tertiary treatment for phosphorus removal. The process is based on electrolysis with inert electrodes, where hydroxide ions generated by the cathode cause the precipitation of calcium phosphate, and oxygen and hydrogen generated by the electrodes perform electroflotation of solids, contributing to sludge thickening. The calcium phosphate generated could be an alternative for phosphorus extraction from natural sources. However, the time required for these processes to occur is significantly different, which may render their joint use impractical. In light of this, the cited authors tested covering the anode with bivalve shells and obtained promising results at the bench scale (Fig. 11.3).

An alternative to the simultaneous development of thickening and digestion of waste activated sludge (WAS) has been

Fig. 11.3 Simultaneous phosphorus precipitation and WWTP sludge thickening by electrolysis (Takabe and Ida, 2023). Reprinted from Takabe, Y., Ida, K., 2023. Simultaneous phosphorus precipitation and sludge thickening by electrolysis with an anode covered by bivalve shells. Water Res. 247, 120789, Copyright (2023), with permission from Elsevier.

proposed by Wang et al. (2008a,b). The method is based on the use of flat sheet membranes in a reactor receiving excess sludge from activated sludge treatment. The main advantages highlighted for the use of this process are higher thickening efficiency compared to gravity-based processes and lower energy consumption compared to dissolved air flotation (Wang et al., 2008a,b). To address one of the limitations of the method, membrane fouling, Kim and Chung (2013) proposed the preliminary application of coagulants in a pilot-scale study (Fig. 11.4), achieving good results for operating conditions of a hydraulic retention time (HRT) of 6.7 days and a flux of 10.7 L/m^2 h. In addition, with the aim of solving problems related to internal concentration polarization (ICP) and fouling phenomena, Zhao et al. (2022) proposed the use of a novel double-skinned Forward Osmosis membrane, achieving satisfactory bench-scale results for sludge thickening.

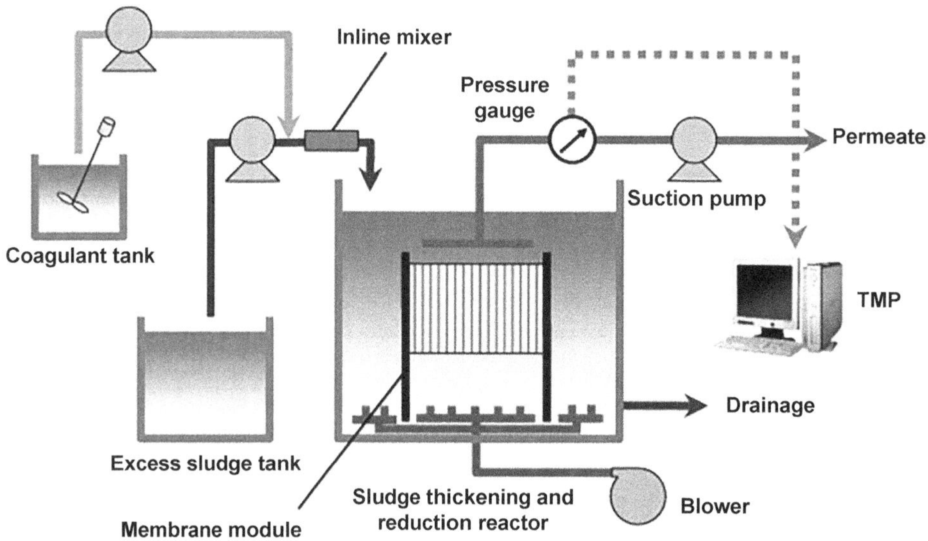

Fig. 11.4 Diagram of sludge thickening by coagulation and membrane module proposed by Kim and Chung (2013). Reprinted from Kim, H., Chung, T., 2013. Performance of the sludge thickening and reduction at various factors in a pilot-scale MBR. Sep. Purif. Technol. 104, 297–306, Copyright (2013), with permission from Elsevier.

11.3 Sludge stabilization

The sludge stabilization stage aims to stabilize the biodegradable fraction of the remaining organic matter, reduce the risk of material putrefaction, and eliminate part of the pathogen concentration. However, it is important to note that the sludge stabilization stage depends on the type of treatment employed in municipal wastewater. Some employed technologies may already provide stabilized sludge, rendering this stage unnecessary (Luduvice, 2007). In such cases, dewatering can be proceeded directly, or combined thickening and dewatering techniques can be used.

In general, anaerobic technologies such as UASB (Upflow Anaerobic Sludge Blanket), or mixed systems like facultative ponds generate stabilized sludges. Conversely, aerobic technologies like activated sludge systems generate unstabilised sludge (Luduvice, 2007). Among the alternatives for sludge stabilization directly in the sewage line, it is possible to opt for combined aerobic and anaerobic systems (e.g., a UASB followed by activated sludge), optimizing the treatment processes.

Another relevant point to mention is the importance of the stabilization process for the final disposal of treated sludge (Tomei et al., 2016). Depending on the intended use of the sludge—agriculture, for example (Major et al., 2022)—or the chosen final destination—such as incineration—certain characteristics need to be controlled to optimize processes and to avoid problems during storage and transportation.

The sludge stabilization processes can be divided into three categories (Luduvice, 2007): (i) biological stabilization—which is the most commonly used option, with anaerobic digestion being more widely used than aerobic digestion due to its superior performance in stabilization; (ii) chemical stabilization—involving the oxidation of organic matter, typically occurring in conjunction with other processes, with alkaline oxidation being the most common; and (iii) thermal stabilization—which involves stabilization by heating the volatile fraction of the sludge, often in conjunction with thermal drying, which is a dewatering process.

New research into sludge stabilization processes is focusing on multifunctional approaches that have the combined ability to stabilize and condition, dewater or sanitize the sludge. In these cases, stabilization is often a secondary process. In addition, sludge stabilization solutions can often be integrated directly into the wastewater treatment line using combined systems. On the other hand, the influence of stabilization processes on the quality of the sludge produced on a real scale is the subject of ongoing research.

Tomei et al. (2016) evaluated the impact of two upgrades involving the use of sludge from a wastewater treatment plant in Italy on the quality of treated sludge sent to two final destinations: agricultural use or incineration. The first upgrade involved the separation of primary sludge from secondary sludge, while the second upgrade involved the addition of a post-anaerobic digestion stage after aerobic digestion for sludge stabilization.

With regard to the second upgrade evaluated by Tomei et al. (2016), the results obtained demonstrated the feasibility of the solution and enabled its implementation at a real scale. Technically, a significant increase in the volatile solids (VS) removal efficiency was achieved and nitrogen removal from the aerobic reactor supernatant was introduced, reducing the nitrogen load recirculated to the WWTP. The economic analysis showed that the investment in the anaerobic reactor was economically sustainable. The life cycle analysis carried out by the authors showed that the option would increase nitrous oxide emissions to air from the WWTP; however, spreading the sludge for agricultural use would reduce methane, nitrous oxide and ammonia emissions, as well as reducing fertilizer consumption, demonstrating an environmental benefit.

Although the results obtained by Tomei et al. (2016) are specific to a local context, it can be inferred that the implementation of a post-aerobic stabilization stage may be an interesting alternative for denitrification and increased VS removal. Major et al. (2022) investigated the influence of the stabilization system on the microbial community composition in 10 WWTPs in Croatia, 4 of which used a combined anaerobic-aerobic system. The results showed that the stabilization methods used did not affect the microbial communities, with the quality of the raw wastewater being the determining factor.

The quality of stabilized sludge is another area of concern. Quality standards for agricultural application should be more stringent and should be accompanied by methods to remove pathogens. Mei et al. (2020) proposed a more detailed assessment of the potential of stabilized sludge as a soil conditioner, heavy metal fixer and nutrient store. In a study conducted in 16 Chinese WWTPs treating sludge by anaerobic digestion followed by composting, the authors concluded that the use of two parameters—the content of humic-like substances in the final products (greater than 150 mg/g VS) and the fluorescence complexity index (greater than 5.0)—allows a more accurate assessment of the land use potential of sludge than the limits of organic degradation rates.

On the other hand, the presence of emerging contaminants is another aspect that can affect the quality of stabilized sludge. The presence of microplastics in various environmental compartments has recently become a major concern. Some protocols for the assessment of microplastics in stabilized and dewatered sludge (Fig. 11.5) have been evaluated by Bretas Alvim et al.

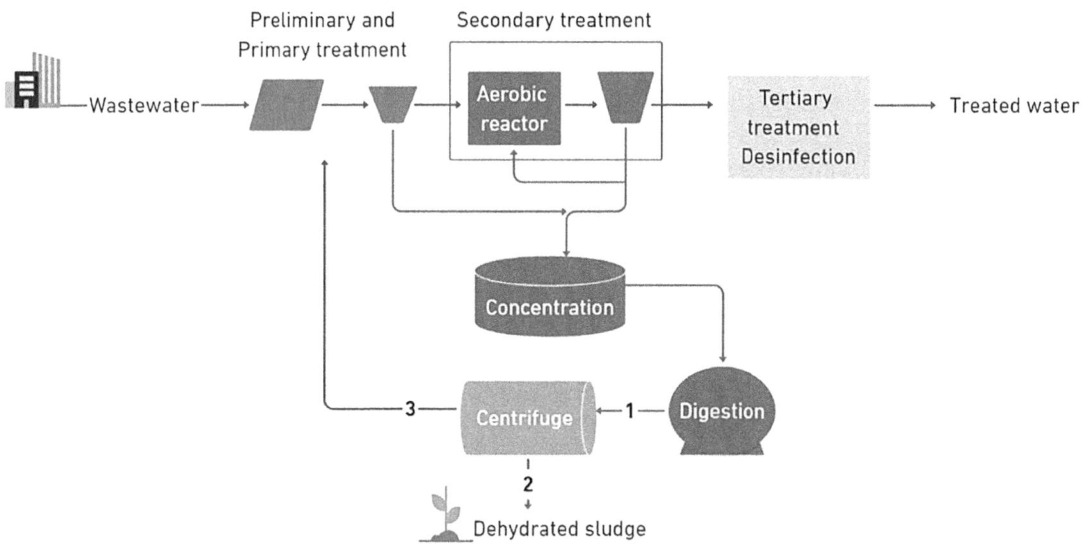

Fig. 11.5 Sampling points of the study by Bretas Alvim et al. (2022) for the evaluation of microplastics in digested sludge (1), dewatered sludge (2), and centrifuged effluent (3). Reprinted from Bretas Alvim, C., Navajas Valiente, S., Bes-Piá, M.A., Mendoza-Roca, J.A., 2022. Methodology for removing microplastics and other anthropogenic microparticles from sludge dewatering system. J. Environ. Manage. 314, 115010, Copyright (2022), with permission from Elsevier.

(2022), with peroxidation followed by density separation using NaI proving to be the most effective. The protocol can be used for sludge quality monitoring and may become part of new requirements for the removal of microplastics from stabilized and dewatered sludge.

11.4 Sludge dewatering

Sludge dewatering is often evaluated in conjunction with sludge thickening because of the similar objective of removing excess water. However, in cases where wastewater treatment does not produce stabilized sludge, dewatering is carried out after the stabilization stage to reduce the volume of sludge to be transported for final disposal or to increase its calorific value for an incineration stage. Since the sludge has already undergone moisture reduction processes, techniques with higher energy consumption can be used at this stage, allowing to some extent the removal of interstitial or capillary water from the sludge. After this stage of treatment, the sludge is referred to as "cake" and has a semi-solid appearance, with a solids content that can reach up to 30%. The volume reduction of the sludge after the moisture removal stages can reach 90–95% of its original volume (Gonçalves et al., 2007).

11.4.1 Sludge conditioning

Sludge dewatering is usually preceded by a conditioning stage. Sludge conditioning involves the application of flocculants to facilitate the destabilization of colloidal particles and their aggregation into flocs. The aim of this stage is to optimize the dewatering process, reduce energy consumption and increase efficiency. Inorganic chemical salts such as ferric chloride, lime or others, as well as organic polymers—neutral, cationic or anionic—can be used for this purpose (Gonçalves et al., 2007). Further information on the mechanisms and details of flocculation used for municipal sludge dewatering can be found in Hyrycz et al. (2022).

Research into sludge conditioning aims to optimize the use of these products or to test and develop options that are technically, environmentally or economically more effective.

Alternatives currently being investigated for sludge conditioning include new products, the use of waste from other processes, or new combinations, such as the combined use of bentonite and polyaluminium chloride (PAC) (Hatamifar and Gholikandi, 2024), red mud with Fenton reagent (Zhang et al., 2014), the combined use of cement kiln dust and fly ash (Elbaz et al., 2021), solidified

CO_2 application (Kazimierowicz and Dębowski, 2023), Ion Exchange Resins (IERs) application (Geng et al., 2022), $KMnO_4$[sbnd]Fe(II) utilization (Ai et al., 2019), Polyhexamethylene biguanide (PHMB) usage (Y. Wang et al., 2022), lignin-based flocculants application (Wang and Chen, 2023), modified areca husk fiber biochar (MAFB-$AlCl_3$) use (Chopade and Devatha, 2024), coal fly ash modified by sulfuric acid utilization (Chen et al., 2010), semi-coke powder from lignite pyrolysis application (Zhang et al., 2017), oxidizing agents—Fenton's reagent and Fe^{2+}/persulfate (Yu et al., 2017), or iron-based advanced oxidation processes (Liang and Zhou, 2022; Wang et al., 2023).

Furthermore, the use of natural products for sludge conditioning has also been the subject of studies, such as *Opuntia ficus indica* juice application (Betatache and Aouabed, 2015), potato starch, rice stover utilization, or the application of a polymeric chitosan quaternary ammonium salt (Liu et al., 2024). However, none of these options are yet widely used on a large scale to replace traditional chemical coagulants or polyelectrolytes.

11.4.2 Sludge drying beds

Sludge drying beds are the oldest and simplest method of sludge dewatering (Guo et al., 2017). Their operation is based on evaporation and percolation processes. These drying beds can be constructed with masonry and filled with layers of draining material, supporting media (usually sand), and a drainage system (Gonçalves et al., 2007). The dewatering time is approximately 6 weeks, and depending on the climatic conditions, after this period, the sludge can reach up to about 40% solids content (Hyrycz et al., 2022). Drying beds can be built in open areas or covered to protect against rain (Gonçalves et al., 2007). Disadvantages of this method include the need for large areas and the possibility of odor generation (Rumky et al., 2023).

An alternative to traditional drying beds is the use of Sludge Drying Reed Beds (SDRB) (Fig. 11.6), which are constructed wetlands structures using plants such as *Phragmites australis* (common reed) (Uggetti et al., 2009). This is not a new technology (De Maeseneer, 1997), and has been used successfully for over 40 years in countries such as Denmark, Poland, and Belgium (Uggetti et al., 2009). However, for its use in different climatic zones, it is necessary to use plants adapted to the specific environment. A list of plants used in different climatic conditions can be found in Gholipour et al. (2022). Another advantage of using SDRB is its ability to stabilize sludge (Peruzzi et al., 2017), which makes it an alternative for this process.

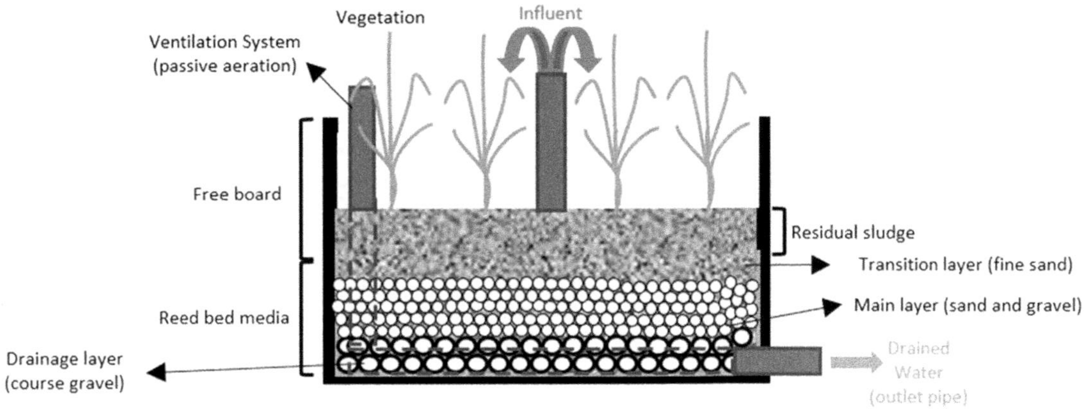

Fig. 11.6 Typical layout of a sludge treatment reed bed (Gholipour et al., 2022). Reprinted from Gholipour, A., Fragoso, R., Duarte, E., Galvão, A., 2022. Sludge treatment reed bed under different climates: a review using meta-analysis. Sci. Total Environ. 843, 156953, Copyright (2022), with permission from Elsevier.

11.4.3 Mechanical sludge dewatering

The use of mechanical systems is emerging as an alternative that requires less space, is faster and offers a higher level of operational control than drying beds. However, it typically achieves up to 35% dry solids (Rumky et al., 2023) and is associated with higher energy consumption. The most common mechanical systems include centrifuges and filter presses (Gonçalves et al., 2007; Mamais et al., 2009).

Centrifugation is a process in which solid-liquid separation is induced by the application of centrifugal forces (Gonçalves et al., 2007). Centrifugal decanters (Fig. 11.7) offer several advantages, including: (i) the use of high forces ranging from 2000 to 4000 $\times g$; (ii) the ability to handle sludges with high solid contents; (iii) the potential for continuous operation (Wakeman, 2007). Moreover, centrifuges can be used for both sludge thickening and dewatering and can even be installed in series (Gonçalves et al., 2007). Some alternative equipment has been developed to optimize the unit operation of centrifugation, such as a centrifuge coupled with thermal drying (Wakeman, 2007).

Filter presses were originally developed for industrial use and later adapted for sludge dewatering purposes. They are known for their reliability and operate in batch mode. Their main advantages are: (i) cake with a higher solids content compared to other mechanical dewatering options; (ii) high quality effluent (filtrate); and (iii) low consumption of conditioning products (Gonçalves et al., 2007).

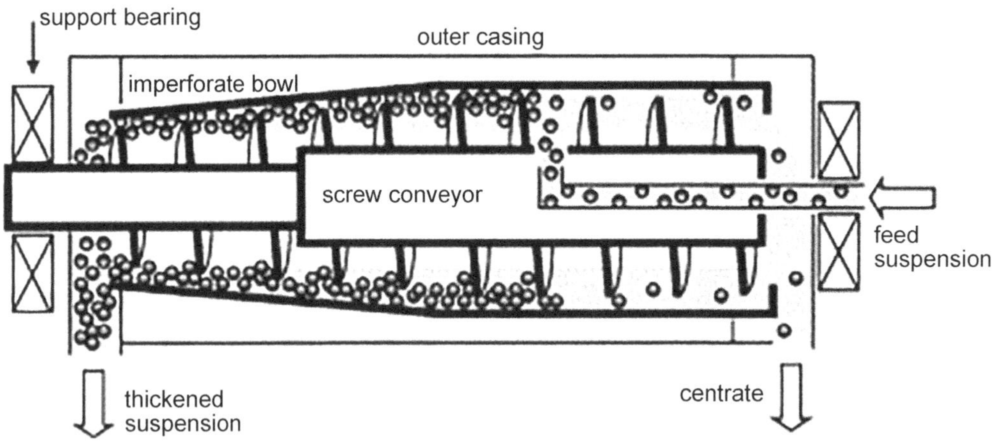

Fig. 11.7 Schematic illustration of a centrifuge decanter (Wakeman, 2007). Reprinted from Wakeman, R., 2007. Separation technologies for sludge dewatering. J. Hazard. Mater. 144, 614–619, Copyright (2007), with permission from Elsevier.

Plate and frame, recessed plate, and membrane plate options are commonly used for sludge dewatering with filter presses. Filter plates (Fig. 11.8) are the most commonly used equipment (Wakeman, 2007) and their operating cycle lasts between 3 and 5 h and consists of three phases: (i) charging; (ii) filtration under maximum pressure; and (iii) cake discharge (Gonçalves et al., 2007). Some modern improvements the development of filter press equipment are presented by Wakeman (2007) highlighting automation and control of the plate assembly opening, optimisation of agitation mechanisms for cake removal, and improvement of the cloth washing system.

Another option for pressure filtration is the belt filter press (Fig. 11.9). Its operation is based on a system of rollers and conveyors, where a tensioned cloth presses the sludge to reduce its moisture content (Wakeman, 2007).

Fig. 11.8 Picture of a side-bar filter press (Wakeman, 2007). Reprinted from Wakeman, R., 2007. Separation technologies for sludge dewatering. J. Hazard. Mater. 144, 614–619, Copyright (2007), with permission from Elsevier.

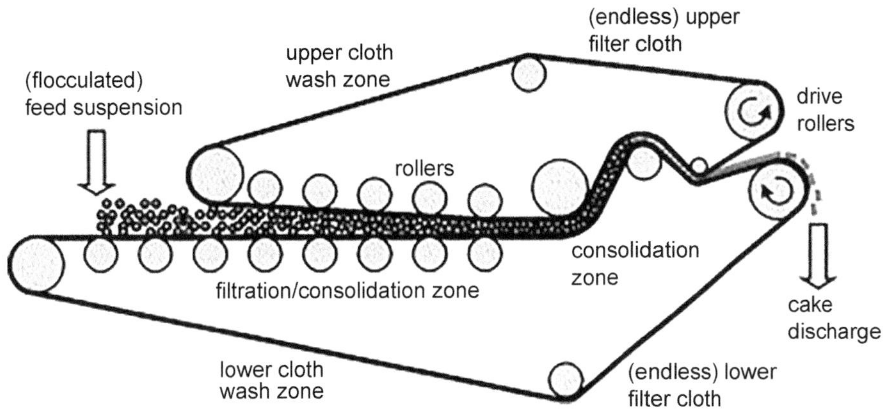

Fig. 11.9 Schematic illustration of a belt filter press (Wakeman, 2007). Reprinted from Wakeman, R., 2007. Separation technologies for sludge dewatering. J. Hazard. Mater. 144, 614–619, Copyright (2007), with permission from Elsevier.

The selection of one or another mechanical dewatering equipment depends on the characteristics of the sludge and the options available to the designers. Although there is a perception that filter presses achieve higher solids content than other equipment, results from a study conducted by Mamais et al. (2009) indicated that centrifuges produced a cake with higher solids content, using lower amounts of conditioners than belt filter presses. It is therefore important to evaluate the choice of equipment to be used on a case-by-case basis.

11.4.4 Other options for sludge dewatering

Several alternatives to drying beds and mechanical dewatering equipment are currently being investigated to improve process performance or reduce energy consumption. These alternatives may include new methods or combinations with existing dewatering options.

Feng et al. (2022) proposed a novel mechanism for sludge dewatering called piezo dewatering. In this process, a piezo material, tetragonal $BaTiO_3$ (*t*-BT), was incorporated into the sludge to act as a piezo conditioning agent. The process was carried out on a bench scale under the same conditions as a filter press operation. Although further studies on a larger scale are needed, the mechanism showed promising results by optimizing the dewatering process without increasing energy consumption. There are still uncertainties regarding the mechanisms that enable sludge conditioning, but the authors suggest a likely mechanism based on the usual principles of electricity (Fig. 11.10).

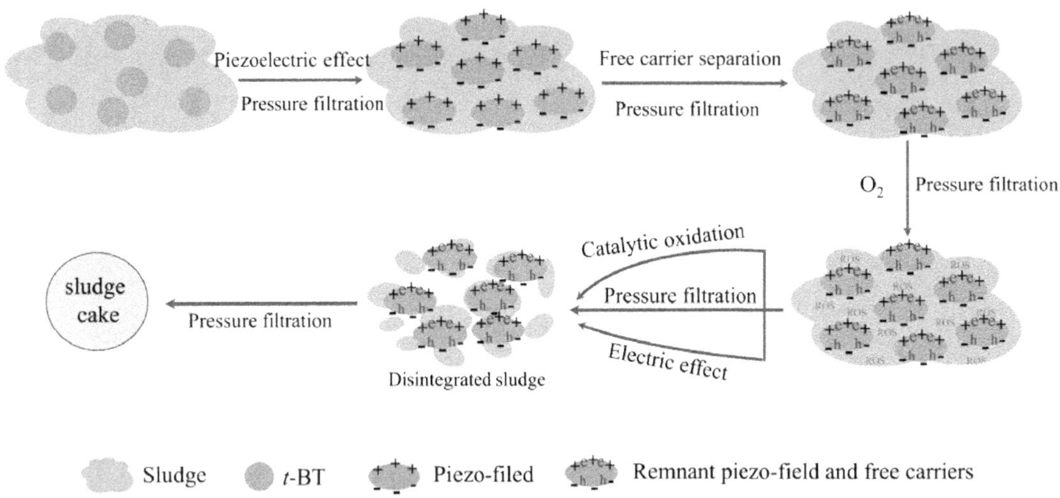

Fig. 11.10 Schematic representation of the piezo-dewatering mechanism (Feng et al., 2022). Reprinted from Feng, J., Zhang, T., Sun, J., Zhu, J., Yan, W., Tian, S., Xiong, Y., 2022. Improvement of sewage sludge dewatering by piezoelectric effect driven directly with pressure from pressure filtration: towards understanding piezo-dewatering mechanism. Water Res. 209, 117922, Copyright (2022), with permission from Elsevier.

Another promising alternative for sludge dewatering is the electro-dewatering (EDW) process. This process is capable of achieving a solids content of up to 60% while using less than a quarter of the energy required by a mechanical dewatering process combined with thermal drying. EDW combines electrochemical cells with a mechanical pressure system. Its mechanism of action involves breaking down the sludge structure by direct electrical force and indirect oxidation (Rumky et al., 2023).

The electro-dewatering process consists of several subprocesses such as water electrolysis, electromigration, electrophoresis and electro-osmosis. In addition, the use of pre-treatments can increase the efficiency of the process and has promising applications, such as the application of a constant negative pressure (i.e. vacuum) or the use of an ultrasonic method. On the other hand, the electro-Fenton process is emerging as an alternative electrochemical advanced oxidation process (EAOP) that can contribute to the oxidation of organic pollutants in addition to the dewatering process (Rumky et al., 2023). Another modification resulted in a protocol called pressurized electro-dewatering with constant voltage gradient mode (G-PEDW), which shows superior results compared to conventional EDW (Rao et al., 2022). However, most of the research is still at the bench-scale testing stage, and development strategies need to be modified to assess the potential for scale-up (Rumky et al., 2023).

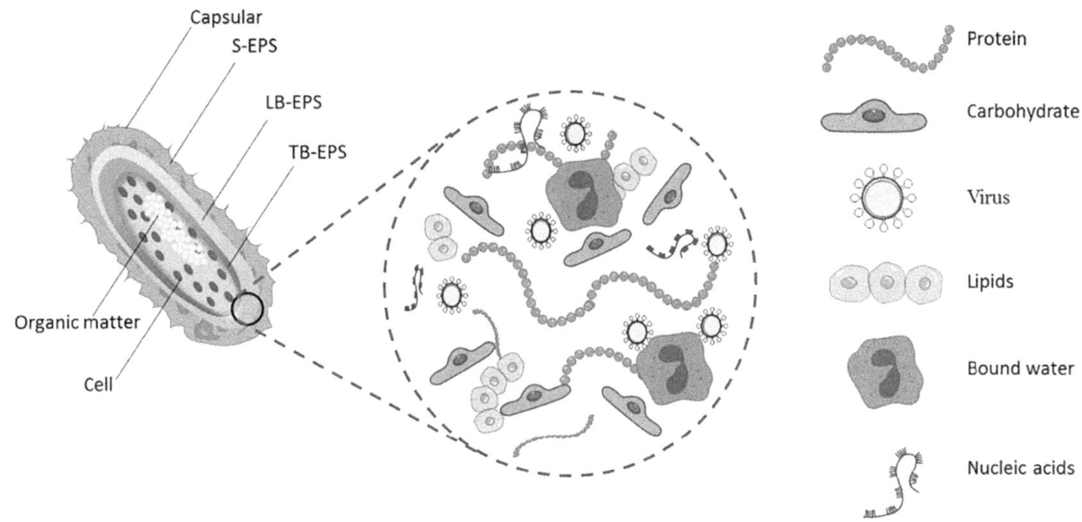

Fig. 11.11 Schematic representation of the components of extracellular polymeric substances (EPS) (Qi et al., 2024). Adapted from Lin, W., Liu, X., Ding, A., Ngo, H.H., Zhang, R., Nan, J., Ma, J., Li, G., 2022. Advanced oxidation processes (AOPs)-based sludge conditioning for enhanced sludge dewatering and micropollutants removal: a critical review. J. Water Process Eng. 45, 102468, ISSN 2214-7144, https://doi.org/10.1016/j.jwpe.2021.102468. Reprinted from Qi, Y., Chen, J., Xu, H., Wu, S., Yang, Z., Zhou, A., Hao, Y., 2024. Optimizing sludge dewatering efficiency with ultrasonic treatment: insights into parameters, effects, and microstructural changes. Ultrason. Sonochem. 102, 106736, Copyright (2024), with permission from Elsevier.

Ultrasonic treatment is another emerging technology that can be applied to sludge dewatering. Ultrasound works by disrupting the structure of the sludge's extracellular polymeric substances (EPS) (Fig. 11.11) through cavitation and hydrodynamic shear mechanisms, releasing the bound water encapsulated in the EPS. At the same time, the sponge effect caused by these mechanisms facilitates the flow of free water through the particles and contributes to the agglomeration of sludge particles. In addition, ultrasound has the ability to eliminate pathogenic microorganisms and some hazardous substances from the sludge (Qi et al., 2024). Again, the challenge is to scale up the technique for application to large volumes of sludge.

Other interesting alternatives worth mentioning are: mechanical compaction pre-treatment using hydrothermal carbonization (Senfter et al., 2021) and the use of a hydrate-based process (Sun et al., 2023).

11.5 Disposal alternatives

Commonly, the final destination of sewage sludge, after treatment stages, is a Sanitary Landfill. Although this is considered to be a safe disposal method, landfilling a material that has undergone several treatment stages, with energy and input consumption, and that has potential properties for reuse in other processes, is not an option that aligns with the principles of Circular Economy and Sustainable Development.

Therefore, emphasis will be given here to the beneficial uses of sewage sludge, even though it is acknowledged that landfill disposal or incineration of sludge are relevant options and should be used when no other alternative is available.

Various applications for sewage sludge have been investigated and implemented over time. These include the restoration of degraded land, use as raw material for organic compost, application in green roofs, erosion control, utilization in forestry soil, use in cement kilns, manufacturing of construction materials, pavement, daily and final landfill cover, and utilization as raw material in biodiesel production (Urban et al., 2019).

However, due to the characteristics of sewage sludge, with high levels of organic matter and nutrients, the most common application that maximizes these aspects is its use in agricultural soils (Urban et al., 2019).

11.5.1 Logistical aspects of the sludge management process

Another aspect that cannot be overlooked when choosing between beneficial use or disposal alternatives for sludge is logistics. The costs and environmental footprints associated with transportation, the need for storage facilities, as well as the acceptance by recipients, are fundamental to make the beneficial use of sludge viable (Urban et al., 2019). These aspects make the sludge disposal decision a complex task that can affect the entire sludge treatment process or even the entire wastewater treatment line (Urban et al., 2023).

To address logistical challenges and optimize sludge management at the basin level, Urban and Isaac (2018) proposed a method based on Geographic Information Systems (GIS) and linear optimisation to optimize the disposal of sewage sludge and water treatment plants (WTPs) sludge in a large basin in Brazil, directing the sludge to priority destinations such as ceramics production and agricultural use. This basin includes 76 municipalities, over 5 million urban water consumers, approximately 116 sewage treatment plants (STPs), and 100 WTPs. The integrated management approach has proved effective in avoiding competition between final destinations, optimizing distances within the system, and preventing overloading of receiving sites. However, given the complexity of the system, with various stakeholders and treatment companies with their own interests, it is essential to reach a broad agreement to ensure that the points set out in the sludge management plan are adhered to by all participants.

11.6 Biosolids and agricultural recycling of WWTP sludge

Biosolids is a term commonly used to refer to treated sewage sludge, with potential applications for agriculture and revegetation of degraded areas. Biosolids contains phosphorus and nitrogen, essential nutrients for plant growth, and thus can replace inorganic fertilizers, contributing to reduce the greenhouse gases emissions associated to their production, and moving towards a circular economy approach.

Annual reports for 2022 show that in the United States, 56% of biosolids produced are destined for land application, of which 31% is used for agricultural purposes (Fig. 11.12).

Recycling sludge in agriculture is the most suitable option because it reduces pressure on natural resources and avoids less appropriate options that have a greater impact on the population and the environment (Fig. 11.13).

In a literature review conducted by Ma and Rosen (2021), the recovery of phosphorus from sewage sludge incinerator ash (SSA) was addressed, once this byproduct contains significant amounts of phosphorus (up to 14% total P), which are not beneficially reused when the ashes are disposed in landfills.

When sewage sludge is incinerated, most carbon and nitrogen contents are lost, while phosphorus content remains in a significantly reduced volume and weight, making sewage sludge ash the most efficient and economical form for phosphorus recovery.

Beneficial reuse of SSA includes agricultural or urban land application, daily landfill cover, land reclamation, athletic field footing, brick, tile, and glass-ceramics production, and application as road subbase material.

Although SSA contain significant amounts of phosphorus in terms of total P, the amount available for plant uptake (bioavailable P) varies widely, from approximately 10% to 68% of total P.

Thermochemical treatments and other techniques such as co-incineration of sewage sludge with corn or cotton stalks, and microbial inoculation with P-solubilizing microbes have been reported as successful approaches to increasing P solubility.

As environmental degradation increases worldwide, slow-release fertilizers, such as struvite or SSA, have been identified as potential sources of phosphorus that are relatively concentrated

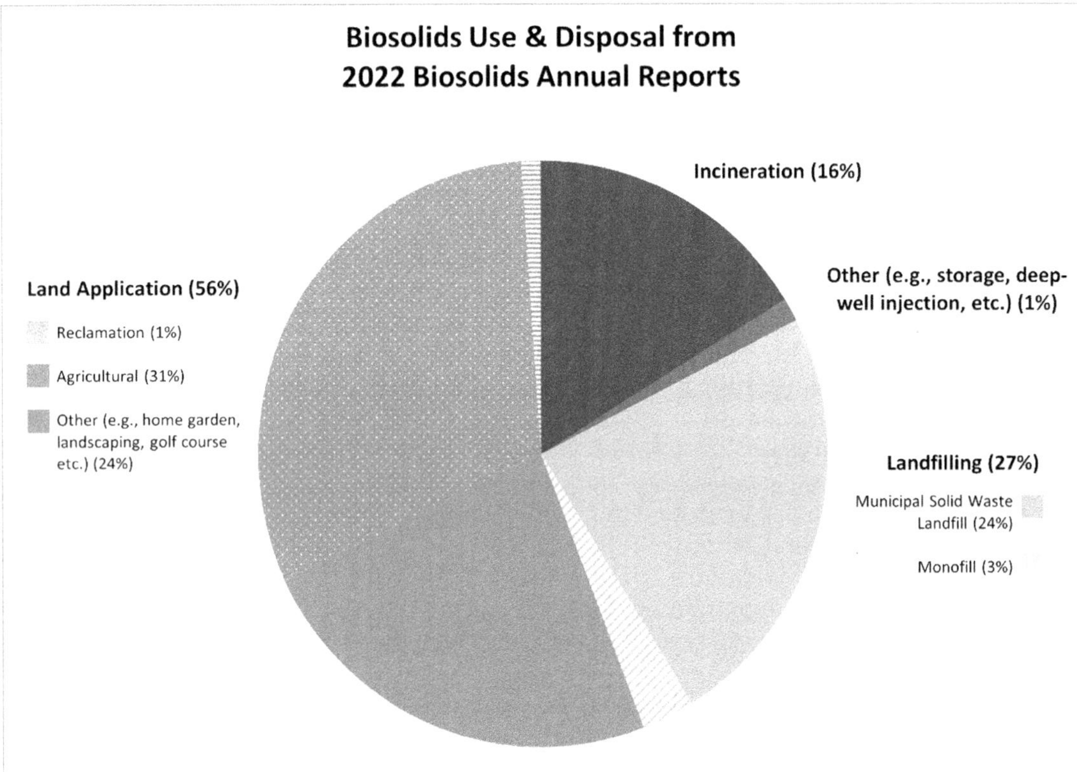

Fig. 11.12 Distribution of biosolids use and disposal from biosolids annual reports covering 2022 available in Enforcement and Compliance History Online (ECHO). https://www.epa.gov/biosolids/basic-information-about-biosolids#:~:text=Biosolids%20are%20divided%20into%20%E2%80%9CClassgeneral%20requirements%20and%20management%20practices.

and insoluble in water, requiring root exudates to access P as needed throughout the season (Rahman et al., 2014; Talboys et al., 2016).

Research and practical experience worldwide have demonstrated the benefits of controlled sludge application in agriculture. These benefits include increased productivity, improved crop quality, reduced costs, and enhanced physical, chemical, and biological soil properties.

When biosolids are applied to land for agricultural purposes, an appropriate agronomic rate must be followed, which is the application rate of sludge designed to provide the required amount of nitrogen for the crop or vegetation grown on the land. The agronomic rate depends on the type of crop and soil characteristics.

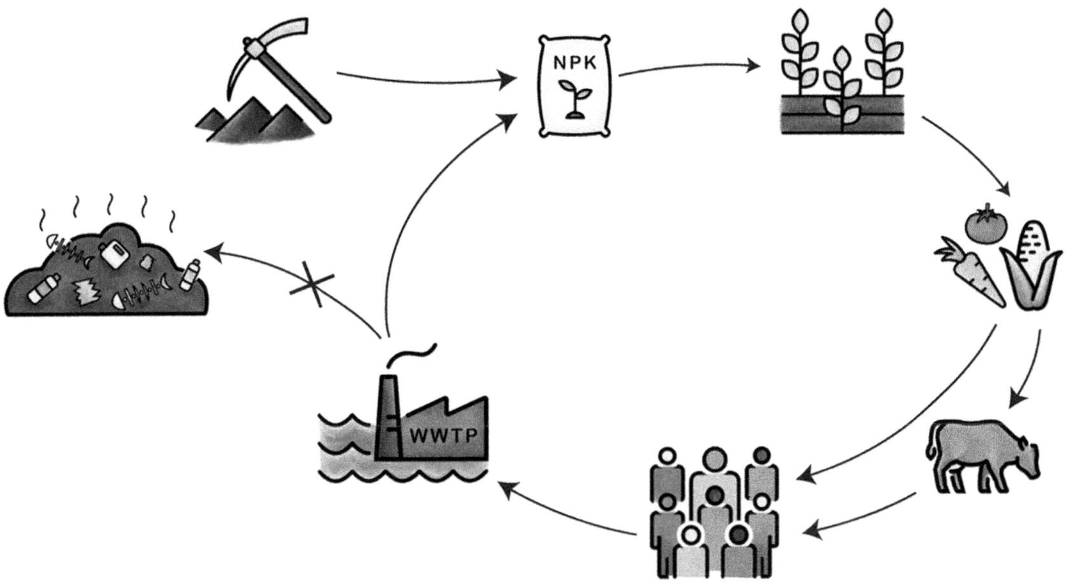

Fig. 11.13 Nutrient recovery from biosolids (Ma and Rosen, 2021). Reprinted from Ma, P., Rosen, C., 2021. Land application of sewage sludge incinerator ash for phosphorus recovery: a review. Chemosphere 274, 129609, Copyright (2021), with permission from Elsevier.

Andreoli et al. (2007a) argue that recycling sludge in agriculture is the best alternative when it meets the necessary requirements regarding heavy metal concentration and pathogens. They explain that sewage sludge should not be regarded as a mere waste, as its physicochemical characteristics make it an excellent soil conditioner. Sewage sludge is therefore redefined as biosolids, which are sludges from the biological treatment of liquid waste that have been processed to allow safe handling in agriculture.

According to von Sperling and Gonçalves (2007) and Gonçalves et al. (2007), important characteristics of sewage sludge are the nutrient content, due to its nutritional properties; the content of heavy metals and organic pollutants, which can be a threat to the balance of the food chain and consequently human health; and the organic matter content, which determines the stability of the material and its potential to emit odors and attract vectors.

The most commonly used inferential parameter for assessing organic matter content in sludge is its concentration of volatile solids. In treatment processes with short cell retention times, the sludge presents low levels of total solids (TS) and high percentages of volatile solids (VS), requiring additional stabilization treatment steps (Saqqar and Pescod, 1995).

The physicochemical characteristics of sludge depend on the composition of the wastewater and the processes involved in both the liquid and solid phases of treatment. It is known that these characteristics can vary annually, seasonally, or even daily, as a result of variations in the characteristics of the wastewater. These variations are more pronounced in systems receiving large quantities of industrial discharges (Metcalf and Eddy, 1991).

The concentration of metals in sludge depends on the type and quantity of industrial effluents discharged into sewage treatment systems and is of fundamental importance in defining alternatives for final disposal. At low concentrations, metals serve as nutrient sources for plants; however, at high concentrations, they can be harmful. It is known that elements such as iron, copper, cobalt, molybdenum, and zinc play important roles in organisms: they compose the electron donor system and function as bridges in enzymatic compounds. However, when present in high concentrations, they can become toxic and even lethal (Gonçalves, 1999).

Andreoli et al. (1999) explain that during the sewage sludge formation process, along with bacterial colonies, the main pathogens and heavy metals present in sewage co-precipitate. The removal processes of metals from the generated sludge are expensive, complex, and inefficient; for this reason, this is a problem to be preventively solved through the identification of the sources generating these materials and their isolation from the public sewage network associated with a highly restrictive policy for receiving industrial sewage in networks, capable of guaranteeing the quality of the produced sludge.

In addition to agricultural applications, biosolids have been successfully used as a sustainable means of reclaiming degraded land, serving as a medium for establishing vegetation, reducing the bioavailability of toxic substances often found in soils, controlling soil erosion, and regenerating soil layers.

More recently, certain challenges associated with emerging contaminants have emerged, including the presence of microplastics and *per-* and polyfluoroalkyl substances (PFAS). Because of the environmental risks related to the persistent and ubiquitous nature of PFAS, these compounds are considered of significant concern. PFAS are synthetic fluorinated substances with strong C—F bonds, being extremely resistant to degradation, and potentially bioaccumulating.

High concentrations of PFAS have been found in biosolids as well as in effluents and influents of wastewater treatment plants. Therefore, PFAS contamination increases the cost of biosolids management. Pyrolysis and gasification have been implemented on a commercial scale after they were reported to be effective thermochemical techniques for degrading PFAS in biosolids, while maintaining the carbon and nutrient levels, and producing PFAS-free biochar. Both technologies are sustainable in terms of energy, as no energy supply is required because the syngas generated by the processes could be used to produce energy (Kumar et al., 2023).

The vermiremediation of sewage sludge containing various micropollutants, including pharmaceuticals, personal care products, endocrine disruptors, and per/polyfluoroalkyl substances by *Eisenia andrei* earthworms was recently studied (Grasserová et al., 2024). The authors reported that a wide range of micropollutants was vermiaccumulated and observed that vermicomposted material contained significantly lower contents of diclofenac, metoprolol, telmisartan, and triclosan than the control (without earthworms). The authors concluded that, as expected, no substantial removal of PFAS was observed, except for the transformation of compounds that are not fully fluorinated. Moreover, the authors identified an interesting trend for the overall bioaccumulation of the detected micropollutants, as although the sewage sludge samples contained different concentrations of micropollutants, both groups of pharmaceuticals and personal care products (PPCPs) and PFAS vermiaccumulated to a similar extent, suggesting that organic micropollutants are absorbed via very general mechanisms controlled by earthworm physiology rather than by specific transporters, as previously suggested for PFAS in the literature.

11.6.1 Microplastics in sewage sludge and biosolids

Recently, Harley-Nyang et al. (2023) conducted a literature review comprising 65 studies from 25 countries on the presence of microplastics in sludge, biosolids, and soils. The authors analyzed microplastic concentrations, and found a wide variation, ranging from 6.44×10^8 to 1.67×10^{12} microplastics per 1000 t of sludge. It is estimated that biosolids recycling releases up to 6430 t of microplastics to land globally per year, representing a significant source of microplastics in soil (Fig. 11.14).

The analysis of the data using the Kruskal-Wallis test indicated that there was no significant difference in the concentrations of microplastics in the sludge and biosolids samples subjected to different treatment levels (Fig. 11.15).

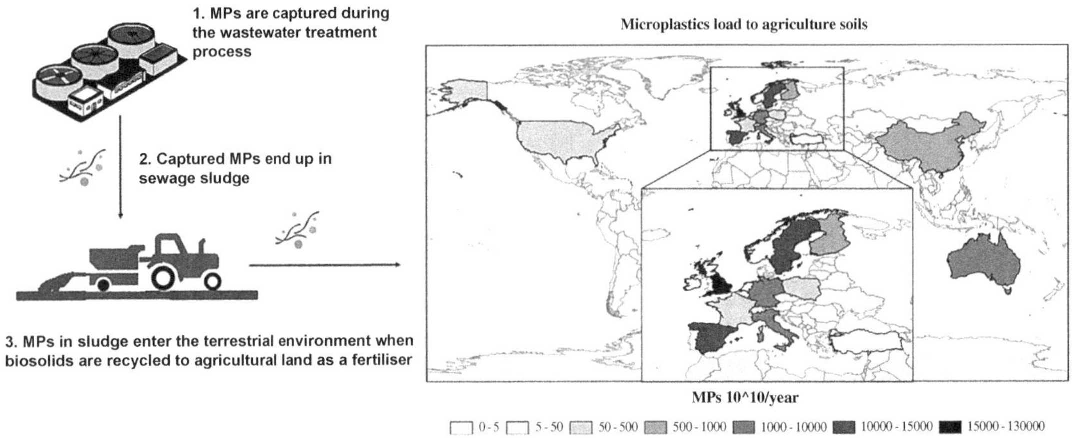

Fig. 11.14 Agricultural recycling of sludge from wastewater treatment plants as a source of microplastics in the soil (Harley-Nyang et al., 2023). Reproduction is permitted under a Creative Commons CC-BY license. Original available at: https://doi.org/10.1016/j.scitotenv.2023.164068.

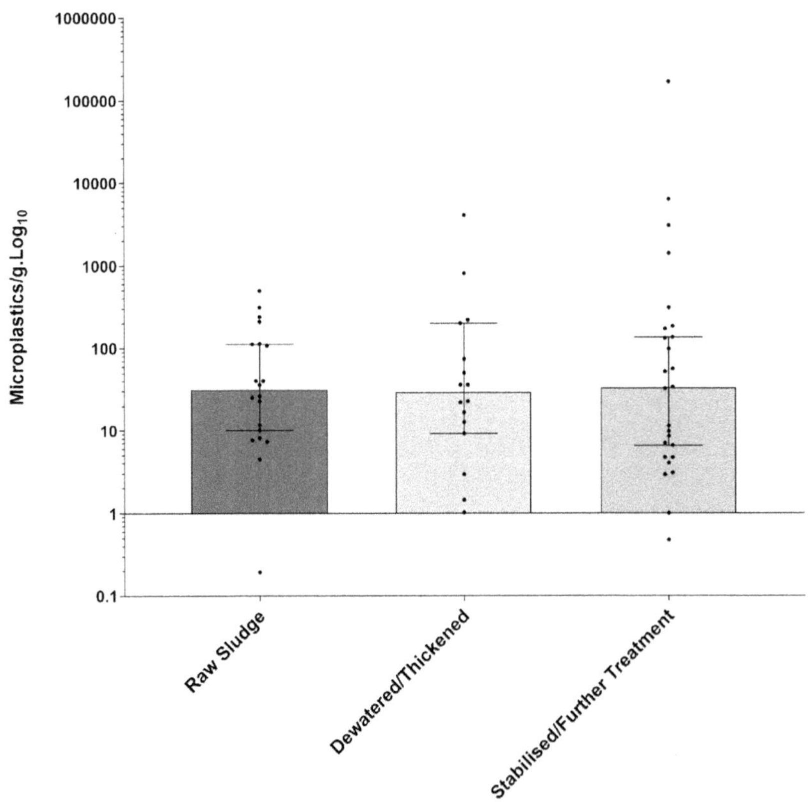

Fig. 11.15 Microplastic concentration in raw and treated sewage sludge (Harley-Nyang et al., 2023). Reproduction is permitted under a Creative Commons CC-BY license. Original available at: https://doi.org/10.1016/j.scitotenv.2023.164068.

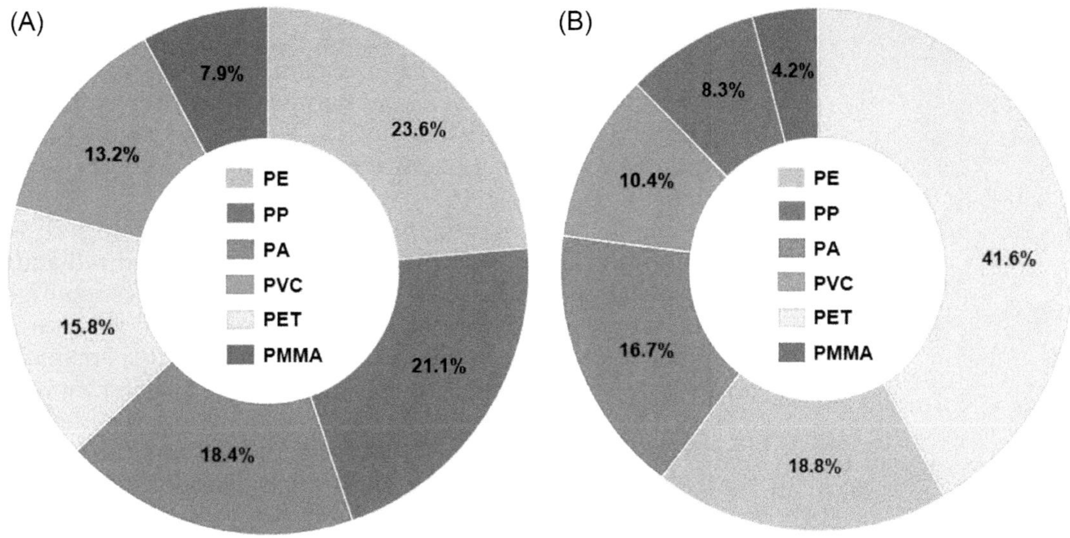

Fig. 11.16 Microplastic polymers identified in: (A) soil, and (B) biosolids Rashti et al. (2023). *PA*, polyamide; *PE*, polyethylene; *PET*, polyethylene terephthalate; *PMMA*, polymethyl methacrylaten; *PP*, polypropylene; *PVC*, polyvinyl chloride. Reprinted from Rezaei Rashti, M., Hintz, J., Esfandbod, M., Bahadori, M., Chen, C., 2023. Detecting microplastics in organic-rich materials and their potential risks to earthworms in agroecosystems. Waste Manage. 166, 96–103, Copyright (2023), with permission from Elsevier.

Rashti et al. (2023) analyzed soil samples from agricultural land amended with biosolid and biosolid samples from wastewater treatment plants and reported that the total number of microplastics in soil samples (1000–3100 MPs per kg dry mass) was on average 30 times lower than in biosolid samples (55,400–73,800 MPs per kg dry mass), and samples showed lower concentrations of fragments than fibers. The results indicated that the polymers polyethylene (PE, 23.6%) and polypropylene (PP, 21.1%) were the major sources of microplastic contamination in biosolid amended soil samples, while polyethylene terephthalate (PET, 41.6%), PE (18.8%) and polyamide (PA, 16.7%) were the most abundant polymers in biosolid samples (Fig. 11.16).

Jose et al. (2024) reviewed the transformation of plastic waste in the environment and the fate of microplastics in wastewater treatment plants, describing the chemical, physicochemical and biodegradation mechanisms that break plastics down into smaller fragments. The main mechanisms of plastic degradation include hydrolysis, oxidation, photodegradation, thermal degradation and the four-step microbial degradation process consisting of biodeterioration, depolymerisation, assimilation and mineralization.

11.7 Pathogen removal

Sanitisation aims to eliminate or reduce the density of pathogenic microorganisms and can be carried out by chemical, physical, or biological processes. It is a treatment stage of great importance when the alternative for final disposal is agricultural recycling (Gonçalves, 1999).

Among the various principles that can promote sludge disinfection, three factors stand out: temperature, pH and radiation. Under certain environmental conditions, these factors represent ranges within which organisms remain present or develop in the sludge, and once this balance is disturbed, the organisms are destroyed. The intensity and duration of these factors applied to the sludge mass determine the effectiveness of disinfection. The state of these factors in the sludge is therefore the basis of the disinfection methods. The most economical disinfection processes are based on changing the parameters that favor the inviability or destruction of pathogens by changing, at least for some time, the chemical and physical conditions of the material, mainly pH and temperature. It is important to emphasize that the most important element of sludge from an agronomic point of view is nitrogen, which is lost in two ways: (i) by volatilization as ammonia; (ii) by leaching as nitrate. Both temperature and pH increase accelerate volatilization losses, which increase over time (Andreoli et al., 1999).

The United States Environmental Protection Agency (USEPA) uses a classification of biosolids based on treatment methods[a]:

- Class A: Biosolids that have been treated to non-detectable metals and vector attraction reduction options.
- Class A EQ (sale/give away): Exceptional Quality biosolids that meet low pollutant and Class A pathogen reduction limits and have reduced levels of degradable vector attractants.
- Class B: Biosolids that contain detectable levels of pathogens but have been reduced to pathogen levels that do not pose a threat to public health and the environment as long as measures are taken to prevent exposure to the biosolids after use or disposal.

Table 11.1 shows the limits for microbial indicators for Class A biosolids treatment.

Table 11.2 shows the description of the Processes to Significantly Reduce Pathogens (PSRP).

[a]https://echo.epa.gov/help/reports/biosolids-facility-report-help#biosolidsAnnualReport.

Table 11.1 Microbial indicators for Class A biosolids treatment (Boczek et al., 2023).

Microbial indicator	Class A regulatory limit
Fecal coliforms or	Less than 1000 MPN per g total solids
Salmonella sp.	Less than 3 MPN per 4 g total solids
Enteric viruses	Less than 1 PFU per 4 g total solids
Viable helminth ova	Less than 1 viable helminth ova per 4 g total solids

MNP, most probable number; *PFU*, plaque-forming units. Total solids in dry weight basis.

Table 11.2 Description of processes to significantly reduce pathogens (PSRP) (Boczek et al., 2023).

PSRP	Regulatory conditions to maintain PSRP
Aerobic digestion	Sewage sludge is agitated with air or oxygen to maintain aerobic conditions for a specific mean cell residence time (i.e., solids retention time) at a specific temperature. Values for the mean cell residence time and temperature shall be between 40 days at 20 °C and 60 days at 15 °C
Air drying	Sewage sludge is dried on sand beds or on paved or unpaved basins. The sewage sludge dries for a minimum of 3 months. During 2 of the 3 months, the ambient average daily temperature is above 0 °C
Anaerobic digestion	Sewage sludge is treated in the absence of air for a specific mean cell residence time (i.e., solids retention time) at a specific temperature. Values for the mean cell residence time and temperature shall be between 15 days at 35–55 °C and 60 days at 20 °C
Composting	Using either the within-vessel, static aerated pile, or windrow composting methods, the temperature of the sewage sludge is raised to 40 °C or higher and remains at 40 °C or higher for 5 days. For 4 h during the 5-day period, the temperature in the compost pile exceeds 55 °C
Lime stabilization	Sufficient lime is added to the sewage sludge to raise the pH of the sewage sludge to 12 for more than 2 h of contact

Table 11.3 shows a comparison of the log reductions in bacteria, viruses, protozoa and helminths following different sludge treatments. A Process to Significantly Reduce Pathogens (PSRP) is defined as a process capable of consistently reducing the density of microorganisms in sludge by 2 log or more, or by a factor of 100. PSRP processes set treatment parameters for Class B biosolids (Boczek et al., 2023).

Table 11.3 Summary of log reductions in pathogens after sludge treatment (Boczek et al., 2023).

Process to significantly reduce pathogens (PSRP)	Bacteria	Viruses	Parasites[a]
Anaerobic digestion	0.5–4.0	0.5–2.0	0.5
Aerobic digestion	0.5–4.0	0.5–2.0	0.5
Composting	2.0–4.0	2.0–4.0	2.0–4.0
Air drying	0.5–4.0	0.5–4.0	0.5–4.0
Lime stabilization	0.5–4.0	4.0	0.5

[a]Protozoa and helminths.

11.7.1 Lime stabilization

Liming is one of the most efficient processes for pathogen elimination, sludge stabilization and deodorisation (Gonçalves, 1999). Defined as a process of chemical stabilization and disinfection, liming results from the addition and mixing of lime to the sludge, causing a rapid alkalisation of the environment, raising the pH to around 12, making the environment unsuitable for the survival and development of pathogens present in the sludge.

According to Andreoli et al. (1999), the liming process involves mixing quicklime (CaO) with sludge in proportions ranging from 30% to 50% of the dry weight of the sludge. In this case, three factors intervene in the disinfection process: first, the change in (i) the temperature; (ii) the pH of the resulting mass (sludge-lime mixture); and finally, (iii) the action of ammonia formed from the nitrogen in the sludge under conditions of high temperature and pH. These authors explain that the contact of the lime with the water contained in the sludge leads to an exothermic reaction that causes the temperature to rise for several days until the mixture stabilizes. This effect depends on the quality and proportion of lime used in relation to the amount of sludge. If the lime is not processed properly, part of the product will not be lime (CaO) but limestone ($CaCO_3$), which has a much slower reaction rate. Raising the pH of the mixture is another disinfection factor. Sludge limed to 50% of its dry weight has a pH slightly above 12.0. At this pH all pathogens are eliminated, provided the exposure time is sufficient. The contact of the ammonia produced from the nitrogen in the sludge is also a disinfection factor, which is more effective with longer exposure times of the pathogens, which is why it is advisable to cover the storage heaps with plastic sheeting, which reduces ammonia losses and thus increases the exposure time of

Table 11.4 Description of processes to further reduce pathogens (PFRP) (Boczek et al., 2023).

PFRP	Description
Composting	Using either the within-vessel composting method or the static aerated pile composting method, the temperature of sewage sludge is maintained at 55 °C or higher for 3 consecutive days Using the windrow composting method, the temperature of the sewage sludge is maintained at 55 °C or higher for 15 consecutive days or longer. During the period when the compost is maintained at 55 °C or higher, there shall be a minimum of five turnings of the windrow
Heat drying	Sewage sludge is dried by direct or indirect contact with hot gases to reduce the moisture content of the sewage sludge to 10% or lower. Either the temperature of the sewage sludge particles exceeds 80 °C or the wet bulb temperature of the gas in contact with the sewage sludge as the sewage sludge leaves the dryer exceeds 80 °C
Heat treatment	Liquid sewage sludge is heated to a temperature of 180 °C or higher for 30 min
Thermophilic aerobic digestion	Liquid sewage sludge is agitated with air or oxygen to maintain aerobic conditions and the mean cell residence time (i.e., the solids retention time) of the sewage sludge is 10 days at 55–60 °C
Beta ray irradiation	Sewage sludge is irradiated with beta rays from an electron accelerator at dosages of at least 1.0 Mrad at room temperature (ca. 20 °C)
Gamma ray irradiation	Sewage sludge is irradiated with gamma rays from certain isotopes, such as Cobalt 60 and Cesium 137, at dosages of at least 1.0 Mrad at room temperature (ca. 20 °C)
Pasteurization	The temperature of the sewage sludge is maintained at 70 °C or higher for 30 min or longer

the pathogens to the action of the ammonia and, consequently, the effectiveness of the disinfection. This covering also prevents the sludge from being exposed to atmospheric precipitation, which can moisten the material, and mitigates the unpleasant odors common to sludge that has not been properly stabilized.

Table 11.4 presents the description of the Processes to Further Reduce Pathogens (PFRP), which are defined as processes used to consistently reduce the density of the microbial population, including all bacterial, viral, and protozoan pathogens, to below detectable levels. PFRP meet Class A criteria for pathogens, vector attraction reduction and land application (Boczek et al., 2023).

11.7.2 Composting

Composting can produce either Class A or Class B biosolids, depending on the process operation, which results in certain temperatures over time, as shown in Tables 11.2 and 11.4.

According to Andreoli et al. (1999), composting is a biological process in which microorganisms degrade the organic matter contained in sludge, either alone or mixed with other organic residues (such as straw, sawdust, garden waste and prunings, and the organic fraction of municipal waste) in exothermic processes, generating heat and consequently increasing the temperature of the piles. For composting, sludge should be mixed with organic residues (chopped vegetable residues, straw, sugar cane bagasse), ensuring that the components are thoroughly mixed. It is advisable to chop the organic material into pieces between 0.5 and 4.0 cm in length to allow good aeration, which is essential for microbial activity. Microbial activity also depends on adequate moisture, ideally between 55% and 65%. Microbial activity consumes nitrogen in the breakdown and resynthesis of organic matter. Carbon is extracted from the residues and nitrogen is provided by the sludge. The ideal balance, or good nutritional status of the compost, is a C/N ratio between 20 and 30, i.e. 20–30 units of carbon for one unit of nitrogen. A temperature above 60 °C should be maintained for at least 10 days. As the biological activity decreases, the temperature also decreases. At this stage, the compost pile should be turned to promote aeration and mixing of the materials. Heating of the compost indicates that the process has not reached stability and a stable temperature indicates that the compost is ready.

It is believed that nitrogen losses are low with this method as there is a cycling of the element, mainly from nitrate and ammoniacal forms to various organic forms which slowly release nitrogen into the soil.

Rashti et al. (2023) investigated the effects of microplastics in biosolid-amended soils and reported that the application of biosolids containing microplastics significantly reduced the survival rate of *Eisenia Fetida* to 81% and showed no significant effect on *Amynthas Gracilis*. In addition, the biosolids amendment significantly reduced the growth rate of earthworms, with a more pronounced negative effect on *Eisenia Fetida*. Given that *Eisenia Fetida* is the most common earthworm in vermicomposting, these results can be considered of significant concern.

11.8 Conclusions and future trends

Sludge management is a complex, costly, and essential stage of municipal wastewater treatment. However, it is often neglected. The common line of sludge treatment consists of the application of thickening, stabilization, dewatering, disinfection, and final disposal. The configuration of this treatment depends on the characteristics of the sludge to be treated and the requirements of the final destination. For example, sludge from anaerobic treatment is already stabilized and the stabilization stage can be eliminated. If the sludge is to be landfilled or incinerated, it does not need to be safe from pathogens and the pathogen removal stage can be eliminated. Some treatment options dewater the sludge directly, disregarding the thickening stage.

Emerging technologies for thickening and dewatering include intensive research into new ways of conditioning sludge. The development of new products, the use of waste and the use of alternative techniques such as electrolysis and ultrasound are emerging as viable alternatives to the traditional metal salts and polyelectrolytes used, with economic and environmental benefits. However, the major challenge for the future is to scale up these new technologies to assess which of them are really viable for full-scale application.

Other new alternatives for sludge management include: the development of mixed technologies that allow two processes to run simultaneously, saving space and time on sludge treatment lines; and optimizing the agricultural use of sludge, in this case called biosolids.

Despite the challenges posed by emerging contaminants in biosolids, land application of sewage sludge remains a viable alternative to biosolids use in a circular economy approach. This requires treatment processes that produce biosolids of exceptional quality.

The vermiremediation of sludge containing micropollutants is emerging as a potentially promising treatment to improve the quality of biosolids, and further studies are needed to support this practice.

References

Ai, J., Zhang, W., Chen, F., Liao, G., Li, D., Hua, X., Wang, D., Ma, T., 2019. Catalytic pyrolysis coupling to enhanced dewatering of waste activated sludge using $KMnO_4Fe(II)$ conditioning for preparing multi-functional material to treat groundwater containing combined pollutants. Water Res. 158, 424–437. 00431354 https://doi.org/10.1016/j.watres.2019.04.044.

Andreoli, C.V., Lara, Ferreira, A.C., Pegorini, E.S., Ihlenfeld, R.G.K., 1999. Uso e Manejo Do Lodo de Esgoto na Agricultura. SANEPAR/PROSAB.

Andreoli, C.V., Pegorini, E.S., Fernandes, F., dos Santos, H.F., 2007a. Land application of sewage sludge. In: Andreoli, C.V., von Sperling, M., Fernandes, F. (Eds.), Sludge Treatment and Disposal, first ed. vol. 6. IWA Publishing, pp. 162–207.

Andreoli, C.V., von Sperling, M., Fernandes, F., 2007b. Sludge Treatment and Disposal, first ed. vol. 6 IWA Publishing.

Betatache, H., Aouabed, A., 2015. Sewage sludge conditioning using *Opuntia Ficus Indica* juice. Rev. Sci. Eau. 28 (1), 67–72. 1718-8598 https://doi.org/10.7202/1030009ar.

Boczek, L., Herrmann, R., Resek, E., Richman, T., 2023. Pathogens and vector attraction in sewage sludge. https://cfpub.epa.gov/si/si_public_record_Report.cfm?dirEntryId=356976%26Lab=CESER.

Bretas Alvim, C., Navajas Valiente, S., Bes-Piá, M.A., Mendoza-Roca, J.A., 2022. Methodology for removing microplastics and other anthropogenic microparticles from sludge dewatering system. J. Environ. Manage. 314, 115010. 03014797 https://doi.org/10.1016/j.jenvman.2022.115010.

Chen, C., Zhang, P., Zeng, G., Deng, J., Zhou, Y., Lu, H., 2010. Sewage sludge conditioning with coal fly ash modified by sulfuric acid. Chem. Eng. J. 158 (3), 616–622. 13858947 https://doi.org/10.1016/j.cej.2010.02.021.

Chopade, G., Devatha, C.P., 2024. Experimental investigation on sludge conditioning and dewatering using an agricultural biomass coupled with resource recovery. J. Environ. Manage. 352, 120098. 03014797 https://doi.org/10.1016/j.jenvman.2024.120098.

De Maeseneer, J.L., 1997. Constructed wetlands for sludge dewatering. Water Sci. Technol. 35 (5), 80–82. https://doi.org/10.1016/S0273-1223(97)00080-2.

Elbaz, A.A., Aboulfotoh, A.M., Dohdoh, A.M., Wahba, A.M., 2021. Reuse of by-products material as a conditioner to improve thickening and dewatering of sludge. Desalin. Water Treat. 225, 104–111. 19443986 https://doi.org/10.5004/dwt.2021.27224.

Feng, J., Zhang, T., Sun, J., Zhu, J., Yan, W., Tian, S., Xiong, Y., 2022. Improvement of sewage sludge dewatering by piezoelectric effect driven directly with pressure from pressure filtration: towards understanding piezo-dewatering mechanism. Water Res. 209, 117922. 00431354 https://doi.org/10.1016/j.watres.2021.117922.

Geng, H., Xu, Y., Dai, X.H., Yang, D.H., 2022. Overview and prospect of the application of ion exchange resin in treatment of sewage sludge. Zhongguo Huanjing Kexue/China Environ. Sci. 42 (11), 5220–5228. Chinese Society for Environmental Sciences, China, 10006923. Available from: http://www.zghjkx.com.cn/EN/volumn/current.shtml.

Gholipour, A., Fragoso, R., Duarte, E., Galvão, A., 2022. Sludge treatment reed bed under different climates: a review using meta-analysis. Sci. Total Environ. 843, 156953. 00489697 https://doi.org/10.1016/j.scitotenv.2022.156953.

Gonçalves, R.F., 1999. Gerenciamento de Lodo de Lagoas de Estabilização não Mecanizadas. PROSAB.

Gonçalves, R.F., Luduvice, M., von Sperling, M., 2007. Sludge thickening and dewatering. In: Andreoli, C.V., von Sperling, M., Fernandes, F. (Eds.), Sludge Treatment and Disposal, first ed. vol. 6. IWA Publishing.

Grasserová, A., Pacheco, N.I.N., Semerád, J., Filipová, A., Innemanová, P., Hanč, A., Procházková, P., Cajthaml, T., 2024. New insights into vermiremediation of sewage sludge: the effect of earthworms on micropollutants and vice versa. Waste Manag. 174, 496–508. Elsevier Ltd, Czech Republic, 18792456. Available from: https://www.sciencedirect.com/science/journal/0956053X. https://doi.org/10.1016/j.wasman.2023.12.016.

Guo, J., Jiali, D., Chen, P., Tan, X., Huang, X., Gan, P., Lin, F., 2017. Enhanced efficiencies of sludge dewatering and domestic wastewater treatment by using the bioflocculant from rice stover. Water Environ. J. 31 (1), 120–126. 1747–6585 https://doi.org/10.1111/wej.12221.

Harley-Nyang, D., Memon, F.A., Baquero, A.O., Galloway, T., 2023. Variation in microplastic concentration, characteristics and distribution in sewage sludge & biosolids around the world. Sci. Total Environ. 891, 164068. 00489697 https://doi.org/10.1016/j.scitotenv.2023.164068.

Hatamifar, A., Gholikandi, G.B., 2024. Synergistic effect of innovative bentonite and poly aluminum chloride (PAC) combination for excess and digested sludge thickening and dewatering. Desalin. Water Treat. 317, 100131. 19443986 https://doi.org/10.1016/j.dwt.2024.100131.

Hyrycz, M., Ochowiak, M., Krupińska, A., Włodarczak, S., Matuszak, M., 2022. A review of flocculants as an efficient method for increasing the efficiency of municipal sludge dewatering: mechanisms, performances, influencing factors and perspectives. Sci. Total Environ. 820, 153328. 00489697 https://doi.org/10.1016/j.scitotenv.2022.153328.

Jose, S., Lonappan, L., Cabana, H., 2024. Prevalence of microplastics and fate in wastewater treatment plants: a review. Environ. Chem. Lett. 22 (2), 657–690. Springer Science and Business Media Deutschland GmbH, Canada, 16103661. Available from: https://www.springer.com/journal/10311. https://doi.org/10.1007/s10311-023-01679-y.

Kazimierowicz, J., Dębowski, M., 2023. Characteristics of solidified carbon dioxide and perspectives for its sustainable application in sewage sludge management. Int. J. Mol. Sci. 24 (3), 2324. 1422-0067 https://doi.org/10.3390/ijms24032324.

Kim, H.G., Chung, T.H., 2013. Performance of the sludge thickening and reduction at various factors in a pilot-scale MBR. Sep. Purif. Technol. 104, 297–306. 13835866 https://doi.org/10.1016/j.seppur.2012.11.008.

Kumar, R., Dada, T.K., Whelan, A., Cannon, P., Sheehan, M., Reeves, L., Antunes, E., 2023. Microbial and thermal treatment techniques for degradation of PFAS in biosolids: a focus on degradation mechanisms and pathways. J. Hazard. Mater. 452, 131212. 03043894 https://doi.org/10.1016/j.jhazmat.2023.131212.

Liang, J., Zhou, Y., 2022. Iron-based advanced oxidation processes for enhancing sludge dewaterability: state of the art, challenges, and sludge reuse. Water Res. 218, 118499. 00431354 https://doi.org/10.1016/j.watres.2022.118499.

Liu, M., Rashid, S., Wang, W., Zhang, H., Zhao, Y., Xingmin, F., Zhaoyang, S., Graham, N., Yu, W., 2024. The application of chitosan quaternary ammonium salt to replace polymeric aluminum ferric chloride for sewage sludge dewatering. Water Res. 256, 121539. 00431354 https://doi.org/10.1016/j.watres.2024.121539.

Luduvice, M., 2007. Sludge stabilization. In: Andreoli, C.V., von Sperling, M., Fernandes, F. (Eds.), Sludge Treatment and Disposal, first ed. vol. 6. IWA Publishing, pp. 48–76.

Ma, P., Rosen, C., 2021. Land application of sewage sludge incinerator ash for phosphorus recovery: a review. Chemosphere 274, 129609. 00456535 https://doi.org/10.1016/j.chemosphere.2021.129609.

Major, N., Jechalke, S., Nesme, J., Goreta Ban, S., Černe, M., Sørensen, S.J., Ban, D., Grosch, R., Schikora, A., Schierstaedt, J., 2022. Influence of sewage sludge stabilization method on microbial community and the abundance of antibiotic resistance genes. Waste Manage. 154, 126–135. Elsevier Ltd, Croatia, 18792456. Available from: www.elsevier.com/locate/wasman. https://doi.org/10.1016/j.wasman.2022.09.033.

Mamais, D., Tzimas, A., Efthimiadou, A., Kiissandrakis, J., Andreadakis, A., 2009. Evaluation of different sludge mechanical dewatering technologies. J. Residuals Sci. Technol. 6 (1), 27–34. 15448053.

Mei, X., Tang, J., Zhang, Y., 2020. Sludge stabilization: characteristics of the end-products and an alternative evaluative methodology. Waste Manag. 105, 355–363. 0956053X https://doi.org/10.1016/j.wasman.2020.02.027.

Metcalf, Eddy, 1991. Wastewater Engineering: Treatment, Disposal, and Reuse. McGraw Hill.

Peruzzi, E., Macci, C., Doni, S., Iannelli, R., Masciandaro, G., 2017. Stabilization process in reed bed systems for sludge treatment. Ecol. Eng. 102, 381–389. 09258574 https://doi.org/10.1016/j.ecoleng.2017.02.017.

Qi, Y., Chen, J., Xu, H., Wu, S., Yang, Z., Zhou, A., Hao, Y., 2024. Optimizing sludge dewatering efficiency with ultrasonic treatment: insights into parameters, effects, and microstructural changes. Ultrason. Sonochem. **102**, 106736. 13504177 https://doi.org/10.1016/j.ultsonch.2023.106736.

Rahman, M.M., Salleh, M.A.M., Rashid, U., Ahsan, A., Hossain, M.M., Ra, C.S., 2014. Production of slow release crystal fertilizer from wastewaters through struvite crystallization—a review. Arab. J. Chem. 7 (1), 139–155. 18785352 https://doi.org/10.1016/j.arabjc.2013.10.007.

Rao, B., Jiangge, S., Xu, J., Xu, S., Pang, H., Zhang, Y., Xu, P., Wu, B., Lian, J., Deng, L., 2022. Coupling mechanism and parameter optimization of sewage sludge dewatering jointly assisted by electric field and mechanical pressure. Sci. Total Environ. 817, 152939. 00489697 https://doi.org/10.1016/j.scitotenv.2022.152939.

Rashti, M.R., Hintz, J., Esfandbod, M., Bahadori, M., Lan, Z., Chen, C., 2023. Detecting microplastics in organic-rich materials and their potential risks to earthworms in agroecosystems. Waste Manag. 166, 96–103. 0956053X https://doi.org/10.1016/j.wasman.2023.04.047.

Romero-Güiza, M.S., Flotats, X., Asiain-Mira, R., Palatsi, J., 2022. Enhancement of sewage sludge thickening and energy self-sufficiency with advanced process control tools in a full-scale wastewater treatment plant. Water Res. 222, 118924. 00431354 https://doi.org/10.1016/j.watres.2022.118924.

Rumky, J., Deb, A., Shim, M.J., Laakso, E., Repo, E., 2023. A review on the recent advances in electrochemical treatment technologies for sludge dewatering and alternative uses. J. Hazard. Mater. Adv. 11, 100341. https://doi.org/10.1016/J.HAZADV.2023.100341.

Saqqar, M.M., Pescod, M.B., 1995. Modelling sludge accumulation in anaerobic wastewater stabilization ponds. Water Sci. Technol. 31 (12), 185–190. 02731223 https://doi.org/10.1016/0273-1223(95)00505-H.

Senfter, T., Fritsch, L., Berger, M., Kofler, T., Mayerl, C., Pillei, M., Kraxner, M., 2021. Sludge thickening in a wastewater treatment plant using a modified hydrocyclone. Carbon Resour. Convers. 4, 132–141. 25889133 https://doi.org/10.1016/j.crcon.2021.03.001.

Sun, L., Hassanpouryouzband, A., Sun, H., Wang, T., Zhang, L., Yang, L., Zhao, J., Song, Y., 2023. Advancement in sewage sludge dewatering with hydrate crystal phase change: unveiling the micro-moisture migration and dewaterability mechanisms. ACS Sustain. Chem. Eng. 11 (32), 12075–12083. 2168-0485 https://doi.org/10.1021/acssuschemeng.3c02718.

Takabe, Y., Ida, K., 2023. Simultaneous phosphorus precipitation and sludge thickening by electrolysis with an anode covered by bivalve shells. Water Res. 247, 120789. 00431354 https://doi.org/10.1016/j.watres.2023.120789.

Talboys, P.J., Heppell, J., Roose, T., Healey, J.R., Jones, D.L., Withers, P.J.A., 2016. Struvite: a slow-release fertiliser for sustainable phosphorus management? Plant Soil 401 (1–2), 109–123. Springer International Publishing, United Kingdom, 15735036. Available from: http://www.wkap.nl/journalhome.htm/0032-079X. https://doi.org/10.1007/s11104-015-2747-3.

Tomei, M.C., Bertanza, G., Canato, M., Heimersson, S., Laera, G., Svanström, M., 2016. Techno-economic and environmental assessment of upgrading alternatives for sludge stabilization in municipal wastewater treatment plants. J. Clean. Prod. 112, 3106–3115. https://doi.org/10.1016/j.jclepro.2015.10.017.

Uggetti, E., Llorens, E., Pedescoll, A., Ferrer, I., Castellnou, R., García, J., 2009. Sludge dewatering and stabilization in drying reed beds: characterization of three full-scale systems in Catalonia, Spain. Bioresour. Technol. 100 (17), 3882–3890. 09608524 https://doi.org/10.1016/j.biortech.2009.03.047.

Urban, R.C., Isaac, R.L., 2018. WTP and WWTP sludge management: a case study in the metropolitan area of Campinas, southeastern Brazil. Environ. Monit. Assess. 190 (10). Springer International Publishing, Brazil, 15732959. Available from: www.wkap.nl/journalhome.htm/0167-6369. https://doi.org/10.1007/s10661-018-6972-0.

Urban, R.C., Isaac, R.d.L., Morita, D.M., 2019. Uso benéfico de lodo de estações de tratamento de água e de tratamento de esgoto: estado da arte. Revista DAE 67 (219), 128–158. https://doi.org/10.4322/dae.2019.050.

Urban, R.C., Nakada, L.Y.K., Isaac, R.L., 2023. A system dynamics approach for large-scale water treatment plant sludge management: a case study in Brazil. J. Clean. Prod. 419. Elsevier Ltd, Brazil, 09596526. Available from: https://www.journals.elsevier.com/journal-of-cleaner-production. https://doi.org/10.1016/j.jclepro.2023.138105.

von Sperling, M., Andreoli, C.V., 2007. Introduction to sludge management. In: Andreoli, C.V., von Sperling, M., Fernandes, F. (Eds.), Sludge Treatment and Disposal, first ed. vol. 6. IWA Publishing, pp. 1–4.

von Sperling, M., Gonçalves, R.F., 2007. Sludge characteristics and production. In: Andreoli, C.V., von Sperling, M., Fernandes, F. (Eds.), Sludge Treatment and Disposal, first ed. vol. 6. IWA Publishing, pp. 4–30.

Wakeman, R.J., 2007. Separation technologies for sludge dewatering. J. Hazard. Mater. 144 (3), 614–619. 03043894 https://doi.org/10.1016/j.jhazmat.2007.01.084.

Wang, S., Chen, H., 2023. Enhanced dewaterability of sewage sludge by grafted cationic lignin-based flocculants. Sci. Total Environ. 903, 166958. 00489697 https://doi.org/10.1016/j.scitotenv.2023.166958.

Wang, Z., Wu, Z., Hua, J., Wang, X., Du, X., Hua, H., 2008a. Application of flat-sheet membrane to thickening and digestion of waste activated sludge (WAS). J. Hazard. Mater. 154 (1–3), 535–542. https://doi.org/10.1016/J.JHAZMAT.2007.10.057.

Wang, X., Wu, Z., Wang, Z., Du, X., Hua, J., 2008b. Membrane fouling mechanisms in the process of using flat-sheet membrane for simultaneous thickening and digestion of activated sludge. Sep. Purif. Technol. 63 (3), 676–683. https://doi.org/10.1016/J.SEPPUR.2008.07.013.

Wang, Y., Gu, M.X., Ge, D., Dong, Y., Bai, L., Han, Y., Zhu, N., 2022. Polyhexamethylene biguanidine used as a new type sewage sludge conditioning agent: effect on sludge dewaterability and mechanism. J. Environ. Manag. 315, 115146. https://doi.org/10.1016/J.JENVMAN.2022.115146.

Wang, S., Sadhukhan, J., Xuan, J., Yu, Y., Mao, X., Wang, M., Chen, X., Zhou, X., Xing, L., Wang, X., 2023. Life cycle assessment and life cycle cost of sludge dewatering, conditioned with Fe^{2+}/H_2O_2, $Fe^{2+}/Ca(ClO)_2$, $Fe^{2+}/Na_2S_2O_8$, and Fe^{3+}/CaO based on pilot-scale study data. ACS Sustain. Chem. Eng. 11 (20). https://doi.org/10.1021/acssuschemeng.3c00505.

Xiao, T., Zhang, L., Chen, S., Dai, X., Dong, B., 2022. A novel application of dissolved ozone flotation on sewage sludge thickening: performance and mechanism investigation. Sci. Total Environ. 842, 156874. 00489697 https://doi.org/10.1016/j.scitotenv.2022.156874.

Yu, W., Yang, J., Tao, S., Shi, Y., Jiangwei, Y., Lv, Y., Liang, S., Xiao, K., Liu, B., Hou, H., Jingping, H., Xu, W., 2017. A comparatively optimization of dosages of oxidation agents based on volatile solids and dry solids content in dewatering of sewage sludge. Water Res. 126, 342–350. 00431354 https://doi.org/10.1016/j.watres.2017.09.044.

Zhang, H., Yang, J., Yu, W., Luo, S., Peng, L., Shen, X., Shi, Y., Zhang, S., Song, J., Ye, N., Li, Y., Yang, C., Liang, S., 2014. Mechanism of red mud combined with Fenton's reagent in sewage sludge conditioning. Water Res. 59. https://doi.org/10.1016/j.watres.2014.04.026.

Zhang, L., Wang, W., Chen, Y., Liu, Q., Li, Q., Long, Q., 2017. Sewage sludge conditioning and mechanism with semi-coke powder. Chin. J. Environ. Eng. 11 (3), 1831–1836. Science Press, China, 16739108. Available from: http://www.cjee.ac.cn/teepc_en/ch/index.aspx. 10.12030/j.cjee.201511207.

Zhao, L., Liu, Z., Soyekwo, F., Liu, C., Hu, Y., Niu, Q.J., 2022. Exploring the feasibility of novel double-skinned forward osmosis membranes with higher flux and superior anti-fouling properties for sludge thickening. Desalination 523, 115410. 00119164 https://doi.org/10.1016/j.desal.2021.115410.

12

Forward osmosis for municipal wastewater

S.M. Sadati Tilebon[a], Kamran Ghasemzadeh[b,c], Alfredo Cassano[d], and Angelo Basile[e,f]

[a]Seawater Reverse Osmosis (SWRO) Desalination Plant Operator, Neka Power Generation Management Co., Shahid Salimi Neka Power Plant, Neka, Mazandaran, Iran. [b]Chemical Engineering Department, Urmia University of Technology, Urmia, Iran. [c]Institutes for Materials and Processes, University of Edinburgh, Scotland, United Kingdom. [d]Institute of Membrane Technology (ITM) of the Italian National Research Council (CNR), Rende, Italy. [e]Institute of Membrane Technology (ITM) of the Italian National Research Council (CNR), Cosenza, Italy - Retired. [f]General TAG, Ligornetto, Switzerland

Abbreviations

AeOMBR	aerobic osmosis membrane bioreactor
AnOMBR	anaerobic osmotic membrane bioreactor
COD	chemical oxygen demand
CP	concentration polarization
CTA	cellulose triacetate
DOC	direct osmotic concentration
ED	electrodialysis
FO	forward osmosis
MD	membrane distillation
MF	microfiltration
MRC	manual reconcentration
NASA	U.S. National Aeronautics and Space Administration
NF	nanofiltration
OD	osmotic dilution
OMBR	osmosis membrane reactor
PRO	pressure related osmosis
RO	reverse osmosis
TFC	thin film composite
TDS	total dissolved solids
TMWRF	Truckee Meadows Water Reclamation Facility
TN	total nitrogen
TOC	total organic content
TP	total phosphorus
TrOC	trace organic pollutants
TSS	total suspended solid
UF	ultrafiltration

Municipal Wastewater Treatment. https://doi.org/10.1016/B978-0-443-24826-9.00012-7

List of symbols

A water membrane permeability constant
pH potential of hydrogen
J_w water flux
σ reflection coefficient
ΔP applied pressure
$\Delta\pi$ osmotic pressure difference
π osmotic pressure

12.1 Introduction

Forward osmosis (FO) is a new technology based on the osmosis process that humans have been familiar with since ancient times. Osmosis is a spontaneous diffusion of solvent molecules through a selectively-permeable membranes from a region of lower solute concentration to one with higher concentration. FO is an emerging membrane process that can greatly reduce energy consumption for separation. Also, many purification and separation processes can be performed using this concept with very high efficiency (Chung et al., 2012; Basile et al., 2020; Wibisono and Bilad, 2020). Currently, many applications for forward osmosis have been considered, but a limited number of them have been extended from laboratory scale to commercial scale. Municipal wastewater treatment is one of the most promising applications of forward osmosis, which can lead to the scaling up of this process from laboratory scale to industrial scale.

In recent years, municipal wastewater has been considered for producing water, clean energy, helping to reduce the consumption of fossil fuels, and underground water resources, as well as reducing the rate of global warming. Effluent from municipal wastewater treatment plants can be converted into renewable energy and the energy produced from wastewater can be used as a clean energy source. On the other hand, other contaminants in municipal wastewater can also be converted into other by-products. Municipal wastewater is the most abundant type of wastewater that falls into the category of low-strength waste streams, characterized by low organic strength and high particulate organic matter content (Sikosana et al., 2019).

Government institutions of different Countries are trying to prevent the destruction of the environment by reducing the emission of waste and the discharge of untreated sewage into the environment. Many environmental protection processes are costly, while the processes with proper economic justification that lead to environmental protection can also be ideal. Municipal wastewater and its treatment using aerobic and anaerobic treatment processes or combined processes can be considered as an environmentally friendly process that, in addition to preventing the spread of pollution in the environment, also increases profitability by producing by-products (biomass, biogas, water, etc.) (Qu et al., 2019; Sikosana et al., 2019).

In general, aerobic and anaerobic treatment systems are considered as two basic processes for municipal wastewater treatment. Usually, when biodegradable Chemical Oxygen Demand (COD) concentrations are less than 1000 mg/L, the use of aerobic wastewater treatment methods is preferred. Meanwhile, anaerobic methods are also used for municipal wastewater containing biodegradable COD concentrations over 4000 mg/L. Anaerobic wastewater treatment usually consumes much less energy and the activity of microorganisms in this method requires a high concentration of organic substances. On the other hand, despite the fact that more energy is consumed in the aerobic method, usually the organic substances in the wastewater are removed better and the biomass clots are well formed and can be separated from the wastewater. Hence, much less suspended solids are seen in the aerobic wastewater treatment product. Generally, the effluent quality produced from an aerobic treatment system is higher than the anaerobic system. It can be said that the choice of method is completely dependent on the quality of municipal wastewater and the desired product produced from the wastewater treatment process (Daelman et al., 2012; Gerba and Pepper, 2019).

Municipal wastewater treatment as one of the important applications of forward osmosis process can produce high-quality effluent and pre-concentrated wastewater for anaerobic treatment to facilitate the recovery of nutrients and energy. Anyway, it seems that at least the FO process can be considered as an excellent pretreatment for complex wastewater. Additionally, the FO process can be combined with other separation technologies so designing hybrid processes optimized for the treatment of municipal wastewater (Ismail et al., 2018; Rastogi, 2016, 2020). In this chapter, the fundamentals of FO along with typical membranes and membrane modules used in this technology are previously introduced. Then, typical applications of FO and their latest achievements in the treatment of municipal wastewater are illustrated and discussed.

12.2 Fundamental of forward osmosis

The human has been familiar with the phenomenon of osmosis since the beginning of his life on earth, or at least it can be said that he used it without having a proper understanding of this phenomenon (Marbach and Bocquet, 2019; Ng et al., 2006). Years ago, drying food using salt was used for long-term preservation (Borg, 2003; Qasim et al., 2015). Also, we know that bacteria, microorganisms, fungi, and many pathogenic organisms lose their water and eventually die when exposed to a salty environment. This phenomenon is similar to osmosis, a phenomenon in which water passes through a semi-permeable membrane from a purer environment to an environment with a lower purity of water, while the membrane does not allow the passage of other components in water such as molecules and ions (Kiil, 1989; Maddah and Chogle, 2017). Osmotic pressure is known as the driving force of water transfer from semi-permeable membrane, which is currently used in different applications such as water purification, drug release, and food processing (Cath et al., 2006; Guell and Brenner, 1996).

Nowadays, reverse osmosis (RO) is the most well-known process that is used in industries based on the phenomenon of osmosis. The spontaneous path of water movement is from pure water to impure water through a semi-permeable membrane. Meanwhile, by applying hydraulic pressure on impure water, the natural movement path of water is reversed, and the process of RO is formed. It is clear that hydraulically pressure applied on the impure water is the driving force of mass transfer in the RO process (Joo and Tansel, 2015; Kucera, 2023; Altaee et al., 2014).

In general, chemical potential difference is known as the driving force of mass transfer. Pure water has a higher chemical potential and passes through the semi-permeable membrane to impure water with a lower chemical potential. Applying hydraulic pressure on impure water can prevent mass transfer from semi-permeable membrane, which is known as osmotic pressure (π) (Tow et al., 2018; Altaee et al., 2014). If the osmotic pressure difference ($\Delta\pi$) leads to the transfer of mass through the semi-permeable membrane, the process of FO is obtained, and the path of mass transfer in this process is the opposite of the process of RO. The purpose of the FO is to concentrate the feed stream by using a very concentrated stream as a draw solution (Kwan et al., 2015; Siddiqui et al., 2018).

In addition to RO and FO processes, a pressure related osmosis (PRO) process has also been introduced, which is known as an intermediate process between FO and RO. In this process, similar to the RO process, hydraulic pressure is applied to impure water, but the mass transfer path is from pure water to impure water, like the FO process (Blandin et al., 2016; Lee et al., 2010). The mass transfer in all three mentioned processes can be defined using Eq. (1) (Cath et al., 2006; Shaffer et al., 2015):

$$J_w = A \cdot (\sigma \Delta \pi - \Delta P) \quad (1)$$

in which J_w is the water flux, A is the water permeability constant of the membrane, σ the reflection coefficient, $\Delta\pi$ the osmotic pressure difference. Obviously, for the FO process, ΔP is equal to zero, and for the RO process, $\Delta P > \Delta\pi$ (Cath et al., 2006; Kucera, 2023). Naturally, the relation $\Delta\pi > \Delta P$ is also established for the PRO process. Schematic of FO, RO, and PRO processes along with water flow path is shown in Fig. 12.1. In Fig. 12.2 also graphically shows the relations between the applied pressure and osmotic pressure in these processes (der Bruggen and Luis, 2015; Shaffer et al., 2015; Wibisono and Bilad, 2020).

The FO process has received considerable attention in recent years due to the favorable benefits that exist in this process. In the FO process, due to any requirement to high pressure, current energy supply costs will be lower. Also, due to the lack of pressure in the FO process, the propensity of fouling/scaling is much lower than pressure-driven membrane systems (Holloway et al., 2007). In addition, the equipment used in this process is very simple and requires much less maintenance. Processes which require high pressures or temperatures are limited for a range of industries such as pharmaceutical and food industries. Adversely, the FO process has the potential to be used in several industries. Also, the principles of FO process can be exploited in medicine for slow and precise drug release (Bhatt, 2004).

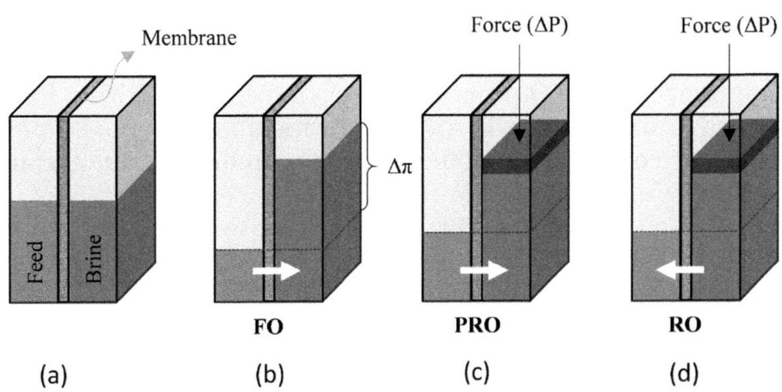

Fig. 12.1 Schematic of: (A) membrane module, (B) FO where ΔP is almost zero and water is transferred from the pure side to the impure side (or to the saltier side), (C) PRO where $\Delta\pi > \Delta P$ is maintained but water still diffuses to the impure (saltier) side, (D) RO where $\Delta P > \Delta\pi$ and water is transferred to the pure side (Rastogi, 2020).

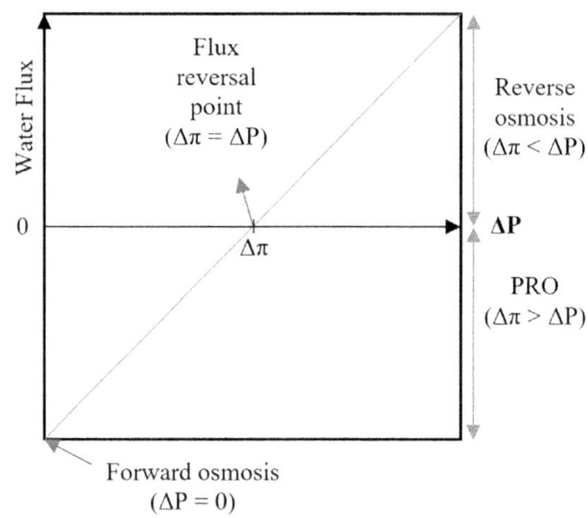

Fig. 12.2 The diagram for determining the range and conditions of each of the PRO, FO and RO processes, in which the case that the hydraulic pressure is greater than the osmotic pressure difference, the RO process will be formed. When the hydraulic pressure is zero, the FO process will take place. Naturally, the PRO process will also be when the applied hydraulic pressure is greater than zero and less than the hydraulic pressure difference (Rastogi, 2020).

12.3 Membranes and modules for forward osmosis

Membranes, draw solutions and membrane modules are the three main pillars of the FO process; their properties for specifical applications in FO are illustrated in the following.

12.3.1 Draw solutions

The draw solution, also indicated as osmotic media, osmotic agent, osmotic engine, driving solution or brine, plays a very important rule in FO processes. The source of the driving force that leads to mass transfer through the membrane in the FO process is the concentration difference between the feed stream and the concentrated stream.

Choosing a draw solution is the first step to develop a FO process for designing a separation process for a specific application. The basic point of choosing draw solution is that this solution must have a higher osmotic pressure than the feed solution to provide the mass transfer from the feed to this solution (Hilal et al., 2021; Basile et al., 2020; Rastogi et al., 2015).

The FO process is designed based on the pull of the solvent from a feed to the draw solution, which realizes the required driving force by the difference in the chemical potential between the two sides of the membrane. In the meantime, in addition to choosing a draw solution, choosing a draw solution re-concentration method is also very important (Awad et al., 2019; Nguyen et al., 2023; Rastogi, 2016). With the start of the FO process, the feed solvent is drawn toward the draw solution, and over time, the concentration of this solution decreases, so that the draw solution must be concentrated again for the proper operation of the FO system. Concentrated sodium chloride solution is known as one of the most widely used draw solutions. The large use of sodium chloride as draw solution in FO is its high solubility in water and the possibility of re-concentrating it in the RO process with limited possibility of scaling on the membrane surface. Of course, for special applications in which high rejection is usually considered, polyvalent ionic solutions are preferable to sodium chloride (Rastogi, 2016; Suwaileh et al., 2021; Zhao et al., 2012).

Various solutions have been proposed as draw solutions in the past, including the use of sea water. Sea water is available in a large volume in many Countries of the world and if the use of separation processes using sea water is expanded, this issue will be a fundamental advantage for the FO methodology. (Batchelder, 1965) first proposed an FO separation process for water demineralization by using sulfur dioxide as a volatile draw solute. (Glew, 1965) used a mixture of water and gas such as sulfur dioxide or a mixture of water with a liquid such as aliphatic alcohols as a draw solution in the FO process. Glew was also the first to propose the recycling of the draw solution in conjunction with FO.

(Frank, 1972; Kravath and Davis, 1975; Kessler and Moody, 1976; Stache, 1989) respectively used aluminum sulfate solution, solution containing glucose, mixed solution of glucose and concentrated fructose, and a solution containing concentrated fructose to purify seawater and produce a nutritious drink. (McGinnis, 2002) also used a two-step process to use the FO process for seawater treatment. In this study, the property of temperature changes has also been used to change the solubility of substances. Also, in this study, KNO_3 and SO_2 solutions were used as draw solution for desalination of seawater. It was also confirmed that merging ammonia and carbon dioxide gases in specific ratios formed highly concentrated draw solutions of thermally removable ammonium salts (McCutcheon et al., 2005, 2006; McGinnis, 2002). This method produced FO draw solutions with excess of 250 atm osmotic pressures, allowing unprecedented great recoveries of drinkable water from concentrated saline feeds and substantial decreases in brine releases from desalination. In a new nanotechnological method, naturally non-toxic magnetoferritin is being experimented as a possible solute for draw solutions (Cath et al., 2006; Rastogi, 2020). Magnetoferritin can be rapidly separated from aqueous streams by a magnetic field.

12.3.2 Membranes for FO processes

A wide range of membranes can be candidates for use in the FO process. In other words, any material that has a selective permeability, is non-porous and has a suitable density to create a phase against the transfer of part of the particles can be used as a FO membrane. Early membrane researchers experimented with every type of membrane material available, including bladders of pigs, cattle, and fish; collodion (nitrocellulose); rubber porcelain and goldbeaters' skin (Anderson, 1977). In the 1960s, the science of membrane manufacturing has made significant advances and modern membranes were produced, which was due to the discovery of the Loeb-Sourirajan process, which led to the production of RO membranes with defect-free, high-flux, and anisotropic properties (Baker, 2023; Loeb, 1995; Loeb et al., 1997) investigated asymmetric aromatic polyamide membranes for use in FO and PRO processes.

In the 1970s, flat sheet or tubular membranes were used as the main membranes for the application in FO processes. Of course, most of the experiments in these years were done on the PRO process. Obviously, the flat sheet and tubular modules produced much less flux than expected. Votta (1974) and Anderson (1977) used commercially available RO membranes as well as synthetic cellulose acetate membranes in the FO process. They used the mentioned membranes to treat the diluted wastewater in the FO process by a simulated seawater draw solution. Kravath and Davis (1975) used flat-sheet RO membranes for seawater desalination by FO by using glucose solution as draw solution. Goosens and Van Haute (1978) also investigated the effect of changes by adding mineral fillers to the membrane manufacturing process. They tried to prove that membrane changes during FO process can predict RO process changes.

Also in the 1970s, Mehta and Loeb (1978) investigated the DuPont B-9 (flat sheet) and B-10 (hollow fiber) Permasep RO membranes made of an aromatic polyamide polymer. Flux through the B-10 Permasep membrane was comparatively low (approximately 8 $L\,m^{-2}\,h^{-1}$ at a pressure of 80 bar). During the

1990s, a singular membrane for FO was developed by Osmotek Inc. (Albany, Oregon) (currently Hydration Technologies Inc. (HTI)). This membrane has been experimented in a wide range of applications (Beaudry and Herron, 1997; Beaudry and Lampi, 1990; Cath et al., 2005; McCutcheon et al., 2006; Wrolstad et al., 1993; York et al., 1999). Also, it was successfully used commercially for many applications such as water purification for military and recreational purposes and showed good efficiency (Goosens and Van Haute, 1978). Cellulose triacetate (CTA) membrane was synthesized for use in the FO process (McCutcheon et al., 2005). The thickness of this membrane, based on the results obtained from the cross-sectional SEM images, was less than 50 μm, while the thickness of current RO membranes, placed on a thick membrane support, is less than 1 μm. CTA FO membranes were designed and synthesized without the need for a support layer and were used with suitable efficiency. The success factors of this membrane are probably the relative thickness of the membrane and the absence of a thick support layer.

To sum up, we can name certain characteristics for a membrane that can lead to the improvement of the membrane performance in the FO process. The high density of the active layer of the membrane can lead to a high rejection rate, in which case, the manufactured product will pass through the membrane with greater purity. On the other hand, the membrane active layer thickness minimization can lead to a decrease in internal concentration polarization (CP) and an increase in water permeation flux. Increasing membrane hydrophilicity can reduce membrane fouling and increase water permeation flux. Increasing the mechanical strength of FO membranes can lead to the possibility of applying higher hydraulic pressure to perform the PRO process. However, it seems that the development of FO membranes is needed to achieve efficient processes. Optimum membranes for FO processes should have high water flux and salt rejection rate as well minimal internal concentration polarization. High mechanical strength to support high hydraulic pressures is another suitable feature for optimal FO membrane (Hilal et al., 2021; Cath et al., 2006; Linares et al., 2017).

12.3.3 Membrane modules and devices

Many designs have been investigated to configure FO modules, each of which has advantages and disadvantages. In laboratory scale, flat or tubular/capillary membranes are usually used. In larger scales and for industrial applications, flat sheet membranes in plate-and-frame configurations are usually evaluated to check their efficiency for scaling up. Obviously, any design can lead to certain benefits. Different configurations also have their own limitations. In addition to the difference between plate-and-frame, spiral-wound, tubular, and bag configurations and the specific advantages and limitations of each, the differences between continuous and batch processes should be considered (Shaffer et al., 2015).

When the FO process is used as a continuous process, the draw solution is continuously fed into the concentration/regeneration process. In this type of process, a material cycle is formed on both sides of the membrane and the feed solution is continuously recirculated on the retentate side and the draw solution on the permeate side (see Fig. 12.3). Due to the fact that the flow rotation process is done continuously on both sides of the membrane, the exploitation of the FO process is more complicated compared to the pressure-driven methods. For example, the spiral-wound module cannot be used for FO because it is not possible to create a forced flow on the permeate side of the membrane (Rastogi, 2020).

In the batch application of the FO process, unlike the continuous process, the draw solution is not regenerated and is not used after concentration. In this system, usually after setting up the process, the primary solutions continue until the mass transfer process is completed. Therefore, it can be said that batch FO can be considered a one-time process and is not reused. Applications using this mode of operation include hydration bags for water purification and osmotic pumps for drug delivery (Suwaileh et al., 2021).

After examining the batch and continuous systems, we can discuss about the membrane modules that can be used in the

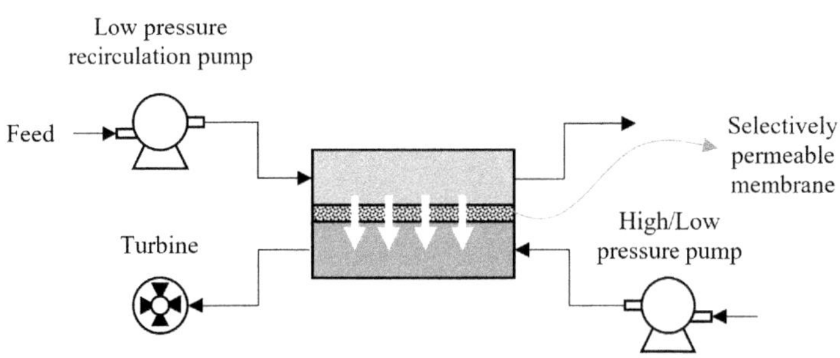

Fig. 12.3 Schematic of the flow path in FO and PRO processes where the feed solution flows on the active layer side of the membrane and the draw solution flows counter-currently on the permeate side. Of course, when the PRO process is done, a hydraulic pressure is applied to the draw solution, which can be fed into a turbine after the process is done and generate electricity to optimize energy consumption.

FO process. There are many choices for making membranes and membrane modules for FO based separation systems. Membranes can be made in many configurations such as flat sheet, tubular, or hollow fiber. Despite many disadvantages of flat sheet configuration, these membranes are currently considered as the best option for FO applications. Flat sheet membranes can be used in plate-and-frame or unique spiral-wound modules (der Bruggen and Luis, 2015; Hilal et al., 2021).

The simplest module used for flat sheet membranes is the *plate-and-frame module*, which is available in various shapes and sizes for laboratory and industrial applications. Plate-and-frame modules can accommodate one to several thousand flat sheet membranes and provide a suitable efficiency for the use. Of course, this type of design is more known for plate-and-frame heat exchangers, but the use of plate-and-frame configuration, despite its much greater complexity than the design of heat exchangers of the same type, has also been taken into consideration (Wibisono and Bilad, 2020). Like any other design, many advantages can be mentioned for plate-and-frame modules; on the other hand, the lack of adequate membrane support and the low packing density are the two main disadvantages of these modules.

The hydraulic or osmotic pressure difference (or a combination of both) is known as the driving force of separation in FO, RO, and PRO processes, and the ability to change the pressure parameter is very important to achieve the desired permeate production flux and the appropriate rejection rate. Lack of adequate membrane support eliminates the ability to apply large changes in the process or create high pressure differences. Therefore, the process must be carried out at low or similar pressures on both sides of the membrane, in which case very strict process control is required (Suwaileh et al., 2021). On the other hand, high packing density can lead to lower investment cost, smaller system footprint, and lower process costs (labor for membrane replacement). Therefore, despite the low packing density that exists for flat sheet membranes, an increase in the footprint of the membrane process, an increase in investment costs, and an increase in operational costs are expected. There are other limitations for plate-and-frame modules, which include external and internal sealing problems, the difficulty of quality control in the membrane process, such as monitoring the integrity of the membrane, and serious limitations in the operating conditions of the process, such as flow rate, pressure and temperature (Zhao et al., 2012).

Spiral-wound modules are currently known as the most important membrane configuration that is used in many membrane separation processes such as RO. The control of this system is done only by controlling the feed flow and the permeate flow is only collected and removed from the permeate side. Meanwhile, in order to exploit FO membranes, it is necessary to be able to circulate the flow of draw solution as a forced flow in the system. Therefore, due to the fact that the permeate flow in the spiral-wound system collects very slowly in a collector tube and leaves the system, this system cannot be used as a module for the FO process (Chung et al., 2012; der Bruggen and Luis, 2015; Suwaileh et al., 2021).

In order to use the spiral-wound module in the FO process, certain changes must be made to it. An innovative design for the use of spiral-wound modules in the FO process is shown in Fig. 12.4 (Mehta, 1982). The draw solution flows through the spacers and among the rolled membranes, in the similar method that a feed stream flows in a spiral-wound element for RO. However, unlike RO elements, the central collecting tube is blocked halfway through so that the feed solution cannot flow to the other side. Instead, an additional glue line at the center of the membrane envelope provides a path for the feed to flow inside the envelope. In this configuration, the feed flows into the first half of the perforated central pipe, is then forced to flow into the

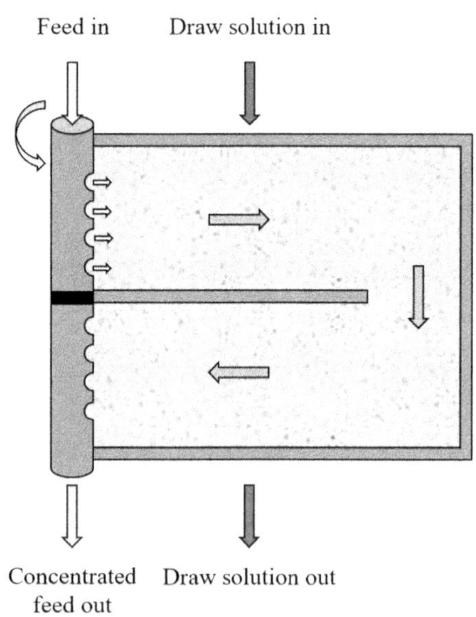

Fig. 12.4 Schematic of a specific spiral-wound membrane module intended for use in the FO process.

envelope, and then flows out through the second half of the perforated central pipe. Obviously, the presented smart design can also be used for the PRO system. In addition, the draw solution can be pressurized, similar to what happens in the RO process, with the aim of achieving greater efficiency and increasing the driving force of mass transfer to perform separation in the process (Rastogi, 2016; Basile et al., 2020; Cath et al., 2006).

One of the important points about spiral-wound membranes is the inability to reverse the process flow path (reversing the flow path through the membrane) for the purpose of membrane washing. In the RO process, pressure reversal with the aim of creating a permeate flux from the permeate side to the feed side is one of the most important problems that can lead to the membrane active layer damaging. This phenomenon, which is also known as back-pressure, can cause the formation of ballooning or active layer swelling. During this phenomenon, the active layer is separated from the membrane support and the volume of production increases and salt rejection is greatly reduced. However, due to the lack of membrane support in many FO membranes, flow reversal can be tested in spiral-wound membranes. Sagiv and Semiat (2005) tested flow reversal for backwashing FO membranes in a seawater desalination process. They observed that over the time of operation in the FO process with spiral-wound membranes, a large $CaCO_3$ precipitate forms on the membrane surface. As a result, in certain periods of time, the flow direction was reversed to wash the scales from the surface of the membrane. They found that backwashing the spiral-wound membrane can lead to the complete recovery of the water permeation rate through the membrane.

The use of *tubular membranes* consisting of tubes or hollow fibers can be considered the most important type of membrane to operationalize the use of the FO process. The first reason for this issue is the appropriate mechanical strength of tubular membranes, which leads to the ability to withstand higher hydraulic pressure of these membranes. This issue can lead to the easier packing of these membranes in membrane modules without considering the support. No need for membrane support reduces the internal CP and increases the efficiency of the membrane in the FO process. As a second advantage, the process of making tubular membranes is almost simple and they can be easily synthesized. Also, good packing density can be obtained from these membranes. The ability to be close to the flow on both sides of the membrane is also the third advantage that can be mentioned about tubular membranes for use in the FO process (Linares et al., 2017; Rastogi, 2016, 2020).

Hollow fiber membrane modules are largely used at industrial level. These membranes are usually made with a diameter of less than 1 mm. This issue leads to the reduction of the possibility of turbulent flow on the membrane surface and the bore side. Therefore, the possibility of fouling and scaling in these membranes increases. The same phenomenon reduces mass transfer in the membrane and increases concentration polarization. On the other hand, tubular membranes usually have a diameter higher than 2 mm. At this scale, it is possible to reach turbulent flow conditions and fouling and scaling are reduced. Therefore, this configuration is characterized by higher mass transfer and lower CP (Nguyen et al., 2023; Rastogi, 2020; Rastogi et al., 2015).

Today, the most important industrial application of membranes is in the RO process. Therefore, the membranes used in the RO process are more commonly seen in the market, which are generally semi-permeable flat sheet membranes in spiral-wound modules. Therefore, the tubular membranes did not expand much. The spiral-wound configuration is recognized as an ideal configuration for RO membranes, which has eliminated the need to produce tubular membranes. Thus, expanding the use of the FO process requires the development of tubular membranes specified to this process (Eyvaz et al., 2018; Hilal et al., 2021; Nguyen et al., 2023).

Another widely used application of the FO process is the use of a FO-based membrane in *hydration bag*. This system is a double lined bag, in which the external bag is a completely sealed plastic bag that contains feed water (water that needs to be purified) and FO bag. The internal bag is also the location of draw solution inside a FO membrane, typically a sucrose solution (Awad et al., 2019; Hilal et al., 2021; Wibisono and Bilad, 2020).

12.4 Forward osmosis applications

FO can be considered an ideal process for liquid separation, because if it is designed properly and an acceptable product production rate is reached, the separation process is done with very little energy consumption. However, the FO process has been tested in several separation processes to date. Industrial wastewater treatment has been tested as a bench-scale FO method (Haupt and Lerch, 2018; Kerusha Lutchmiah et al., 2014; Mahto et al., 2021). In addition, the FO process has been used for landfill leachate treatment and its concentration and has been expanded from pilot scale to full-scale (Iskander et al., 2017; Li et al., 2017; Zhang et al., 2021). Due to the lack of harsh conditions and the elimination of the possibility of damage to food, the food industry is one of the most popular targets of the FO process, which has led to the use of this method especially to purify liquid foods (Cath et al., 2006; Rastogi, 2016; Sant'Anna et al., 2012). Reclaiming wastewater for potable reuse in life support systems (Butler et al., 2013; Hancock et al., 2013; Kessler and Moody, 1976) is another application of the FO process. Besides that, we can mention seawater desalination (Akther et al., 2015; Phuntsho et al., 2013; Qin et al., 2012) and water purification in emergency relief situations (Butler et al., 2013; Miller and Evans, 2006; Saiful et al., 2020). Electricity generation from the PRO process is another application of this process that uses the osmotic pressure difference between seawater and fresh water (Zhang et al., 2011; Zhu et al., 2015).

12.4.1 Wastewater treatment and water purification

Considering the global concern about the supply of drinking water needed by humans, water purification can be considered the most important application of the FO process. This process has been used in different configurations for water and wastewater treatment. Some of the tests carried out for the use of these methods have also led to favorable results, and others need to be developed in order to achieve widespread use. Some of the most important applications of the FO process for water and wastewater treatment are discussed below (Awad et al., 2019; Chung et al., 2012).

The concentration of dilute industrial wastewater is one of the first applications considered for the FO process. The purpose of this process was to obtain a method to treat wastewater containing very low concentration of heavy metals for potential reuse. The use of commercial cellulosic membranes has shown that internal CP can lead to a decrease in production flux in the membrane, which must be eliminated to achieve maximum permeate flux production (der Bruggen and Luis, 2015; Suwaileh et al., 2021).

Seawater has been identified as a very ideal source for purification due to the fact that it is a cheap and infinite resource fully available in coastal areas. Different approaches have been considered to improve the efficiency of the FO process for use in seawater desalination, including membrane modifications to improve its performance. However, in order to expand the use of membranes in the FO process, membranes specific to this process, with the purpose of water purification, should be further investigated (der Bruggen and Luis, 2015; Shaffer et al., 2015).

Landfill leachate is known as one of the most important environmental concerns related to landfilling, which can lead to severe soil pollution. Landfill leachate usually contains four types of important pollutants, including total dissolved solids (TDS), dissolved heavy metals, organic and inorganic nitrogen, and organic compounds. Today, the wastewater treatment industry has developed many methods for landfill leachate treatment, which can effectively eliminate the problems caused by landfill leachate. In the meantime, suitable methods for removing organic materials, heavy metals, and nitrogen-containing materials have been provided, but TDS is usually not considered in common wastewater treatment methods that FO-based concentration of landfill leachate can be used efficiently for TDS removal. In other words, the efficient method to remove TDS has not been properly addressed in conventional treatment methods. It seems that FO method can concentrate TDS with good efficiency (Eyvaz et al., 2018; Hilal et al., 2021; Nguyen et al., 2023).

Usually, there are two main methods for effectively removing TDS from landfill leachate: mechanical processes/evaporation and membrane processes. A combination of mechanical, evaporation, RO and FO methods have been investigated for TDS removal, and the appropriate efficiency of the FO process has been confirmed. The results have shown that the combination of membrane and mechanical methods can effectively reduce TDS to a level less than 100 mg/l so that the treated effluent can be used for irrigation (Nguyen et al., 2023; Rastogi, 2020; Rastogi et al., 2015).

Salty water recovery can be done by FO-based processes. CTA membranes produced by Osmotek have been used in FO processes to separate water containing NaCl; in addition to water recovery of about 95%, these membranes provided high rejections for water contaminant. It seems that the reduction of water flow is one of the most important drawbacks of the FO process. Due to the fact that the mass transfer in FO is carried out spontaneously and without applying external pressure, the deposition of materials is carried out along with the water flow, and it is not possible to separate water using the high speed of water movement on the surface of the membrane. Since the increase in the rate of water passage leads to a decrease in the time of proximity of the two streams, as a result, the amount of mass transfer decreases. A reduction of about 30–50% of the flux through the FO membrane has been reported during operation. Of course, fouling of FO membranes usually has less adhesion to the membrane surface than processes such as RO, and experimental results have shown that usually the entire efficiency is restored by chemical washing of the membrane. However, studies have shown that the FO process can be successfully used to concentrate landfill leachate. Therefore, the scale of these units was increased from pilot to plant (Cath et al., 2006; Rastogi, 2020).

Today, space and space missions have become one of the most important strategic issues of different Countries in the world. Space missions are usually difficult and long-term missions, so it is not possible to transport the water needed by astronauts for such a long time. Therefore, carrying out space missions requires continuous and low-cost supply of water needed for various purposes such as sanitary and drinking purposes. Direct potable reuse for advanced life support systems is another application of FO-based membrane process. In the space missions, it is necessary that a purification system can convert the generated wastewater into purified water. Moisture condensate, urine, and sanitary sewage are known as the three main sources of wastewater production in space missions. Therefore, a reliable wastewater treatment system is needed that can convert the generated wastewater into usable water. This system should be reliable, light, durable, and have high recovery capability. Also, the maintenance of this system should be simple with reduced energy consumption. The use of side chemicals is also a limitation of this system. The system without the need to use additional chemicals for water purification is known as an ideal system for use in spacecraft (Cath et al., 2006; Rastogi, 2020; Wibisono and Bilad, 2020).

One smart designed system that have been developed to use the FO process in space is called forward osmotic concentration (FOC), which has the task of purifying wastewater produced in space missions. Among the various methods developed by the U.S. National Aeronautics and Space Administration (NASA) for wastewater treatment in space missions, FOC is considered as an ideal efficient method based on FO process. This system consists of two stages named FOC 01 and FOC 02, as illustrated in Fig. 12.5. Before entering these steps, an RO system is considered. The concentrate stream of this stage enters a two-stage process that initially includes a FO process. The second step is a combination of FO and osmotic distillation (Eyvaz et al., 2018; Nguyen et al., 2023; (Wibisono and Bilad, 2020)).

NASA has been able to achieve many advantages by using the FO process instead of RO. The energy consumption in the FO process is much less, which can eliminate the problem of supplying pressure to perform the RO process. FO membranes have less internal CP than RO membranes, which leads to better performance of FO membranes compared to RO. Also, RO membranes are usually very hard to regenerate by chemical cleaning

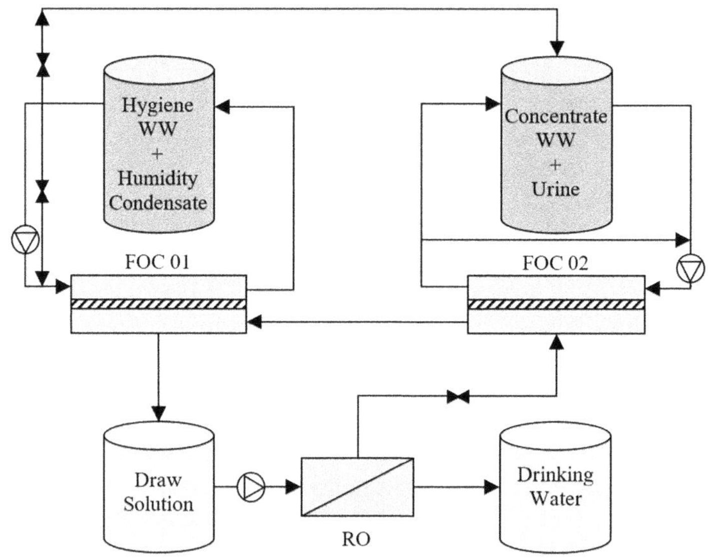

Fig. 12.5 A schematic of the FOC process developed by NASA, where the wastewater stream is pretreated in two stages, FOC 01 and FOC 02, and the draw solution stream enters the RO process. By selecting water and separating it from the draw solution, in addition to producing the pure water required by the spacecraft, the flow of the draw solution is re-concentrated.

processes. On the other hand, in addition to less fouling in FO membranes, their cleaning process is also easy. Therefore, the raw wastewater flow is connected to the FO membranes and the flow containing draw solution and water is sent to the RO membranes. Choosing the type of draw solution is being studied with the aim of increasing the osmotic pressure of water transfer from the wastewater stream to the draw solution and also minimizing fouling on RO membranes in the presence of draw solution (Eyvaz et al., 2018; Nguyen et al., 2023).

Another application of FO in water and wastewater treatment is the concentration of digested sludge liquids. Sludges produced in wastewater treatment facilities are often treated in anaerobic digesters with the aim of further destroying resistant organic solids and further stabilizing the sludge. Dewatering of sludge is usually done after digestion, and the output of this process is a thick liquid stream that contains many organic pollutants. A solid stream (biosolid) is also produced. The concentrated liquid stream produced from the dewatering process contains many pollutants such as nutrients (e.g., ortho-phosphate, ammonia, and organic nitrogen), TDS, TOC, color, TSS, and heavy metals. Currently, it is usually recommended to return this flow back to the primary treatment process, in which case many costs such as re-transfer and re-operation of the facilities are imposed. Also, some of the materials in this stream include recalcitrant nitrogen and phosphorus species that can enter the wastewater stream.

Therefore, the use of a suitable treatment method should replace the return of the concentrated liquid stream to the treatment process (Kucera, 2023; Rastogi et al., 2015).

Holloway et al. (2007) used the FO process to concentrate anaerobic digester centrate at the Truckee Meadows Water Reclamation Facility (TMWRF) in Reno (Nevada) and investigated the advantages and disadvantages of this system. The main purpose of this study was to investigate the FO process as an alternative method for centrate concentration, and the use of this method was successfully tested. Also, the concentrated flow produced from this process was considered for use as fertilizer. In addition, the economic model and the cost effectiveness of the FO process were also investigated.

Other applications of FO in water purification are hydration bags. Hydration bags are usually used in special situations and have military, relief, and recreational applications. When reliable water sources are not available in emergency situations, it is recommended to use hydration bags. These bags are known as one of the few applications of FO membranes, which has become very important today. Compared to other water purification devices, these bags are much slower, but they do not need to use electricity, and even muddy water can be turned into suitable water. FO membranes currently have good selectivity. Therefore, the filtered stream in hydration bags is usually free of microorganisms, most ions, and most macromolecules (Shaffer et al., 2015; Zhao et al., 2012).

12.4.2 Seawater desalination

Desalination of sea water is one of the most important methods that can solve the problem of water shortage in the future. Many patents have been registered in recent years for the use of FO systems for desalination of seawater, but almost all of these patents have faced problems in increasing the scale and reaching the industrial scale. Currently, there are limited studies that have investigated the use of FO in seawater desalination. However, Kravath and Davis (1975) used cellulose acetate flat sheet and hollow fiber membranes for desalination of Atlantic Ocean water. In this study, glucose solution was used as draw solution. Kessler and Moody (1976) conducted experiments on the FO process for desalination. They pointed out that in special and sensitive conditions, this system can be used to supply the required water, but the FO process cannot be used as a sustainable process to supply the required drinking water using sea water and during the desalination process.

Laboratory-scale and bench-scale studies have shown that, under certain conditions, the FO process can be used efficiently for seawater desalination (McCutcheon et al., 2005, 2006; McGinnis, 2002). The efficiency of the process is largely dependent on the choice of membrane type and draw solution. For example, the use of CTA membranes and a very strong draw solution, such as highly soluble ammonia and carbon dioxide gases, led to the creation of an efficient process for desalination of sea water. In another study, ammonium carbonate and ammonium hydroxide were introduced as suitable draw solutions (McCutcheon et al., 2006). However, draw solutions have great effect on the cost effectivity of the FO process with the aim of desalination of sea water, and studies have shown that very high osmotic pressures as driving force ($\Delta\pi$) can also be obtained for 0.05 and 2 M NaCl solutions (about 238 bar and 127 bar, respectively) with an appropriate combination of ammonia/carbon dioxide as draw solution (Chung et al., 2012; Suwaileh et al., 2021).

A simple schematic of a FO process using a special ammonia-carbon dioxide solution as draw solution is given in Fig. 12.6. During this process, pure water from seawater enters the draw solution through a semi-permeable membrane. Therefore, the draw solution is diluted and the water flow must be separated from it for re-concentration. Here, the draw solution is heated to a temperature of about 60 °C. Therefore, the draw solution decomposes and a mixture of ammonia and carbon dioxide is produced. Now, the process of separating draw solution and absorbed water from seawater is done by processes such as column distillation or membrane distillation (MD). During this process, the water stream is separated from the draw solution and the concentrated draw solution stream is recovered and can be used again in the FO process (McCutcheon et al., 2005).

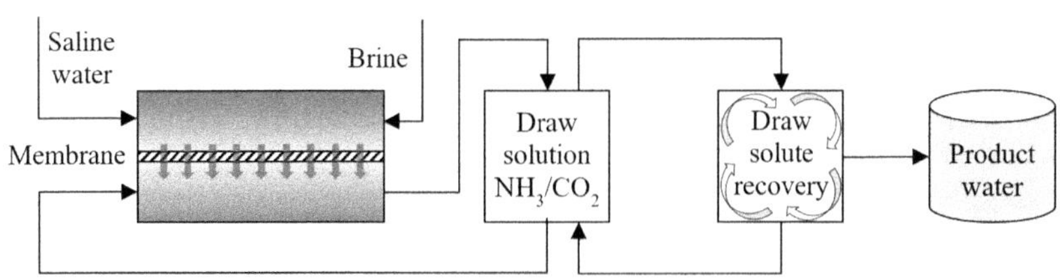

Fig. 12.6 Schematic of seawater desalination process using FO process and a special draw solution containing ammonia and carbon dioxide.

12.4.3 Other applications of FO processes

In addition to all studied applications of FO process, food processing systems can efficiently use this process. Osmotic treatment is a common and well-known process in food products for materials such as meats and fruits (Rosa and Giroux, 2001). However, the FO process, despite its high potential, has only been investigated at laboratory scale (Beaudry and Lampi, 1990; Wrolstad et al., 1993). It should be noted that the FO process, unlike many other processes, does not require harsh operation conditions and is performed at acceptable temperatures and pressures in food applications. Therefore, it can preserve the taste, aroma, and color of the food and also preserve the nutritional value of the products. This process is characterized by high removal, less membrane fouling, and favorable performance; therefore, it can be a suitable option for the concentration of beverages and liquid foods.

Regarding the use of FO process in the pharmaceutical industry, we can also discuss about osmotic pumps, which have many advantages compared to common drug delivery methods. Common methods of drug delivery, such as the oral method, have many limitations to achieve the desired results. In situations where drug delivery requires a specific, targeted dose, or long-term release of the drug in the patient's body, the oral method cannot be effective. Regular drug concentration, reduction of medication frequency, controlled and accurate release of drug in the blood environment, increased bioavailability, improved patient compliance, and reduced side effects are some of the positive features that can be obtained by using osmotic pumps. Of course, in some cases, such as patients suffering from gastrointestinal problems, it is generally not possible to use some medicines orally. Therefore, osmotic pumps are an ideal procedure for these patients (Hilal et al., 2021; Nguyen et al., 2023).

12.5 Municipal wastewater treatment with forward osmosis

Nowadays, sustainable water supply, less energy consumption, and environmental protection are among the most important concerns of human life. It is possible to solve some of these concerns by using an efficient method for municipal wastewater treatment. This solution should be economical as well to prevent the spread of pollution in the environment as an environmentally friendly process. In addition to supplying the required water, the recovery and rehabilitation of municipal wastewater can provide a concentrated flow of organic matter as well biogas, during the anaerobic treatment process, as a green energy source. Recovery of nutrients from wastewater also deserves special attention due to the increasing stringency of effluent discharge regulations and uncertainties associated with minable phosphorus supply for food security (Elser and Bennett, 2011; Koppelaar and Weikard, 2013; Morse et al., 1998).

Activated sludge treatment is a well-known biological treatment process, which basically aims to remove organic matter, nutrients, and pathogens from wastewater. Of course, nutrients and energy cannot be effectively recycled in this method. On the other hand, this method requires aeration, which consumes large amounts of energy. Also, the remaining sludge from this process can cause problems for the environment. In activated sludge treatment process, the carbon content of the wastewater as chemical energy source and the nitrogen content as nutrient materials source are converted into biomass, carbon dioxide, and nitrogen during the activity of microorganisms. It can be said that during this process, large amounts of extractable nutrients and energy are wasted, and also large amounts of energy are consumed to carry out this process. Anaerobic treatment can be considered as an alternative to activated sludge treatment, which is carried out in the absence of oxygen, and the water contaminants are converted into methane-rich gas. Anaerobic treatment also converts phosphorus into more common materials through a chemical change so that it can be extracted in subsequent recovery processes (Smith et al., 2012). Due to the fact that there is no need for aeration in the anaerobic process, the amount of energy consumption is significantly reduced. Also, the anaerobic process can be converted into a process without the need for external energy by using the energy from biogas consumption (Batstone et al., 2015; Frijns et al., 2013; Sills et al., 2016).

Anaerobic treatment of wastewater and biogas production as a suitable opportunity to obtain a sustainable source of water as well as renewable energy production has shown a very good potential to reduce the environmental stress of wastewater. The results of studies have shown that the content of chemical energy in municipal wastewater is about 9 times higher than the energy required to operate an activated sludge unit. Despite all the advantages of the anaerobic treatment process, a major problem has led to the stop of the development of this process (Meerburg et al., 2015; Wan et al., 2016). The concentration of organic pollutants in municipal wastewater is usually low. This issue leads to a decrease in the activity of the anaerobic digester, which ultimately significantly reduces the amount of gas produced per unit volume of wastewater, and the removal of wastewater pollutants will not be done properly. On the other hand, methane can be dissolved in a small amount in water (about 22.7 mg/L), which leads to the lack of biogas production and low efficiency of the anaerobic treatment process (Smith et al., 2012). Membrane technologies and membrane bioreactors are considered as an attractive technology to solve the problem of lack of organic matter concentration in municipal wastewater.

The hybridization of anaerobic treatment process with membrane technology such as RO, nanofiltration, MD and FO have led to the development of anaerobic osmotic membrane bioreactors (AnOMBR). Among the different membrane methods, the FO method is known as the most promising choice for the hybridization of anaerobic treatment of municipal wastewater and membrane processes due to the appropriate separation efficiency, reversibility of membrane fouling and low energy consumption (Achilli et al., 2009; Lay et al., 2010; Lee et al., 2010). As a major advantage, FO technology operates with minimal external energy requirements (Cath et al., 2006). Kerusha Lutchmiah et al. (2014) used FO for water treatment and pointed out that development of new membranes and optimization of operating conditions of wastewater treatment based on FO process should be considered in order to achieve a scalable efficiency. Holloway et al. (2015) conducted a comprehensive study on the application of membrane bioreactors for the production of potable water using municipal wastewater. They reported that using osmotic membrane bioreactors, municipal wastewater treatment can be done with high efficiency. Xie et al. (2016) mentioned FO as a new technology with good potential to recover municipal wastewater nutrients. Previous studies consider the FO process as an important technology for the future of municipal wastewater treatment (Kerusha Lutchmiah et al., 2014; Subramani and Jacangelo, 2015; Li et al., 2015). However, despite all the advantages, membrane fouling, concentration polarization, and anaerobic treatment integration require further investigation and more appropriate development.

By summarizing the available studies, it can be concluded that using the FO process for municipal wastewater treatment can be done with three types of configurations (Fig. 12.7). Currently, the most common approach for wastewater treatment using the FO process is the aerobic osmosis membrane reactor (AeOMBR). AnOMBR has the potential to be used as a simultaneous process of wastewater treatment and biogas production. A submerged FO membrane is used to separate the water in the wastewater stream and concentrate its solutes. The third proposed process for

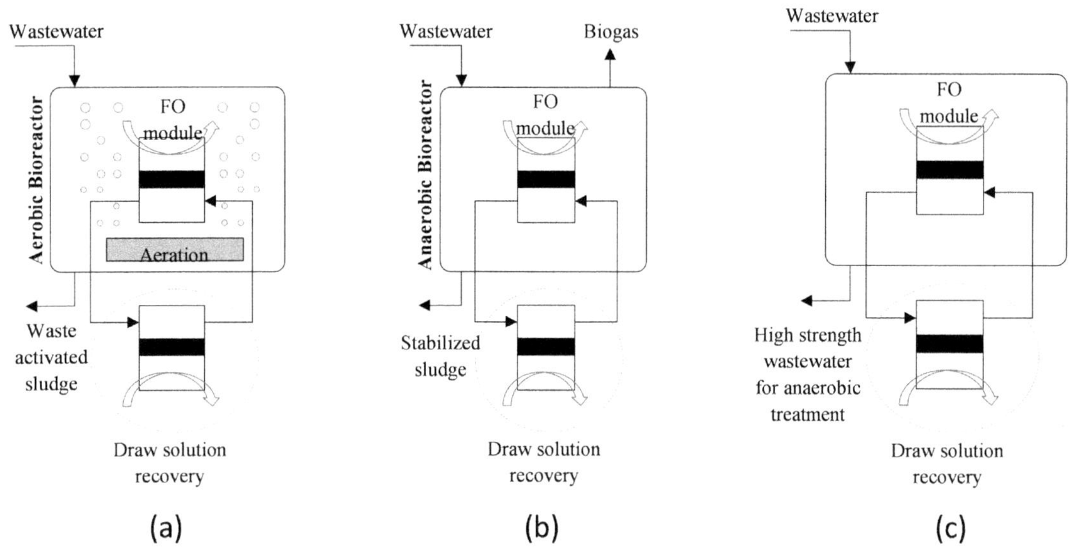

Fig. 12.7 Schematic of the three main processes containing the osmotic bioreactor: (A) AeOMBR; (B) AnOMBR; (C) the wastewater concentration process with the aim of providing concentrated wastewater to enter the anaerobic digestion reactor.

combining FO membranes with bioreactors for wastewater treatment involves performing the FO process before anaerobic digestion. This system has a similar concept to the AnOMBR process. Obviously, in this method, the FO membrane is in contact with concentrated wastewater, which has a lower tendency to cause fouling on the FO membrane compared to mixed liquor inside an AnOMBR. However, it seems that concentration of the wastewater flow before entering the OMBR is the best choice to increase the efficiency of municipal wastewater treatment (Ansari et al., 2017; Suwaileh et al., 2021).

The type and extent of separation in the FO process can be completely different based on the type of biological treatment. In other words, membrane type, biological treatment process, and operation conditions can affect the level of municipal wastewater treatment. The removal of nitrogen, organic matter, phosphorus, and trace organic pollutants (TrOCs) are known as the basic parameters for determining the efficiency of the separation process. If the concentrated flow of these pollutants enters the biological treatment process, a suitable removal efficiency can be expected. It should be noted that the FO process can effectively separate the mentioned pollutants and obtain a concentrated flow from them. Therefore, a high removal efficiency can be expected.

Compared to other methods, such as microfiltration (MF) and ultrafiltration(UF), osmotic pretreatment of wastewater has better efficiency in the rejection of phosphorus, organic compounds, colloidal particles, and microbes due to the difference in the separation mechanism (Gu et al., 2015).

Nitrogen separation in FO-based processes can be very different, which stems from the different structures of nitrogen-containing materials in wastewater (Zhang et al., 2014). FO membranes cannot completely provide rejection of neutral ammonia and usually this parameter is reported to be less than 80% (Gu et al., 2015; Linares et al., 2013). Of course, the rejection rate for positively charged ammonium ions is much higher using FO membranes (Xue et al., 2015). Potential of hydrogen (pH) is one of the influencing parameters on the efficiency of AeOMBR process. At neutral pH, biological degradation of nitrogenous materials is achieved during processes such as nitrification/denitrification with appropriate rejection of FO membrane. As can be seen in Table 12.1, the amount of nitrogenous substances removal in the process of anaerobic digestion is very low and compared to the aerobic method, it is not very effective in removing nitrogenous substances. However, it is noteworthy that nitrogen removal could be achieved via struvite (i.e., magnesium ammonium phosphate) recovery, by deploying a dedicated ammonia recovery process, or by converting ammonia microbiologically into nitrous oxide for enhanced biogas utilization (Scherson et al., 2014). Very good TrOC removal using the FO process is a significant advantage of municipal wastewater treatment processes based on these methods (Alturki et al., 2013; Coday et al., 2014). Treated wastewater can be easily used in irrigation and agricultural applications, but the use of this flow as drinking water requires attention to the removal level of a wide range of TrOCs such as pharmaceutical residues, steroid hormones, phytoestrogens, UV-blockers, and pesticides (Luo et al., 2014; Siegrist and Joss, 2012; Wijekoon et al., 2013). AeOMBR is known as the most efficient process for the removal of TrOCs, which simultaneously biodegradation and membrane rejection processes can provide a suitable level of removal, while for AnOMBRs methods, this parameter is less reported (Coday et al., 2014; Song et al., 2016).

Choosing the draw solution recovery method is one of the most important part of municipal wastewater treatment, and currently three types of technology can be used for this task: MD as temperature derived process, RO or NF as pressure derived processes, and ED as electrical energy derived process. Choosing the type of re-concentration process is related to many parameters, among which we can mention draw solution rejection ability and draw

Table 12.1 A selection of the results presented for the use of the combined process of FO and aerobic/anaerobic biological removal and the removal capabilities of various pollutants in wastewater.

FO system	Aerobic/ anaerobic	Membrane	Arrangement	Removal efficiency	Reference
Osmotic membrane bioreactor	Aerobic	CTA	Cross-flow	98% for TOC and 99% for NH_4-N	Wan et al. (2016)
Osmotic membrane bioreactor	Aerobic	TFC	Cross-flow	96% for TOC and 99% for NH_4-N	Wan et al. (2016)
Osmotic membrane bioreactor	Aerobic	CTA	Submerged plate-and-frame	>99% for COD, >99% for TP, and > 82% for TN	Holloway et al. (2014)
Osmotic membrane bioreactor	Aerobic	CTA	Submerged plate-and-frame	98% for TOC, >99% for PO_4, and 80–90% for NH_4-N	Qiu and Ting (2014)
Osmotic membrane bioreactor	Aerobic	CTA	Submerged plate-and-frame	98% for TOC and 98% for NH4-N	Qiu and Ting (2013)
Osmotic membrane bioreactor	Aerobic	CTA	Submerged plate-and-frame	>98% for TOC and > 98% for NH_4-N	Achilli et al. (2009)
Osmotic membrane bioreactor	Anaerobic	CTA	Submerged plate-and-frame	>95% for COD, >99% for TP and 70–80% for ammonia	Gu et al. (2015)
Osmotic membrane bioreactor	Anaerobic	CTA	Submerged plate-and-frame	96.7% for COD, 99% for TP, and 60% for NH_4-N	Chen et al. (2014)
Osmotic membrane bioreactor	Anaerobic	CTA	Submerged plate-and-frame	92.9% for TOC	Tang and Ng (2014)
Wastewater pre-conc.	-	CTA	Submerged plate-and-frame	99% for COD, 99% for PO_4, 67–68% for ammonia, and 56–59% for TN	Linares et al. (2013)
Wastewater pre-conc.	-	CTA	Pilot-scale spiral wound	99.8% for COD, 99.7% for TP, 48.1% for NH_4-N, and 67.8% for TN	Wan et al. (2016)

COD, chemical oxygen demand; TOC, total organic carbon; TP, total phosphorus; TN, total nitrogen; CTA, cellulose triacetate.

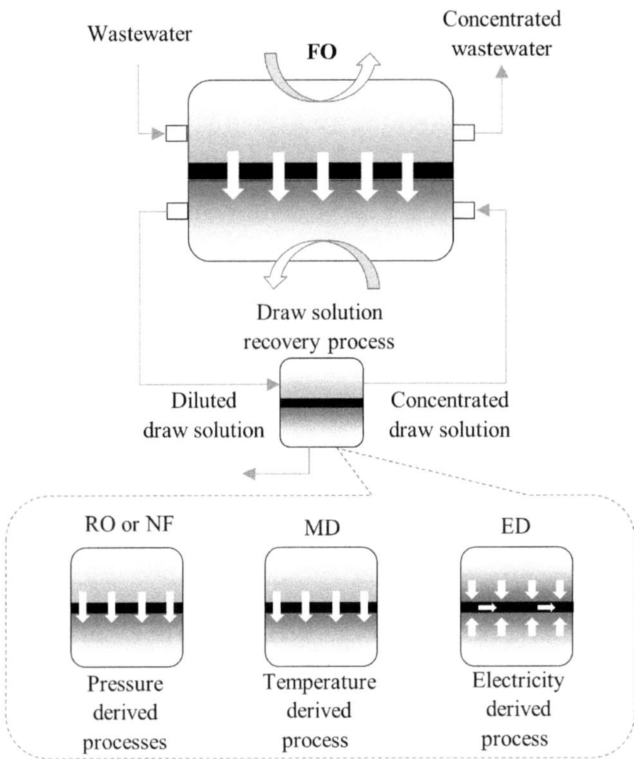

Fig. 12.8 A schematic of the important processes that can be used in re-concentration of draw solution, such as MD (as a temperature driven method), RO or NF (as pressure driven methods), and ED (as an electrical energy driven method).

solution nature. A schematic of the mentioned processes as candidates for draw solution re-concentration is given in Fig. 12.8.

Recent studies have been able to prove the effectiveness of the FO method to recover nutrients and produce energy from municipal wastewater. Some of these results are given in Table 12.2 which shows the suitable resource recovery of combining the FO method with other methods for municipal wastewater treatment. As mentioned before, the combination of FO process and anaerobic treatment with the aim of producing biogas is known as a promising method, but more studies are needed to solve problems such as CP and membrane fouling. In addition, the investigation of long-term operation and the investigation of the degree of fouling in the long term are among the issues that require further investigation regarding the FO process.

Table 12.2 Summary of FO-based resource recovery processes.

Feed solution	FO-based process	Recovered resource	Draw solution	Draw solute recovery process	Performance	Reference
Digested sludge centrate	FO-RO	Nutrient concentrate (i.e., ammonia and phosphate) Fresh water	NaCl	RO	Ammonia removal 82.9–92.1% Phosphate removal 99.6–99.9% Optimum water recovery 70%	Holloway et al. (2007)
Synthetic wastewater	An-OMBR	Biogas	NaCl	MRC	Methane yield 0.21 L CH_4/g COD	Chen et al. (2014)
Synthetic wastewater	An-OMBR	Biogas	NaCl and Na2SO4	MRC	NaCl AnOMBR had a higher biogas methane composition than Na_2SO_4 An-OMBR	Tang and Ng (2014)
Urine	FO pretreatment	Nutrient concentrate (i.e., ammonium, phosphate, and potassium)	Synthetic seawater and brine	OD	Ammonium removal 50–80% Phosphate removal >90% Potassium removal >90%	Zhang et al. (2014)
Digested sludge centrate	FO-MD	Struvite ($MgNH_4PO_4{\cdot}6H_2O$) Fresh water	$MgCl_2$	MD	Ammonium removal >90% Phosphate removal >97% Bidirectional diffusion of Mg^{2+} and protons improved struvite recovery	Xie et al. (2014)
Activated sludge	Ae-OMBR	Calcium phosphate	$MgCl_2$ and NaCl	MRC	Phosphorus content >11%	Qiu and Ting (2014)
Synthetic wastewater	An-OMBR	Biogas	NaCl	MRC	Methane yield 0.3 L CH_4/g COD	Gu et al. (2015)
Activated sludge	MF-AeOMBR	Calcium phosphate	Seawater brine	OD	MF extracted dissolved nutrients Phosphorus content 11–13%	Qiu et al. (2015)
Secondary treated effluent	FO pretreatment	Nutrient concentrate (i.e., ammonia and phosphate)	Synthetic seawater	OD	Ammonia removal 66.7% Phosphate removal 92.1	Xue et al. (2015)
Activated sludge	MF-Ae-OMBR-RO	Calcium or magnesium phosphate Fresh water	NaCl	RO	Precipitate 15–20% phosphorus	Luo et al. (2016)
Digested sludge centrate	FO pretreatment	Calcium phosphate	Seawater	OD	Phosphate removal >98% Bidirectional diffusion of protons improved calcium phosphate recovery	Ansari et al. (2016)

OD, Osmotic dilution; MRC; Manual reconcentration.

12.6 Conclusions and future trends

The FO process is considered as a suitable process for municipal wastewater treatment. This process transforms the municipal wastewater flow into two streams including the purified water flow and the concentrated flow of the wastewater pollutants, which in addition to supplying the required water, can be used to produce biogas. Despite all the advantages of FO, this process is still underdeveloped and in its early stages. Realization of commercial scale implementation of this process requires development to reach evolution. Currently, economic barriers are considered as the first problem of the FO process. The low production flux in the FO process leads to an increase in the required volume of equipment and significantly increases the footprint of the process. It is obvious that increasing the footprint increases the cost of capital investment and operation of the unit. On the other hand, despite all the environmental and operational advantages of the FO process, it is not possible to use this process in centralized applications and it can only be implemented for decentralized applications. In general, increasing the scale of municipal wastewater treatment using the FO process requires more research to evaluate the feasibility of increasing the scale of FO in terms of technical and economic viability. It seems that increasing studies aimed at improving FO membranes, designing high-efficiency membrane modules, and optimizing solute recovery processes can increase the scale of this process and make it possible to use FO systems at a centralized level. In addition, the design of hybrid systems of FO and anaerobic bioreactors have been able to eliminate some of the concerns about the ability to increase the scale of FO. Also, it seems that the use of FO process together with processes such as aerobic bioreactors, anaerobic bioreactors, RO, nanofiltration, electrodialysis, and MD can lead to solving the drawbacks of these methods. For example, anaerobic bioreactors cannot be used directly for municipal wastewater treatment because of the very low concentration of organic substances in it. While the use of the FO process can create a concentrated stream so that the municipal wastewater treatment cycle can be completed by using the anaerobic bioreactor.

References

Achilli, A., Cath, T.Y., Marchand, E.A., Childress, A.E., 2009. The forward osmosis membrane bioreactor: a low fouling alternative to MBR processes. Desalination 239 (1–3), 10–21. https://doi.org/10.1016/j.desal.2008.02.022. 00119164.

Akther, N., Sodiq, A., Giwa, A., Daer, S., Arafat, H.A., Hasan, S.W., 2015. Recent advancements in forward osmosis desalination: a review. Chem. Eng. J. 13858947 (281), 502–522. https://doi.org/10.1016/j.cej.2015.05.080.

Altaee, A., Zaragoza, G., van Tonningen, H.R., 2014. Comparison between forward osmosis-reverse osmosis and reverse osmosis processes for seawater desalination. Desalination 336, 50–57.

Alturki, A.A., McDonald, J.A., Khan, S.J., Price, W.E., Nghiem, L.D., Elimelech, M., 2013. Removal of trace organic contaminants by the forward osmosis process. Sep. Purif. Technol. 103, 258–266. https://doi.org/10.1016/j.seppur.2012.10.036. 13835866.

Anderson, D.K., 1977. Concentration of Dilute Industrial Wastes by Direct Osmosis. University of Rhode Island. https://digitalcommons.uri.edu/theses/1164/.

Ansari, A.J., Hai, F.I., Price, W.E., Nghiem, L.D., 2016. Phosphorus recovery from digested sludge centrate using seawater-driven forward osmosis. Elsevier, Australia, Sep. Purif. Technol. 163, 1–7. 18733794. Available from: http://www.journals.elsevier.com/separation-and-purification-technology/. https://doi.org/10.1016/j.seppur.2016.02.031.

Ansari, A.J., Hai, F.I., Price, W.E., Drewes, J.E., Nghiem, L.D., 2017. Forward osmosis as a platform for resource recovery from municipal wastewater—A critical assessment of the literature. Elsevier B.V., Australia, J. Membr. Sci. 529, 195–206. 18733123. Available from: www.elsevier.com/locate/memsci. https://doi.org/10.1016/j.memsci.2017.01.054.

Awad, A.M., Jalab, R., Minier-Matar, J., Adham, S., Nasser, M.S., Judd, S.J., 2019. The status of forward osmosis technology implementation. Elsevier B.V., Qatar, Desalination 461, 10–21. 00119164. Available from: https://www.journals.elsevier.com/desalination. https://doi.org/10.1016/j.desal.2019.03.013.

Baker, R.W., 2023. Membrane Technology and Applications. Wiley, United States, pp. 1–539. Available from: https://onlinelibrary.wiley.com/doi/book/10.1002/9781119686026. https://doi.org/10.1002/9781119686026.

Basile, A., Cassano, A., Rastogi, N.K., 2020. Current Trends and Future Developments on (Bio-)Membranes: Reverse and Forward Osmosis: Principles, Applications, Advances, first ed. Elsevier, Amsterdam, Netherlands.

Batchelder, G.W., 1965. Process for the Demineralization of Water. US Patent. **3**,.

Batstone, D.J., Hülsen, T., Mehta, C.M., Keller, J., 2015. Platforms for energy and nutrient recovery from domestic wastewater: a review. Chemosphere 140, 2–11. https://doi.org/10.1016/j.chemosphere.2014.10.021. 00456535.

Beaudry, E.G., Herron, J.R., 1997. SAE Technical Papers. In: Direct Osmosis for Concentrating Wastewater. SAE International, United States, https://doi.org/10.4271/972270 26883627. http://papers.sae.org/.

Beaudry, E.G., Lampi, K.A., 1990. Membrane technology for direct-osmosis concentration of fruit juices. Food Technol. 44, 121. https://pascal-francis.inist.fr/vibad/index.php?action=getRecordDetail&idt=19381215.

Bhatt, P.P., 2004. Osmotic Drug Delivery Systems for Poorly Water Soluble Drugs. Pharmaventures Ltd, pp. 26–29.

Blandin, G., Verliefde, A.R.D., Comas, J., Rodriguez-Roda, I., Le-Clech, P., 2016. Efficiently combining water reuse and desalination through forward osmosis-reverse osmosis (FO-RO) hybrids: a critical review. MDPI AG, Spain, Membranes 6 (3). 20770375. Available from: http://www.mdpi.com/2077-0375/6/3/37/pdf. https://doi.org/10.3390/membranes6030037.

Borg, F.G., 2003. What is osmosis? Explanation and understanding of a physical phenomenon. arXiv: General Phys. https://doi.org/10.48550/arXiv.physics/0305011.

Butler, E., Silva, A., Horton, K., Rom, Z., Chwatko, M., Havasov, A., McCutcheon, J.R., 2013. Point of use water treatment with forward osmosis for emergency relief. Desalination (312), 23–30. https://doi.org/10.1016/j.desal.2012.12.013. 00119164.

Cath, T.Y., Gormly, S., Beaudry, E.G., Flynn, M.T., Adams, V.D., Childress, A.E., 2005. Membrane contactor processes for wastewater reclamation in space: part I. Direct osmotic concentration as pretreatment for reverse osmosis. J. Membr. Sci. 257 (1–2), 85–98. https://doi.org/10.1016/j.memsci.2004.08.039.

Cath, T.Y., Childress, A.E., Elimelech, M., 2006. Forward osmosis: principles, applications, and recent developments. J. Memb. Sci. 281, 70–87.

Chen, L., Gu, Y., Cao, C., Zhang, J., Ng, J.W., Tang, C., 2014. Performance of a submerged anaerobic membrane bioreactor with forward osmosis membrane for low-strength wastewater treatment. Elsevier Ltd, Singapore, Water Res. 50, 114–123. 18792448. Available from: www.elsevier.com/locate/watres. https://doi.org/10.1016/j.watres.2013.12.009.

Chung, T.S., Zhang, S., Wang, K.Y., Su, J., Ling, M.M., 2012. Forward osmosis processes: yesterday, today and tomorrow. Desalination 287, 78–81. https://doi.org/10.1016/j.desal.2010.12.019. 00119164.

Coday, B.D., Yaffe, B.G.M., Xu, P., Cath, T.Y., 2014. Rejection of trace organic compounds by forward osmosis membranes: a literature review. American Chemical Society, United States, Environ. Sci. Technol. 48 (7), 3612–3624. 15205851. Available from: http://pubs.acs.org/journal/esthag. https://doi.org/10.1021/es4038676.

Daelman, M.R.J., van Voorthuizen, E.M., van Dongen, U.G.J.M., Volcke, E.I.P., van Loosdrecht, M.C.M., 2012. Methane emission during municipal wastewater treatment. Elsevier Ltd, Netherlands, Water Res. 46 (11), 3657–3670. 18792448. Available from: www.elsevier.com/locate/watres. https://doi.org/10.1016/j.watres.2012.04.024.

der Bruggen, B., Luis, P., 2015. Forward osmosis: understanding the hype. Rev. Chem. Eng. 31, 1–12.

Elser, J., Bennett, E., 2011. A broken biogeochemical cycle. Nature 478 (7367), 29–31. https://doi.org/10.1038/478029a. 0028-0836.

Eyvaz, M., Arslan, S., İmer, D., Yüksel, E., Koyuncu, İ., 2018. Forward osmosis membranes—a review: part I. IntechOpen. https://doi.org/10.5772/intechopen.72287.

Frank, B.S., 1972. Desalination of Sea Water. U.S. Patent NO. **3**,.

Frijns, J., Hofman, J., Nederlof, M., 2013. The potential of (waste) water as energy carrier. Energy Convers. Manag. 65, 357–363.

Gerba, C.P., Pepper, I.L., 2019. Municipal Wastewater Treatment. Elsevier BV, pp. 393–418, https://doi.org/10.1016/b978-0-12-814719-1.00022-7.

Glew, D.N., 1965. Process for Liquid Recovery and Solution Concentration. US Patent. **3**,.

Goosens, I., Van Haute, A., 1978. The use of direct osmosis tests as complementary experiments to determine the water and salt permeabilities of reinforced cellulose acetate membranes. Desalination 26 (3), 299–308. https://doi.org/10.1016/s0011-9164(00)82204-3. 00119164.

Gu, Y., Chen, L., Ng, J.W., Lee, C., Chang, V.W.C., Tang, C.Y., 2015. Development of anaerobic osmotic membrane bioreactor for low-strength wastewater treatment at mesophilic condition. Elsevier, Singapore, J. Membr. Sci. 490, 197–208. 18733123. Available from: www.elsevier.com/locate/memsci. https://doi.org/10.1016/j.memsci.2015.04.032.

Guell, D.C., Brenner, H., 1996. Physical mechanism of membrane osmotic phenomena. American Chemical Society, United States, Ind. Eng. Chem. Res. 35 (9), 3004–3014. 08885885. Available from: http://pubs.acs.org/journal/iecred. https://doi.org/10.1021/ie950787f.

Hancock, N.T., Xu, P., Roby, M.J., Gomez, J.D., Cath, T.Y., 2013. Towards direct potable reuse with forward osmosis: technical assessment of long-term process performance at the pilot scale. J. Membr. Sci. 445, 34–46. https://doi.org/10.1016/j.memsci.2013.04.056. 18733123.

Haupt, A., Lerch, A., 2018. Forward osmosis application in manufacturing industries: a short review. Membranes 8 (3), 47. https://doi.org/10.3390/membranes8030047. 2077-0375.

Hilal, N., Ismail, A.F., Khayet, M., Johnson, D., 2021. Osmosis Engineering. Elsevier, United Arab Emirates, pp. 1–367. Available from: https://www.sciencedirect.com/book/9780128210161. https://doi.org/10.1016/B978-0-12-821016-1.00016-4.

Holloway, R.W., Childress, A.E., Dennett, K.E., Cath, T.Y., 2007. Forward osmosis for concentration of anaerobic digester centrate. Elsevier Ltd, United States, Water Res. 41 (17), 4005–4014. 00431354. Available from: www.elsevier.com/locate/watres. https://doi.org/10.1016/j.watres.2007.05.054.

Holloway, R.W., Regnery, J., Nghiem, L.D., Cath, T.Y., 2014. Removal of trace organic chemicals and performance of a novel hybrid ultrafiltration-osmotic membrane bioreactor. American Chemical Society, United States, Environ. Sci. Technol. 48 (18), 10859–10868. 15205851. Available from: http://pubs.acs.org/journal/esthag. https://doi.org/10.1021/es501051b.

Holloway, R.W., Achilli, A., Cath, T.Y., 2015. The osmotic membrane bioreactor: a critical review. Environ. Sci. Water Res. Technol. 1, 581–605.

Iskander, S.M., Zou, S., Brazil, B., Novak, J.T., He, Z., 2017. Energy consumption by forward osmosis treatment of landfill leachate for water recovery. Elsevier Ltd, United States, Waste Manag. 63, 284–291. 18792456. Available from: www.elsevier.com/locate/wasman. https://doi.org/10.1016/j.wasman.2017.03.026.

Ismail, F., Khulbe, K.C., Matsuura, T., 2018. Reverse Osmosis. Elsevier. https://www.sciencedirect.com/book/9780128114681/reverse-osmosis.

Joo, S.H., Tansel, B., 2015. Novel technologies for reverse osmosis concentrate treatment: a review. Academic Press, United States, J. Environ. Manage. 150, 322–335. 10958630. Available from: https://www.sciencedirect.com/journal/journal-of-environmental-management. https://doi.org/10.1016/j.jenvman.2014.10.027.

Kerusha Lutchmiah, A.R.D., Verliefde, K., Roest, L.C., Rietveld, E.R.C., 2014. Forward osmosis for application in wastewater treatment: a review. Water Res. 58, 179–197. https://doi.org/10.1016/j.watres.2014.03.045. 00431354.

Kessler, J.O., Moody, C.D., 1976. Drinking water from sea water by forward osmosis. Desalination 18 (3), 297–306. https://doi.org/10.1016/s0011-9164(00)84119-3. 00119164.

Kiil, F., 1989. Molecular mechanisms of osmosis. Am. J. Physiol. Regul. Integr. Comp. Physiol. 256 (4), R801. https://doi.org/10.1152/ajpregu.1989.256.4.r801. 0363-6119.

Koppelaar, R.H.E.M., Weikard, H.P., 2013. Assessing phosphate rock depletion and phosphorus recycling options. Glob. Environ. Chang. 23 (6), 1454–1466. https://doi.org/10.1016/j.gloenvcha.2013.09.002. 09593780.

Kravath, R.E., Davis, J.A., 1975. Desalination of sea water by direct osmosis. Desalination 16 (2), 151–155. https://doi.org/10.1016/S0011-9164(00)82089-5. 00119164.

Kucera, J., 2023. Reverse Osmosis. John Wiley & Sons.

Kwan, S.E., Bar-Zeev, E., Elimelech, M., 2015. Biofouling in forward osmosis and reverse osmosis: measurements and mechanisms. J. Memb. Sci. 493, 703–708.

Lay, W.C.L., Chong, T.H., Tang, C.Y., Fane, A.G., Zhang, J., Liu, Y., 2010. Fouling propensity of forward osmosis: investigation of the slower flux decline phenomenon. IWA Publishing, Singapore, Water Sci. Technol. 61 (4), 927–936. 02731223. Available from: http://www.iwaponline.com/wst/toc.htm. https://doi.org/10.2166/wst.2010.835.

Lee, S., Boo, C., Elimelech, M., Hong, S., 2010. Comparison of fouling behavior in forward osmosis (FO) and reverse osmosis (RO). J. Membr. Sci. 365 (1–2), 34–39. https://doi.org/10.1016/j.memsci.2010.08.036. 03767388.

Li, W.-W., Yu, H.-Q., Rittmann, B.E., 2015. Chemistry: reuse water pollutants. Nature 528, 29–31.

Li, J., Niu, A., Lu, C.J., Zhang, J.H., Junaid, M., Strauss, P.R., Xiao, P., Wang, X., Ren, Y.W., Pei, D.S., 2017. A novel forward osmosis system in landfill leachate treatment for removing polycyclic aromatic hydrocarbons and for direct fertigation. Elsevier Ltd, China, Chemosphere 168, 112–121. 18791298. Available from: www.elsevier.com/locate/chemosphere. https://doi.org/10.1016/j.chemosphere.2016.10.048.

Linares, R.V., Li, Z., Abu-Ghdaib, M., Wei, C.-H., Amy, G., Vrouwenvelder, J.S., 2013. Water harvesting from municipal wastewater via osmotic gradient: an evaluation of process performance. J. Memb. Sci. 447, 50–56.

Linares, R.V., Li, Z., Elimelech, M., Amy, G., Vrouwenvelder, H., 2017. Recent Developments in Forward Osmosis Processes. IWA Publishing, London, UK.

Loeb, S., 1995. Proceedings of the Euromembrane '95 Conference Pressure-Retarded Osmosis Revisited: The Prospects for Osmotic Power at the Dead Sea.

Loeb, S., Titelman, L., Korngold, E., Freiman, J., 1997. Effect of porous support fabric on osmosis through a Loeb-Sourirajan type asymmetric membrane. J. Membr. Sci. 129 (2), 243–249. https://doi.org/10.1016/S0376-7388(96)00354-7.

Luo, Y., Guo, W., Ngo, H.H., Nghiem, L.D., Hai, F.I., Zhang, J., Liang, S., Wang, X.C., 2014. A review on the occurrence of micropollutants in the aquatic environment and their fate and removal during wastewater treatment. Elsevier B.V., Australia, Sci. Total Environ. 473-474, 619–641. 18791026. Available from: www.elsevier.com/locate/scitotenv. https://doi.org/10.1016/j.scitotenv.2013.12.065.

Luo, W., Hai, F.I., Price, W.E., Guo, W., Ngo, H.H., Yamamoto, K., Nghiem, L.D., 2016. Phosphorus and water recovery by a novel osmotic membrane bioreactor-reverse osmosis system. Elsevier Ltd, Australia, Bioresour. Technol. 200, 297–304. 18732976. Available from: www.elsevier.com/locate/biortech. https://doi.org/10.1016/j.biortech.2015.10.029.

Maddah, H., Chogle, A., 2017. Biofouling in reverse osmosis: phenomena, monitoring, controlling and remediation. Appl. Water Sci. 7 (6), 2637–2651. https://doi.org/10.1007/s13201-016-0493-1. 2190-5487.

Mahto, A., Aruchamy, K., Meena, R., Kamali, M., Nataraj, S.K., Aminabhavi, T.M., 2021. Forward osmosis for industrial effluents treatment—sustainability considerations. Elsevier B.V., India, Sep. Purif. Technol. **254**. 18733794. Available from: http://www.journals.elsevier.com/separation-and-purification-technology/. https://doi.org/10.1016/j.seppur.2020.117568.

Marbach, S., Bocquet, L., 2019. Osmosis, from molecular insights to large-scale applications. Chem. Soc. Rev. 48 (11), 3102–3144. https://doi.org/10.1039/c8cs00420j. 0306-0012.

McCutcheon, J.R., McGinnis, R.L., Elimelech, M., 2005. A novel ammonia–carbon dioxide forward (direct) osmosis desalination process. Desalination 174, 1.

McCutcheon, J.R., McGinnis, R.L., Elimelech, M., 2006. Desalination by ammonia–carbon dioxide forward osmosis: influence of draw and feed solution concentrations on process performance. J. Memb. Sci. 278, 114–123.

McGinnis, R.L., 2002. Osmotic Desalination Process. U.S. Patent. 6.

Meerburg, F.A., Boon, N., Van Winckel, T., Vercamer, J.A.R., Nopens, I., Vlaeminck, S.E., 2015. Toward energy-neutral wastewater treatment: a high-rate contact stabilization process to maximally recover sewage organics. Elsevier Ltd, Belgium, Bioresour. Technol. 179, 373–381. 18732976. Available from: www.elsevier.com/locate/biortech. https://doi.org/10.1016/j.biortech.2014.12.018.

Mehta, G.D., 1982. Further results on the performance of present-day osmotic membranes in various osmotic regions. J. Membr. Sci. 10 (1), 3–19. https://doi.org/10.1016/S0376-7388(00)82183-3. 03767388.

Mehta, G.D., Loeb, S., 1978. Internal polarization in the porous substructure of a semipermeable membrane under pressure-retarded osmosis. J. Membr. Sci. 4 (C), 261–265. https://doi.org/10.1016/S0376-7388(00)83301-3. 03767388.

Miller, J.E., Evans, L.R., 2006. Forward Osmosis: A New Approach to Water Purification and Desalination. Sandia National Laboratories (SNL), Albuquerque, NM/Livermore, CA.

Morse, G.K., Brett, S.W., Guy, J.A., Lester, J.N., 1998. Review: phosphorus removal and recovery technologies. Elsevier Sci B.V., United Kingdom, Sci. Total Environ. 212 (1), 69–81. 00489697. Available from: www.elsevier.com/locate/scitotenv. https://doi.org/10.1016/S0048-9697(97)00332-X.

Ng, H.Y., Tang, W., Wong, W.S., 2006. Performance of forward (direct) osmosis process: membrane structure and transport phenomenon. Environ. Sci. Technol. 40, 2408–2413.

Nguyen, D.V., Nguyen, T.T., Adha, R.S., Zheng, L., Bui, X.T., Ma, X., Vo, H.N.P., 2023. Forward osmosis: principle and applications in sustainable water and energy development. Curr. Dev. Biotechnol. Bioeng., 463–491. Elsevier, South Korea. Available from: https://www.sciencedirect.com/book/9780443191800. https://doi.org/10.1016/B978-0-443-19180-0.00017-1.

Phuntsho, S., Sahebi, S., Majeed, T., Lotfi, F., Kim, J.E., Shon, H.K., 2013. Assessing the major factors affecting the performances of forward osmosis and its implications on the desalination process. Chem. Eng. J. 231, 484–496. https://doi.org/10.1016/j.cej.2013.07.058. 13858947.

Qasim, M., Darwish, N.A., Sarp, S., Hilal, N., 2015. Water desalination by forward (direct) osmosis phenomenon: a comprehensive review. Elsevier B.V., United Arab Emirates, Desalination 374, 47–69. 00119164. Available from: https://www.journals.elsevier.com/desalination. https://doi.org/10.1016/j.desal.2015.07.016.

Qin, J.-J., Lay, W.C.L., Kekre, K.A., 2012. Recent developments and future challenges of forward osmosis for desalination: a review. Desalin. Water Treat. 39, 123–136.

Qiu, G., Ting, Y.P., 2013. Osmotic membrane bioreactor for wastewater treatment and the effect of salt accumulation on system performance and microbial community dynamics. Elsevier Ltd, Singapore, Bioresour. Technol. 150, 287–297. 18732976. Available from: www.elsevier.com/locate/biortech. https://doi.org/10.1016/j.biortech.2013.09.090.

Qiu, G., Ting, Y.P., 2014. Direct phosphorus recovery from municipal wastewater via osmotic membrane bioreactor (OMBR) for wastewater treatment. Elsevier Ltd, Singapore, Bioresour. Technol. 170, 221–229. 18732976. Available from: www.elsevier.com/locate/biortech. https://doi.org/10.1016/j.biortech.2014.07.103.

Qiu, G., Law, Y.M., Das, S., Ting, Y.P., 2015. Direct and complete phosphorus recovery from municipal wastewater using a hybrid microfiltration-forward osmosis membrane bioreactor process with seawater brine as draw solution. American Chemical Society, Singapore, Environ. Sci. Technol. 49 (10), 6156–6163. 15205851. Available from: http://pubs.acs.org/journal/esthag. https://doi.org/10.1021/es504554f.

Qu, J., Wang, H., Wang, K., Yu, G., Ke, B., Yu, H.Q., Ren, H., Zheng, X., Li, J., Li, W.W., Gao, S., Gong, H., 2019. Municipal wastewater treatment in China: development history and future perspectives. Higher Education Press, China, Front. Environ. Sci. Eng. 13 (6). 2095221X. Available from: http://www.springerlink.com/content/2095-2201/. https://doi.org/10.1007/s11783-019-1172-x.

Rastogi, N.K., 2016. Opportunities and challenges in application of forward osmosis in food processing. Taylor and Francis Inc., India, Crit. Rev. Food Sci. Nutr. 56 (2), 266–291. 15497852. Available from: www.tandf.co.uk/journals/titles/10408398.asp. https://doi.org/10.1080/10408398.2012.724734.

Rastogi, N.K., 2020. Forward osmosis: principles, applications, and recent developments. Curr. Trends Futur. Dev. Membr., 3–35.

Rastogi, N.K., Cassano, A., Basile, A., 2015. Water treatment by reverse and forward osmosis. In: Advances in Membrane Technologies for Water Treatment. Elsevier, pp. 129–154.
Rosa, M.D., Giroux, F., 2001. Osmotic treatments (OT) and problems related to the solution management. J. Food Eng. 49 (2–3), 223–236. https://doi.org/10.1016/s0260-8774(00)00216-8. 02608774.
Sagiv, A., Semiat, R., 2005. Backwash of RO spiral wound membranes. Desalination 179 (1–3), 1–9. https://doi.org/10.1016/j.desal.2004.11.050. 00119164.
Saiful, S., Ajrina, M., Wibisono, Y., Marlina, M., 2020. Development of chitosan/starch-based forward osmosis water filtration bags for emergency water supply. MDPI AG, Indonesia, Membranes 10 (12), 1–13. 20770375. Available from: https://www.mdpi.com/2077-0375/10/12/414/pdf. https://doi.org/10.3390/membranes10120414.
Sant'Anna, V., Marczak, L.D.F., Tessaro, I.C., 2012. Membrane concentration of liquid foods by forward osmosis: process and quality view. J. Food Eng. 111, 483–489.
Scherson, Y.D., Woo, S.-G., Criddle, C.S., 2014. Production of nitrous oxide from anaerobic digester centrate and its use as a co-oxidant of biogas to enhance energy recovery. Environ. Sci. Technol. 48, 5612–5619.
Shaffer, D.L., Werber, J.R., Jaramillo, H., Lin, S., Elimelech, M., 2015. Forward osmosis: where are we now? Desalination 356, 271–284. https://doi.org/10.1016/j.desal.2014.10.031. 00119164.
Siddiqui, F.A., She, Q., Fane, A.G., Field, R.W., 2018. Exploring the differences between forward osmosis and reverse osmosis fouling. Elsevier B.V., United Kingdom, J. Membr. Sci. 565, 241–253. 18733123. Available from: www.elsevier.com/locate/memsci. https://doi.org/10.1016/j.memsci.2018.08.034.
Siegrist, H., Joss, A., 2012. Review on the fate of organic micropollutants in wastewater treatment and water reuse with membranes. Water Sci. Technol. 66 (6), 1369–1376. https://doi.org/10.2166/wst.2012.285. 0273-1223.
Sikosana, M.L., Sikhwivhilu, K., Moutloali, R., Madyira, D.M., 2019. Municipal wastewater treatment technologies: a review. Procedia Manuf. 35, 1018–1024. https://doi.org/10.1016/j.promfg.2019.06.051.
Sills, D.L., Wade, V.L., DiStefano, T.D., 2016. Comparative life cycle and technoeconomic assessment for energy recovery from dilute wastewater. Environ. Eng. Sci. 33, 861–872.
Smith, A.L., Stadler, L.B., Love, N.G., Skerlos, S.J., Raskin, L., 2012. Perspectives on anaerobic membrane bioreactor treatment of domestic wastewater: a critical review. Bioresour. Technol. 122, 149–159. https://doi.org/10.1016/j.biortech.2012.04.055. 18732976.
Song, X., McDonald, J., Price, W.E., Khan, S.J., Hai, F.I., Ngo, H.H., Guo, W., Nghiem, L.D., 2016. Effects of salinity build-up on the performance of an anaerobic membrane bioreactor regarding basic water quality parameters and removal of trace organic contaminants. Elsevier Ltd, Australia, Bioresour. Technol. 216, 399–405. 18732976. Available from: www.elsevier.com/locate/biortech. https://doi.org/10.1016/j.biortech.2016.05.075.
Stache, K., 1989. Apparatus For Transforming Sea Water, Brackish Water, Polluted Water or the Like Into a Nutrious Drink by Means of Osmosis. US Patent. **4**,.
Subramani, A., Jacangelo, J.G., 2015. Emerging desalination technologies for water treatment: a critical review. Elsevier Ltd, United States, Water Res. 75, 164–187. 18792448. Available from: www.elsevier.com/locate/watres. https://doi.org/10.1016/j.watres.2015.02.032.
Suwaileh, W., Johnson, D., Hilal, N., 2021. Principles of forward osmosis. In: Osmosis Engineering. Elsevier, pp. 131–148.

Tang, M.K.Y., Ng, H.Y., 2014. Impacts of different draw solutions on a novel anaerobic forward osmosis membrane bioreactor (AnFOMBR). Water Sci. Technol. 69 (10), 2036–2042. https://doi.org/10.2166/wst.2014.116. 0273-1223.

Tow, E.W., Warsinger, D.M., Trueworthy, A.M., Swaminathan, J., Thiel, G.P., Zubair, S.M., Myerson, A.S., Lienhard V, J.H., 2018. Comparison of fouling propensity between reverse osmosis, forward osmosis, and membrane distillation. Elsevier B.V., United States, J. Membr. Sci. 556, 352–364. 18733123. Available from: www.elsevier.com/locate/memsci. https://doi.org/10.1016/j.memsci.2018.03.065.

Votta, F., 1974. Concentration of Industrial Waste by Direct Osmosis: Completion Report.

Wan, J., Jun, G., Zhao, Q., Yu, L., 2016. COD capture: a feasible option towards energy self-sufficient domestic wastewater treatment. Sci. Rep. 6 (1). https://doi.org/10.1038/srep25054. 2045-2322.

Wibisono, Y., Bilad, M.R., 2020. Design of forward osmosis system. In: Current Trends and Future Developments on (Bio-) Membranes. Elsevier, pp. 57–83.

Wijekoon, K.C., Hai, F.I., Kang, J., Price, W.E., Guo, W., Ngo, H.H., Nghiem, L.D., 2013. The fate of pharmaceuticals, steroid hormones, phytoestrogens, UV-filters and pesticides during MBR treatment. Elsevier Ltd, Australia, Bioresour. Technol. 144, 247–254. 18732976. Available from: www.elsevier.com/locate/biortech. https://doi.org/10.1016/j.biortech.2013.06.097.

WROLSTAD, R.E., MCDANIEL, M.R., DURST, R.W., MICHEALS, N., LAMPI, K.A., BEAUDRY, E.G., 1993. Composition and sensory characterization of red raspberry juice concentrated by direct-osmosis or evaporation. J. Food Sci. 58 (3), 633–637. https://doi.org/10.1111/j.1365-2621.1993.tb04344.x. 17503841.

Xie, M., Nghiem, L.D., Price, W.E., Elimelech, M., 2014. Toward resource recovery from wastewater: extraction of phosphorus from digested sludge using a hybrid forward osmosis-membrane distillation process. American Chemical Society, Australia, Environ. Sci. Technol. Lett. 1 (2), 191–195. 23288930. Available from: http://pubs.acs.org/page/estlcu/about.html. https://doi.org/10.1021/ez400189z.

Xie, M., Shon, H.K., Gray, S.R., Elimelech, M., 2016. Membrane-based processes for wastewater nutrient recovery: technology, challenges, and future direction. Elsevier Ltd, Australia, Water Res. 89, 210–221. 18792448. Available from: www.elsevier.com/locate/watres. https://doi.org/10.1016/j.watres.2015.11.045.

Xue, W., Tobino, T., Nakajima, F., Yamamoto, K., 2015. Seawater-driven forward osmosis for enriching nitrogen and phosphorous in treated municipal wastewater: effect of membrane properties and feed solution chemistry. Water Res. 69, 120–130. https://doi.org/10.1016/j.watres.2014.11.007. 00431354.

York, R.J., Thiel, R.S., Beaudry, E.G., 1999. Full-scale experience of direct osmosis concentration applied to leachate management. In: Proceedings of the Seventh International Waste Management and Landfill Symposium (Sardinia'99), S. Margherita Di Pula, Cagliari, Sardinia, Italy. CISA, Environmental Sanitary Engineering Centre, pp. 4–8.

Zhang, F., Brastad, K.S., He, Z., 2011. Integrating forward osmosis into microbial fuel cells for wastewater treatment, water extraction and bioelectricity generation. Environ. Sci. Technol. 45, 6690–6696.

Zhang, J., She, Q., Chang, V.W.C., Tang, C.Y., Webster, R.D., 2014. Mining nutrients (N, K, P) from urban source-separated urine by forward osmosis dewatering. American Chemical Society, Singapore, Environ. Sci. Technol. 48 (6), 3386–3394. 15205851. Available from: http://pubs.acs.org/journal/esthag. https://doi.org/10.1021/es405266d.

Zhang, J., Wang, D., Chen, Y., Gao, B., Wang, Z., 2021. Scaling control of forward osmosis-membrane distillation (FO-MD) integrated process for pre-treated landfill leachate treatment. Desalination 520, 115342. https://doi.org/10.1016/j.desal.2021.115342. 00119164.

Zhao, S., Zou, L., Tang, C.Y., Mulcahy, D., 2012. Recent developments in forward osmosis: opportunities and challenges. J. Membr. Sci. 396, 1–21. https://doi.org/10.1016/j.memsci.2011.12.023. 18733123.

Zhu, X.Z., Zhang, F., Li, W.W., Liu, H.Q., Wang, Y.K., Huang, M.S., 2015. Forward osmosis membrane favors an improved proton flux and electricity generation in microbial fuel cells. Elsevier B.V., China, Desalination 372, 26–31. 00119164. Available from: https://www.journals.elsevier.com/desalination. https://doi.org/10.1016/j.desal.2015.06.011.

13

Emerging wastewater disinfection technologies

Patricia García-Muñoz[*], Jesús Quintana-Terriza[*], Niloofar Arabbaseri[*], Pablo Santiago Espiñeira[*], Cheikh Gaye[*], and Jorge Rodríguez-Chueca[*]

Department of Industrial Chemical & Environmental Engineering, Escuela Técnica Superior de Ingenieros Industriales, Universidad Politécnica de Madrid, Madrid, Spain

Abbreviations

2,4-DCP	2,4-dichlorophenol
AOPs	advanced oxidation process
CB	conduction band
CECs	contaminants of emerging concern
DBPs	disinfection by-products
HAAs	haloacetic acids
HANs	haloketonitriles
HTTP	high temperature plasma
LTP	low temperature plasma
NTP	non-thermal plasma
ORP	oxide-reduction potential
PAW	plasma activated water
PDS or PS	persoxydisulfate or persulfate
PMS	peroxymonosulfate
RONS	reactive nitrogen and oxygen species
ROS	reactive oxygen species
SODIS	solar disinfection
THMs	trihalomethanes
UV-A	ultraviolet radiation A
UV-C	ultraviolet radiation C
VB	valence band
WWTs	wastewater treatment plants

[*]All authors have contributed to the writing of the chapter to an equal degree of participation.

Municipal Wastewater Treatment. https://doi.org/10.1016/B978-0-443-24826-9.00013-9

Symbols

B relationship between the excitation and ionization energies
d electrode distance
e^- electron
E_g energy barrier or gap
h Planck constant
h^+ hole
p pressure
pH hydrogen potential
Te electron temperature
Ti ion temperature
υ wave frequency
V volts
V_B breakdown voltage
Υ townsend coefficient

13.1 Introduction

Water is vital for the development of all living organisms as it is essential for the proper functioning of cells. Additionally, on a human level, water is a critical substance in nearly all human activities, such as industry, agriculture, livestock, power generation, transportation, recreational activities, and more. Therefore, the quality and availability of this resource become crucial in the economic and social development of different territories and societies.

The use of this resource in different human activities will require specific physical-chemical and biological quality for each case. In any case, the presence of pathogenic microorganisms is a limiting factor that poses a risk to human health. For this reason, if water is to be used directly by humans, it must be thoroughly disinfected, and the fact is that infectious diseases caused by pathogenic bacteria, viruses, protozoa, and helminths represent the most widespread health risks associated with drinking water ("Guidelines for drinking-water quality. World Health Organization," 2011).

13.1.1 Fundamentals of disinfection

Disinfection is the process through which all microorganisms present in water, whether pathogenic or not, are eliminated with the aim of preventing the transmission of waterborne diseases to humans. The mechanisms of disinfection follow three fundamental routes: (i) blocking vital enzymatic activities for the microorganism by the action of the oxidizing agent added to the water; (ii) causing direct damage to the organism, such as the rupture of cytoplasmic membranes leading to cell lysis and consequently cell death; (iii) interference with the biosynthesis of regenerative components, hindering cell growth. It was in the year 1854 when an outbreak of cholera in the city of London, United Kingdom, led the English physician John Snow to correlate cholera cases with the water from public sources that supplied the city (Goldstein, 2012). This discovery marked the birth of epidemiology and represented a turning point in human development because it raised awareness of the need to disinfect water as a transmission vector for diseases. Since then, the disinfection of drinking water has become a necessary measure for ensuring safe water consumption. Unfortunately, despite the recognized need for it, still, 2 billion people worldwide lack access to safely managed drinking water services, resulting in around 3.5 million annual deaths caused by waterborne diseases such as dysentery, typhoid, gastroenteritis, and the aforementioned cholera (Chaúque et al., 2023).

The types of conventional water disinfectants are numerous and can be classified into physical methods, such as temperature, UV radiation, or ultrasound, and chemical methods, including chlorination, ozonation, and others like hydrogen peroxide, potassium permanganate, etc. Multiple factors influence the disinfection process, such as contact time, type and concentration of the disinfectant agent, nature and intensity of the physical agent, temperature, number and type of microorganisms, and the physicochemical characteristics of the aqueous medium that needs to be disinfected. Furthermore, an ideal disinfectant must meet all or most of a list of requirements, including factors such as toxicity to microorganisms and non-toxicity to humans, solubility and stability in water, high availability, cost-effectiveness, and easy handling (EPA, 2020). In this regard, chlorination and ozonation are the most common and widely used disinfection processes worldwide, as they fulfil most of the aforementioned characteristics (EPA, 1999).

13.1.2 Drawbacks of conventional disinfectants

However, the aforementioned conventional disinfectants are not infallible and have drawbacks or disadvantages that may restrict their use to certain actions or situations or under specific water characteristics. In this section, an attempt is made to compile the main disadvantages of conventional disinfectants, especially those related to chlorination, ozonation, and the use of UV-C radiation. Table 13.1 summarizes the main advantages and disadvantages of both technologies.

The disinfection process through chlorination had its first application during the cholera epidemic in London, where John Snow discovered the transmission of the bacteria through the drinking water supply. By the early 20th century, chlorination

Table 13.1 Advantages and disadvantages of conventional disinfectants.

Advantages	Disadvantages
Chlorination	
High disinfecting power	Formation of hazardous by-products
Important residual disinfecting action	Degradation of hypochlorite
Low cost	Toxicity
Easy application	Storage issues
Oxidizes Fe, Mn, and sulfides	Microorganisms highly resistant to chlorine (viruses)
Improves color removal	Possible odor and taste issues
Oxidation of suspended particles	Release of metals in the distribution system
Ozonation	
High oxidation potential	High treatment cost. Capital investment and electric power demand
Does not generate THMs	Bromates and aldehydes may be generated
Does not require high concentrations or long contact times (10–30 min)	No residual effect
Does not alter the water's pH and increases the dissolved oxygen concentration	Low dosage may not properly deactivate certain viruses, spores, or cysts
Reduces water's odor, taste, and color	Very reactive and corrosive
UV-C radiation	
Absence of disinfection by-products	High cost
No harmful effects on human health or aquatic organisms	No residual effect
Does not generate any tastes or odors	Photoreactivation of microorganisms

had become widely popular worldwide, marking a turning point in water purification and drastically reducing the incidence of waterborne diseases in more developed countries. Chlorination can be carried out primarily by adding Cl_2 (g) or liquid NaClO, which contains between 5% and 15% active Cl_2. In any case, during the process, the concentration of hypochlorous acid and hypochlorite ions forms the term "free chlorine," and the amount of either depends on the pH of the medium. Despite being one of the main disinfection methods, as it meets virtually all the requirements for an ideal disinfectant, its main drawback is the generation of disinfection byproducts (DBPs). This fact was discovered in 1974 when Rook described that HClO and HBrO reacted with natural organic matter, such as humic and fulvic acids (Formation of Haloforms During Chlorination of Natural Waters, 1974). The list of DBPs formed during chlorination includes trihalomethanes (THMs), haloacetic acids (HAAs), haloketonitriles (HANs), haloketones, chloral hydrate, and chloropicrin. Since this discovery, various studies have been conducted to determine the toxicity, mutagenicity, teratogenic activity, and carcinogenicity of these substances, ultimately leading to their classification as carcinogens for humans (Benítez et al., 2021). Chronic exposure to these substances has been primarily linked to bladder cancer but also to other types of cancer (Krasner et al., 2017). Hence, the chlorination process is not the most suitable for disinfecting wastewater due to the high organic load present in these waters.

Ozonation is a disinfection method widely used, especially in countries like France or Switzerland. It relies on harnessing the high oxidation potential of ozone, making it one of the strongest and most efficient oxidants for the elimination of almost any microorganism. Ozone (O_3) is an allotropic molecule of O_2 that is not stable under ambient conditions and must be generated *in situ*, primarily by applying a high voltage to a stream of dry air (Epelle et al., 2023). Therefore, the economic cost and handling become some of the main disadvantages of this method, especially when compared to chlorination. Additionally, ozone has a very low residual power as it tends to decompose rapidly due to its low stability. Furthermore, ozonation also generates DBPs. Whether directly or indirectly, it can react with bromide to form brominated ozone DBPs, such as bromate ion (Jahan et al., 2021). Moreover, in the presence of natural organic matter, ozonation generates nonhalogenated organic DBPs such as aldehydes, ketoacids, and carboxylic acids (Jahan et al., 2021). Again, we are dealing with the generation of substances that pose a toxicity problem for human health and should be avoided at all costs.

Finally, UV radiation is considered a physical method of disinfection. Specifically, UV-C radiation has a high germicidal capacity because DNA chains have a high absorption of short-wave radiation, primarily at 254 nm. The delivery of this type of radiation has traditionally been done using mercury lamps. Therefore, the main drawback is associated with the energy consumption of these lamps. This drawback has been partially offset by the advent of LED lamps emitting in the UV-C spectrum. However, other problems or disadvantages associated with this type of disinfection stem from the lack of residual effect or the potential regrowth of microorganisms due to partial inactivation and the DNA repair mechanisms of cells. However, concerning the formation of DBPs, there is no evidence that UV-C radiation generates specific DBPs (Kashinkunti et al., 2004).

13.1.3 Microbial pollution in wastewater

Microbiological contamination in wastewater primarily originates from human metabolism, so a wide variety of bacteria, viruses, and protozoa can be found in wastewater effluents. As mentioned earlier, these microorganisms are carriers of a long list of waterborne diseases, and their removal is mandatory in water treatment processes, but not necessarily in wastewater treatment processes. The routes of infection not only include ingestion of contaminated water, but also inhalation or dermal contact as shown in Fig. 13.1.

Wastewater Treatment Plants (WWTPs) are designed in most cases for the removal of suspended solids, organic matter, nitrogen, and phosphorus, but in many countries, legislation does not mandate the disinfection of wastewater (López et al., 2019), as microorganisms do not pose a risk to aquatic ecosystems. However, the situation is different when it comes to the regeneration of wastewater effluents for reuse in various human activities. The reuse of wastewater has become a common practice in countries and regions facing water scarcity due to increased water stress resulting from factors such as population growth and irregular precipitation patterns due to the consequences of climate change. The European Union has specific regulations for the reuse of wastewater in agriculture, known as Regulation 2020/741 ("Regulation (EU) 2020/741, Minimum requirements for water reuse," 2020). This regulation establishes different minimum

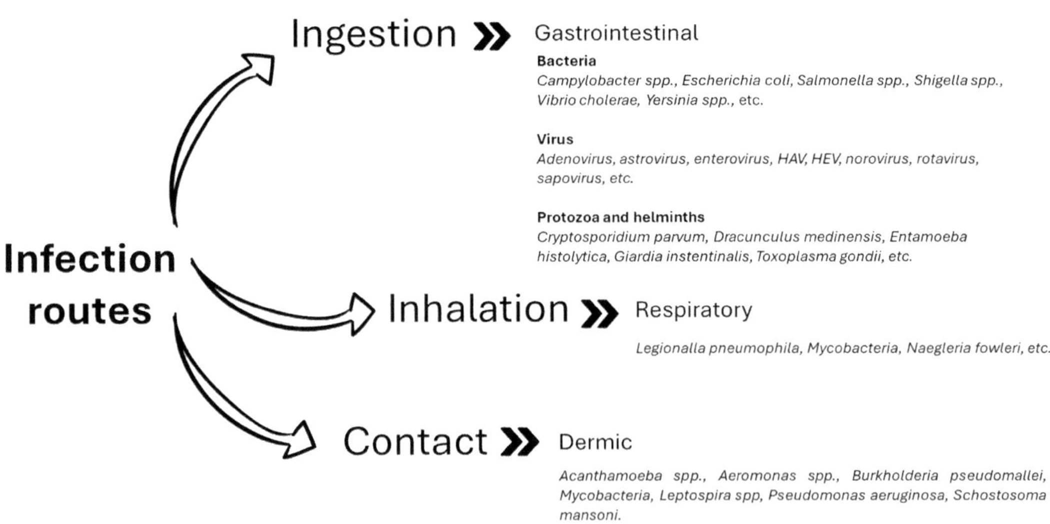

Fig. 13.1 Main routes of infection and microorganisms causing waterborne diseases.

quality classes for regenerated water based on its agricultural use. Among the microbiological indicator parameters are *E. coli*, total coliphages, and spores of *Clostridium perfringens*. Additionally, the regulation requires monitoring the concentration of *Legionella* spp. and helminth eggs as a representation of intestinal nematodes. In addition, it will be necessary to reduce the concentration of BOD5, suspended solids, and water turbidity to specific limits. To achieve this, suitable tertiary treatments must be implemented to meet the required values, especially for microbiological parameters, to prevent potential indirect contamination to humans and the outbreak of diseases. To achieve this, disinfection treatments must be as efficient as possible in the disinfection process, both in terms of investment and operation costs, while minimizing the generation of disinfection byproducts that could pose health risks to humans and ecosystems. That is why the scientific community is exploring emerging disinfection treatments that can address the issues associated with conventional disinfectants.

13.2 Advanced oxidation processes and their applications in wastewater disinfection

As the population grows and consumption patterns evolve to meet society's demands, so does wastewater volume and composition. The presence of new, more complex wastewater matrices, as well as the increasing demand of consumers and governments for cleaner effluents and more sustainable disinfection methods, has drawn attention to new treatment approaches (Krishnan et al., 2023).

Advanced Oxidation Processes (AOPs) have recently gained new interest as a solution to this challenge. In these procedures, highly reactive hydroxyl radicals ($HO^{\cdot}$) are produced, and their interaction with pollutants in water is enhanced, favoring the inactivation of pathogens in wastewater and thus allowing its disinfection (Cuerda-Correa et al., 2020). Hydroxyl and sulfate radicals are characterized by high oxidation potential, higher than conventional oxidants used in chemical disinfection, as it can be seen in Table 13.2.

Even though the first oxidation processes via $HO^{\cdot}$ interaction were studied as early as the 19th century (when H.J.H. Fenton employed a solution of hydrogen peroxide and iron (II) sulfate for the analytic determination of halo-organic compounds), they had little interest until Glaze et al., in the year 1987, suggested their potential in wastewater treatment. It was then when the term 'Advanced Oxidation Processes' was coined (Glaze et al., 1987).

Table 13.2 Oxidation potential (V) of hydroxyl and sulfate radicals compared to conventional oxidants.

F_2	$^{\bullet}OH$	$SO_4^{\bullet-}$	O at.	O_3	H_2O_2	MnO_4^{2-}	Cl_2	ClO_2
3.03	2.80	2.60	2.42	2.07	1.76	1.67	1.36	1.15

The call for stricter wastewater regulations, as well as the necessity of proposing new water reuse applications, makes AOPs a powerful tool to achieve a more sustainable water management paradigm. The challenge of developing and scaling these processes, guaranteeing clean effluents produced under sustainable approaches, is one of the greatest environmental tasks that humanity will face in the 21st century (Armah et al., 2020).

There are many approaches to categorizing AOPs. According to the distribution of the reagent/catalyst at work, the systems may be categorized into homogeneous and heterogeneous processes. Depending on the utilization or omission of radiation, they may belong to the light-driven or dark types of processes (Poyatos et al., 2010). Some of the most common processes are Photolysis, Photocatalysis, the Fenton process and its derivatives, and Ozonation; however, new configurations of these approaches often involve combining two or more methods. Fig. 13.2 summarizes all the types of AOP groups available.

13.2.1 Fundamentals of advanced oxidation processes

Even though all AOPs feature the *in-situ* generation of hydroxyl radicals, their mechanisms of reactive oxygen species (ROS) production differ for each method.

13.2.1.1 UV-driven photolysis

Photolysis processes in the presence of oxidants are based on the cleavage of typically unstable bonds, yielding radical species as represented in Eqs. (1)–(3) (Lee et al., 2021):

$$H_2O_2 + h\nu \rightarrow 2^{\bullet}OH \quad (1)$$

$$NaClO + h\nu \rightarrow Na^+ + Cl^{\bullet} + {}^{\bullet}OH \quad (2)$$

$$HSO_5^- + h\nu \rightarrow SO_4^{\bullet-} + HO^{\bullet} \quad (3)$$

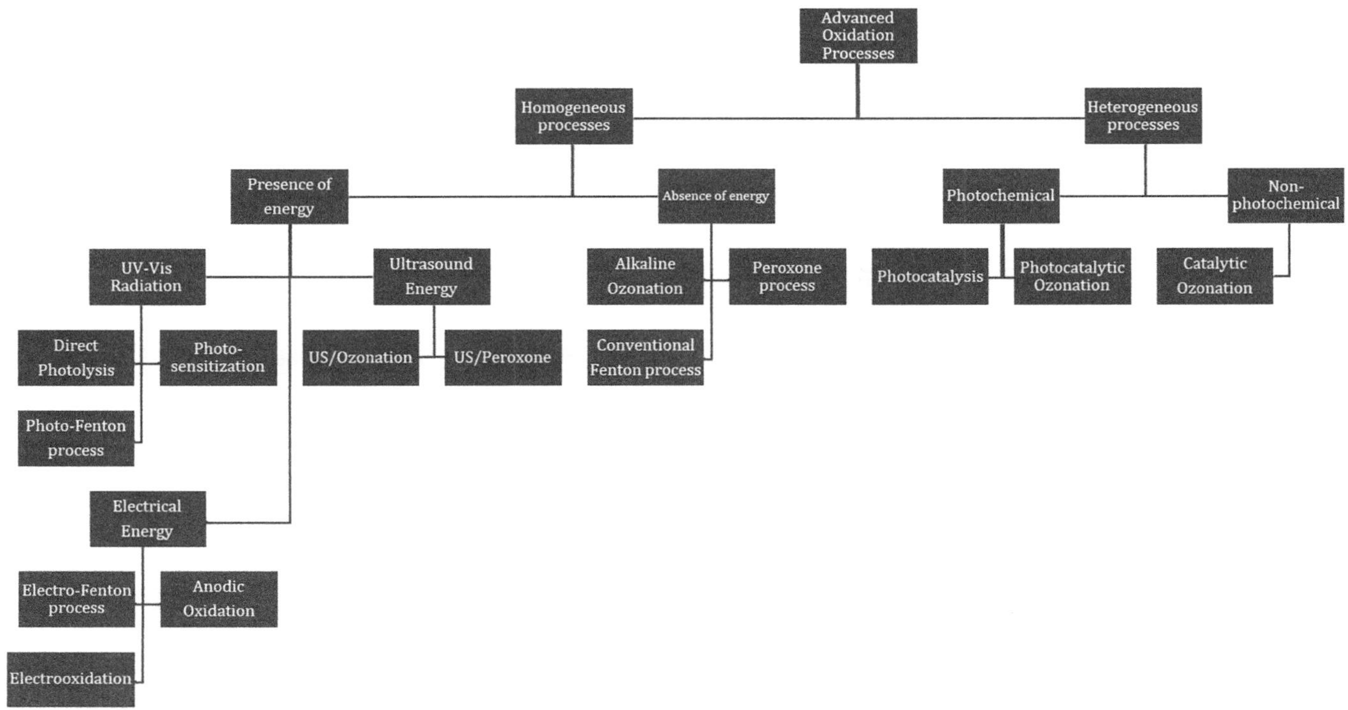

Fig. 13.2 Classification of advanced oxidation processes.

To guarantee the effectiveness of the process, the energy provided by the radiation must be at least equal to the energy of dissociation of the corresponding chemical bonds. This approach, besides enhancing oxidation via radical generation, also promotes disinfection, since UV radiation interacts with cellular tissues and DNA/RNA chains, altering genetic material and ultimately destroying the microorganisms (Amin et al., 2013).

13.2.1.2 Ozonation

Ozone is an unstable oxygen allotrope and a versatile reagent whose reactivity highly depends on the pH of the system. While ozone is relatively stable in the acidic range (Galdeano et al., 2018), its oxidation mechanism as an electrophilic substitution model, its evolution in an alkaline environment consists of a set of radical reactions (Eqs. 4–7) (Beltrán, 2004):

$$O_3 + HO^- \rightarrow HO_2^- + O_2 \quad (4)$$

$$O_3 + HO_2^- \rightarrow HO_2^{\bullet} + O_3^- \quad (5)$$

$$O_3^{-\bullet} + H^+ \rightarrow HO_3^{\bullet} \quad (6)$$

$$HO_3^{\bullet} \rightarrow {}^{\bullet}OH + O_2 \quad (7)$$

This proposed reaction route shows that other reactive oxygen species, such as the hydroperoxyl radical ($HO_2^{\cdot}$), are generated in situ. These species, while having a lower standard reduction potential than $HO^{\cdot}$, are also accountable for microorganism removal (Suganuma et al., 2019).

13.2.1.3 Fenton process and its modifications with other processes

The Fenton process was first developed at the end of the 19th century by Henry John Horstman Fenton as an analytical method for the determination of organic compounds (Fenton, 1894). Its notable oxidation potential has drawn interest for wastewater disinfection and decontamination. The highlight of this system is the increased oxidative potential of hydrogen peroxide (H_2O_2) through the production of hydroxyl radicals under acidic conditions using iron species as catalysts. The reactions that play a role in this process are shown in Eqs. (8)–(12) (Ameta et al., 2018):

$$Fe^{2+} + H_2O_2 \rightarrow Fe^{3+} + HO^{\bullet} + HO^- \quad (8)$$

$$HO^{\bullet} + H_2O_2 \rightarrow HO_2^{\bullet} + H_2O \quad (9)$$

$$Fe^{2+} + HO^{\bullet} \rightarrow Fe^{3+} + HO^- \quad (10)$$

$$Fe^{3+} + HO_2^{\bullet} \rightarrow Fe^{2+} + O_2 + H^+ \tag{11}$$

$$HO^{\bullet} + HO^{\bullet} \rightarrow H_2O_2 \tag{12}$$

One of the limitations of the conventional Fenton process is the depletion of the Fe^{2+} species, crucial for the generation of hydroxyl radicals. The application of radiation to the system, which is the foundation of the photo-Fenton method, may induce photochemical reactions that regenerate the reduced iron species, as well as promote the formation of hydroxyl radicals (Bauer and Fallmann, 1997). The reactivity is shown in Eqs. (13)–(15):

$$H_2O_2 + h\nu \rightarrow 2HO^{\bullet} \tag{13}$$

$$Fe^{3+} + H_2O + h\nu \rightarrow Fe^{2+} + HO^{\bullet} + H^+ \tag{14}$$

$$Fe^{3+} + H_2O_2 + h\nu \rightarrow Fe^{2+} + HO_2^{\bullet} + H^+ \tag{15}$$

Light, as shown in Eqs. (13)–(15), allows the regeneration of Fe^{2+} ions through the reduction of Fe^{3+} via reaction with hydrogen peroxide or water. This process is subject to the optimization of the optimal light wavelength and intensity. More energetic radiations (for example, ultraviolet light) tend to increase disinfection kinetics, but applying UV light is an expensive operation due to the high energetic requirements associated with its generation. Moreover, the mutagenic nature of this radiation (which is the principle of the disinfection process) increases the operational risks of the photo-Fenton process. Safer alternatives, however, are not as efficient in terms of degradation since the applied light is less energetic.

One of the main drawbacks of the Fenton process is the need to maintain a constant addition of hydrogen peroxide. The formation of ferric sludge and its implications in terms of waste transport and treatment are other limitations of the technique (Masood et al., 2023).

The incorporation of electrochemistry into this method is a novel approach, which aims to solve these problems. The electro-Fenton process features a controlled synthesis of hydrogen peroxide in the reaction medium, as well as the electrolytical regeneration of Fe^{2+}(Oturan and Oturan, 2018). The generation of H_2O_2 is based on the incomplete reduction of O_2 (Oxygen Reduction Reaction, ORR) in a carbon-based electrode, according to Eq. (16):

$$O_2 + 2H^+ + 2e^- \rightarrow H_2O_2 \tag{16}$$

The iron source, in the form of ferrous or ferric species, may be externally added to the solution or generated from a complementary electrode (the sacrificial iron anode) (Brillas et al., 2009).

Other approaches, such as the Photoelectro-Fenton process, are based on the combination of both methods and allow for enhanced degradation kinetics, lower reactive dosage, and reduced ferric sludge generation (Brillas, 2020).

13.2.1.4 Photocatalysis

Photocatalysis emerges as an interesting approach for the generation of ROS. This technology relies on the interaction of radiation with semiconductors. These materials possess electrical conductivity due to electron mobility between two energy levels (bands) separated by an energy barrier or gap (*Eg*). The first band, known as the valence band, is completely full of electrons, while the second band, the conduction band, is completely empty. Since the *Eg* associated with these materials is relatively low (usually not greater than 4 eV), relatively low-energy radiations (including sunlight) could satisfy the resonance condition and enable the promotion of electrons from the valence to the conduction band (Fox and Dulay, 1993).

When radiation with greater energy than the electronic gap interacts with the semiconductor, an electron is promoted from the valence to the conduction band, generating a "hole" (h^+, with oxidizing properties) and an "electron" (e^-, with reducing properties). This polarization is, by nature, unstable, resulting in the nearly immediate neutralization of the e^-/h^+ pair and energy liberation in the form of heat. However, some pairs reach the surface of the catalyst and trigger redox reactions. Holes, as oxidants, react with electron donors (^-OH groups, water, etc.). They generate hydroxyl radicals according to Eqs. (17) and (18):

$$h^+ + H_2O_{ads} \rightarrow H_2O^{\bullet +} + HO^{\bullet} \quad (17)$$

$$h^+ + OH^-{}_{ads} \rightarrow HO^{\bullet} \quad (18)$$

On the other hand, electrons in the conduction band are reducing agents and may interact with acceptor species such as molecular oxygen, which is present in solution. Superoxide anion, a ROS, is generated, and its evolution in the presence of the hydroperoxyl ion ultimately yields hydroxyl radicals via hydrogen peroxide homolysis:

$$O_{2ads} + e^- \rightarrow O_2^{\bullet -} \quad (19)$$

$$O_2^{\bullet -} + H^+ \rightarrow HO_2^{\bullet} \quad (20)$$

$$O_2^{\bullet} + HO_2^{\bullet} + H^+ \rightarrow H_2O_2 + O_2 \quad (21)$$

$$H_2O_2 + H^+ + e^- \rightarrow HO^{\bullet} + H_2O \quad (22)$$

As it has been seen, photocatalysis is a versatile tool for wastewater disinfection (Regmi et al., 2018). Catalyst development and dosage, selection of radiation and its application, and operational mode are challenges to be addressed to ensure good quality in the generated effluent as well as to minimize operational costs and environmental impact (Mishra and Sundaram, 2023).

13.2.2 Applications of AOPs in wastewater disinfection

The application of AOPs in pathogen disinfection in wastewater has drawn increasing interest in recent years. Authors have assessed the employment of these processes for the degradation of different pathogens, considering different water matrices and aiming to scale the operative systems.

Moreno-Andrés et al. (2023) addressed the inactivation of *Vibrio alginolyticus* (a waterborne marine pathogen) by the synergetic application of UV-A, UV-B, and UV-C under two common oxidizing agents, H_2O_2 and HSO_5^-. They concluded that shorter-range UV radiation provoked a faster degradation yield, with UV-A being unable to promote inactivation. The application of oxidants, particularly peroxymonosulfate (PMS, HSO_5^-), enhanced disinfection, with this enhancement being greater for long-range radiation (Moreno-Andrés et al., 2023). Guerra-Rodríguez et al. (2023) focused on the interaction of PMS with several types of disinfection agents (H_2O_2/O_3/UV-A) for the inactivation of *Enterococcus faecalis* in the pilot scale. One of their main conclusions was that the combination of 0.25 g/h O_3 and 0.5 mM PMS supposed the fastest treatment, achieving quantitative disinfection in less than 45 min, while the combination of PMS, O_3 and UV-A radiation, while synergic, was slower in terms of disinfection kinetics (Guerra-Rodríguez et al., 2023).

Another example of application of oxidizing agents and electromagnetic radiation can be found in the study performed by Martínez-García et al. (2023). In this work, the inactivation of several pathogenic agents of bacterial and viral nature is achieved by application of H_2O_2 and dissolved oxygen in presence of solar radiation. The study concludes that even low doses of H_2O_2 enhance disinfection kinetics, being the MS2 bacteriophage inactivation the most favored by oxidant application, yielding a 95–99% less radiation energy for quantitative disinfection than in the experiments in absence of oxidant (Martínez-García et al., 2023). Berruti et al. (2021) studied the application of the UV-C/PMS in the pilot scale for the simultaneous disinfection of three bacterial species and three pollutants of emerging concern. In their experiments, they successfully achieved simultaneous disinfection and decontamination in 15 min of treatment while minimizing sulfate dosage (0.5 mM). Applied PMS concentrations above this limit have potential to also remove antibiotic resistant bacteria and genes (ARBs and ARGs), while authors remark that the removal of ARGs requires harsher, more oxidizing conditions (Berruti et al., 2021).

Maniakova and Rizzo (2023) studied the simultaneous removal of pharmaceuticals and inactivation of pathogens in urban wastewater using UV-C photo-Fenton, UV-C/H_2O_2, and ozonation. In their work, they concluded that UV-C treatments, while effective in terms of disinfection, would require extensive optimization in terms of operational cost (Maniakova and Rizzo, 2023). Another example of photo-Fenton mediated disinfection of pathogens is the work performed by Tian et al. (2024), where the inactivation of a bacteriophage virus is achieved via oxidation in alkaline medium in presence of microstructured (with sizes in the nano-micrometer range) Fe oxides. In this study, the presence of natural organic matter and carbonates enhances virus disinfection, diminishing the concentration of bacteria to values under the limit of detection in 2 min for the solar-assisted degradation experiments (Tian et al., 2024).

Cho et al. (2023) focused on the application of UV/peracetic acid for the disinfection of *Pseudomonas syringae* in water subsequently used for irrigation. Their conclusions, both in terms of pathogen inactivation and plant growth after treated water applications, are optimistic (Cho et al., 2023). The potential of UV/PAA was also assessed by Moussavi et al. (2024) for the simultaneous degradation of persistent pollutants and *E. coli*, being this approach able to effectively inactivate a 6.5 log of 1.2×10^9 in 5 min of treatment in a batch system with tap water as the employed matrix (Moussavi et al., 2024). The influence of the matrix and the reactor configuration on the degradation process in studied by Guerra-Rodríguez et al. (2022). In their work, a comparative study between two reactor configurations (batch and tubular reactors) operating in three different matrices (distilled water, saline solution, and simulated wastewater) was carried out. The study concluded that performing treatment in more complex matrices (like simulated wastewater) hinders process performance. Scavenging experiments point to some ions being beneficial to the degradation process (such as bromine), while others may slow it (for example, fluoride and nitrite). The study highlighted the higher effectiveness of the tubular configuration, which may be attributable to the higher radiation intensity associated with this operation mode (Guerra-Rodríguez et al., 2022).

13.2.3 Challenges and drawbacks of AOPs disinfection

As has been seen, AOPs have earned their reputation as the "wastewater treatment of the 21st century" due to their enormous potential and versatility for not only pathogen disinfection but also intensive decontamination of wastewater in terms of micropollutant removal. However, it must be acknowledged that their systematic application on a global scale is yet to be achieved (Suty et al., 2004).

Many AOPs rely on the continuous application of radiation (especially in the UV range of the spectrum), the steady consumption of reagents, and the profuse utilization of specific catalysts. These circumstances considerably increase the operational and technological cost of AOPs, making their systematic application in conventional wastewater treatment plants a challenge yet to be solved (Kumar and Shah, 2021).

Moreover, even though all AOPs aim at the mineralization of the organic load of the effluents they receive, full conversion of the organic matter to inorganic species does not always happen. The occurrence of DBPs, generated after the incomplete degradation of organic matter, may yield effluents with a higher associated toxicity than the influent prior to treatment (Wang and Wang, 2021).

These drawbacks hinder the extensive implementation of AOPs at the global level and are challenges to be addressed in further research and scale-up initiatives.

13.3 Heterogeneous photocatalysis

Throughout this section, we will delve into the fundamentals, mechanisms, and application of heterogeneous photocatalysis in wastewater disinfection.

13.3.1 Fundamentals

As introduced previously in Section 13.2.1.4, a photocatalyst is a substance that in the presence of light not only causes chemical changes in the substances in the reaction, but is also excited after the absorption of light and causes successive interactions with the reaction partners, which result in the formation of reaction intermediates (Tan et al., 2019). Photocatalytic reactions have been used in recent years for multiple purposes, such as environmental care and energy generation. There are multiple factors required to achieve high photocatalytic activity, which will result in high reaction kinetics, such as mass of catalyst and initial concentration, wavelength and high absorption of UV radiation, temperature, radiant flux, quantum yield, high electron conductivity for better separation of charge carriers, and a high active surface for reaction (Faraji et al., 2019; Herrmann, 1999; Samadi et al., 2019). The photocatalyst in heterogeneous photocatalysis is in solid form, and all reactions take place at solid-gas or solid-liquid interfaces. The mechanism that takes place was briefly mentioned in Section 13.2.1.4. and consists of the transfer of holes and electrons from the excited semiconductor (McNaught and Wilkinson, 1997). Heterogeneous photocatalysis is one of the most promising subgroups of treatments within AOPs. Among the most studied heterogeneous photocatalysts, TiO_2 undoubtedly stands out, as it has always been the material that has allowed obtaining the best photocatalytic performances with maximum quantum yield. Within TiO_2, there are different allotropes with photocatalytic activity (anatase, rutile, and brookite), but it is undoubtedly anatase that exhibits the best photocatalytic conditions. By means of photocatalysis, it has been reported significant quantities of contaminants like organic compounds, noxious gases, and heavy metal ions can be disintegrated, converted into minerals, or diminished to inert substances (Gao et al., 2020). But also, in last years, heterogeneous photocatalysis has emerged as a promising method for neutralizing a broad spectrum of pathogenic microorganisms, even those resilient to conventional disinfection methods. One of the early references regarding the use of heterogeneous photocatalysis for disinfection purposes was proposed by Matsunaga et al. (1985), reporting the death of microbial cells using photoelectrochemistry with semiconductor powder (TiO_2/Pt) (Matsunaga et al., 1985). Since then, there have been numerous articles by authors describing the use of heterogeneous photocatalysis for water disinfection at the laboratory scale, being applied to any type of microorganism such as bacteria (Dunlop et al., 2002; García-Muñoz et al., 2022; Martín et al., 2024; Núñez-Salas et al., 2021; Rodríguez-Chueca et al., 2015), spores (Dunlop et al., 2008; Kim et al., 2021; Lanao et al., 2010; Polo-López et al., 2014), viruses (Giannakis et al., 2017; Watts et al., 1995; Xie et al., 2024), protozoa (Abeledo-Lameiro et al., 2016; Sunnotel et al., 2009), fungi (Polo-López et al., 2014; Sichel et al., 2007).

13.3.1.1 Mechanism of heterogeneous photocatalysis

A significant portion of research has specifically delved into the interplay between reactive oxygen species (ROS) and structures within biological microorganisms. These studies have revealed that the semiconductor material used as a photocatalyst in the disinfection process possesses an appropriate band gap, facilitating the separation of its electrons and holes when exposed to light radiation of the correct wavelength. Electrons (e^-) transition to the conduction band (CB), leaving behind holes (h^+) in the valence band (VB) (Herrmann, 1999). These separated electrons and holes may eventually recombine, releasing absorbed energy in the process. The resulting h^+ ions react with water, yielding potent hydroxyl radicals and hydrogen ions. These redox reactions, occurring in the presence of water and oxygen, can generate ROS capable of attacking and deactivating microorganisms. Among these ROS, the hydroxyl radical (HO·) or superoxide radical have been proposed as primary species responsible for microorganism inactivation (Fig. 13.3). Despite their short lifespan, these reactive oxygen species are potent in nature and are lethal to microorganisms, leading to their disinfection (Kelly et al., 2021).

The photocatalytic disinfection mechanism generally revolves around the degradation of microbial cell membranes, followed by the release of cellular contents. The objective of these studies is to

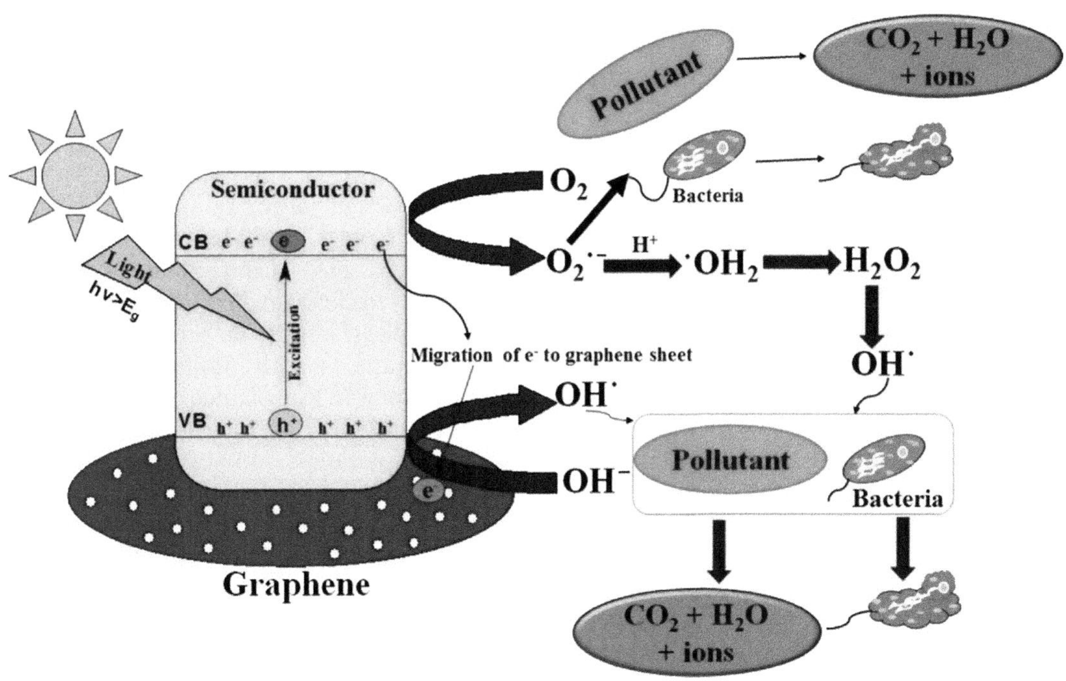

Fig. 13.3 Mechanism of photocatalysis for disinfection and organic pollutant removal with a graphene catalyst in water. Copyright from Dutta, V., Singh, P., Shandilya, P., Sharma, S., Raizada, P., Saini, A.K., Gupta, V.K., Hosseini-Bandegharaei, A., Agarwal, S., Rahmani-Sani, A., 2019. Review on advances in photocatalytic water disinfection utilizing graphene and graphene derivatives-based nanocomposites. J. Environ. Chem. Eng. 7 (3). Available from: http://www.journals.elsevier.com/journal-of-environmental-chemical-engineering/. https://doi.org/10.1016/j.jece.2019.103132.

unravel the mechanisms driving the decline in microbial viability. While a review conducted by Dalrymple et al. (2010), Dalrymple et al. (2010) examined the mechanisms underlying photocatalytic disinfection, the precise sequence of events leading to viability loss remains incompletely understood. Further insights into how ROS interact with microorganisms will empower researchers to optimize material properties and reactor designs, thereby enhancing the efficiency and speed of photocatalytic disinfection (Chong et al., 2010).

13.3.2 Application on wastewater: Advantages and disadvantages

As mentioned earlier, numerous authors have described the use of heterogeneous photocatalysis as a technology capable of removing various microorganisms present in water samples. Furthermore, one of the advantages of this type of technology is the simultaneous removal capacity of pathogenic microorganisms and organic contaminants present in water samples (Xu et al., 2022).

Heterogeneous photocatalysis has become an excellent choice due to its various advantages in most fields. One of the big advantages is that it is eco-friendly, for instance due to the possibility of using sunlight to start chemical reactions and clean up pollution without harming the environment (Giannakis et al., 2017; Polo-López et al., 2014). Also, it is good at getting rid of lots of different kinds of pollution, not only biological, but also organics as dyes, medicines, drugs, antibiotics, etc., from different human activities. So, it is a versatile way to clean up dirty water and fix environmental problems. Nonetheless, amidst its promise, heterogeneous photocatalysis also grapples with limitations (Rengifo-Herrera and Pulgarin, 2023). Notably, the reliance on specific semiconductor materials poses a hurdle, as these may be costly or limited in availability, necessitating exploration into alternative catalysts. There are many factors that depend on the efficiency of photocatalytic reactions, including the type of light, the surface of the catalyst, and what kind of pollutants they are dealing with (Rengifo-Herrera and Pulgarin, 2023). Furthermore, while photocatalysis offers a path to cleaner environments, some processes may inadvertently produce undesirable byproducts or require careful management to avoid secondary pollution incidents (Byrne et al., 2015). Another major issue is the recovery of the catalyst when it is in powder form, and the turbidity it can introduce to the water, causing problems with solar dispersion. For example, in the use of the primary photocatalyst, TiO_2, different methods of utilization have been reported in water treatment: (1) as a suspension of TiO_2 particles in water, and (2) as TiO_2 immobilized on a durable, inert support (Martins et al.,

2022). In that sense, factors such as the configuration of the catalyst in the photoreactor have an effect on the final result for water purification. Particularly when considering the treatment of drinking water for human consumption, utilizing TiO_2 suspended particles as a routine measure to enhance water potability at the household level (point-of-use water treatment) may not always be feasible (Liu et al., 2024). Various techniques have been employed to evaluate and compare the effectiveness of photocatalytic disinfection between a slurry reactor with a suspended catalyst and a structured reactor with an immobilized catalyst in a macroporous foam. The fixed-bed TiO_2 catalyst, which did not involve recovery of the catalyst post-purification, exhibited no particulates. However, this method may result in decreased disinfection rates or require more intricate systems with higher costs. Diverse forms of supported TiO_2, such as cylinders, tablets, balls, meshes, etc., have been studied to enhance the efficiency of solar water disinfection (Herrmann, 1999; Rengifo-Herrera and Pulgarin, 2023). The TiO_2 suspended particle reactor demonstrated the highest efficiency, attributed to improved surface contact between bacteria and catalyst. Conversely, fixed-bed reactors achieved disinfection rates comparable to slurry reactors.

Numerous strategies exist to improve the efficacy of solar disinfection and solar photocatalysis in water treatment. These enhancements should focus on the following aspects:

- **I.** Maximizing the penetration of suitable solar photons into the photo-reactor system.
- **II.** Enhancing treatment efficacy against resilient waterborne pathogens.
- **III.** Scaling up the volume of treated water within a given solar exposure duration.
- **IV.** Simplifying water treatment systems and reducing reliance on user intervention.
- **V.** Implementing cost-effective systems with straightforward designs and locally available materials for constructing water disinfection setups, especially for deployment in developing regions.
- **VI.** Developing photo-reactor systems optimized to prevent post-treatment recovery or regrowth of microorganisms and to minimize recontamination risks due to improper handling or storage.

13.4 Carbocatalytic processes

According to IUPAC, a catalyst is like a compound that can increase the speed of a reaction without altering the energy change. In other words, a catalyst is a compound that provides a lower activation energy for a given chemical reaction, but does not disturb its thermodynamic equilibrium, since it is neither consumed nor formed during the process. This translates into faster rates, so the effect of the catalyst is to accelerate the reaction it catalyzes, even to the point of making otherwise energetically inaccessible transformations possible. From this point of view, catalysis has become an interdisciplinary process, with numerous applications in various water treatment processes. These include some processes previously mentioned as photocatalytic degradation, AOPs, sonocatalytic degradation and carbocatalysis (Quimbaya-Ñañez et al., 2023; Zhao et al., 2021), the latter of which has been extensively studied in the field of water purification, particularly for treating contaminants of concern and biological pollutants. The species produced with carbocatalysis are highly reactive oxidation species (ROS) (Descorme, 2017; Duan et al., 2020), consequently, they can serve as significant oxidizers to achieve swift decomposition and complete mineralization of targeted micropollutants through processes involving electrophilic attack, addition reactions, hydrogen extraction, and electron transfer.

13.4.1 Fundamentals

It is a metal-free heterogeneous catalytic technique with great potential for current and future synthesis processes. Carbocatalysts are heterogeneous materials based on interconnected *sp2*/*sp3* carbons decorated with oxygen functional groups, structural defects and other heteroatomic dopants (N, S, etc.) (Descorme, 2017).

The carbocatalytic process is often considered an AOP, this process specifically uses carbon materials as catalysts for ROS production. This process has garnered significant interest in recent years, primarily due to the potential utilization of biomass as a carbon source. Moreover, biomass is an abundant and renewable source, and the resulting materials can be environmentally friendly and sustainable alternatives to conventional carbocatalysts, such as graphene, carbon nanotubes and active carbon (Qu et al., 2022). Carbocatalysis had already shown its effectiveness in the field of adsorption, so the catalytic properties of carbon materials are used to activate various oxidants such as H_2O_2, PMS and peroxydisulfate (PDS) to obtain radical and no-radical species with greater oxidizing power (Estrada-Flórez et al., 2024; Gao et al., 2022). Several authors have already demonstrated the effectiveness of carbocatalysis in the degradation of contaminants of emerging concern (CEC) and conclude that it is a strategy that can be combined with other AOPs to greatly improve the removal of organic pollutants (Jiang et al., 2023;

Wang et al., 2022). It must be considered and remembered that the radical species generated will be capable not only of eliminating CECs but also microbiological contaminants.

Studies in the literature have demonstrated that in carbocatalytic processes, carbon primarily serves to facilitate electron transfer from the oxidant, thereby aiding in the generation of hydroxyl and sulfate radicals (Fig. 13.4). Duan et al. (2016, 2018) explored the catalytic potential of various forms of carbon in water treatment technology. They emphasized the significance of defect indices as a determining factor for the catalytic activity of any carbonaceous material, with higher defect levels correlating to more effective catalytic properties (Duan et al., 2016, 2018). From this point of view biomass-derived carbons, with their defective structures can be a good alternative to conventional carbons. Given that in the structure of carbon materials, the active sites are in the edges, in the *sp2* hybrid atoms and on the dangling bonds in the six-chain carbon rings, (Schlögl, 2021), the physico-chemical properties of carbonaceous materials (electronic configuration, localized electrons, electron-rich functionalities, heteroatoms, etc.) will play an important role in catalytic processes.

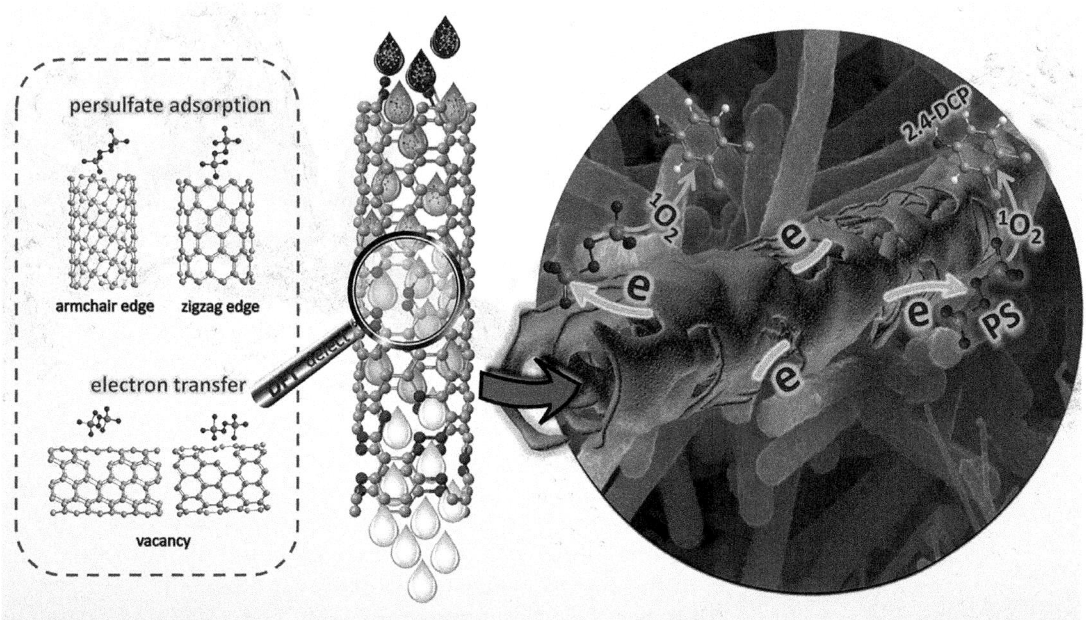

Fig. 13.4 General mechanism of carbocatalytic processes. Copyright from Cheng, X., Guo, H., Li, W., Yang, B., Wang, J., Zhang, Y., Du, E., 2020. Metal-free carbocatalysis for persulfate activation toward nonradical oxidation: enhanced singlet oxygen generation based on active sites and electronic property. Chem. Eng. J. 396, 125107. https://doi.org/10.1016/j.cej.2020.125107.

13.4.2 Application on wastewater

As previously confirmed, carbocatalysis was already promising strategy in the field of adsorption and is now gaining popularity in the catalytic degradation of pollutants (Song et al., 2023; Zhao et al., 2021), particularly in the AOPs, proving to be a good complement for improving water decontamination (Estrada-Flórez et al., 2024). This meanwhile there is not too much information on the role of carbocatalysis in water disinfection in the literature. However, there are a few studies on the combination of carbocatalysts with other materials. In a recent study, the combined effect of $Cu(OH)_2$ and a nanostructured carbon layer was show to be effective in bacteria removal from the flow (Liu et al., 2023). With their method, they successfully inactivated over 99% of bacteria in the water and maintained continuous mechanical disinfection for a 30-day period, showcasing the effectiveness of their process in deactivating pathogenic bacteria. Another study also shows the application of a carbon-based catalytic membrane, focusing on persulfate (PS) activation and continuous flow water disinfection (Li et al., 2022). By studying the cumulative effect of each membrane component, the influence of carbon on PS activation and its effectiveness in inactivating micro-organisms indicative of fecal contamination in a defined suspension of *Escherichia coli* and *Enterococcus* was demonstrated. Without PS, the pristine PVDF membrane retained 1.98 ± 0.051 log, while the rGO- M-PVDF membrane eliminated 3.71 ± 0.66 log of the initial population after 6 h of operation. They then demonstrated the catalytic characteristics of the membrane by adding a PS precursor to the feed suspension. During 24 h of treatment, this system effectively decreased the presence of culturable microorganisms in both specified suspensions and river water to levels significantly lower than the detection threshold (0.5 CFU/mL). Still in the same vein, a more recent study by Chabi et al. (2024) demonstrated the contribution of carbocatalysis to an AOP system, applied to the treatment of wastewater containing a dynamic bacterial community. They employed a multi-stage water purification setup that incorporated an advanced oxidation process (AOP) featuring activated carbon in a pilot reactor to achieve effective drinking water production. This treatment approach led to a significant decrease in bacteria levels, with reductions ranging from 2 to 4 log10. By the conclusion of the treatment, the average count of pathogenic bacteria in the surface water was notably diminished.

All these examples confirm the effectiveness of carbocatalytic processes in water treatment and ensure that treated water complies with established guidelines for microbiological bacteria in drinking water.

13.4.3 Key characteristics of carbocatalytic processes

Due to their stability in the pH range, their tunable chemical properties, and their structural properties with remarkable electrical capacities (Morales-Torres et al., 2012), carbon compounds remain outstanding candidates for the development of innovative low-cost catalysts. It has already been widely reported that, in contrast to metal-based catalysts, carbon-based catalysts offer the benefits of reduced secondary pollution, improved thermal stability, and heightened catalytic efficiency. Some authors have shown that PMS can be activated by graphene, activated carbon, etc., and applied the elimination of organic contaminants (Peng et al., 2020; Wang et al., 2019). The major advantage of oxidant activation is the possibility of producing other products with higher oxidation potentials, more effective in oxidizing pollutants due to their higher redox potential. For example, hydrogen peroxide (1.77 eV), peroxymonosulfate (2.07 eV) and peroxodisulfate (2.01 eV) respectively produce hydroxyl (2.7 eV), sulfate radicals (2.5–3.1 eV) and singlet oxygen (Descorme, 2017). Recent developments in carbocatalysis have emphasized the utilization of biochar for activating oxidants. Biochar's inherent properties, including surface area, pore diameter, surface defects, and the presence of heteroatoms, stem from various biomass precursors with diverse compositions and structures. So chemical modification of biochar can also increase the number of Lewis acid sites, thus broadening the range of active sites (Duan et al., 2015). In this way, biochar doping can considerably improve the catalytic efficiency of materials when activated by oxidants such as PMS, PDS, etc. (Duan et al., 2015; Yang et al., 2016). Within the structure of biochar/PMS system, the heteroatoms present had different roles and the activation was dominated by nitrogen atoms (Wang et al., 2020). By incorporating nitrogen doping, the catalytic efficacy of the biochar experiences significant enhancement during PMS activation, as indicated by Yang et al. (2016). Alongside facilitating PMS surface adsorption, nitrogen atoms also stimulate electron transfer reactions between biochar and PMS by activating the *sp2*-carbon hybrid (Zhao et al., 2021). The graphitic nitrogen presents in the biochar acts as an active site for activating PMS to generate ROS, and as an adsorption site for adsorbing PMS encourage direct electrons transfer to contaminants to prevent their oxidation (Arampatzidou and Deliyanni, 2016). Despite the good results obtained with this carbocatalytic process, previous work has only presented information on kinetic and mechanical aspects. Similarly, non-radical pathways for pollutant removal are not well implicated, and theoretical calculations are not included in this literature. What could be interesting perspectives in this field of research.

13.5 Plasma based disinfection

As previously said conventional disinfection process have limitations on the elimination of some contaminants, being necessary to study some alternative ways. Conventional AOPs are a very useful way of elimination CECs and biological contaminants. However, these processes require the use of different chemical products in order to produce the oxidative species such as hydroxyl or sulfate radicals necessary to oxidize the organic matter. Plasma treatments offer several advantages. On one hand, they provide an alternative method for generating hydroxyl radicals and other reactive species. On the other hand, the intrinsic ionic state of plasma allows for its direct utilization.

13.5.1 Fundamentals

Plasma which constitutes the 99% of the visible matter in the known universe, is the fourth state of the matter, it is the most energized state, and it is composed of ionized matter. In the solid-state particles are linked ones with the others, in the liquid state, not all the particles are united being able to flow. In the gas state, particles are not united ones with the others being separated. Lastly, plasma is reached when a gas is ionized, this means that the particles are no longer united forming molecules but forming a cloud of charged particles generated after the separation of the previous particle. The decomposition of the previous particle generates electrons, positive and negative ions, free radicals, photons and excited and neutral atoms and molecules. The plasma is characterized for being a very good conductor of the electricity and temperature, against normal behavior of gases, that usually are used as insulant. In the universe, plasma is reached in conditions of very high temperature and/or low-pressure conditions. However, is possible to produce it in atmospheric conditions by inducing a big electrical field (Raizer, 1991; Tonks and Langmuir, 1929).

Plasma can be characterized by electron and ion temperatures and densities, gas temperature, pressure, current magnitude, powering mode, thermodynamic equilibrium, and ionization degree. Being possible to categorize different types of plasma. The main classification for this topic, will be the temperature

(T_e for electrons, T_i for ions), being low temperature plasma (LTP) and high temperature plasma (HTP) ($Ti \approx Te \geq 10^7$ K). In the low temperature plasma depending on the thermal equilibrium between electron and ions there can be non-thermal plasmas (NTP) ($Te \approx Ti \approx 300$ K) and thermal plasmas ($Ti \approx Te \leq 2 \cdot 10^4$ K). An example of NTP will be corona discharges, for LTP an arc of plasma and for HTP the fusion plasma (de Cadiz, 2022; Hippler and Kersten, 2008).

It is possible to generate the ionization effect artificially reaching the dielectric resistivity of the medium discharge achieving the breakdown of the discharge. Once it is reached the ionization starts. Being possible to confine the ionization area between electrodes. This phenomenon is modelized by the Paschen's breakdown law (Eq. 23) (Llewellyn-Jones, 1957; Raizer, 1991).

$$V_B(voltios) = \frac{Bpd}{Ln(Apd) - Ln\left[\left(1 + \frac{1}{\gamma}\right)\right]} \tag{23}$$

where V_B is the breakdown voltage, p is the pressure of the ionized gas, d electrode distance, Υ is the Townsend coefficient, B is the relationship between the excitation and ionization energies and A is the maximum ionization in the gas. Varying the pressure, gas and electrode configurations different energy and chemical reactions are achieved.

Any gas can be used to produce plasma, however for industrial application has a lot of importance the atmospheric air plasma (Domonkos et al., 2021; Reema et al., 2022). The ionization of air produces a lot of Reactive Nitrogen and Oxygen Species (RONS) (Lukes et al., 2014; Simoncelli et al., 2019; Thirumdas et al., 2018). When the discharge is produced in air, species derived from nitrogen and oxygen are mainly generated with lifespan of nanoseconds. Due to the elevated reactivity, reactive species quickly react to produce more stable species such as ozone and nitrogen oxides (Simoncelli et al., 2019). When the discharge of plasma is produced in water, the ionized gas is mainly composed of nitrogen, oxygen and water vapor. These discharges can be used for treating water and generate plasma activated water (PAW).

PAW composition and properties depend on the way of generating the plasma discharge and the composition of the ionized gas (Guo et al., 2021; Perinban et al., 2019). In the case of air ionization for producing PAW, generated reactive oxygen species such as hydroxyl radicals ($^{\bullet}OH$), singlet oxygen ($^1O_2$), superoxide (O_2^-), ozone (O_3), hydrogen peroxide (H_2O_2) will be generated. From the nitrogen of the air, in the water will appear the more stable species (Lee et al., 2021; Wang et al., 2022). Also, PAW will be rich in nitrates (NO_3^-), nitrites (NO_2^-) and peroxynitrates species (Guo et al., 2021; Lukes et al., 2014; Thirumdas et al., 2018; Zhou et al., 2018). Due to the generation of this chemical species the pH will decrease, the conductivity will increase, and the Oxide-Reduction Potential (ORP) will increase, becoming an oxidative substance (Zhao et al., 2020).

There are different technologies for obtaining PAW, some of the most studied ones are, Dielectric Barrier Discharge (DBD), spark, jet spark, pulsed corona, corona discharge, gliding arc, etc. Also, it is possible to use direct or alternating current, in positive or negative polarity with different geometries for the electrodes (Cui et al., 2018).

13.5.2 Application on wastewater

Due to the different ways of producing the ionization and the versality of plasma technologies, there are a lot of applications, some of the most described into the bibliography are:

- Elimination of different types of contaminants from wastewaters such microorganism, biofilms, resistant genes, pharmaceutical drugs, dye elimination, microplastics degradation, etc. (Dhivya et al., 2022; Kumar et al., 2022; Ostrikov et al., 2020; Liu et al., 2024; Mai-Prochnow et al., 2021; Gupta and Ayan, 2019; Zhao et al., 2020).
- Treatment of water to obtain a disinfectant with different uses, like medicine (Ran Lee et al., 2022; Wang et al., 2022; Zhou et al., 2018).
- Treatment of water to produce a fertilizer due to the capacity of PAW to increase germination and growth speed of the crops (Bradu et al., 2020; Domonkos et al., 2021; Guo et al., 2021; Liu et al., 2020; Thirumdas et al., 2018).
- Use of PAW into food processing (Calvo et al., 2020; Dhivya et al., 2022; Perinban et al., 2019).

As previously mentioned, conventional disinfection technologies are not perfect and they present some disadvantages as the generation of DBPs after reacting with organic matter. Other technologies such as UV-C could be an alternative, but it does not have residual effect, so there might be bacterial regrowth. Plasma technologies are a very interesting alternative in the wastewater treatments due to:

I. No necessity of using additional chemicals to achieve the remove. It is possible to degrade the contaminants just with the application of the plasma and with the reaction of the RONS generated in the process showing big elimination rates (Cui et al., 2018).

II. The activation of the water increases the oxidative species concentration, most of them have with short life time. However, usually recombine into species with longer lifespan in the waters but with oxidative capacity as well, avoiding the posttreatment bacterial regrowth.

III. The chemistry and composition of the treated water can be easily controlled by the treatment time, ionized gas, and discharge geometry.

IV. When the ionized gas is atmospheric air, after a short plasma treatment time, the concentration of reactive species from oxygen and nitrogen increases considerably.

V. Apart of plasma treatment, the water obtained can be used as PAW, an oxidative substance capable of achieving bacterial reductions and avoid regrowth.

VI. Also, PAW is possible to use as a fertilizer due to the increase of nitrogen species. It is greatly proven that seeds watered with PAW, increase its germination and growth range.

13.5.3 Advantages and disadvantages

The use of plasma technologies is an innovative technology in the wastewater sector and that is because have different advantages with the conventional treatment ways:

- Plasma treatments does not need the addition of any chemical compound in order to make the disinfection. It can be used to induce AOPs with a smaller price than conventional ways.
- Water treated with plasma, have a residual antimicrobial effect due to the presence of oxidative species in the water.
- It is possible to produce changes in the water just inducing an electric field enough strong to ionize the vicinity of the water interphase.
- It has to be produced in-situ reducing the transport impact.
- These technologies have a residual effect, which prevents microbial regrowth post-treatment.
- It is possible to remove different types of contaminants with a high elimination rate.
- Atmospheric plasmas only require of a power source to be able to produce it.

However, there are some problems with the implementation of these plasma technologies:

- For the generation of the plasma, it is necessary to achieve the dielectric breakdown, inducing the proper electric field, this requires the use of a proper power source. Depending on the type of discharge it will require of high voltage or elevated current with high frequency. So, it is necessary to install proper security measures.

- For making the scalation to be able to install it in big surfaces, the two main limitation factors are the proper species diffusion among the reactor to have a homogeneous treatment and to be able to have a source with enough power.
- It is necessary to have the discharge conditions properly controlled to avoid the excessive acidification of the PAW.
- The materials used in the discharge vicinity must be resistant to the degradation in order to increase the usable time of the plasma treatment equipment.

Conclusion and future trends

There is a growing necessity of developing novel disinfection technologies to face the presence of new pollutants and microorganisms in water because they represent a risk to the different living organisms of the trophic chain. In this scenario, scientific community have made efforts to implement high-efficiency/low-cost processes.

Among all the conventional water disinfection processes, chlorination, ozonation and UV-C dosing are the most commonly used due to the inherent advantages like their low cost and simplicity. Nevertheless, in the last decades, several controversial items have emerged. During the treatment of water with chlorination and ozonation, the presence of DBPs and toxic intermediates compounds have been detected and so they are in the spotlight of scientific community. Besides, for those cases using ozone, there is the necessity of the *in situ* generation that rises more the price of the cubic meter of treated water. When using UVC, the detection of DBPs has not been shown, but another of the main disadvantages also is relative to the energy consumption of the lamps which need a continuous, and sometimes expensive, replacement.

For that reason, there is a rising tendency to the development of treatments centred on AOPs. These processes based on the generation of oxidizing agents are gaining attention *vs* the conventional treatments. AOPs benefit from their sustainable and eco-efficient nature to face a new era of water treatment. Despite the low level of TRL demonstrated till now for scaling-up those process, in some cases, concretely, when the mineralization of the effluent to CO_2 and H_2O is not complete, the detection of DBPs has been shown too.

AOPs are classified according to the phase in which the catalyst stays compared to the reagents being homogeneous when it is in the same phase than water and heterogeneous when stays in a different phase, normally in a solid one.

The main advantage of the use of AOPs resides in their potential and their versatility, but the increase in the operational cost due to the reagent consumption (i.e., H_2O_2, O_3, PMS, etc.) as well as the incomplete mineralization of the organic pollutant reached while using some processes have provoked the reluctance of promoting their scaling-up for treating higher volumes of water.

The photocatalytic process involving the employment of a solid material can be lead in two modes of catalyst: (i) directly suspended in a solution and (ii) immobilized onto a support. In the first case, the process is limited to some applications like its use for drinking water despite the higher efficiency demonstrated when working in this mode. Nevertheless, when using a immobilized catalyst the applications rise up at the expense of reducing the disinfection/decontamination rates of the overall process.

The main advantage of photocatalysis *vs* SODIS is that the application of a solid catalyst can enhance the efficiency of the microorganisms removal in terms of rate in exchange of a certain cost of the employed material. However, there are several drawbacks that still must be taken into account in order to be able to scaling-up the process to treat larger volumes of polluted water. In addition to the cost of the catalyst, the type of required light plays a key role in the system and sometimes is correlated to the type and quantity of pollution inside the water.

In immobilized systems, the main drawbacks is related to the chosen support and the leaching of the catalyst to the solution since it must maximize the light penetration into the material to ensure the activation of the full active phase and so the guarantee the photoreactions.

Carbocatalysis have emerged as an alternative to heterogeneous catalytic processes focusing its attention in the activation of an oxidant (PMS, PDS, H_2O_2, O_3, etc.). The main advantage resides in the dual activity of the carbonaceous material involving adsorption of the pollutant and the promotion of additional ways of generation of radicals from the initial oxidants at the surface of the catalyst. This dual role theoretically will improve not only the reaction rate of the decontamination process, but also the yield of the oxidant decomposition toward $HO^{\cdot}$ or $HOO^{\cdot}$ and another species.

The main advantage of the employment of carbonaceous material are related to the presence of defects in their structure which favors the catalytic properties of the materials; these are those involving the electronic configuration and the presence of doping heteroatoms focused mainly on the N element. Till now, the effect of doping has demonstrated an important role in terms of efficiency. But, as always in the heterogeneous processes, there is a still a huge gap to reach the implementation of them at larger scales.

Finally, the last process cited here is relative to the generation of radical species *via* activation of plasma. This process consists of one of a fewer studied for this application indicating that it represents a totally emerging technology despite the number of plasma devices used for water treatment has been increased in the last few decades.

The main advantage resides onto its capacity of eliminating contaminants (such as resistant genes, pharmaceutical waste) and microorganisms without the necessity of the addition of any chemical product, the residual antimicrobial effect and besides, without the generation of DBPs.

There are different ways to produce the ionization of the gas and the composition of the Plasma Activated Water (PAW) depends on the type of ionized gas, the treatment time, the aqueous matrix and the discharge geometry. Another advantage is that PAW can be used for different purposes in different sectors such as wastewater treatment, food industry and agriculture sector.

However, plasma activation needs a power source with energy enough to cover the production of species to treat large volumes of water. The low and incipient knowledge about the reaction mechanisms and the high energy consumption represent the main drawbacks to lead a further implementation. In this sense, it can be said that the scale-up of the process is still far compared to conventional processes.

In terms of future prospects for disinfection, it is important to consider that chlorination is by far the ideal disinfectant, but with the disadvantage of forming DBPs (disinfection by-products), so the first premise should be to obtain a technology capable of retaining the positive aspects of chlorination (efficiency, price, availability, easy application, etc.) but without generating DBPs. Additionally, the scientific community is proposing that future disinfection treatments based on chemical oxidation methods should be capable of simultaneously removing pathogenic microorganisms and emerging contaminants of concern. Technologies that can address this task at a competitively economic level will have gained significant ground. Finally, sustainability must always be considered, and treatments that are environmentally friendly, considering the entire life cycle of the process, should be prioritized.

Acknowledgments

The authors wish to thank the MCIN/AEI/10.13039/501100011033 and the European Union NextGenerationEU/PRTR for funding the DIGIT4WATER project (TED2021-129969A-C-32), and also Spanish MCIN/AEI/10.13039/501100011033 and "ERDF A way of making Europe," for the project PHOTORAS (PID2021-128165OA-I00). J.R.-C. acknowledges the Programme of Excellence for University

Teaching Staff (M190020074BJJRC). P.G.-M. acknowledges Comunidad de Madrid through the call Research Grants for Young Investigators from Universidad Politécnica de Madrid for funding the research project SUNCAT4PLAST (APOYO-JOVENES-21-JV4DEB-3-Q2WGKV).

References

Abeledo-Lameiro, M.J., Ares-Mazás, E., Gómez-Couso, H., 2016. Evaluation of solar photocatalysis using TiO_2 slurry in the inactivation of Cryptosporidium parvum oocysts in water. J. Photochem. Photobiol. B 163, 92–99. https://doi.org/10.1016/j.jphotobiol.2016.08.016.

Ameta, R., Chohadia, A.K., Jain, A., Punjabi, P.B., 2018. Chapter 3—fenton and photo-fenton processes. Advanced Oxidation Processes for Waste Water TreatmentEmerging Green Chemical Technology. Academic Press.

Amin, M.M., Hashemi, H., Bovini, A.M., 2013. A review on wastewater disinfection. Int. J. Environ. Health Eng. 2 (1), 22. https://doi.org/10.4103/2277-9183.113209.

Arampatzidou, A.C., Deliyanni, E.A., 2016. Comparison of activation media and pyrolysis temperature for activated carbons development by pyrolysis of potato peels for effective adsorption of endocrine disruptor bisphenol-A. Academic Press Inc., Greece, J. Colloid Interface Sci. 466, 101–112. 10957103. Available from: http://www.elsevier.com/inca/publications/store/6/2/2/8/6/1/index.htt. https://doi.org/10.1016/j.jcis.2015.12.003.

Armah, E.K., Chetty, M., Adedeji, J.A., Kukwa, D.T., Mutsvene, B., Shabangu, K.P., Bakare, B.F., 2020. Emerging trends in wastewater treatment technologies: the current perspective. In: Promising Techniques for Wastewater Treatment and Water Quality Assessmen. Intechopen. https://doi.org/10.5772/intechopen.93898.

Bauer, R., Fallmann, H., 1997. The photo-fenton oxidation—a cheap and efficient wastewater treatment method. Res. Chem. Intermed. 23, 341–354.

Beltrán, F.J., 2004. Ozone Reaction Kinetics for Water and Wastewater Systems. Lewis Publishers.

Benítez, J.S., Rodríguez, C.M., Casas, A.F., 2021. Disinfection byproducts (DBPs) in drinking water supply systems: a systematic review. Elsevier Ltd, Colombia, Phys. Chem. Earth 123. 14747065. Available from: http://www.journals.elsevier.com/physics-and-chemistry-of-the-earth/. https://doi.org/10.1016/j.pce.2021.102987.

Berruti, I., Nahim-Granados, S., Abeledo-Lameiro, M.J., Oller, I., Polo-López, M.I., 2021. Uv-c peroxymonosulfate activation for wastewater regeneration: simultaneous inactivation of pathogens and degradation of contaminants of emerging concern. MDPI AG, Spain, Molecules 26 (16). 14203049. Available from: https://www.mdpi.com/1420-3049/26/16/4890/pdf. https://doi.org/10.3390/molecules26164890.

Bradu, C., Kutasi, K., Magureanu, M., Puač, N., Živković, S., 2020. Reactive nitrogen species in plasma-activated water: generation, chemistry and application in agriculture. J. Phys. D. Appl. Phys. 53 (22), 223001. https://doi.org/10.1088/1361-6463/ab795a. 0022-3727.

Brillas, E., Sirés, I., Oturan, M.A., 2009. Electro-fenton process and related electrochemical technologies based on fenton's reaction chemistry. Chem. Rev. 109 (12), 6570–6631. 15206890. Available from: http://pubs.acs.org/doi/pdfplus/10.1021/cr900136g. https://doi.org/10.1021/cr900136gSpain.

Brillas, E., 2020. A review on the photoelectro-Fenton process as efficient electrochemical advanced oxidation for wastewater remediation. Treatment with UV light, sunlight, and coupling with conventional and other photo-assisted advanced technologies. Chemosphere 250, 126198. https://doi.org/10.1016/j.chemosphere.2020.126198. 00456535.

Byrne, J.A., Dunlop, P.S.M., Hamilton, J.W.J., Fernández-Ibáñez, P., Polo-López, I., Sharma, P.K., Vennard, A.S.M., 2015. A review of heterogeneous photocatalysis for water and surface disinfection. MDPI AG, United Kingdom, Molecules 20 (4), 5574–5615. 14203049. Available from: http://www.mdpi.com/1420-3049/20/4/5574/pdf. https://doi.org/10.3390/molecules20045574.

de Cadiz, C., 2022. Research and Industrial Development of Innovative Electrode Configurations for Electrokinetic Cooling Devices. Doctorado industrial en ingeniería de máquinas, Universidad politécnica de Madrid.

Calvo, T., Prieto, M., Alvarez-Ordóñez, A., López, M., 2020. Effect of non-thermal atmospheric plasma on food-borne bacterial pathogens on ready-to eat foods: morphological and physico-chemical changes occurring on the cellular envelopes. Food Secur. 9 (12), 1865. https://doi.org/10.3390/foods9121865. 2304-8158.

Chabi, K., Ye, C., Yan, W., Li, J., Kiki, C., Xianzhong, L., Shikan, Z., Gad, M., Xin, Y., 2024. Removing bacteria contaminants from raw surface water using multiple barriers pilot-scale reactor: sand filtration, activated carbon, and advanced oxidation processes. Biochem. Eng. J. 204, 109246. https://doi.org/10.1016/j.bej.2024.109246. 1369703X.

Chaúque, B.J.M., Issufo, M., Benitez, G.B., Cossa, V.C., Chaúque, L.G.H., Stauber, C.-E., Benetti, A.D., Rott, M.B., 2023. Why do low-cost point-of-use water treatment technologies succeed or fail in combating waterborne diseases in the field? A systematic review. Elsevier Ltd, Brazil, J. Environ. Chem. Eng. 11 (5). 22133437. Available from: http://www.journals.elsevier.com/journal-of-environmental-chemical-engineering/. https://doi.org/10.1016/j.jece.2023.110575.

Cho, M., Kim, S.R., Hwang, I., Kim, K., 2023. Ultraviolet-activated peracetic acid treatment-enhanced Arabidopsis defense against Pseudomonas syringae pv. tomato DC3000. Korean Society of Plant Biotechnology, South Korea, J. Plant Biotechnol. 50, 215–224. 12292818. Available from: https://www.kspbtjpb.org/journal/view.html?pn=vol%26uid=2354%26vmd=Full. https://doi.org/10.5010/JPB.2023.50.027.215.

Chong, M.N., Jin, B., Chow, C.W.K., Saint, C., 2010. Recent developments in photocatalytic water treatment technology: a review. Elsevier Ltd, Australia, Water Res. 44 (10), 2997–3027. 00431354. Available from: https://www.sciencedirect.com/science/journal/00431354. https://doi.org/10.1016/j.watres.2010.02.039.

Cuerda-Correa, E.M., Alexandre-Franco, M.F., Fernández-González, C., 2020. Advanced oxidation processes for the removal of antibiotics from water. An overview. MDPI AG, Spain, Water (Switzerland) 12 (1). 20734441. Available from: https://res.mdpi.com/d_attachment/water/water-12-00102/article_deploy/water-12-00102-v2.pdf. https://doi.org/10.3390/w12010102.

Cui, Y., Cheng, J., Chen, Q., Yin, Z., 2018. The types of plasma reactors in wastewater treatment. In: IOP Conference Series: Earth and Environmental Science. Institute of Physics Publishing, China, p. 208, https://doi.org/10.1088/1755-1315/208/1/012002. 17551315 1 https://iopscience.iop.org/journal/1755-1315.

Dalrymple, O.K., Stefanakos, E., Trotz, M.A., Goswami, D.Y., 2010. A review of the mechanisms and modeling of photocatalytic disinfection. Appl. Catal. B Environ. 98 (1–2), 27–38. https://doi.org/10.1016/j.apcatb.2010.05.001. 09263373.

Descorme, C., 2017. Catalytic wastewater treatment: oxidation and reduction processes. Recent studies on chlorophenols. Catal. Today 297, 324–334. https://doi.org/10.1016/j.cattod.2017.03.039. 09205861.

Dhivya, R., Rajakrishnapriya, V.C., Sruthi, K., Chidanand, D.V., Sunil, C.K., Rawson, A., 2022. Biofilm combating in the food industry: overview, non-thermal approaches, and mechanisms. J. Food Process. Preserv. 46 (10). https://doi.org/10.1111/jfpp.16282. 0145-8892.

Domonkos, M., Tichá, P., Trejbal, J., Demo, P., 2021. Applications of cold atmospheric pressure plasma technology in medicine, agriculture and food industry. MDPI AG, Czech Republic, Appl. Sci. (Basel) 11 (11). 20763417. Available from: https://www.mdpi.com/2076-3417/11/11/4809/pdf. https://doi.org/10.3390/app11114809.

Duan, X., Indrawirawan, S., Sun, H., Wang, S., 2015. Effects of nitrogen-, boron-, and phosphorus-doping or codoping on metal-free graphene catalysis. Catal. Today 249, 184–191. https://doi.org/10.1016/j.cattod.2014.10.005. 09205861.

Duan, X., Chao, S., Zhou, L., Sun, H., Suvorova, A., Odedairo, T., Zhu, Z., Shao, Z., Wang, S., 2016. Surface controlled generation of reactive radicals from persulfate by carbocatalysis on nanodiamonds. Appl. Catal. B Environ. 194, 7–15. https://doi.org/10.1016/j.apcatb.2016.04.043. 09263373.

Duan, X., Indrawirawan, S., Kang, J., Tian, W., Zhang, H., Duan, X., Zhou, X., Sun, H., Wang, S., 2020. Synergy of carbocatalytic and heat activation of persulfate for evolution of reactive radicals toward metal-free oxidation. Catal. Today 355, 319–324. https://doi.org/10.1016/j.cattod.2019.02.051. 09205861.

Dunlop, P.S.M., McMurray, T.A., Hamilton, J.W.J., Byrne, J.A., 2008. Photocatalytic inactivation of Clostridium perfringens spores on TiO2 electrodes. J. Photochem. Photobiol. A Chem. 196 (1), 113–119. https://doi.org/10.1016/j.jphotochem.2007.11.024. 10106030.

Duan, X., Sun, H., Wang, S., 2018. Metal-free carbocatalysis in advanced oxidation reactions. Acc. Chem. Res. 51 (3), 678–687. https://doi.org/10.1021/acs.accounts.7b00535.

Dunlop, P.S.M., Byrne, J.A., Manga, N., Eggins, B.R., 2002. The photocatalytic removal of bacterial pollutants from drinking water. J. Photochem. Photobiol. A Chem. 148 (1–3), 355–363. https://doi.org/10.1016/s1010-6030(02)00063-1. 10106030.

EPA, 1999. Alternative Disinfectants and Oxidants. Guidance Manual. United States Environmental Protection Agency.

EPA, 2020. Disinfection Profiling and Benchmarking Technical Guidance Manual. United States Environmental Protection Agency.

Epelle, E.I., Macfarlane, A., Cusack, M., Burns, A., Okolie, J.A., Mackay, W., Rateb, M., Yaseen, M., 2023. Ozone application in different industries: a review of recent developments. Elsevier B.V., United Kingdom, Chem. Eng. J. 454. 13858947. Available from: www.elsevier.com/inca/publications/store/6/0/1/2/7/3/index.htt. https://doi.org/10.1016/j.cej.2022.140188.

Estrada-Flórez, S.E., Serna-Galvis, E.A., Lee, J., Torres-Palma, R.A., 2024. Unraveling kinetic and synergistic effects during ultrasound-enhanced carbocatalysis for water remediation as a function of ultrasonic frequency. Academic Press, Colombia, J. Environ. Manag. 350. 10958630. Available from: https://www.sciencedirect.com/science/journal/03014797. https://doi.org/10.1016/j.jenvman.2023.119548.

Faraji, M., Yousefi, M., Yousefzadeh, S., Zirak, M., Naseri, N., Jeon, T.H., Choi, W., Moshfegh, A.Z., 2019. Two-dimensional materials in semiconductor photoelectrocatalytic systems for water splitting. Royal Society of Chemistry, Iran, Energy Environ. Sci. 12 (1), 59–95. 17545706. Available from: http://pubs.rsc.org/en/journals/journal/ee. https://doi.org/10.1039/c8ee00886h.

Fenton, H.J.H., 1894. LXXIII—oxidation of tartaric acid in presence of iron. J. Chem. Soc. Trans. 65, 899–910. https://doi.org/10.1039/CT8946500899. 03681645.

Rook, J.J., 1974. Formation of Haloforms During Chlorination of Natural Waters. Water Treatment and Examination. vol. 23 Society for Water Treatment and Examination, p. 234.

Fox, M.A., Dulay, M.T., 1993. Heterogeneous photocatalysis. Chem. Rev. 93 (1), 341–357. https://doi.org/10.1021/cr00017a016. 15206890.

Galdeano, M.C., Wilhelm, A.E., Goulart, I.B., Tonon, R.V., Freitas-Silva, O., Germani, R., Chávez, D.W.H., 2018. Effect of water temperature and pH on the concentration and time of ozone saturation. Instituto de Tecnologia de Alimentos—ITAL, Brazil, Braz. J. Food Technol. 21. 19816723. Available from: http://www.scielo.br/pdf/bjft/v21/1981-6723-bjft-21-e2017156.pdf. https://doi.org/10.1590/1981-6723.15617.

Gao, C., Low, J., Long, R., Kong, T., Zhu, J., Xiong, Y., 2020. Heterogeneous single-atom photocatalysts: fundamentals and applications. Chem. Rev. 120 (21), 12175–12216. https://doi.org/10.1021/acs.chemrev.9b00840.

Gao, Y., Wang, Q., Ji, G., Li, A., 2022. Degradation of antibiotic pollutants by persulfate activated with various carbon materials. Chemical Engineering Journal 429, 132387. https://doi.org/10.1016/J.CEJ.2021.132387.

García-Muñoz, P., López-Maxías, C., Guerra-Rodríguez, S., Carbajo, J., Casas, J.A., Rodríguez-Chueca, J., 2022. Photocatalytic activation of peroxymonosulfate using ilmenite (FeTiO3) for Enterococcus faecalis inactivation. Elsevier Ltd, Spain, J. Environ. Chem. Eng. 10 (5). 22133437. Available from: http://www.journals.elsevier.com/journal-of-environmental-chemical-engineering/. https://doi.org/10.1016/j.jece.2022.108231.

Giannakis, S., Liu, S., Carratalà, A., Rtimi, S., Amiri, M.T., Bensimon, M., Pulgarin, C., 2017. Iron oxide-mediated semiconductor photocatalysis vs. heterogeneous photo-Fenton treatment of viruses in wastewater. Impact of the oxide particle size. J. Hazard. Mater. 339, 223–231. https://doi.org/10.1016/j.jhazmat.2017.06.037. 03043894.

Glaze, W.H., Kang, J.W., Chapin, D.H., 1987. The chemistry of water treatment processes involving ozone, hydrogen peroxide and ultraviolet radiation. Ozone Sci. Eng. 9 (4), 335–352. https://doi.org/10.1080/01919518708552148. 15476545.

Goldstein, B.D., 2012. John Snow, the broad street pump and the precautionary principle. Environ. Dev. 1 (1), 3–9. https://doi.org/10.1016/j.envdev.2011.12.002. 22114645.

Guerra-Rodríguez, S., Rodríguez, E., Moreno-Andrés, J., Rodríguez-Chueca, J., 2022. Effect of the water matrix and reactor configuration on Enterococcus sp. inactivation by UV-A activated PMS or H_2O_2. J. Water Process Eng. 47, 102740. https://doi.org/10.1016/j.jwpe.2022.102740.

Guerra-Rodríguez, S., Rodríguez, E., Rodríguez-Chueca, J., 2023. Pilot-scale regeneration of wastewater through intensified sulfate radical-based advanced oxidation processes (PMS/UV-A, PMS/H2O2/UV-A, and PMS/O3): inactivation of bacteria and mechanistic considerations. Chem. Eng. J. 469, 143859. https://doi.org/10.1016/j.cej.2023.143859. 13858947.

Guo, D., Liu, H., Zhou, L., Xie, J., He, C., 2021. Plasma-activated water production and its application in agriculture. J. Sci. Food Agric. 101 (12), 4891–4899. https://doi.org/10.1002/jsfa.11258.

Gupta, T.T., Ayan, H., 2019. Application of non-thermal plasma on biofilm: a review. Appl. Sci. 9 (17), 3548. https://doi.org/10.3390/app9173548. 2076–3417.

Herrmann, J.M., 1999. Heterogeneous photocatalysis: fundamentals and applications to the removal of various types of aqueous pollutants. Elsevier, France, Catal. Today 53 (1), 115–129. 09205861. Available from: http://www.sciencedirect.com/science/journal/09205861. https://doi.org/10.1016/S0920-5861(99)00107-8.

Hippler, R., Kersten, H., 2008. Low Temperature Plasmas. Wiley-VCH.

Jahan, B.N., Li, L., Pagilla, K.R., 2021. Fate and reduction of bromate formed in advanced water treatment ozonation systems: a critical review. Elsevier Ltd, United States, Chemosphere 266. 18791298. Available from: www.elsevier.com/locate/chemosphere. https://doi.org/10.1016/j.chemosphere.2020.128964.

Jiang, T., Wang, B., Gao, B., Cheng, N., Feng, Q., Chen, M., Wang, S., 2023. Degradation of organic pollutants from water by biochar-assisted advanced oxidation processes: mechanisms and applications. J. Hazard. Mater. 442, 130075. https://doi.org/10.1016/j.jhazmat.2022.130075. 03043894.

Kashinkunti, R.D., Linden, K.G., Shin, G.A., Metz, D.H., Sobsey, M.D., Moran, M.C., Samuelson, A.M., 2004. Investigating multibarrier inactivation for Cincinnati—UV, by-products, and biostability. American Water Works Association, United States, J. Am. Water Works Assoc. 96 (6), 114–127. 0003150X, Available from: http://onlinelibrary.wiley.com/journal/10.1002/(ISSN)1551-8833. https://doi.org/10.1002/j.1551-8833.2004.tb10785.x.

Kelly, J., McDonnell, C., Skillen, N., Manesiotis, P., Robertson, P.K.J., 2021. Enhanced photocatalytic degradation of 2-methyl-4-chlorophenoxyacetic acid (MCPA) by the addition of H2O2. Elsevier Ltd, United Kingdom, Chemosphere 275. https://doi.org/10.1016/j.chemosphere.2021.130082. 18791298. Available from: www.elsevier.com/locate/chemosphere.

Kim, J.U., Shahbaz, H.M., Cho, J., Lee, H., Park, J., 2021. Inactivation of Bacillus cereus spores using a combined treatment of UV-TiO2 photocatalysis and high hydrostatic pressure. Elsevier Ltd, South Korea, Innov. Food Sci. Emerg. Technol. 70. 14668564. Available from: http://www.elsevier.com/locate/ifset. https://doi.org/10.1016/j.ifset.2021.102676.

Krasner, S.W., Cantor, K.P., Weyer, P.J., Hildesheim, M., Amy, G., 2017. Case study approach to modeling historical disinfection by-product exposure in Iowa drinking waters. Chinese Academy of Sciences, United States, J. Environ. Sci. (China) 58, 183–190. 18787320. Available from: http://www.journals.elsevier.com/journal-of-environmental-sciences/. https://doi.org/10.1016/j.jes.2017.03.007.

Krishnan, R.Y., Manikandan, S., Subbaiya, R., Karmegam, N., Kim, W., Govarthanan, M., 2023. Recent approaches and advanced wastewater treatment technologies for mitigating emerging microplastics contamination—a critical review. Sci. Total Environ. 858, 159681. https://doi.org/10.1016/j.scitotenv.2022.159681. 00489697.

Kumar, V., Shah, M.P., 2021. Advanced oxidation processes for complex wastewater treatment. Advanced Oxidation Processes for Effluent Treatment Plants. Elsevier.

Kumar, A., Škoro, N., Gernjak, W., Povrenović, D., Puač, N, 2022. Direct and indirect treatment of organic dye (acid blue 25) solutions by using cold atmospheric plasma jet. Front. Phys. 10. https://doi.org/10.3389/FPHY.2022.835635.

Lanao, M., Ormad, M.P., Goñi, P., Miguel, N., Mosteo, R., Ovelleiro, J.L., 2010. Inactivation of Clostridium perfringens spores and vegetative cells by photolysis and TiO2 photocatalysis with H2O2. Sol. Energy 84 (4), 703–709. https://doi.org/10.1016/j.solener.2010.01.023. 0038092X.

Lee, B.C.Y., Lim, F.Y., Loh, W.H., Ong, S.L., Hu, J., 2021. Emerging contaminants: an overview of recent trends for their treatment and management using light-driven processes. MDPI, Singapore, Water (Switzerland) 13 (17). 20734441. Available from: https://www.mdpi.com/2073-4441/13/17/2340/pdf. https://doi.org/10.3390/w13172340.

Li, W., Zhou, R., Zhou, R., Weerasinghe, J., Zhang, T., Gissibl, A., Cullen, P.J., Speight, R., Ostikov, K., 2022. Insights into amoxicillin degradation in water by non-thermal plasmas. Chemosphere 291, 132757. https://doi.org/10.1016/J.CHEMOSPHERE.2021.132757.

Liu, K., Yang, Z., Liu, S., 2020. Study of the characteristics of DC multineedle-to-water plasma-activated water and its germination inhibition efficiency: the effect of discharge mode and gas flow. IEEE Trans. Plasma Sci. 48 (4), 969–979. https://doi.org/10.1109/TPS.2020.2980040.

Liu, F., Hou, Y., Wang, S., Li, Z., Zhang, B., Tong, M., 2023. Periodate activation by pyrite for the disinfection of antibiotic-resistant bacteria: Performance and mechanisms. Water Res. 230, 119508. https://doi.org/10.1016/J.WATRES.2022.119508.

Liu, L., Shen, Z., Wang, C., 2024. Recent advances and new insights on the construction of photocatalytic systems for environmental disinfection. Academic Press, China, J. Environ. Manag. 353. 10958630. Available from: https://www.sciencedirect.com/science/journal/03014797. https://doi.org/10.1016/j.jenvman.2024.120235.

Llewellyn-Jones, F., 1957. Ionization and Breakdown in Gases. John Wiley & Sons.

López, A., Rodríguez-Chueca, J., Mosteo, R., Gómez, J., Rubio, E., Goñi, P., Ormad, M.P., 2019. How does urban wastewater treatment affect the microbial quality of treated wastewater? Institution of Chemical Engineers, Spain, Process Saf. Environ. Prot. 130, 22–30. 09575820. Available from: http://www.elsevier.com/wps/find/journaldescription.cws_home/713889/description#description. https://doi.org/10.1016/j.psep.2019.07.016.

Lukes, P., Dolezalova, E., Sisrova, I., Clupek, M., 2014. Aqueous-phase chemistry and bactericidal effects from an air discharge plasma in contact with water: evidence for the formation of peroxynitrite through a pseudo-second-order post-discharge reaction of H 2 O 2 and HNO 2. Plasma Sources Sci. Technol. 23 (1), 015019. https://doi.org/10.1088/0963-0252/23/1/015019. 0963-0252.

Mai-Prochnow, A., Zhou, R., Zhang, T., Ostrikov, K., Mugunthan, S., Rice, S.A., Cullen, P.J., 2021. Interactions of plasma-activated water with biofilms: inactivation, dispersal effects and mechanisms of action. Nature Research, Australia, NPJ Biofilms Microbiomes 7 (1). 20555008. Available from: https://www.nature.com/npjbiofilms/. https://doi.org/10.1038/s41522-020-00180-6.

Maniakova, G., Rizzo, L., 2023. Pharmaceuticals degradation and pathogens inactivation in municipal wastewater: a comparison among UVC photo-Fenton with chelating agents, UVC/H2O2 and ozonation. J. Environ. Chem. Eng. 11 (6), 111356. https://doi.org/10.1016/j.jece.2023.111356. 22133437.

Martín, I., Reñones, P., Rodríguez-Chueca, J., de la Peña O'Shea, V.A., Fresno, F., García-Muñoz, P., 2024. Independent and simultaneous photocatalytic decontamination and disinfection of water over In2O3/TiO2 heterojunctions. Elsevier B.V., Spain, Catal. Today 426. 09205861. Available from: http://www.sciencedirect.com/science/journal/09205861. https://doi.org/10.1016/j.cattod.2023.114397.

Martínez-García, A., Nahim-Granados, S., Berruti, I., Oller, I., Polo-López, M.I., 2023. Assessment of solar water disinfection enhancement with H2O2 and dissolved oxygen on inactivating different waterborne pathogens. J. Environ. Chem. Eng. 11 (6), 111145. https://doi.org/10.1016/j.jece.2023.111145. 22133437.

Martins, R.C., Sacras, Â., Jovanovic, S., Alves, P., Ferreira, P., Gomes, J., 2022. Solar energy for liquid wastewater treatment with novel TiO2 supported catalysts. Elsevier Ltd, Portugal, Energy Rep. 8, 489–494. 23524847. Available from: http://www.journals.elsevier.com/energy-reports/. https://doi.org/10.1016/j.egyr.2022.01.196.

Masood, A.S., Ali, M.S., Manzar, M.S., Khan, N.A., Khan, A.H., 2023. Current situation of pharmaceutical wastewater around the globe. In: The Treatment of Pharmaceutical Wastewater: Innovative Technologies and the Adaptation of

Treatment Systems. Elsevier, Italy, pp. 19–52. Available from: https://www.sciencedirect.com/book/9780323991605. https://doi.org/10.1016/B978-0-323-99160-5.00013-8.

Matsunaga, T., Tomoda, R., Nakajima, T., Wake, H., 1985. Photoelectrochemical sterilization of microbial cells by semiconductor powders. FEMS Microbiol. Lett. 29 (1–2), 211–214. https://doi.org/10.1111/j.1574-6968.1985.tb00864.x. 03781097.

McNaught, A., Wilkinson, A., 1997. Compendium of Chemical Terminology: Gold Book. IUPAC International Union of Pure and Applied Chemistry.

Mishra, S., Sundaram, B., 2023. A review of the photocatalysis process used for wastewater treatment. Mater. Today Proc. https://doi.org/10.1016/j.matpr.2023.07.147. 22147853.

Morales-Torres, S., Pastrana-Martínez, L.M., Figueiredo, J.L., Faria, J.L., Silva, A.M.-T., 2012. Design of graphene-based TiO2 photocatalysts—a review. Environ. Sci. Pollut. Res. 19 (9), 3676–3687. https://doi.org/10.1007/s11356-012-0939-4. 09441344.

Moreno-Andrés, J., Tierno-Galán, M., Romero-Martínez, L., Acevedo-Merino, A., Nebot, E., 2023. Inactivation of the waterborne marine pathogen Vibrio alginolyticus by photo-chemical processes driven by UV-A, UV-B, or UV-C LED combined with H2O2 or HSO5−. Water Res. 232, 119686. https://doi.org/10.1016/j.watres.2023.119686. 00431354.

Moussavi, G., Bakhtiarynasab, M., Shekoohiyan, S., Mohammadi, S., Giannakis, S., Li, M., 2024. High potential of the vacuum UV-activated peracetic acid (VUV/PAA) process in eliminating recalcitrant contaminants and waterborne pathogens: assessing the efficacy of annular and helical reactor configurations. J. Water Process Eng. 60, 105143. https://doi.org/10.1016/j.jwpe.2024.105143. 22147144.

Núñez-Salas, R.E., Rodríguez-Chueca, J., Hernández-Ramírez, A., Rodríguez, E., Maya-Treviño, M.D.L., 2021. Evaluation of B-ZnO on photocatalytic inactivation of Escherichia coli and Enterococcus sp. Elsevier Ltd, Mexico, J. Environ. Chem. Eng. 9 (1). 22133437. Available from: http://www.journals.elsevier.com/journal-of-environmental-chemical-engineering/. https://doi.org/10.1016/j.jece.2020.104940.

Ostrikov, K. (k.), Zhou, R., Zhou, R., Wang, P., Xian, Y., Mai-Prochnow, A., Lu, X., Cullen, P.J., Ostrikov, K. (k.), Bazaka, K., 2020. Plasma-activated water: generation, origin of reactive species and biological applications. Institute of Physics Publishing, Australia, J. Phys. D. Appl. Phys. 53 (30). 13616463. Available from: https://iopscience.iop.org/article/10.1088/1361-6463/ab81cf. https://doi.org/10.1088/1361-6463/ab81cf.

Oturan, N., Oturan, M.A., 2018. Electro-Fenton Process: Background, New Developments, and Applications Electrochemical Water and Wastewater Treatment. Elsevier, France, pp. 193–221. Available from: http://www.sciencedirect.com/science/book/9780128131602. https://doi.org/10.1016/B978-0-12-813160-2.00008-0.

Peng, Z., Liu, X., Zhang, W., Zeng, Z., Liu, Z., Zhang, C., Liu, Y., Shao, B., Liang, Q., Tang, W., Yuan, X., 2020. Advances in the application, toxicity and degradation of carbon nanomaterials in environment: a review. Environ. Int. 134, 105298. https://doi.org/10.1016/j.envint.2019.105298. 01604120.

Perinban, S., Orsat, V., Raghavan, V., 2019. Nonthermal plasma–liquid interactions in food processing: a review. Compr. Rev. Food Sci. Food Saf. 18 (6), 1985–2008. https://doi.org/10.1111/1541-4337.12503. 1541–4337.

Polo-López, M.I., Castro-Alférez, M., Oller, I., Fernández-Ibáñez, P., 2014. Assessment of solar photo-Fenton, photocatalysis, and H2O2 for removal of phytopathogen fungi spores in synthetic and real effluents of urban wastewater. Chem. Eng. J. 257, 122–130. https://doi.org/10.1016/j.cej.2014.07.016. 13858947.

Poyatos, J.M., Muñio, M.M., Almecija, M.C., Torres, J.C., Hontoria, E., Osorio, F., 2010. Advanced oxidation processes for wastewater treatment: state of the art. Water Air Soil Pollut. 205 (1–4), 187–204. https://doi.org/10.1007/s11270-009-0065-1. 0049–6979.

Qu, S., Yuan, Y., Yang, X., Xu, H., Mohamed, A.K., Zhang, J., Zhao, C., Liu, L., Wang, B., Wang, X., Rinklebe, J., Li, Y.C., Wang, S., 2022. Carbon defects in biochar facilitated nitrogen doping: the significant role of pyridinic nitrogen in peroxymonosulfate activation and ciprofloxacin degradation. Elsevier B.V., China, Chem. Eng. J. 441. 13858947. Available from: www.elsevier.com/inca/publications/store/6/0/1/2/7/3/index.htt. https://doi.org/10.1016/j.cej.2022.135864.

Quimbaya-Ñañez, C., Serna-Galvis, E.A., Silva-Agredo, J., García-Rubio, I., Torres-Palma, R.A., Ávila-Torres, Y.P., 2023. Improvement of the carbocatalytic degradation of pharmaceuticals in water by the use of ultrasound waves. Multidisciplinary Digital Publishing Institute (MDPI), Colombia, Water (Switzerland) 15 (20). 20734441. Available from: http://www.mdpi.com/journal/water. https://doi.org/10.3390/w15203679.

Raizer, Y.P., 1991. Gas Discharge Physics. Springer.

Ran Lee, H., Sang Lee, Y., Suk You, Y., Young Huh, J., Kim, K., Cheol Hong, Y., Kim, C.-H., 2022. Antimicrobial effects of microwave plasma-activated water with skin protective effect for novel disinfectants in pandemic era. *Sci. Rep.* 12, 5968. https://doi.org/10.1038/s41598-022-10009-1.

Reema, R.R., Khanikar, H., Bailung, K.S., 2022. Review of the cold atmospheric plasma technology application in food, disinfection, and textiles: a way forward for achieving circular economy. Frontiers Media SA, India, Front. Phys. 10. https://doi.org/10.3389/fphy.2022.942952. 2296424X. Available from: journal.frontiersin.org/journal/physics.

Regmi, C., Joshi, B., Ray, S.K., Gyawali, G., Pandey, R.P., 2018. Understanding mechanism of photocatalytic microbial decontamination of environmental wastewater. Frontiers Media S.A., South Korea, Front. Chem. 6. 22962646. Available from: http://journal.frontiersin.org/journal/chemistry. https://doi.org/10.3389/fchem.2018.00033.

Regulation (EU) 2020/741, 2020. Minimum requirements for water reuse. OJEU 177, 32–55.

Rengifo-Herrera, J.A., Pulgarin, C., 2023. Why five decades of massive research on heterogeneous photocatalysis, especially on TiO2, has not yet driven to water disinfection and detoxification applications? Critical review of drawbacks and challenges. Elsevier B.V., Argentina, Chem. Eng. J. 477. https://doi.org/10.1016/j.cej.2023.146875. 13858947. Available from: www.elsevier.com/inca/publications/store/6/0/1/2/7/3/index.htt.

Rodríguez-Chueca, J., Ormad Melero, M.P., Mosteo Abad, R., Esteban Finol, J., Ovelleiro Narvión, J.L., 2015. Inactivation of Escherichia coli in fresh water with advanced oxidation processes based on the combination of O3, H2O2, and TiO2. Kinetic modeling. Springer Verlag, Spain, Environ. Sci. Pollut. Res. 22 (13), 10280–10290. 16147499. Available from: http://www.springerlink.com/content/0944-1344. https://doi.org/10.1007/s11356-015-4222-3.

Samadi, M., Zirak, M., Naseri, A., Kheirabadi, M., Ebrahimi, M., Moshfegh, A.Z., 2019. Design and tailoring of one-dimensional ZnO nanomaterials for photocatalytic degradation of organic dyes: a review. Springer Netherlands, Iran, Res. Chem. Intermed. 45 (4), 2197–2254. 15685675. Available from: http://www.

springer.com/chemistry/journal/11164. https://doi.org/10.1007/s11164-018-03729-5.

Schlögl, R., 2021. Quo vadis carbocatalysis? J. Energy Chem. 61, 219–227. https://doi.org/10.1016/j.jechem.2021.02.024. 20954956.

Sichel, C., de Cara, M., Tello, J., Blanco, J., Fernández-Ibáñez, P., 2007. Solar photocatalytic disinfection of agricultural pathogenic fungi: Fusarium species. Appl. Catal. B Environ. 74 (1–2), 152–160. https://doi.org/10.1016/j.apcatb.2007.02.005. 09263373.

Simoncelli, E., Schulpen, J., Barletta, F., Laurita, R., Colombo, V., Nikiforov, A., Gherardi, M., 2019. UV–VIS optical spectroscopy investigation on the kinetics of long-lived RONS produced by a surface DBD plasma source. Plasma Sources Sci. Technol. 28 (9), 095015. https://doi.org/10.1088/1361-6595/ab3c36. 1361–6595.

Song, T., Gao, Y., Ye, J., Zhang, X., Rui, S., Luo, J., 2023. Insight into enhanced degradation of tetracycline over peroxymonosulfate activated via biochar-based nanocomposite: performance and mechanism. Environ. Sci. Pollut. Res. 30 (10), 27394–27408. https://doi.org/10.1007/s11356-022-24102-5. 1614–7499.

Suganuma, R., Kawasaki, S., Yasuoka, K., 2019. Disinfection by HO 2 radicals generated by negative corona discharge on acidic solution. IEEE Trans. Plasma Sci. 47 (1), 346–349. https://doi.org/10.1109/tps.2018.2879384. 0093–3813.

Sunnotel, O., Verdoold, R., Dunlop, P.S.M., Snelling, W.J., Lowery, C.J., Dooley, J.S.G., Moore, J.E., Byrne, J.A., 2009. Photocatalytic inactivation of Cryptosporidium parvum on nanostructured titanium dioxide films. J. Water Health 8 (1), 83–91. https://doi.org/10.2166/wh.2009.204.

Suty, H., De Traversay, C., Cost, M., 2004. Applications of advanced oxidation processes: present and future. Water Sci. Technol. 49 (4), 227–233. https://doi.org/10.2166/wst.2004.0270. 0273–1223.

Tan, H.L., Abdi, F.F., Ng, Y.H., 2019. Heterogeneous photocatalysts: an overview of classic and modern approaches for optical, electronic, and charge dynamics evaluation. Royal Society of Chemistry, Australia, Chem. Soc. Rev. 48 (5), 1255–1271. 14604744. Available from: http://pubs.rsc.org/en/journals/journal/cs. https://doi.org/10.1039/c8cs00882e.

Thirumdas, R., Kothakota, A., Annapure, U., Siliveru, K., Blundell, R., Gatt, R., Valdramidis, V.P., 2018. Plasma activated water (PAW): chemistry, physico-chemical properties, applications in food and agriculture. Trends Food Sci. Technol. 77, 21–31. https://doi.org/10.1016/j.tifs.2018.05.007.

Tian, N., Giannakis, S., Oji-Okoro, O.C., Farinelli, G., Garcia-Muñoz, P., Pulgarin, C., 2024. Photo-Fenton inactivation of MS2 bacteriophage at alkaline pH by Fe salts or nm to μm-sized oxides, and the Janus-faced effects of natural organic matter in surface waters. Catal. Today 430, 114536. https://doi.org/10.1016/j.cattod.2024.114536. 09205861.

Tonks, L., Langmuir, I., 1929. A general theory of the plasma of an arc. Phys. Rev. J. 34, 876. https://doi.org/10.1103/PhysRev.34.876.

Wang, J., Wang, S., 2021. Toxicity changes of wastewater during various advanced oxidation processes treatment: an overview. J. Clean. Prod. 315, 128202. https://doi.org/10.1016/J.JCLEPRO.2021.128202.

Wang, S., Xu, L., Wang, J., 2019. Nitrogen-doped graphene as peroxymonosulfate activator and electron transfer mediator for the enhanced degradation of sulfamethoxazole. Chem. Eng. J. 375, 122041. https://doi.org/10.1016/J.CEJ.2019.122041.

Wang, H., Guo, W., Liu, B., Si, Q., Luo, H., Zhao, Q., Ren, N., 2020. Sludge-derived biochar as efficient persulfate activators: sulfurization-induced electronic structure modulation and disparate nonradical mechanisms. Appl. Catal. B Environ. 279, 119361. https://doi.org/10.1016/J.APCATB.2020.119361.

Wang, J., Li, W., Zhao, Z., Musoke, F.S.N., Wu, X., 2022. Ultrasonic activated biochar and its removal of harmful substances in environment. MDPI, China, Microorganisms 10 (8). 20762607. Available from: www.mdpi.com/journal/microorganisms. https://doi.org/10.3390/microorganisms10081593.

Watts, R.J., Kong, S., Orr, M.P., Miller, G.C., Henry, B.E., 1995. Photocatalytic inactivation of coliform bacteria and viruses in secondary wastewater effluent. Water Res. 29 (1), 95–100. https://doi.org/10.1016/0043-1354(94)E0122-M. 00431354.

World Health Organization, 2011. Guidelines for Drinking-Water Quality. World Health Organization.

Xie, Y., Zhang, Z., Zhao, Y., Han, Y., Liu, C., Sun, Y., 2024. Effect of dissolved organic matter on the inactivation of bacteriophage MS2 by graphitic carbon nitride-based photocatalysis. J. Environ. Chem. Eng. 12 (2), 112025. https://doi.org/10.1016/j.jece.2024.112025. 22133437.

Xu, X., Wu, C., Guo, A., Qin, B., Sun, Y., Zhao, C., Zhang, F., Cai, A., 2022. Visible-light photocatalysis of organic contaminants and disinfection using biomimetic-synthesized TiO2-Ag-AgCl composite. Appl. Surf. Sci. 588, 152886. https://doi.org/10.1016/j.apsusc.2022.152886. 01694332.

Yang, S., Li, L., Xiao, T., Zhang, Y., Zheng, D., 2016. Promoting effect of ammonia modification on activated carbon catalyzed peroxymonosulfate oxidation. Sep. Purif. Technol. 160, 81–88. https://doi.org/10.1016/j.seppur.2016.01.020. 13835866.

Zhao, Y.M., Patange, A., Sun, D.W., Tiwari, B., 2020. Plasma-activated water: physicochemical properties, microbial inactivation mechanisms, factors influencing antimicrobial effectiveness, and applications in the food industry. Blackwell Publishing Inc., Ireland, Compr. Rev. Food Sci. Food Saf. 19 (6), 3951–3979. 15414337. Available from: http://onlinelibrary.wiley.com/journal/10.1111/(ISSN)1541-4337. https://doi.org/10.1111/1541-4337.12644.

Zhao, C., Shao, B., Yan, M., Liu, Z., Liang, Q., He, Q., Wu, T., Liu, Y., Pan, Y., Huang, J., Wang, J., Liang, J., Tang, L., 2021. Activation of peroxymonosulfate by biochar-based catalysts and applications in the degradation of organic contaminants: a review. Chem. Eng. J. 416, 128829. https://doi.org/10.1016/J.CEJ.2021.128829.

Zhou, R., Zhou, R., Prasad, K., Fang, Z., Speight, R., Bazaka, K., Ostrikov, K.(K.), 2018. Cold atmospheric plasma activated water as a prospective disinfectant: the crucial role of peroxynitrite. Green Chem. 20 (23), 5276–5284. https://doi.org/10.1039/c8gc02800a. 1463–9262.

14

Modeling of municipal biological wastewater treatment

Elham Jalilnejad, Salar Fakhraddinfakhriazar, and Mehran Alizadeh
Faculty of Chemical Engineering, Urmia University of Technology, Urmia, Iran

Abbreviations

AI	artificial intelligence
ANFIS	adaptive neuro fuzzy inference system
ANN	artificial neural network
ANN-GA	artificial neural networks that employ genetic algorithms
ARIMA	autoregressive integrated moving average
AS	activated sludge
ASM	activated sludge model
ASM1	activated sludge model no. 1
ASM2	activated sludge model no. 2
ASM3	activated sludge model no. 3
ASP	activated sludge plants
BOD	biochemical oxygen demand
CFD	computational fluid dynamics
CNNs	convolutional neural networks
COD	chemical oxygen demand
DL	deep learning
DO	dissolved oxygen
DTs	decision trees
FFNN	feed forward neural network
FIS	fuzzy inference system
FL	fuzzy logic
GA	genetic algorithm
HRT	hydraulic retention time
IFAS	integrated fixed-film activated sludge process
IWA	International Water Association
MAPE	mean absolute percentage error
MBR	membrane bioreactors
MIA SG	Modeling and Integrated Assessment Specialist Group
ML	machine learning
MLCM	machine learning cost modeling
MLP	multi-layer perceptron
PAO	phosphorus-accumulating organism
RMSE	root mean square error
RNNs	recurrent neural networks
SBR	sequencing batch reactors

Municipal Wastewater Treatment. https://doi.org/10.1016/B978-0-443-24826-9.00014-0

SRT	solids retention time
SVMs	support vector machines
SVR	support vector regression
TUDP	metabolic model developed at the Delft University of Technology
WWTP	wastewater treatment plants

Symbols

CA	concentration of microorganism A
CB	concentration of microorganisms B
Ci	mass concentration
G	exponential growth rate
kd	intangible and indeterminate substances
kr	first-order endogenous decay rate
Ks	saturation constant
pH	potential of hydrogen
Ri	overall rate
S	substrate concentration
SALK	alkalinity of the wastewater
SI	soluble inert organic matter
SN2	dinitrogen, N2
SND	soluble biodegradable organic N
SNH	free and ionized ammonia
SNH4	ammonium plus ammonia nitrogen
SNO	nitrite and nitrate nitrogen
SNOX	nitrate plus nitrite nitrogen
SO	dissolved oxygen
SS	readily biodegradable substrate
t	time
X	microbial population
XA	nitrifying organisms
Xb	biochemical process
XBA	active autotrophic biomass
XBH	active heterotrophic biomass
XH	heterotrophic organisms
XI	particulate inert organic matter
XND	particulate biodegradable organic N
XP	particulate products from biomass decay
XS	slowly biodegradable substrates
XSS	suspended solids
XSTO	a cell internal storage product of heterotrophic organisms
Y	yield coefficient
μmax	maximum specific growth rate
μs	specific growth rate

14.1 Introduction

Addressing issues related to water scarcity and growth in population is proving to be challenging for many countries (Jalilnejad et al., 2023). Wastewater treatment is an intricate procedure that utilizes an integration of physical, chemical, and biological methods to eliminate impurities to an acceptable level before releasing it into the surrounding environment (Alizadeh et al., 2019; Jalilnejad et al., 2018). Biological remediation technologies provide a more sustainable alternative for treating and reusing organic waste and wastewater (Hedaoo et al., 2012). By conducting thorough investigation and implementing effective environmental control measures, it is possible to easily treat nearly all wastewaters that contain biodegradable components with a biochemical oxygen demand (BOD) to chemical oxygen demand (COD) ratio of 0.5 or higher using biological methods (Chan et al., 2009). Compared to other wastewater treatment systems, they offer the benefits of reduced treatment expenses and no additional pollution (Wei et al., 2003). Both aerobic and anaerobic processes can be employed for waste treatment. Aerobic processes utilize oxygen to convert organic wastes into biomass and carbon dioxide through the activity of microorganisms known as aerobes (Tang et al., 2022). On the other hand, anaerobic processes occur in the absence of oxygen and break down complex organic wastes into methane, carbon dioxide, and water through three distinct steps including hydrolysis, acidogenesis, and methanogenesis (Arslan et al., 2024). Typically, aerobic systems are appropriate for treating low strength effluents with biodegradable COD concentrations below 1000 mg/L, whereas anaerobic systems are suited for treating high strength sewage with biodegradable COD concentrations exceeding 4000 mg/L (Chan et al., 2009).

The Activated Sludge (AS) process is widely used as a biological method for treating municipal wastewater. This process involves the use of a suspension of bacterial biomass (known as activated sludge) to eliminate contaminants (Zhou et al., 2024). An activated sludge wastewater treatment facility can achieve biological nitrogen removal, biological phosphorus removal, and removal of organic carbon compounds, depending on its design and specialized application (Lu et al., 2024). Clearly, numerous activated sludge process configurations have developed over time (Tang et al., 2025). After the sedimentation of solid particles in the initial phase of a wastewater treatment plant, the process advances to the secondary treatment phase. During this stage, elimination of dispersed organic compounds from effluent occurs using activated sludge type microorganisms (Altowayti et al., 2022). The activated sludge technique is highly cost-effective, safe, and adaptable, since it can be customized for any type of wastewater. Additionally, it has the capability to produce high-quality effluent (Lu et al., 2024).

Microbial growth in a biological wastewater treatment system result in an increase in the number of cells and the mass of bacteria through cell division (Clagnan et al., 2024). Both extracellular and intracellular factors can be influential in microbial growth. Cells can be influenced by several exterior environmental elements such as pH, heat, oxygen levels, availability of water, and food supply. Simultaneously, intracellular factors encompass the interior design, metabolic activities, and the cell's genetic matter (Stavrakakis et al., 2024). In order for biomass growth to occur in a culture, microorganisms require a suitable environment for their survival and development, which entails meeting several requirements including energy source, adequate nutrition, absence of growth-inhibiting substances, and appropriate physicochemical factors (Horikoshi et al., 2024).

Following the introduction of the activated sludge process, it became imperative for plant engineers to anticipate the overall performance of wastewater treatment plants (Karpinska et al., 2013). This was done to gain a deeper understanding of the biological processes involved and to ensure the reliability and flexibility of the treatment process in response to varying operational circumstances (Ruiz et al., 2023). To address these issues, modeling technique has arisen as a promising tool for simulating and predicting the behavior of processes. During the design phase of a wastewater treatment plant, mathematical models (such as Activated sludge models, Monod model, and Contois model) are utilized to elucidate the dynamic changes in carbon oxidation, nitrification, denitrification, and other processes (Altowayti et al., 2022; Pomiès et al., 2013; 21 AD). These mathematical models elucidate the kinetics of each process, which refers to the relationship between the rate of a reaction and the concentration of its reactants, as well as their stoichiometry (Hauduc et al., 2013). Typically, the modeling techniques rely on a substantial collection of experimental datasets, which is a time-consuming and costly endeavor. Furthermore, in order to obtain compliant and calculable results on a large scale, numerous simplifications and assumptions must be made. Because process parameters calculations vary at diverse treatment levels, it is therefore impractical to integrate multiple mechanistic models for distinct phases of a wastewater treatment plant (Vanhooren et al., 2003).

Increased model complexity, overparameterization, and challenges in model calibration are the primary limitations of the mathematical models. Due to such drawbacks, mechanistic models necessitate the utilization of robust computational instruments to resolution. Presently, CFD (Karpinska and Bridgeman, 2016), AI (Singh et al., 2023), and ML (Tejaswini et al., 2024) techniques have become influential tools for monitoring, simulating, and designing intricate non-linear practical issues in wastewater treatment plants with the aim of enhancing efficiency and sustainability. The AI and ML models rely entirely on establishing a correlation among the input and output datasets. This enables them to accurately predict and manipulate the intended outcomes, hence facilitating informed decision-making (Duarte et al., 2024). The utilization of AI and ML techniques can obviate the necessity for experimental measurements by methodically acquiring knowledge of the connections between variables through analysis of past data (Singh et al., 2023). Implementing these approaches in the municipal biological wastewater treatment sector can effectively demonstrate sustainability by eliminating the need for analyzing efficiency and operational variables. This, in turn, reduces maintenance requirements, the need for sample gathering and shipping to laboratories, and the costs associated with time-consuming analysis of plants (Shahbeik et al., 2022). Till date, several AI-based models such as Artificial Neural Network (ANN) (Matheri et al., 2021), Fuzzy Logic (FL) (Tejaswini et al., 2024), and Genetic Algorithm (GA) (Zhang et al., 2020) have been used in biological wastewater treatment facilities which are demonstrated in Fig. 14.1.

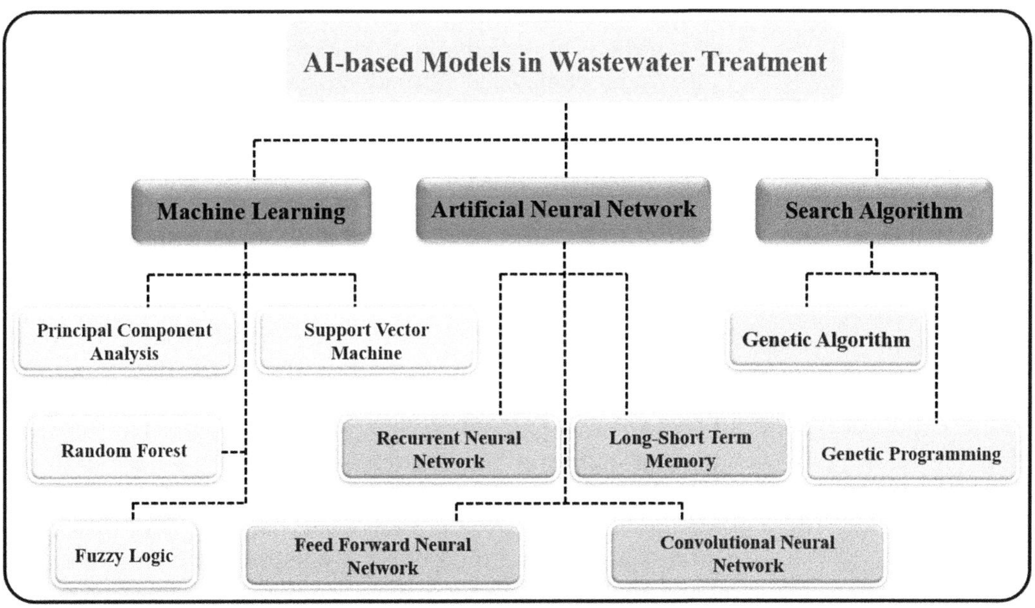

Fig. 14.1 Different AI-based models utilized in wastewater treatment facilities.

This chapter explores the application of both mechanistic and data-driven models in municipal biological wastewater treatment systems. Mechanistic models, based on experimental information, simulate the behavior of these complex systems, while data-driven models leverage historical data to make predictions. Additionally, we delve into the practical applications of these models and discuss the challenges involved in implementing them. Subsequent sections also provide insights into the existing gaps and future research directions in this domain.

14.2 Wastewater treatment modeling types

The modeling of biological wastewater systems has been evolutionary: while the first models was based on organic loadings and volumes of bioreactors without considering biological reactions in the system, in later works the kinetic equations (mainly Monod-type equations) was included in design; this is later followed by current trend of solving mathematical models using computers (Ruiz et al., 2023). Due to its flexibility, cost effectiveness, and safety, the activated sludge process is still the most commonly used for the removal of biological organic matter and nutrients (Orhon, 2015). Nevertheless, other technologies like Membrane Bioreactors (MBR), Sequencing Batch Reactors (SBR), or Integrated Fixed-film Activated Sludge Process (IFAS) may be also used as complementary to further enhance the quality of treated water and compliance to the standards and regulations or decreasing the energy consumption (Waqas et al., 2020; Ahmad and Guria, 2022).

Predicting the behavior of Wastewater Treatment Plants (WWTPs) has been important for engineers and operators to well comprehend the biological processes and examine the steadiness of the process under fluctuating operational conditions (Holloway et al., 2021). The models have been improved since their first development where only two state variables were incorporated to account for decay of substrate and the biomass growth as a primary order equation. In later studies, Monod-type equations were included to include saturation and inhibition (Ruiz et al., 2023). Nowadays, the developed models encompass various state variables and processes that cover mass and energy balances, hydraulic equations and physical-chemical processes in addition to biological processes. By this means, a complete set of differential and algebraic equations are achieved that accurately describe real systems and enable theoretical predictions and facilitate experiments that would otherwise be impractical (Ferrer et al., 2008; Amrutha and Haseena, 2020). In this regard, several deterministic, stochastic, and data-driven based models have been developed to monitor WWTP performance (Duarte et al., 2024).

14.3 Mechanistic models (deterministic/white box models)

Mechanistic models (deterministic or white-box models) are based on the understanding of the system (basic engineering principles) and use the differential equations that represent the balance equations (Gernaey et al., 2004). Although deterministic models are widely used in wastewater treatment field to design, learning, management and process optimization purposes, simplification in these models cause uncertainties (Quaghebeur et al., 2022). In this type of modeling the following steps are implemented which is described in depth by Petersen et al. (2002):

Defining model purpose
Selecting model
Collecting data (including hydraulics and wastewater and biomass characterization)
Data matching
Model calibration
Model validation
Scenario evaluations

Water quality modeling has progressed since the early 1900s, with the work of Streeter and Phelps in 1925 (Streeter and Phelps, 1925). Activated Sludge Models (ASMs), which are the most typical models for the activated sludge process, were introduced by International Water Association (IWA) that provided the basic standard models (Gernaey et al., 2004). After issuing the first ASM model in 1987, several versions have been published and additions have been applied to meet the requirement of different technologies and process configurations (Ruiz et al., 2023). While the ASM consists of differential equations corresponding to biological and chemical reactions in AS tank, WWTP model includes ASM as well as hydraulic model, oxygen transfer model and sedimentation tank model that represents real treatment plant (Gernaey et al., 2004).

14.3.1 Monod model

Degradable solids, often known as substrates, undergo a process of decomposition into smaller molecules that are soluble. Microorganisms flourish on many substrates and generate biomass. Multiple studies have established a correlation between the growth of biomass and the breakdown of substrate. Instances of this include biomass raise, substance hydrolysis and fermentation throughout the acidic stage of anaerobic breakdown. The state of a continuous flow bioreactor can be characterized by the factors that constitute the microorganism and the substrate that limits its growth. Exploration of substrate degradation during the acid-phase digestion of municipal primary sludge has involved the investigation of various operative and ecological factors, including pH, solids retention time (SRT), hydraulic retention time (HRT), and membrane bioreactor (MBR) (Altowayti et al., 2022).

The relationship among the microbiological organisms' concentrations and the material or media that the organisms live on is illustrated by the Monod model (Monod, 1949). The Monod model can be utilized to characterize the growth of bacterial cultures. The specific growth rate (μ_s) of the Monod model is computed using Eq. (1):

$$\mu_s = \frac{\mu_{max} S}{K_s + S} \tag{1}$$

The variable S represents the substrate concentration. K_s is the saturation constant. μ_{max} indicates the maximum specific growth rate. The half-maximal specific rate of growth is the concentration of biodegradable substance where the rate of specific growth is 50% of the greatest specific growth rate. In this model, the growth rate is believed to be in relation to the quantity of accessible nutrients. All these lead to an interrelated structure of two differential equations, one governing the population of microbes and the other controlling the concentration of nutrients (Cheyns et al., 2010).

The reaction mixture is presumed to have been combined properly. The Monod model describes the growth of microorganisms in a liquid environment where bacteria and nutrients are present. In other words, the Monod equation states that growth of biomass is proportional to biomass concentration in a first order manner and to substrate concentration in a mixed order manner (Stanescu and Chen-Charpentier, 2009). This model which is illustrated by Eq. (2) assumes the presence of a single nutrient:

$$\frac{d}{dt}(C_i) = R_i(C_A, C_B) \tag{2}$$

The variables C_A and C_B represent concentration of microorganisms, which are small organisms such as mold, yeasts, and bacteria. R_i denotes the total rate where genus i are created via reactions, while C_i denotes the mass content of reacting agent over unit volume of the liquid component. The Monod model (Eq. 3) provides a mathematical expression for the pace at which microbiological proliferation occurs (Ramesh et al., 2019).

$$\mu(C_i) = \frac{\mu_{max} C_i}{K_s + C_i}, i = A, B \quad (3)$$

The ultimate configuration of the regulatory framework of differential equations, under the assumption that the microbial death rate is directly proportional to the size of the population, can be expressed as Eqs. (4) and (5).

$$\frac{dC_A}{dt} = \frac{\mu_{max} C_B}{K_s + C_B} C_A - k_r C_A \quad (4)$$

$$\frac{dC_B}{dt} = -\frac{1}{Y} \frac{\mu_{max} C_B}{K_s + C_B} C_A \quad (5)$$

The yield coefficient is represented by Y, while the first-order endogenous decay rate is denoted as k_r. The growth of species C_A in response to nutrients C_B is indicated through the initial factor on the right part of Eq. (5). This equation's second term represents the decay of microorganisms. The depletion of nutrients caused by microbial consumption is denoted through the initial term on the right part of Eq. (5). This expressions state that biomass decay is first order with respect to biomass concentration (Cheyns et al., 2010).

The Monod model, which remains the most widely utilized approach, is sufficiently flexible and resilient to be systematically investigated by quantitative, numerical, and experimental methods in the field of environmental engineering processes. The fundamental notion was employed to investigate the models of a continuous flow bioreactor and membrane reactor.

Eq. (6) represents the biochemical process (X_b), where substrate (S) is consumed to make products (P, not shown). Death triggers the generation of intangible and indeterminate substances, which are not visible (k_d). The reactor releases residual substrate, microorganisms, and the final product. The schematic design of the basic model illustrates the passage of unused substrate, microorganisms, and the output (Altowayti et al., 2022).

$$S \overset{\mu_s}{\rightarrow} X_b \overset{k_d}{\rightarrow} \quad (6)$$

14.3.2 Contois model

An essential element of kinetic modeling through various biochemical reactions is the pace where the population of microorganisms (X) grows and consumes the substrate (S). Historically, scientists have frequently operated on the erroneous belief that the specific rate of growth remains constant regardless of the amount of organism's present. Furthermore, empirical findings indicate that in specific systems, the specific growth rate model should incorporate both substrate and microbial concentrations. Research on nutrient-constrained batch systems led Contois to propose a similar theory of growth kinetics (Moser, 1958; CONTOIS, 1959). The evidence suggested that the specific growth rate may rely on the quantity of microbes. The Contois growth model is illustrated by Eq. (7).

$$\mu(S,X) = \mu_{max} + \left(\frac{S}{K_s X + S}\right) \tag{7}$$

The specific growth rate μ (S, X) is determined by the concentrations of microbe (X) and substrate (S). The maximum specific growth rate is written as μ_{max} while the saturation constant is indicated as K_s. In specific circumstances, their values remain constant. Contois's model indicates that the rate of growth μ (S, X) diminishes with increasing microbe content. The exponential rate of growth (G) is the pace at which a quantity increases exponentially over time and is represented by Eqs. (8) and (9).

$$G = \mu_{max} + \left(\frac{S}{XK_S + S}\right) \tag{8}$$

$$X = \mu_{max} + \left(\frac{S/X}{K_S + S/X}\right) \tag{9}$$

As the concentration of bio-mass increases, the Contois model can be seen as having two contrasting scenarios. When this happens, the Contois model for biomass or substrate concentration becomes simplified to first-order kinetics (Ramirez et al., 2009). As accurate as possible, these estimations cab be stated as below:

1 Biomass growth adheres to a first-order kinetics model
2 The kinetics of substrate usage is first order

In situations where the content of biomass is elevated, the Contois rate of growth is employed like a model that accounts for the limitations in mass transport due to the small surface area (Myint et al., 2007). Eq. (10) is a detailed expression of the rate of increase (Altowayti et al., 2022).

$$\mu = \mu_{max} + \left(\frac{S/X}{K_S + S/X}\right) \tag{10}$$

14.3.2.1 The Contois kinetics of growth and inhibition

Suppressing the activity of substances is a vital stage in the decomposition of organic material by bacteria. The Monod model does not consider substrate inhibition. The Monod model, enhanced by include substrate inhibition, has been extensively employed in the investigation of biological reactions, as devised by Andrew. Therefore, incorporating substrate restraint into the Contois modeling could significantly enhance its applicability (Ruel et al., 2002). In order to accomplish this, it is need to modify the growth rate model in the form of Eq. (11).

$$\mu = \mu_{max} + \left(\frac{S}{XK_S + S + S^2/K_1} \right) \tag{11}$$

14.3.2.2 The Contois kinetics model's microbiological implementations

In certain instances, it has been found that the Contois model has an excellent correlation with test data compared to other models which disregard substrate content, like the Monod model (Fujimoto, 1963). Scientists from various fields have conducted investigations into the causes of this phenomena. When the rate of mass transfer or the rate of enzyme kinetics is the limiting factor for growth, the Contois model can be created using a mechanistic approach that is based on saturation kinetics (Lobry and Harmand, 2006). This approach has been detailed by Fujimoto and Characklis. Moreover, a crucial aspect in implementing the Contois model is the existence of a heterogeneous regional media around the cell, which restricts the biomass's ability to access the substrate. Arditi and Ginzburg explored the procedure of determining the growth rate. When determining population growth-rate processes, it was shown that predator-prey ratios are more significant than only considering the prey alone. More studies have provided more evidence supporting this hypothesis (Sharma et al., 2019; Cao et al., 2022).

Although the Monod expression has traditionally been used to describe the pace at which microorganisms grow, certain scientists have utilized Contois kinetics to model the dynamics of growth in wastewater treatment systems (Ajbar et al., 2011). For the first time, growth rate forecasts in one reactor including recycling were made using a Contois model by Nelson et al. (2008). They proved that by increasing the duration of residence, it is possible to reduce the concentration of the effluent to any desired level. However, in processes regulated by Monod kinetics, there is a maximum concentration for the effluent, as indicated by previous research.

A stability analysis and steady-state responses were given by Nelson and Holder (2009) after they investigated the dynamics of a non-recyclable N-reactant cascade device. The objective of their study was to determine the optimal time of residency and number of tanks in order to minimize the concentration of effluent. An increase in the total quantity of reactors in the reactor cascading system resulted in a substantial decrease in effluent concentration in most of the analyzed circumstances. Based on their research, the value of organic substance content that leaves a N reactor cascade drops.

Ajbar et al. (2011) conducted another investigation on the limitations of oxygen delivery. The reactor's performance was assessed by calculating the productivity and efficiency of the process in order to recognize the steady-state resolutions. The effectiveness of the reactor increased, but the productivity decreased when the concentration of the influent became reduced. By the decreasing of oxygen transport coefficient, the effectiveness of reactor also decreased. A variable yield coefficient is necessary to achieve a periodic solution, while it has a negligible effect on reactor performance.

14.3.3 Activated sludge models

The following section provides a summary of ASMs. In this regard, the ASMs developed by IWA and the metabolic model developed at the Delft University of Technology (TUDP model) are described. The comparison between these models is summarized in Table 14.1.

14.3.3.1 ASM1

The Activated Sludge Model No. 1 (ASM1) was introduced in 1986 by IWA and widely accepted in industrial and research community and extensively used in the literature (Henze et al., 1987). The ASM1 is considered as the reference model, as it led to the widespread acceptance of WWTP modeling in research and later in industrial society (Gernaey et al., 2004).

The elementary principles of ASM1 are based on the model developed by Dold and colleagues in 1980, where Monod kinetics is used to describe biological reactions. The ASM1 explains the elimination of organic compounds and nitrogen, while simultaneously employing oxygen and nitrate as electron acceptors. This model links the conservation of COD with stoichiometry and kinetics, which is achieved by representing substances transformation rates using mathematical derivatives (Dold et al., 1981). The model provides a clear description of the sludge formation and estimates the concentration of organic matter, known as COD (Brdjanovic et al., 2015). Fundamentally, ASM1 includes four main processes: (Jalilnejad et al., 2023) biomass production, (Alizadeh et al., 2019) biomass decay, (Jalilnejad et al., 2018) organic nitrogen ammonification, and (Hedaoo et al., 2012)

Table 14.1 Summary of activated sludge models (Gernaey et al., 2004).

Model	Nit	Den	Het/Aut decay	Hyd	Bio-P	Den. PAOs	Lysis of PAO/PHA	Fermentation	Chemical P removal	Reactions	State variables
ASM1	X	X	DR, Cst	EA						8	13
ASM2	X	X	DR, Cst	EA	X		Cst	X	X	19	19
ASM2d	X	X	DR, Cst	EA	X	X	Cst	X	X	21	19
ASM3	X	X	ER, EA	Cst						12	13
ASM3-bio-p	X	X	ER, EA	Cst	X	X	EA			23	17
TUDP	X	X	DR, Cst	EA	X	X	EA	X		21	17

Nit = nitrification; Den = denitrification; Het/Aut decay = heterotrophic/autotrophic decay; Hyd = hydrolysis; Den. PAO = denitrifying PAO activity included in the model; DR = death regeneration concept; EA = electron acceptor depending; ER = endogenous respiration concept; Cst = not electron acceptor depending.

organic particulate matter hydrolysis. Moreover, heterotrophic and autotrophic microorganisms are responsible for the reaction in the model (Henze et al., 1987). The broad range of organic carbon and nitrogenous components in the model are subdivided into a limited number of fractions considering their solubility and biodegradability (Gernaey et al., 2004). The basic processes occurring in this model alongside the state variables are summarized in Tables 14.2 and 14.3. The material and energy balances, as

Table 14.2 The state variables in the ASM1 (Henze et al., 1987).

State variable	ASMI notation	
Soluble inert organic matter	S_I	$gCOD/m^3$
Readily biodegradable substrate	S_s	
Particulate inert organic matter	X_I	
Slowly biodegradable substrate	X_s	
Active heterotrophic biomass	X_{BH}	
Active autotrophic biomass	X_{BA}	
Particulate products from biomass decay	X_P	
Dissolved oxygen	S_O	gO_2/m^3
Nitrite and nitrate nitrogen	S_{NO}	gN/m^3
Free and ionized ammonia	S_{NH}	
Soluble biodegradable organic N	S_{ND}	
Particulate biodegradable organic N	X_{ND}	
Alkalinity of the wastewater	S_{ALK}	Molar units

Table 14.3 The processes in the ASM1 (Henze et al., 1987).

Process	Basic reaction
Aerobic growth of heterotrophs	$S_S + S_O + S_{NH} \rightarrow X_{BH}$
Anoxic growth of heterotrophs	$S_S + S_{NO} + S_{NH} \rightarrow X_{BH}$
Aerobic growth of autotrophs	$S_O + S_{NH} \rightarrow X_{BA} + S_O$
Decay of heterotrophs	$X_{BH} \rightarrow X_P + X_S + X_{ND}$
Decay of autotrophs	$X_{BA} \rightarrow X_P + X_S + X_{ND}$
Ammonification of soluble organic N	$S_{ND} \rightarrow S_{NH}$
Hydrolysis of entrapped organics	$X_S \rightarrow S_S$
Hydrolysis of entrapped organic N	$X_{ND} \rightarrow S_{ND}$

well as these equations, are resolved using the precise Euler method. ASM1 has been used in most commercial software for modeling and simulating nitrogen removal in WWTPs and has served as a guide in many academic and real-world projects (Van Loosdrecht et al., 2015).

While the ASM1 has been in use for a long time, this model has a few drawbacks (Hauduc et al., 2010). Even though the limitation terms can be incorporated easily, the model does not consider the impact of nitrogen, phosphorus, and other inorganic nutrient limitations on the removal of organic substrate and cell growth. Also, the nitrification coefficients are considered constant and exhibit inhibitory effects on the components of the wastewater. Since in this model the composition of heterotrophic biomass is homogeneous and alteration of species diversity is not time-dependent, some parameters such as settleability of sludge, concentration gradients, and configuration of reactor are not taken into consideration. Moreover, in this model the organic matter and organic nitrogen hydrolysis is merged and takes place at an identical rate. Additionally, it cannot deal with high nitrite levels. As this model is designed for municipal WWT, it is not suitable to apply ASM1 to industrial systems (Savun-Hekimoğlu, 2021). The other limitations of ASM1 are as follows:

This model cannot be employed with AS systems characterized by very high loads or low sludge retention times (<1 day).

The heterotrophic yield coefficient remains unaffected by the type of electron acceptor.

It is assumed that the concentration of organic particles in the biomass is instant.

As ASM1 does not involve mechanisms that simulate anaerobic conditions, errors may occur when simulating systems with a large-volume anaerobic reactor.

14.3.3.2 ASM2 and ASM2-d

Further studies on enhanced biological phosphorus removal and development of ASM models resulted in Activated Sludge Model No. 2 (ASM2), which was published in 1995 (Duarte et al., 2024). ASM2 include biological and chemical phosphorus removal processes in addition to organic substances and nitrogen removal (incorporating simultaneous nitrification-denitrification processes) and thus is a more complicated model than ASM1 (Henze et al., 1999; Gujer et al., 1995). This model also accounts for both biological (phosphorus-accumulating organisms) and chemical (chemical precipitation of phosphates) processes (Savun-Hekimoğlu, 2021). Following the release of ASM2, it was remarked that Phosphorus-Accumulating Organisms (PAOs) can denitrify (Drewnowski et al., 2020). This observation resulted in

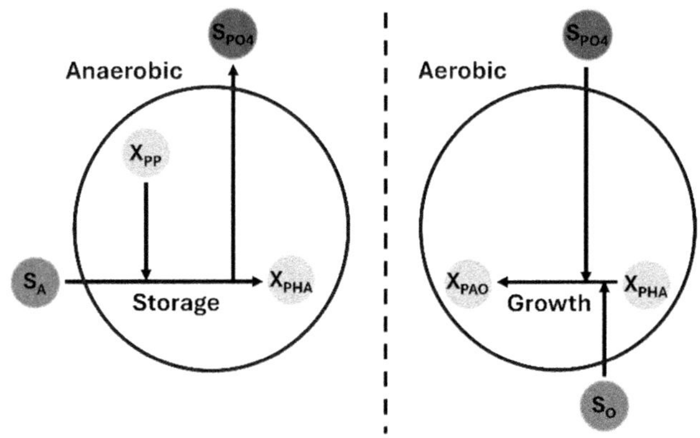

Fig. 14.2 Substrate flows for storage and growth of PAOs in the ASM2 model (Gernaey et al., 2004).

publishing ASM2d in 1999m which considers the denitrification caused by PAOs (Makinia and Zaborowska, 2020).

The modeling of Bio-P in ASM2 is depicted in Fig. 14.2. In this model PAOs are represented with cell internal structure and all the organic storage products are grouped into single element, X_{PHA}. The growth of PAOs solely occurs on cell internal organic storage material. Though storage is not influenced by the conditions of the electron acceptor, it can occur when fermentation products like acetate are present. Therefore, from a practical point of view, storage is typically examined only in the anaerobic activated sludge tanks of the WWT system (Gernaey et al., 2004).

By adding new parameters, ASM2 became more complicated, making it practically more difficult to handle (Gujer, 2006). Also, ASM2d appeared to be overparameterized requiring more systematic calculation method (Brun et al., 2002). The prediction of the steady-state condition of a Bio-P activated sludge plant requires the definition of 131 independent parameters, which shows the high-level complexity of the model that has been reached (Gujer, 2006). The number of independent parameters to predict the steady-state condition of plant in different models are summarized in Table 14.4.

14.3.3.3 ASM3

AMS3 is another IWA activated sludge model that is frequently cited and has extensive acceptance (Gujer et al., 1999). This model was developed in 1999 to enhance the practical drawbacks of ASM1 by adding new coefficients and variables to include storage of organic compounds and describing decay processes more realistically (Duarte et al., 2024). Similar to ASM1, Activated Sludge Model No. 3 (ASM3) only accounts for carbon and nitrogen

Table 14.4 The number of independent parameters to predict the steady-state condition of plant in different models (Gujer, 2006).

Parameter type	ASM1	ASM2	ASM2-d	ASM3
Composition factors	2	13	13	8
Stoichiometric parameters	3	9	9	7
Kinetics parameters	14	43	45	21
Temperature coefficients	14	43	45	21
Influent concentrations	>9	19	19	13
Plant layout definition	?	?	?	?
Total parameter number	>42	>127	>131	>70

removal. In general, ASM3 consists of following main processes: (Jalilnejad et al., 2023) biomass growth, (Alizadeh et al., 2019) endogenous respiration, (Jalilnejad et al., 2018) storage of easily biodegradable organic substrates, (Hedaoo et al., 2012) respiration of stored material and (Chan et al., 2009) hydrolysis (Savun-Hekimoğlu, 2021). The basic processes included in this model alongside the state variables are summarized in Tables 14.5 and 14.6.

Even though ASM3 is enhanced model with introducing new understandings of the systems, ASM1 is still the extensively used

Table 14.5 Processes state variables in ASM3 (Savun-Hekimoğlu, 2021).

1	Hydrolysis
2	Aerobic storage of X_{STO}
3	Anoxic storage of X_{STO}
4	Aerobic growth of X_H
5	Anoxic growth of X_H (denitrification)
6	Aerobic endogenous respiration of X_H
7	Anoxic endogenous respiration of X_H
8	Aerobic respiration of X_{STO}
9	Anoxic respiration of X_{STO}
10	Growth of X_A (nitrification)
11	Aerobic endogenous respiration of X_A
12	Anoxic endogenous respiration of X_A

Table 14.6 State variables in ASM3 (Savun-Hekimoğlu, 2021).

S_{ALK}	Alkalinity of the wastewater
S_I	Particulate inert organic matter
S_{N2}	Dinitrogen, N_2
S_{NH4}	Ammonium plus ammonia nitrogen
S_{NOX}	Nitrate plus nitrite nitrogen
S_O	Dissolved oxygen
S_S	Readily biodegradable substrate
X_H	Heterotrophic organisms
X_I	Inert particulate organic material
X_S	Slowly biodegradable substrates
X_{STO}	A cell internal storage product of heterotrophic organisms
X_A	Nitrifying organisms
X_{SS}	Suspended solids

model (Hauduc et al., 2013). In ASM3, the significant effect of storage polymers in heterotrophic processing of activated sludge is considered, which is the major difference from ASM1. In this model, instead of using circular growth–decay–growth model (or death–regeneration concept) in the conversion of heterotrophic activated sludge, a growth-endogenous respiration model is employed (Gernaey et al., 2004). In ASM1 it is assumed that the entire biomass is produced by external substrate and oxygen consumption is described through biomass decay after external substrate depletion (Savun-Hekimoğlu, 2021). In ASM3, all readily biodegradable organic substrates (S_s) are transformed directly into stored material (X_{STO}) under festive conditions and these stored components serve as the carbon and energy source for growth during subsequent famine period (Gernaey et al., 2004). The COD flows and differences between these two models are illustrated in Fig. 14.3.

In ASM1 the autotrophic and heterotrophic decay regeneration cycles are strongly interrelated, whereas in ASM3, the transformation processes of all types of organisms are not interconnected. The modifications in the decay model, along with the consideration of the storage stage, provide multiple access points for oxygen utilization. This occasionally enhances the separation process and improves process characterization. Moreover, since solely the hydrolysis of X_S in the influent is included in ASM3, oxygen consumption rates of this model are of lower importance in hydrolysis (Savun-Hekimoğlu, 2021).

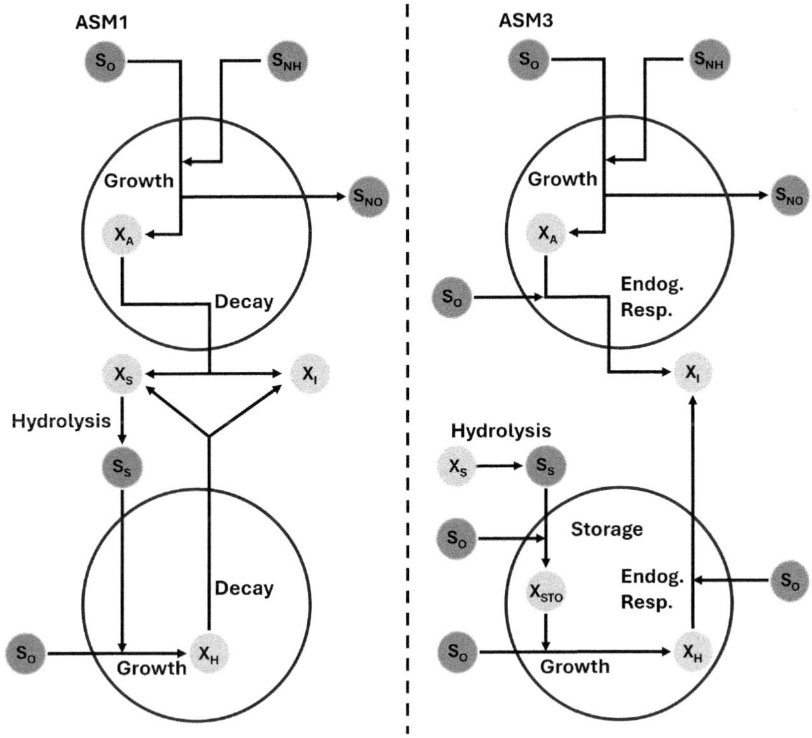

Fig. 14.3 Substrate flows for autotrophic and heterotrophic biomass in the ASM1 and ASM3 models (Gujer et al., 1999).

According to Koch and coworkers, the dynamic behavior of conventional municipal WWTPs can be described by both ASM1 and ASM3. However, ASM3 demonstrates better performance in scenarios where the storage of easily biodegradable substrates is significant, such as in industrial wastewater or WWTPs with substantial non aerated zones (Koch et al., 2000).

14.3.3.4 TUDP model

TUDP model, which was developed at the Delft University of Technology, brings together the metabolic model for denitrifying and non-denitrifying bio-P described by Murnleitner et al. (1997) and ASM1 that contains both autotrophic and heterotrophic reactions (Brdjanovic et al., 2000). In contrast to ASM2 or ASM2d, this model fully accounts for the metabolism of PAOs and specifically includes all organic storage components that are exhibited as X_{PHA} and X_{GLY} in Fig. 14.4 (Gernaey et al., 2004). The validation of the TUDP model was performed in a series of tests in enriched

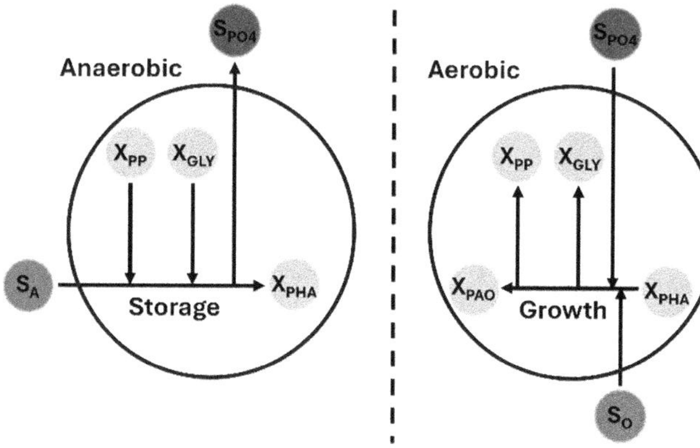

Fig. 14.4 Substrate flows for storage and aerobic growth of PAOs in the TUDP model (Gernaey et al., 2004).

Bio-P SBR laboratory systems. These tests have been carried out with various SRTs (Smolders et al., 1995), aerobic phase durations (Kuba et al., 1997) as well as oxygen and nitrate as electron acceptor (Murnleitner et al., 1997). It should be noted that in cases where the pH (7.5) and Ca^{2+} concentrations are high, incorporation of biologically induced P precipitation to the Bio-P model may be essential (Maurer and Boller, 1999; Maurer et al., 1999).

14.3.4 Selection and application of models

The selection of the model depends on several factors, such as the objective of the modeling study, the targeted biological processes in the WWTP to be modeled and the desired precision (Hauduc et al., 2010). While ASM2/ASM2d models are implemented in the cases where biological phosphorus removal is present (Gujer et al., 1995), in absence this process, ASM1 (Henze et al., 1987) or ASM3 (Gujer et al., 1999) are applicable, since these two models are solely describing organic matter and nitrogen removal processes.

In addition to be a useful tool for aiding system design, control, predicting system, and plant optimization, with the advantage of saving time and money, the modeling and simulation works can also be useful for training plant operators, as they can show the influence of day-to-day decisions on plant performance (Fig. 14.5) (Ferrer et al., 2008). Furthermore, better decisions can be made thanks to modeling studies that lead to both ecological and economic influences on plant performance. Nevertheless, engineers and operators are not usually willing to employ these computational methods, due to the complexity of the mathematical models and required programming expertise (Ruiz et al., 2023).

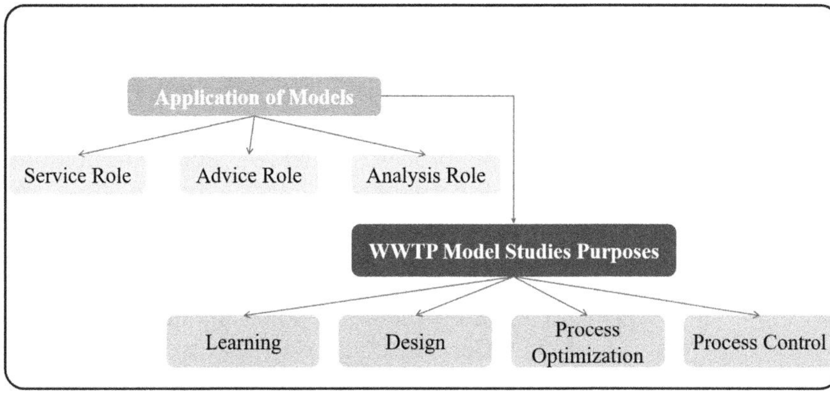

Fig. 14.5 Application and objectives of WWTP models.

Despite development of more complex models by the academic community, engineering based on model integration is not widely applied in the design and operation of wastewater treatment plants, due to a lack of acquaintance with modeling and its foundations among wastewater treatment specialists. This knowledge gap may result in inaccurate model application, and disapproving perspective on the simulation tools that are resolved by the frequent enhancement and updating of more user-friendly simulators such as WEST (Garrido-Baserba et al., 2020; Zahorodny et al., 2022).

14.3.5 Simulator

The simulator, the environment in which the WWTP configuration is simulated, can be classified into three main categories: (Jalilnejad et al., 2023) coding, (Alizadeh et al., 2019) general purpose simulators, and (Jalilnejad et al., 2018) specific commercial simulators (Vanhooren et al., 2003). The modeler can either code manually in a computational language or, alternatively, use general-purpose simulators like MATLAB or Simulink, which offer great flexibility. Nevertheless, their correct application requires expertise and advanced competency (Donoso-Bravo et al., 2013). Eventually, specific commercial simulators can be used, which can be either open or closed-source software in accordance with the possibilities for incorporating new models: (Ruiz et al., 2023) GPS-X, SUMO, AQUASIM, SIMBA, WEST, and BIOWIN. In the WEST simulator in addition to an incorporated library of all ASMs and other developed models, users can create their own models (Ruiz et al., 2023).

14.3.6 Model calibration

The model calibration, defined as determining parameters to fit specific data from the full-scale WWTP under observation, can be carried out for leaning objectives or evaluation and optimization of the process (Petersen, 2000). Calibration generally begins with the use of initial parameters in ASMs, and two main approaches can be implemented: mathematical optimization and process engineering approach. While in a mathematical approach calibration is based only on mathematical optimization, the process engineering method applies model structure and process knowledge (Gernaey et al., 2004).

The mathematical optimization approach can be helpful in model calibration when it is coupled with competence and understanding of experts. Besides being complicated, the mathematical optimization of the treatment plant model results in naturally complex and highly non-linear activated sludge models. The model parameters in calibration of treatment plant models cannot be determined easily. As different combinations of influent characteristics and model parameters can deliver similar results, it is critical to collect detailed data to accurately define the model parameters within realistic limits. Nevertheless, getting these informative data of an actual plant is not straightforward because of overdesign of WWTP, which results in negligible effluent dynamics useful for calibration. It is therefore useful to take measurements at different parts of the bioreactor during the process. Another drawback of this approach is the inability in distinguishing between well- and poorly- defined model parameters, which results in small modifications to a large number of parameters (Weijers and Vanrolleghem, 1997).

In contrast to mathematical method, in model calibration using process engineering approach, which is carried out based on modeler's process knowledge and experience, the model parameters are adjusted one after another until the model fits the data from the treatment plant. By using this approach, instead of single set of parameters a range of parameters are created that can estimate main WWTP performance indicators including sludge production, effluent concentration, and the dynamics of internal concentration (when data are existing). In the case where no data is provided on internal concentration dynamics, it is difficult to correctly calibrate the gains (sensitivities) of the model (Gernaey et al., 2004).

Although mathematical optimization allows more parameters to be modified, the process engineering approach is more sensitive to parameters (Van Veldhuizen et al., 1999). Nonetheless, this requires expertise and understanding of the process. In wastewater treatment plant modeling studies, to confirm the model's sensitivity to parameter changes applied throughout calibration, the mathematical approach is integrated with the process engineering method via sensitivity analysis (Petersen et al., 2002; Van Veldhuizen et al., 1999; Meijer et al., 2001). The calibration approach is defined by the objectives of the modeling work illustrating required data and the degree of detail necessary for calibration. Thus, data collection, that could be either nondynamic or dynamic data, is also an essential step in model calibration (Hulsbeek et al., 2002).

14.3.7 ASM models drawbacks

Despite the decades-long history of ASMs and their effective wide employment in the field, these models have a number of drawbacks that have reduced their popularity. To obtain more accurate predictions, the model needs to include many different parameters, which leads to a more complex over-parameterized model that is also difficult to calibrate. Hence, current ASMs have high level of complexity and solving these complicated models requires higher-capacity computational tools (Sin and Al, 2021). Besides, the ASMs face a high level of uncertainty caused by several assumptions and simplifications. Limited adaptability, poor model validation, and data requirements are other drawbacks of these models (Schneider et al., 2022; Hauduc et al., 2011). As mentioned, model calibration refers to modifying model parameters from a set of default parameters and updating to fit simulation results to experimental data. This step requires a long time that is an obstacle to extensive implementation of ASMs (Hauduc et al., 2011).

14.4 From conventional models to AI-based innovations

A growing population and increased urbanization have led to a heightened demand for effective municipal wastewater treatment systems, necessitating proactive efforts to develop efficient models for these systems. Conventional mathematical models, while initially developed for predicting the behavior of biological wastewater treatment systems, have demonstrated limitations in terms of precision and reliability. In order to forecast the efficiency of wastewater treatment plants and to predict how key operations and processes in municipal wastewater treatment plants will behave, a variety of time series-based, stochastic, and deterministic models have been created (Pérez-Beltrán et al., 2024). Predictive models that incorporate both physical and experiential information are primarily mechanistic models. Nevertheless, these models are an oversimplification of reality, leading to inconsistency in the model's structure and a continuous requirement for calibration. An example of this is the activated sludge model, which operates effectively within a limited order of pH, temperature, and other necessary factors (Fenu et al., 2010). Additionally, it is not practical to combine manifold mathematical models for different stages of wastewater treatment facilities due to variations in the calculation of process parameters at different treatment levels (Cairone et al., 2024).

However, advancements in AI have paved the way for the development of more sophisticated models that can accurately predict system behavior, optimize operational parameters, and enhance overall system efficiency. AI-based models have emerged as a revolutionary force in the subject area of wastewater treatment, bringing forth a significant shift in how municipal biological wastewater treatment systems are managed and optimized. By leveraging the power of machine learning algorithms and intricate mathematical principles, these advanced models have demonstrated a remarkable potential to surpass traditional models in various dimensions (Li et al., 2024). In terms of speed, AI-based models can process complex data sets rapidly, ensuring that they are capable of providing real-time insights and recommendations for wastewater treatment operational parameters. Moreover, by incorporating sophisticated mathematical techniques, these models are well-equipped to handle intricate, non-linear dynamics frequently encountered in biological wastewater treatment processes, thus improving the overall accuracy of predictions and simulations. In addition, AI-based models boast remarkable generalizability, which allows them to adapt and perform effectively even when faced with new, unforeseen situations or data sets, thus further highlighting their superiority over conventional mathematical models (Matheri et al., 2021).

14.5 Artificial intelligence and machine learning modeling

The global research community has shown significant interest in utilizing AI and ML techniques to make groundbreaking advancements in strategies for treating wastewater. These techniques have shown to be a powerful tool for streamlining the modeling of processes in wastewater treatment plants. By employing these methods, machines can gain expertise to carry out jobs often performed by human, taking into account past experiences (Zhao et al., 2020). Therefore, the creation of models that rely solely on input-output data, without incorporating knowledge of physical, chemical, or biological processes in the model structure, can be utilized to detect potential issues in wastewater treatment plants. This allows for the timely implementation of appropriate control measures when necessary (Duarte et al., 2024). Data-driven models do not explicitly specify the structure, but instead determine it by searching for relationships in the available data. These models especially focus on training, validating, and testing the data (Fig. 14.6) (Singh et al., 2023).

Machine Learning involves developing algorithms and instruments that can enhance their performance by continuously collecting data, gaining experience and proficiency in the process (Mohri et al., 2018). Machine learning encompasses three major paradigms: supervised learning, unsupervised learning, and reinforcement learning (Jordan and Mitchell, 2015). Supervised learning is a machine learning strategy that utilizes accumulated experiences to enhance the training set. The system has been programmed or trained using a predetermined and well-categorized dataset. Upon analyzing a sequence of data and assimilating knowledge from it, the program can determine the appropriate timing for accepting additional data recordings (Sarker, 2021). Supervised machine learning models are the most commonly utilized, as they may be applied in scenarios where analyzing past

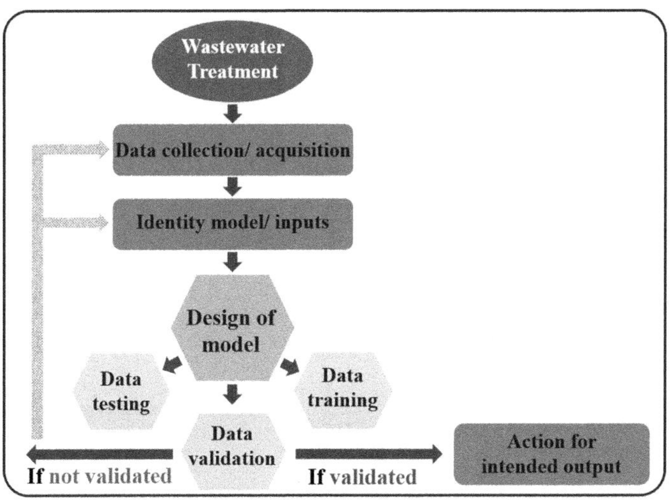

Fig. 14.6 Conceptual diagram of a data-driven model creation.

data enables the prediction of future actions. Decision trees (DTs), Support Vector Machines (SVMs), and ANNs are examples of models used in this form of learning (Zhao et al., 2024).

Artificial Neural Networks are widely employed in simulating and predicting the efficiency of biological treatment in wastewater treatment plants (Essahlaoui et al., 2023). ANNs consist of a variety of artificial neurons that are interconnected by links of variable weights. These networks serve as black box illustrations of pseudo-neurological systems (Pimenta et al., 2016). Fig. 14.7 demonstrates the conceptual diagram of ANN model structure. Every neuron takes input signals from other neurons, analyzes them, and transmits the output, which is then forwarded as input to following neurons (Wei et al., 2023). Artificial neural networks acquire knowledge from training data and effectively model the connections between data points. This acquired knowledge may then be applied for tasks such as simulation, forecasting, and optimization. Artificial Neural Networks are a class of computational models that attempt to imitate the functioning of the human brain in processing information (Yao et al., 2024).

In contrast, unsupervised learning does not rely on feedback or known results to guide the model's learning process. Consequently, it is difficult to determine the accuracy of the obtained results, ensuring supervised models more suitable for practical applications (Hastie et al., 2008). Unsupervised learning relies mainly on observation and discovery, and it is particularly useful in situations where information about intended results is unavailable. The patterns identified through unsupervised learning are potentially valuable when applying supervised machine learning techniques. For instance, unsupervised techniques can be utilized to perform cluster analysis and employ the resulting clusters as an extra asset in a supervised learning model (Dike et al., 2018).

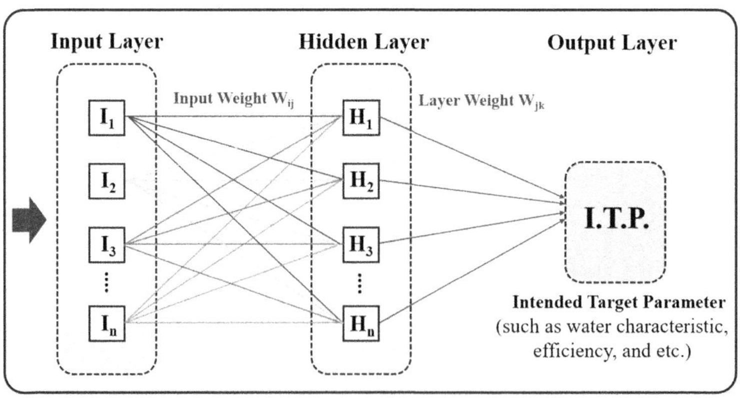

Fig. 14.7 Common architecture of neural network based models.

Deep learning (DL) has arisen as a subarea of ML due to the growing availability of data (Janiesch et al., 2021). The advancement of technology and hardware has facilitated the utilization of DL models to address intricate challenges and enhance current approaches, for example image recognition or tomography analysis. DL is employed to replicate the operations of the human mind in data analysis, acquire knowledge with no human guidance, and utilize unorganized and unlabeled data, employing two potential methods: supervised and unsupervised (Michelucci, 2018). Autoencoders and generative modeling exemplify the unsupervised methodology, whereas Multi-Layer Perceptron (MLP), Recurrent Neural Networks (RNNs), and Convolutional Neural Networks (CNNs) exemplify the supervised methodology (Kim, 2016; Lecun et al., 2015).

Fuzzification, defuzzification, fuzzy rules, and the Fuzzy Inference System (FIS) make up the FL algorithms, which were created to represent intricate and imprecise systems (Zadeh, 1983). Among these compounds, the most commonly utilized is the FIS, which consists of four functioning blocks: the fuzzification unit, the decision-making unit, the knowledge base (comprising the regulations and the information database), and the defuzzification unit (Tejaswini et al., 2024). In order to reach the lowest or maximum objective function, evolutionary algorithms known as GAs replicate the natural evolutionary process using Darwin's theory (Vafaie and De Jong, 1992). The primary principles for implementing genetic algorithms to chromosomal population are selection, crossover, and variation (Schmitt, 2001). To improve the neural network's capacity to solve problems, Artificial Neural Networks that employ Genetic Algorithms (ANN-GA) utilize a GA to continuously optimize the neural network's parameters (Reeves, 2010).

14.5.1 Real-world implementations of AI in biological wastewater treatment

Several AI-based models have been utilized to forecast the properties of effluent from wastewater treatment plants. The ANN and M5 model tree were used by Asami et al. (2021) to model the biochemical oxygen demand in a water sample. Model trees are decision tree structures characterized by their lack of class labels at their terminal nodes, instead including linear regression models. The primary algorithm for constructing model trees is called M5. Although both the ANN and M5 model tree demonstrated dependability, robustness, and high generalization capacity, the ANN outperformed the M5 model tree when implemented to the WWTP of Ramin thermal power in Ahvaz, Iran. The ANN achieved an R^2 value of 0.95, 0.95, and 0.97 for BOD, COD, and total suspended solids respectively, during model validation. In comparison, the M5 model tree achieved an R^2 value of 0.88, 0.90, and 0.83 for BOD, COD, and TSS respectively, during model validation. This evaluation was conducted using a daily dataset spanning 3 years from 2013 to 2015.

It was shown by Nourani et al. (2021) that the black box AI model, which is made up of Feed Forward Neural Network (FFNN), Support Vector Regression (SVR), and Adaptive Neuro Fuzzy Inference System (ANFIS), could make predictions up to 20% more accurate during the verification phase. They did this by using jittering data preprocessing and postprocessing ensemble models to guess model parameters through an Autoregressive Integrated Moving Average (ARIMA) model. Based on data from 2016 to 2018, the model was employed to forecast the BOD and COD concentrations in the effluent of Tabriz wastewater treatment plant. The authors reached the conclusion that AI models are superior to ARIMA in predicting wastewater treatment parameters.

An extensive dataset containing historical data from 2010 to 2020 of a wastewater treatment plant was utilized as input for a machine learning ensemble model. This model integrates ANN, ANFIS, and SVR to accurately forecast 15 process parameters. Using a multistage model structure made it possible to predict the process's intermediary factors that are affected by the characteristics of the influent. This can help explain the overall efficiency of the process (Zaghloul and Achari, 2022).

Some of these prediction models have also been used in systems that remediate industrial wastewater. Picos-Benítez et al. (2020) conducted a study to evaluate and optimize the treatment of wastewaters including sulfate and bromophenol blue dye utilizing an electro-oxidation method. They employed an ANN-GA model to analyze the success of this approach. The ANN model underwent training using 51 electrolytic trials. The input variables included electrolysis time, flow, current density, pH, and dye concentration. The output variable was the discoloration efficiency. The performance of the ANN was evaluated using the Root Mean Square Error (RMSE) and Mean Absolute Percentage Error (MAPE) metrics. The RMSE value was found to be 10.73%, while the MAPE value was 8.81%. These values were determined based on the comparison between the actual and predicted values. The application of a Genetic Algorithm allowed for the identification of the most favorable conditions, leading to a discoloration efficiency of 88.8%. This value closely aligns with the model's prediction of 95.5%. Furthermore, the employed ANN model accurately forecasts the electrochemical nitrate removal, with a maximum coefficient of determination of 0.9020.

AI-based models have also been employed to enhance the energy efficiency in wastewater treatment plants by optimizing energy consumption. Torregrossa et al. (2018) utilized ML techniques, specifically neural networks and random forest models, to create high efficiency energy cost models for wastewater treatment plants. Hence, a database comprising 317 plants located in northwest Europe was utilized. Whenever the Machine Learning Cost Modeling (MLCM) techniques were used, the model performance indicators consistently outperformed those reported in the literature. This study determines that the pollutant load (measured by COD, TP, and TN) in the input is the most influential factor affecting the energy cost of wastewater treatment plants, while the energy price has no effect on the energy consumption cost model.

Furthermore, Bernardelli et al. (2020) provided a detailed account of the development and practical evaluation of the energy way model predictive controller using ANFIS and a heuristic search. The controller was tested on a municipal wastewater treatment plant serving a population of around 500,000. The model successfully forecasted the occurrence of TN peaks with a 30-min lead time, enabling the adjustment of air flow to meet effluent discharge requirements and achieve energy savings.

14.6 Computational fluid dynamics modeling of biological wastewater treatment systems

CFD was initially applied to mechanics and aerospace engineering as a result of improvements in computational capabilities in the 70s and software developments to address fluid flow problems. CFD later found its applications in chemical engineering, being used for reactor design and currently it is widely implemented in different fields including civil and environmental engineering (Kochevsky, 2004). One of these domains is water and wastewater treatment applications, where CFD is applied to variety of unit processes, including hydraulic elements (e.g., flow splitting), physical and chemical as well as biological processes like suspended growth nutrient removal and anaerobic digestion (Samstag et al., 2016). Application of CFD in water and wastewater treatment industries have been accelerated by developing multiphase flow models with focus on different design aspects such as, but not limited to, AS tanks, digesters and clarifiers, pumping stations (Karpinska and Bridgeman, 2016).

CFD studies facilitate the enhancement of treatment efficiency and reduction of energy consumption. Numerous single and multiphase approaches were used in CFD studies to model aeration tank operation and AS tanks. CFD modeling can be used to study the behavior of multiphase flows and assess the impact of different operating parameters on local phenomena (such as the interaction between flow fields, interfacial mass transfer, and chemical reactions) (Karpinska and Bridgeman, 2016). Due to its high accuracy and the ability to evaluate systems safely and cost-effectively, CFD has also became more popular than conventional wastewater treatment models. Given these points, CFD-assisted modeling can be helpful in designing new facilities as well as retrofitting or optimizing existing Activated Sludge Plants (ASP), which in return not only enhances the utility performance and energy consumption but also reducing required time, costs, and labor (Laurent et al., 2014; De Gussem et al., 2014). The applications of CFD-assisted modeling are further discussed in the following section.

14.6.1 Application of CFD in ASP

Enhanced oxygen mass transfer is essential in AS tanks of biological wastewater treatment systems to ensure aerobic processes in biodegrading and nitrifying biomass. In typical ASP configurations, i.e., channel or enclosed bubbly flow AS reactors using diffuser aeration, the maximized interface between the dispersed and continuous phase (air/oxygen bubbles and mixed liquor, respectively) provides the enhanced oxygen mass transfer. Due to the design of wastewater utilities, aeration systems, and the AS tank based on static models in most cases, and the design of the control system using PID controllers, the correct characterization of tank hydrodynamics and its effect on Dissolved Oxygen (DO) profiles in the basin is challenging. Advanced understanding of the bioreactors' behavior, including mixing in micro and macro scales, is required for enhanced efficiency of biochemical conversion, hydraulic and process design, and robust operation of AS systems. CFD modeling provides outstanding visualization features, enabling detailed characterization of localized phenomena, even under variable conditions. In this regard, CFD is a powerful tool that is used in developing and designing new energy efficient and effective unit processes at WWTPs as well as performance optimization or retrofitting of existing AS systems (Karpinska and Bridgeman, 2016). CFD can be also used in wastewater treatment systems to develop new aeration devices in wastewater treatment tanks and optimize and modify their spatial arrangement (Karpinska et al., 2013; Morchain et al., 2000; Bhuyar et al., 2010).

14.6.2 Approaches to CFD analysis

To ensure reliable and energy-efficient performance of multifaceted AS reactor systems, it is important to consider and understand multiphase dynamics of the AS tank and study how efficiency of biological treatment and mixing as well as related energy consumption are influenced by aeration system. Furthermore, it is essential to take into account the mixing and aeration devices and their quantity, type and spatial distribution. The performance of the existing aeration systems also can be evaluated using CFD-assisted tools using different approaches to define the dynamic behavior of aeration tanks.

One of these approaches in analyzing fluid flow within an aeration tank is gas-liquid models, which provides relatively quick and simple prediction of the velocity distribution of multiple phases caused by mixers and aerators, as well as bubble sizes and local gas hold-up. These are process parameters that highly influence the oxygen mass transfer in bioreactors. CFD method can be useful to predict these parameters as their analysis requires expensive or time-consuming experimental methods or they are overlooked in investigations of experimental results (Fayolle et al., 2007; Gillot et al., 2005; Gillot, 2000; Le Moullec et al., 2008; Vermande et al., 2007).

Gas-liquid neutral density simulations have been also used to examine mixing and aeration. These density-coupled models were implemented due to their ease of validation on a laboratory scale. Nonetheless, overestimation of mixing degree can be deemed as the drawback as it has been reported in the literature (Samstag et al., 2012). Even though using Lagrangian multiphase approach with particle tracking is uncommon to simulate the aeration process, it can be employed to assess the performance of the aeration system by evaluating macromixing and the overall transport phenomena within a bioreactor. Several works have been published studying the aeration and agitation systems that cause distinct flow dynamics in the tank and affect both micro and macro mixing in AS reactors (Karpinska et al., 2013, 2015; Kjellstrand, 2006).

Given the complicated hydrodynamics and biochemical conversions reactions in liquid-solid AS system, a full CFD study of this system is difficult, which demands high computing capacities (high memory and CPU capacity use with longer calculation times). Moreover, more complicated models and increased mesh resolution, as well as targeting more precise results raised the cost of CFD study. The complicated and long studies can be prevented by discretely modeling the AS process and subsequent coupling of the results (Karpinska et al., 2013; Pereira et al., 2012). Additionally, the performed CFD study may require further calibration through data acquisition and model validation using advanced measurement methods, depending on the objectives of the CFD analysis (Nopens et al., 2012).

Comprehensive three-phase CFD models have been studied in the literature to conduct simulations of real physico-chemical-biological processes in a variety of AS system configurations. In order to integrate biokinetics with hydrodynamics, an ASM model has been incorporated into a CFD model by implementing a new transport species with source terms that include bioreaction rates. By this means, transport phenomena and interactions between phases in the AS reactor (namely between the water-gas, water-sludge and gas-sludge phases) could be quantified. In this respect, CFD-ASM simulations take several key factors into account: flow behavior in the reactor, oxygen transfer and distribution, biomass increase and decline (growth and decay), biochemical activities of the microorganisms in the AS (metabolic activity), sludge settling. As a result, by simultaneous prediction of the hydrodynamics of the system and its effects on the reactions taking place in the biomass, which are represented by the concentration profiles, the system design, mixing or aeration can be optimized that in turn increases process efficiency and reduces operating costs (Glover et al., 2006; Le Moullec et al., 2010, 2011; Lei and Ni, 2014).

14.6.3 Drawbacks of CFD simulations

The effectiveness of CFD simulations is different in various process units of wastewater treatment. While CFD is frequently used in investigating final sedimentation, activated sludge basin modeling, and disinfection, more work has to be carried out on primary sedimentation and anaerobic digestion units. Even though CFD approaches are thorough with the capability of embedding biokinetics, non-Newtonian fluids, and multiphase systems, they have to be extended to deal with the diversity of process concepts (Samstag et al., 2016). As mentioned, CFD simulation is essential for studying WWTP not only from a financial point of view, but also from a risk management perspective (Wicklein et al., 2016). Nevertheless, there is a limited number of research groups on this subject and lack of guidelines that compelled IWA to form the Modeling and Integrated Assessment Specialist

Group (MIA SG) to support the development of this kind of community (Samstag et al., 2016).

Besides inappropriate application of CFD originated from improper model selection, incorrect setup assumptions or results analysis (Nopens et al., 2012), there are also some modeling challenges that have been not addressed in CFD modeling of AS tanks such as the AS basin coupling with secondary clarifier (Karpinska and Bridgeman, 2016). Consequently, tackling these problems will improve the robustness and reliability of the models. SRT, concentration of suspended solids in mixed liquids (MLSS) and biomass composition are criteria on which the efficiency of BOD and nutrient removal in the AS process depends. Therefore, the operation of AS tanks is strongly linked to the performance of secondary decanters. Dynamic modeling of the combined aeration tank and clarifier system enables the prediction of flow and mass transport behavior under varying flow conditions. As a result, the effect of changes in return sludge flow rates on the distribution of concentrations within the system can be assessed (Patziger et al., 2012).

As several physical and biochemical phenomena have to be taken into account in parallel, full CFD modeling of a clarifier is not feasible. This modeling is further difficult by the fact that the settleability of AS cannot be predicted, and data on the particulate fraction are incomplete. Moreover, the sludge settling described in the advanced clarifier models is based on empirical equations. This dynamic settleability behavior that is not incorporated in models impacts actual SRTs of AS system, which could be an error source for the WWTP models (Plósz et al., 2012).

14.7 Conclusions and future trends

Computational modeling has emerged as a valuable instrument for managing municipal biological wastewater treatment plants, moving from conventional mechanistic models to data-driven models grounded in artificial intelligence and machine learning. Although mechanistic models have a significant role in process design, data-driven models demonstrate superior prediction capabilities and lower errors. The primary obstacle to achieving broader adoption of AI models is data collecting. The most extensively researched applications are the forecasting of wastewater parameters and the energy conservation. The effective optimization and control of specific unit operations or processes are often overlooked. Hence, it is advisable to do additional laboratory and field experiments in order to comprehend the intricate dynamics of wastewater treatment plants, taking into account the diverse range of inputs, hidden layers, and neurons in the model's architecture. Future research should focus on the implementation of combined mechanistic and data-driven models (Hybrid models), which have a vast potential for anomaly detection and improving the monitoring and design of wastewater treatment systems, while also reducing operational costs and saving energy. Moreover, the urgent focus of future research investigations should be on the predictive control of specific contamination and abrupt shocks in both conventional and developing technologies. Finally, a more extensive model should be developed to consider technical problems related to the system, economic variables, social concerns, and atmospheric interference in order to effectively solve the monitoring and design challenges of biological wastewater treatment systems.

References

Ahmad, T., Guria, C., 2022. Progress in the modification of polyvinyl chloride (PVC) membranes: a performance review for wastewater treatment. Elsevier Ltd, India, J. Water Proc. Eng. 45. 22147144. Available from: http://www.journals.elsevier.com/journal-of-water-process-engineering/. https://doi.org/10.1016/j.jwpe.2021.102466.

Ajbar, A.H., AlAhmad, M., Ali, E., 2011. On the dynamics of biodegradation of wastewater in aerated continuous bioreactors. Math. Comput. Model. 54 (9–10), 1930–1942. https://doi.org/10.1016/j.mcm.2011.04.035. 08957177.

Alizadeh, M., Jalilnejad, E., Rafiee, R., 2019. Application of hydrogels in adsorptive removal of aqueous pollutants. Iran. J. Polym Sci. Technol. 31 (6), 499–518.

Altowayti, W.A.H., Shahir, S., Eisa, T.A.E., Nasser, M., Babar, M.I., Alshalif, A.F., AL-Towayti, F.A.H., 2022. Smart modelling of a sustainable biological wastewater treatment technologies: a critical review. Sustainability 14 (22), 15353. https://doi.org/10.3390/su142215353. 2071-1050.

Amrutha, M.C.V., Haseena, P., 2020. EasyChair preprint waste water treatment plant analysis and simulation using computational tools: a review. In: International Conference on Green Energy for Environmental Sustainability (ICGEES).

Arslan, A., Topkaya, E., Sezer, M., Aksan, S., Veli, S., 2024. Investigation of microplastics in advanced biological wastewater treatment plant effluent. Mar. Pollut. Bull. 203, 116486. https://doi.org/10.1016/j.marpolbul.2024.116486. 0025326X.

Asami, H., Golabi, M., Albaji, M., 2021. Simulation of the biochemical and chemical oxygen demand and total suspended solids in wastewater treatment plants: data-mining approach. J. Clean. Prod. 296, 126533. https://doi.org/10.1016/j.jclepro.2021.126533. 09596526.

Bernardelli, A., Marsili-Libelli, S., Manzini, A., Stancari, S., Tardini, G., Montanari, D., Anceschi, G., Gelli, P., Venier, S., 2020. Real-time model predictive control of a wastewater treatment plant based on machine learning. IWA Publishing, Italy, Water Sci. Technol. 81 (11), 2391–2400. 19969732. https://doi.org/10.2166/wst.2020.298.

Bhuyar, L.B., Thakre, S.B., Ingole, N.W., 2010. Design characteristics of Curved Blade Aerator w.r.t. aeration efficiency and overall oxygen transfer coefficient and comparison with CFD modeling. Int. J. Eng. Sci. Technol. 1 (1). https://doi.org/10.4314/ijest.v1i1.58055. 2141-2839.

Brdjanovic, D., Van Loosdrecht, M.C.M., Versteeg, P., Hooijmans, C.M., Alaerts, G.J., Heijnen, J.J., Modeling, C.O.D., 2000. N and P removal in a full-scale wwtp Haarlem Waarderpolder. Elsevier Science Ltd, Netherlands, Water Res. 34 (3), 846–858. 00431354. Available from: www.elsevier.com/locate/watres https://doi.org/10.1016/S0043-1354(99)00219-5.

Brdjanovic, D., Meijer, S.C.F., Lopez-Vazquez, C.M., Hooijmans, C.M., 2015. Applications of Activated Sludge Models. Iwa Publishing.

Brun, R., Kühni, M., Siegrist, H., Gujer, W., Reichert, P., 2002. Practical identifiability of ASM2d parameters—systematic selection and tuning of parameter subsets. Water Res. 36 (16), 4113–4127. https://doi.org/10.1016/s0043-1354(02)00104-5. 00431354.

Cairone, S., Hasan, S.W., Choo, K.H., Li, C.W., Zarra, T., Belgiorno, V., Naddeo, V., 2024. Integrating artificial intelligence modeling and membrane technologies for advanced wastewater treatment: research progress and future perspectives. Elsevier B.V., Italy, Sci. Total Environ. 944. 18791026. Available from: https://www.sciencedirect.com/science/journal/00489697. https://doi.org/10.1016/j.scitotenv.2024.173999.

Cao, Q., Cai, Y., Luo, Y., 2022. Nonconstant positive solutions to the ratio-dependent predator-prey system with prey-taxis in one dimension. Discrete Continuous Dyn. Syst. B 27 (3), 1397. https://doi.org/10.3934/dcdsb.2021095. 1531–3492.

Chan, Y.J., Chong, M.F., Law, C.L., Hassell, D.G., 2009. A review on anaerobic-aerobic treatment of industrial and municipal wastewater. Chem. Eng. J. 155 (1–2), 1–18. https://doi.org/10.1016/j.cej.2009.06.041. 13858947.

Cheyns, K., Mertens, J., Diels, J., Smolders, E., Springael, D., 2010. Monod kinetics rather than a first-order degradation model explains atrazine fate in soil mini-columns: implications for pesticide fate modelling. Environ. Pollut. 158 (5), 1405–1411. https://doi.org/10.1016/j.envpol.2009.12.041. 02697491.

Clagnan, E., Petrini, S., Pioli, S., Piergiacomo, F., Chowdhury, A.A., Brusetti, L., Foladori, P., 2024. Conventional activated sludge vs. photo-sequencing batch reactor for enhanced nitrogen removal in municipal wastewater: microalgal-bacterial consortium and pathogenic load insights. Elsevier Ltd, Italy, Bioresour. Technol. 401. 18732976. Available from: https://www.sciencedirect.com/science/journal/09608524. https://doi.org/10.1016/j.biortech.2024.130735.

CONTOIS, D.E., 1959. Kinetics of bacterial growth: relationship between population density and specific growth rate of continuous cultures. J. Gen. Microbiol. 21 (1), 40–50. https://doi.org/10.1099/00221287-21-1-40. 0022–1287.

De Gussem, K., Fenu, A., Wambecq, T., Weemaes, M., 2014. Energy saving on wastewater treatment plants through improved online control: case study wastewater treatment plant Antwerp-South. Water Sci. Technol. 69 (5), 1074–1079. https://doi.org/10.2166/wst.2014.015. 0273–1223.

Dike, H.U., Zhou, Y., Deveerasetty, K.K., Wu, Q., 2018. Unsupervised learning based on artificial neural network: a review. In: IEEE International Conference on Cyborg and Bionic Systems, CBS 2018. Institute of Electrical and Electronics Engineers Inc., China, https://doi.org/10.1109/CBS.2018.8612259. 9781538673553 322 327 http://ieeexplore.ieee.org/xpl/mostRecentIssue.jsp?punumber=8595708.

Dold, P.L., Ekama, G.A., Marais Van, G.R., 1981. A General Model for the Activated Sludge Process. Water Pollution Research and Development, pp. 47–77.

Donoso-Bravo, A., Mailier, J., Ruiz-Filippi, G., Vande Wouwer, A., 2013. Identification in an anaerobic batch system: global sensitivity analysis, multi-start strategy and optimization criterion selection. Bioprocess Biosyst. Eng. 36 (1), 35–43. https://doi.org/10.1007/s00449-012-0758-5. 16157605.

Drewnowski, J., Szeląg, B., Xie, L., Lu, X., Ganesapillai, M., Deb, C.K., Szulzyk-Cieplak, J., Łagód, G., 2020. The influence of COD fraction forms and molecules size on hydrolysis process developed by comparative OUR studies in activated sludge modelling. MDPI AG, Poland, Molecules 25 (4). 14203049. Available from: https://www.mdpi.com/1420-3049/25/4/929/pdf. https://doi.org/10.3390/molecules25040929.

Duarte, M.S., Martins, G., Oliveira, P., Fernandes, B., Ferreira, E.C., Alves, M.M., Lopes, F., Pereira, M.A., Novais, P., 2024. A review of computational modeling in wastewater treatment processes. American Chemical Society, Portugal, ACS EST Water 4 (3), 784–804. 26900637. Available from: https://pubs.acs.org/page/aewcaa/about.html. https://doi.org/10.1021/acsestwater.3c00117.

Essahlaoui, F., Elhajrat, N., Halimi, M., El Abbassi, A., 2023. New approach to monitoring a wastewater irrigation system controlled by the artificial neural network (ANN). Groundw. Sustain. Dev. 2352801X (23), 100999. https://doi.org/10.1016/j.gsd.2023.100999.

Fayolle, Y., Cockx, A., Gillot, S., Roustan, M., Héduit, A., 2007. Oxygen transfer prediction in aeration tanks using CFD. Chem. Eng. Sci. 62 (24), 7163–7171. https://doi.org/10.1016/j.ces.2007.08.082. 00092509.

Fenu, A., Guglielmi, G., Jimenez, J., Spèrandio, M., Saroj, D., Lesjean, B., Brepols, C., Thoeye, C., Nopens, I., 2010. Activated sludge model (ASM) based modelling of membrane bioreactor (MBR) processes: a critical review with special regard to MBR specificities. Water Res. 44 (15), 4272–4294. https://doi.org/10.1016/j.watres.2010.06.007. 00431354.

Ferrer, J., Seco, A., Serralta, J., Ribes, J., Manga, J., Asensi, E., Morenilla, J.J., Llavador, F., 2008. DESASS: a software tool for designing, simulating and optimising WWTPs. Environ Model Softw. 23 (1), 19–26. https://doi.org/10.1016/j.envsoft.2007.04.005. 13648152.

Fujimoto, Y., 1963. Kinetics of microbial growth and substrate consumption. J. Theor. Biol. 5 (2), 171–191. https://doi.org/10.1016/0022-5193(63)90058-4. 00225193.

Garrido-Baserba, M., Corominas, L., Cortés, U., Rosso, D., Poch, M., 2020. The fourth-revolution in the water sector encounters the digital revolution. Environ. Sci. Technol. 54 (8), 4698–4705. https://doi.org/10.1021/acs.est.9b04251. 15205851.

Gernaey, K.V., Van Loosdrecht, M.C.M., Henze, M., Lind, M., Jørgensen, S.B., 2004. Denmark Activated sludge wastewater treatment plant modelling and simulation: state of the art. In: Environmental Modelling and Software, p. 19, https://doi.org/10.1016/j.envsoft.2003.03.005. 9 763 783.

Gillot, S., 2000. Effect of air flow rate on oxygen transfer in an oxidation ditch equipped with fine bubble diffusers and slow speed mixers. Water Res. 34 (5), 1756–1762. https://doi.org/10.1016/s0043-1354(99)00323-1. 00431354.

Gillot, S., Capela-Marsal, S., Roustan, M., Héduit, A., 2005. Predicting oxygen transfer of fine bubble diffused aeration systems—model issued from dimensional analysis. Water Res. 39 (7), 1379–1387. https://doi.org/10.1016/j.watres.2005.01.008. 00431354.

Glover, G.C., Printemps, C., Essemiani, K., Meinhold, J., 2006. Modelling of wastewater treatment plants—how far shall we go with sophisticated modelling tools? Water Sci. Technol. 53 (3), 79–89. https://doi.org/10.2166/wst.2006.078. 0273–1223.

Gujer, W., 2006. Activated sludge modelling: past, present and future. Water Sci. Technol. 53 (3), 111–119. https://doi.org/10.2166/wst.2006.082. 0273–1223.

Gujer, W., Henze, M., Mino, T., Matsuo, T., Wentzel, M.C., Marais, G.v.R., 1995. The activated sludge model No. 2: biological phosphorus removal. Water Sci. Technol. 31 (2), 1–11. https://doi.org/10.2166/wst.1995.0061. 0273–1223.

Gujer, W., Henze, M., Mino, T., van Loosdrecht, M., 1999. Activated sludge model no. 3. Water Sci. Technol. 39 (1), 183–193. https://doi.org/10.2166/wst.1999.0039. 0273–1223.

Hastie, T., Tibshirani, R., Friedman, J., 2008. Unsupervised Learning. Springer Nature, pp. 485–585, https://doi.org/10.1007/978-0-387-84858-7_14.

Hauduc, H., Rieger, L., Takács, I., Héduit, A., Vanrolleghem, P.A., Gillot, S., 2010. A systematic approach for model verification: application on seven published activated sludge models. Water Sci. Technol. 61 (4), 825–839. https://doi.org/10.2166/wst.2010.898. 0273–1223.

Hauduc, H., Rieger, L., Ohtsuki, T., Shaw, A., Takács, I., Winkler, S., Héduit, A., Vanrolleghem, P.A., Gillot, S., 2011. Activated sludge modelling: development and potential use of a practical applications database. Water Sci. Technol. 63 (10), 2164–2182. https://doi.org/10.2166/wst.2011.368. 0273–1223.

Hauduc, H., Rieger, L., Oehmen, A., van Loosdrecht, M.C.M., Comeau, Y., Héduit, A., Vanrolleghem, P.A., Gillot, S., 2013. Critical review of activated sludge modeling: state of process knowledge, modeling concepts, and limitations. Biotechnol. Bioeng. 110 (1), 24–46. https://doi.org/10.1002/bit.24624. 0006–3592.

Hedaoo, M.N., Bhole, A.G., Ingole, N.W., Hung, Y.T., 2012. Biological Wastewater Treatment Handbook of Environment and Waste Management: Air and Water Pollution Control. World Scientific Publishing Co., India, pp. 431–473. Available from: http://www.worldscientific.com/worldscibooks/10.1142/7971#t=toc. https://doi.org/10.1142/9789814327701_0012.

Henze, M., Grady, C.P.L., Gujer, W., Marais, G., Matsuo, R., 1987. Activated Sludge Model No. 1. International Association on Water Pollution Research and Control, London.

Henze, M., Gujer, W., Mino, T., Matsuo, T., Wentzel, M.C., Marais, G.V.R., Van Loosdrecht, M.C.M., 1999. Water Science and Technology., https://doi.org/10.1016/S0273-1223(98)00829-4. 02731223 1 165 182 Denmark Activated Sludge Model No.2d, ASM2d 39.

Holloway, T.G., Williams, J.B., Ouelhadj, D., Cleasby, B., 2021. Process stress, stability and resilience in wastewater treatment processes: a novel conceptual methodology. Elsevier Ltd, United Kingdom, J. Clean. Prod. 282. 09596526. Available from: https://www.journals.elsevier.com/journal-of-cleaner-production. https://doi.org/10.1016/j.jclepro.2020.124434.

Horikoshi, S., Hidaka, T., Nishimura, F., Kishimoto, N., 2024. Development of separation and recovery technology of cultured Euglena gracilis using activated sludge for anaerobic digestion. Bioresour. Technol. Rep. 25, 101790. https://doi.org/10.1016/j.biteb.2024.101790. 2589014X.

Hulsbeek, J.J.W., Kruit, J., Roeleveld, P.J., Van Loosdrecht, M.C.M., 2002. A practical protocol for dynamic modelling of activated sludge systems. In: Water Science and Technology. IWA Publishing, Netherlands, p. 45, https://doi.org/10.2166/wst.2002.0100. 02731223 6 127 136 https://iwaponline.com/wst.

Jalilnejad, E., Alizadeh, M., Fakhriazar, S.F., 2018. Application of biological methods in decolorization of azo dye containing wastewaters. J. Stud. Color World 8 (3), 27–40.

Jalilnejad, E., Alizadeh, M., Jabbari, B., 2023. Ion-exchange membranes in microbial fuel cell systems. In: Current Trends and Future Developments on (Bio-) Membranes: Recent Achievements for Ion-Exchange Membranes. Elsevier, Iran, pp. 229–263. Available from: https://www.sciencedirect.com/book/9780323885096. https://doi.org/10.1016/B978-0-323-88509-6.00004-6.

Janiesch, C., Zschech, P., Heinrich, K., 2021. Machine learning and deep learning. Springer Science and Business Media Deutschland GmbH, Germany, Electron. Mark. 31 (3), 685–695. 14228890. Available from: http://www.springer.com/business/business+information+systems/journal/12525?detailsPage=aimsAndScopes. https://doi.org/10.1007/s12525-021-00475-2.

Jordan, M.I., Mitchell, T.M., 2015. Machine learning: trends, perspectives, and prospects. American Association for the Advancement of Science, United States, Science 349 (6245), 255–260. 10959203. Available from: http://www.sciencemag.org/content/349/6245/255.full.pdf. https://doi.org/10.1126/science.aaa8415.

Karpinska, A.M., Bridgeman, J., CFD-aided modelling of activated sludge systems—a critical review. Elsevier Ltd, United Kingdom Water Res. 18792448. 88 (2016), 861–879, Available from: www.elsevier.com/locate/watres. https://doi.org/10.1016/j.watres.2015.11.008.

Karpinska, P.S., Nogueira, R.J., Madalena, S., De Queiroz, G., Rui, M., et al., 2013. New Design Tools for Activated Sludge Process (Doctoral dissertation). Universidade do Porto.

Karpinska, A.M., Dias, M.M., Boaventura, R.A.R., Santos, R.J., 2015. Modeling of the hydrodynamics and energy expenditure of oxidation ditch aerated with hydrojets using CFD codes. IWA Publishing, Portugal, Water Qual. Res. J. Can. 50 (1), 83–94. 12013080. Available from: http://www.iwaponline.com/wqrjc/050/0083/0500083.pdf. https://doi.org/10.2166/wqrjc.2014.036.

Kim, K.G., 2016. Book review: deep learning. Healthc. Inform. Res. 22 (4), 351. https://doi.org/10.4258/hir.2016.22.4.351. 2093–3681.

Kjellstrand, R., 2006. THESIS FOR THE DEGREE OF LICENTIATE OF ENGINEERING, Hydraulic Behaviour in an Activated Sludge Tank From Tracer Test through Hydraulic Modelling to Full-Scale Implementation.

Koch, G., Kühni, M., Gujer, W., Siegrist, H., 2000. Calibration and validation of activated sludge model no. 3 for Swiss municipal wastewater. Elsevier Science Ltd, Switzerland, Water Res. 34 (14), 3580–3590. 00431354. Available from: https://www.sciencedirect.com/science/journal/00431354. https://doi.org/10.1016/S0043-1354(00)00105-6.

Kochevsky, A.N., 2004. Possibilities of simulation of fluid flows using the modern CFD software tools. arXiv:physics.

Kuba, T., van Loosdrecht, M.C.M., Murnleitner, E., Heijnen, J.J., 1997. Kinetics and stoichiometry in the biological phosphorus removal process with short cycle times. Water Res. 31 (4), 918–928. https://doi.org/10.1016/s0043-1354(96)00371-5. 00431354.

Laurent, J., Samstag, R.W., Ducoste, J.M., Griborio, A., Nopens, I., Batstone, D.J., Wicks, J.D., Saunders, S., Potier, O., 2014. A protocol for the use of computational fluid dynamics as a supportive tool for wastewater treatment plant modelling. IWA Publishing, France, Water Sci. Technol. 70 (10), 1575–1584. 02731223. Available from: http://www.iwaponline.com/wst/07010/1575/070101575.pdf. https://doi.org/10.2166/wst.2014.425.

Le Moullec, Y., Potier, O., Gentric, C., Pierre Leclerc, J., 2008. Flow field and residence time distribution simulation of a cross-flow gas-liquid wastewater treatment reactor using CFD. Chem. Eng. Sci. 63 (9), 2436–2449. https://doi.org/10.1016/j.ces.2008.01.029. 00092509.

Le Moullec, Y., Gentric, C., Potier, O., Leclerc, J.P., 2010. CFD simulation of the hydrodynamics and reactions in an activated sludge channel reactor of wastewater treatment. Chem. Eng. Sci. 65 (1), 492–498. https://doi.org/10.1016/j.ces.2009.03.021. 00092509.

Le Moullec, Y., Potier, O., Gentric, C., Leclerc, J.P., 2011. Activated sludge pilot plant: comparison between experimental and predicted concentration profiles using three different modelling approaches. Water Res. 45 (10), 3085–3097. https://doi.org/10.1016/j.watres.2011.03.019. 00431354.

Lecun, Y., Bengio, Y., Hinton, G., 2015. Deep learning. Nature Publishing Group, United States, Nature 521 (7553), 436–444. 14764687. Available from: http://www.nature.com/nature/index.html. https://doi.org/10.1038/nature14539.

Lei, L., Ni, J., 2014. Three-dimensional three-phase model for simulation of hydrodynamics, oxygen mass transfer, carbon oxidation, nitrification and denitrification in an oxidation ditch. Elsevier Ltd, China, Water Res. 53, 200–214. 18792448. Available from: https://www.sciencedirect.com/science/journal/00431354. https://doi.org/10.1016/j.watres.2014.01.021.

Li, X., Su, J., Wang, H., Boczkaj, G., Mahlknecht, J., Singh, S.V., Wang, C., 2024. Bibliometric analysis of artificial intelligence in wastewater treatment: current status, research progress, and future prospects. Elsevier Ltd, China, J. Environ. Chem. Eng. 12 (4). 22133437. Available from: https://www.sciencedirect.com/science/journal/22133437. https://doi.org/10.1016/j.jece.2024.113152.

Lobry, C., Harmand, J., 2006. A new hypothesis to explain the coexistence of n species in the presence of a single resource. C. R. Biol. 329 (1), 40–46. https://doi.org/10.1016/j.crvi.2005.10.004. 16310691.

Lu, J., Qiaoling, L., Qian, H., Qiu, B., 2024. Recovery of organic matters by activated sludge from municipal wastewater: performance and characterization. Environ. Res. 252, 118829. https://doi.org/10.1016/j.envres.2024.118829. 00139351.

Makinia, J., Zaborowska, E., 2020. Mathematical Modelling and Computer Simulation of Activated Sludge Systems. IWA Publishing.

Matheri, A.N., Ntuli, F., Ngila, J.C., Seodigeng, T., Zvinowanda, C., 2021. Performance prediction of trace metals and cod in wastewater treatment using artificial neural network. Elsevier Ltd, South Africa, Comput. Chem. Eng. 149. 00981354. Available from: https://www.journals.elsevier.com/computers-and-chemical-engineering. https://doi.org/10.1016/j.compchemeng.2021.107308.

Maurer, M., Boller, M., 1999. Modelling of phosphorus precipitation in wastewater treatment plants with enhanced biological phosphorus removal. Water Sci. Technol. 39 (1), 147–163. https://doi.org/10.2166/wst.1999.0033. 0273–1223.

Maurer, M., Abramovich, D., Siegrist, H., Gujer, W., 1999. Kinetics of biologically induced phosphorus precipitation in waste-water treatment. Water Res. 33 (2), 484–493. https://doi.org/10.1016/s0043-1354(98)00221-8. 00431354.

Meijer, S.C.F., van Loosdrecht, M.C.M., Heijnen, J.J., 2001. Metabolic modelling of full-scale biological nitrogen and phosphorus removing wwtp's. Water Res. 35 (11), 2711–2723. https://doi.org/10.1016/s0043-1354(00)00567-4. 00431354.

Michelucci, U., 2018. Applied Deep Learning. Apress.

Mohri, M., Rostamizadeh, A., Talwalkar, A., 2018. Foundations of Machine Learning. MIT Press.

Monod, J., 1949. The growth of bacterial cultures. Annu. Rev. Microbiol. 3 (1), 371–394. https://doi.org/10.1146/annurev.mi.03.100149.002103. 0066–4227.

Morchain, J., Maranges, C., Fonade, C., 2000. CFD modelling of a two-phase jet aerator under influence of a crossflow. Water Res. 34 (13), 3460–3472. https://doi.org/10.1016/S0043-1354(00)00080-4. 00431354.

Moser, H., 1958. The Dynamics of Bacterial Populations Maintained in the Chemostat. Carnegie Institution of Washington.

Murnleitner, E., Kuba, T., Van Loosdrecht, M.C.M., Heijnen, J.J., 1997. An integrated metabolic model for the aerobic and denitrifying biological phosphorus removal. Biotechnol. Bioeng. 54 (5), 434–450. https://doi.org/10.1002/(SICI)1097-0290(19970605)54:5<434::AID-BIT4>3.0.CO;2-F.

Myint, M., Nirmalakhandan, N., Speece, R.E., Anaerobic fermentation of cattle manure: modeling of hydrolysis and acidogenesis. Elsevier Ltd, United States

Water Res. 00431354. 41 (2) (2007), 323–332, Available from: www.elsevier.com/locate/watres. https://doi.org/10.1016/j.watres.2006.10.026.

Nelson, M.I., Holder, A., 2009. A fundamental analysis of continuous flow bioreactor models governed by Contois kinetics. II. Reactor cascades. Chem. Eng. J. 149 (1–3), 406–416. https://doi.org/10.1016/j.cej.2009.01.028. 13858947.

Nelson, M.I., Balakrishnan, E., Sidhu, H.S., Chen, X.D., 2008. A fundamental analysis of continuous flow bioreactor models and membrane reactor models to process industrial wastewaters. Chem. Eng. J. 140 (1–3), 521–528. https://doi.org/10.1016/j.cej.2007.11.035. 13858947.

Nopens, I., Batstone, D.J., Griborio, A., Samstag, R., Wicklein, E., Wicks, J., 2012. Water environment federation Belgium computational fluid dynamics (CFD): what is good CFD-modeling practice and what can be the added value of CFD models to WWTP modeling? In: WEFTEC 2012—85th Annual Technical Exhibition and Conference, p. 12, https://doi.org/10.2175/193864712811704161. 9781510870437 7400 7405.

Nourani, V., Asghari, P., Sharghi, E., 2021. Artificial intelligence based ensemble modeling of wastewater treatment plant using jittered data. J. Clean. Prod. 291, 125772. https://doi.org/10.1016/j.jclepro.2020.125772. 09596526.

Orhon, D., 2015. Evolution of the activated sludge process: the first 50 years. John Wiley and Sons Ltd, Turkey, J. Chem. Technol. Biotechnol. 90 (4), 608–640. 10974660. Available from: http://onlinelibrary.wiley.com/journal/10.1002/(ISSN)1097-4660. https://doi.org/10.1002/jctb.4565.

Patziger, M., Kainz, H., Hunze, M., Józsa, J., 2012. Influence of secondary settling tank performance on suspended solids mass balance in activated sludge systems. Elsevier Ltd, Hungary, Water Res. 46 (7), 2415–2424. 18792448. Available from: https://www.sciencedirect.com/science/journal/00431354. https://doi.org/10.1016/j.watres.2012.02.007.

Pereira, J.P., Karpinska, A., Gomes, P.J., Martins, A.A., Dias, M.M., Lopes, J.C.B., Santos, R.J., 2012. Activated Sludge Models Coupled to CFD Simulations Single and Two-Phase Flows on Chemical and Biomedical Engineering. Bentham Science Publishers Ltd., Portugal, pp. 153–173. Available from: http://www.eurekaselect.com/openurl/content.php?genre=book%26isbn=9781608052950. https://doi.org/10.2174/978160805295011201010153.

Pérez-Beltrán, C.H., Robles, A.D., Rodriguez, N.A., Ortega-Gavilán, F., Jiménez-Carvelo, A.M., 2024. Artificial intelligence and water quality: from drinking water to wastewater. TrAC Trends Anal. Chem. 172. https://doi.org/10.1016/j.trac.2024.117597. 01659936. 117597.

Petersen, B., 2000. Calibration Experimental of Sludge Models (Doctoral dissertation). Ghent University.

Petersen, B., Gernaey, K., Henze, M., Vanrolleghem, P.A., 2002. Evaluation of an ASM1 model calibration procedure on a municipal-industrial wastewater treatment plant. IWA Publishing, Belgium, J. Hydroinf. 4 (1), 15–38. 14651734. Available from: https://iwaponline.com/jh. https://doi.org/10.2166/hydro.2002.0003.

Picos-Benítez, A.R., Martínez-Vargas, B.L., Duron-Torres, S.M., Brillas, E., Peralta-Hernández, J.M., 2020. The use of artificial intelligence models in the prediction of optimum operational conditions for the treatment of dye wastewaters with similar structural characteristics. Institution of Chemical Engineers, Mexico, Process Saf. Environ. Prot. 143, 36–44. 09575820. Available from: http://www.elsevier.com/wps/find/journaldescription.cws_home/713889/description#description. https://doi.org/10.1016/j.psep.2020.06.020.

Pimenta, A., Carneiro, D., Neves, J., Novais, P., A neural network to classify fatigue from human-computer interaction. Elsevier, Portugal Neurocomputing

18728286. 172 (2016), 413–426, Available from: www.elsevier.com/locate/neucom. https://doi.org/10.1016/j.neucom.2015.03.105.

Plósz, B.G., Nopens, I., Rieger, L., Griborio, A., Clercq, J., Vanrolleghem, P.A., et al., 2012. A critical review of clarifier modelling: state-of-the-art and engineering practices. In: 3rd IWA/WEF Wastewater Treatment Modelling Seminar.

Pomiès, M., Choubert, J.M., Wisniewski, C., Coquery, M., 2013. Modelling of micropollutant removal in biological wastewater treatments: a review. Sci. Total Environ. 443, 733–748. https://doi.org/10.1016/j.scitotenv.2012.11.037. 18791026.

Quaghebeur, W., Torfs, E., De Baets, B., Nopens, I., Hybrid differential equations: integrating mechanistic and data-driven techniques for modelling of water systems. Elsevier Ltd, Belgium Water Res. 18792448. 213 (2022), Available from: www.elsevier.com/locate/watres. https://doi.org/10.1016/j.watres.2022.118166.

Ramesh, C., Vinithkumar, N., Kirubagaran, R., Venil, C., Dufossé, L., 2019. Multifaceted applications of microbial pigments: current knowledge, challenges and future directions for public health implications. Microorganisms 7 (7), 186. https://doi.org/10.3390/microorganisms7070186. 2076–2607.

Ramirez, I., Mottet, A., Carrère, H., Déléris, S., Vedrenne, F., Steyer, J.P., Modified ADM1 disintegration/hydrolysis structures for modeling batch thermophilic anaerobic digestion of thermally pretreated waste activated sludge. Elsevier Ltd, France Water Res. 00431354. 43 (14) (2009), 3479–3492, Available from: www.elsevier.com/locate/watres. https://doi.org/10.1016/j.watres.2009.05.023.

Reeves, C.R., 2010. Genetic Algorithms. Springer Science and Business Media LLC, pp. 109–139, https://doi.org/10.1007/978-1-4419-1665-5_5.

Ruel, S.M., Comeau, Y., Ginestet, P., Héduit, A., 2002. Modeling acidogenic and sulfate-reducing processes for the determination of fermentable fractions in wastewater. Biotechnol. Bioeng. 80 (5), 525–536. https://doi.org/10.1002/bit.10410. 0006–3592.

Ruiz, L.M., Pérez, J.I., Gómez, M.A., 2023. Practical review of modelling and simulation applications at full-scale wastewater treatment plants. Elsevier Ltd, Spain, J. Water Proc. Eng. 56. 22147144. Available from: http://www.journals.elsevier.com/journal-of-water-process-engineering/. https://doi.org/10.1016/j.jwpe.2023.104477.

Samstag, R.W., Wicklein, E.A., Reardon, R.D., Leetch, R.J., Parks, R.M., Groff, C.D., 2012. Water Environment Federation United States Field and CFD analysis of jet aeration and mixing. In: WEFTEC 2012—85th Annual Technical Exhibition and Conference, p. 7, https://doi.org/10.2175/193864712811708301. 9781510870437 4113 4139.

Samstag, R.W., Ducoste, J.J., Griborio, A., Nopens, I., Batstone, D.J., Wicks, J.D., Saunders, S., Wicklein, E.A., Kenny, G., Laurent, J., 2016. CFD for wastewater treatment: an overview. IWA Publishing, United States, Water Sci. Technol. 74 (3), 549–563. 02731223. Available from: http://wst.iwaponline.com/content/74/3/549.full.pdf. https://doi.org/10.2166/wst.2016.249.

Sarker, I.H., 2021. Machine learning: algorithms, real-world applications and research directions. Springer, Australia, SN Comput. Sci. 2 (3). 26618907. Available from: https://www.springer.com/journal/42979. https://doi.org/10.1007/s42979-021-00592-x.

Savun-Hekimoğlu, B., 2021. On the use of mathematical models for wastewater treatment: a review and analysis of activated sludge models ASM1 and ASM3. Int. J. Environ. Geoinform. 8 (1), 1–18. https://doi.org/10.30897/ijegeo.794643. 2148–9173.

Schmitt, L.M., 2001. Theory of genetic algorithms. Elsevier, Japan, Theor. Comput. Sci. 259 (1–2), 1–61. 03043975. Available from: http://www.journals.elsevier.com/theoretical-computer-science/. https://doi.org/10.1016/S0304-3975(00)00406-0.

Schneider, M.Y., Quaghebeur, W., Borzooei, S., Froemelt, A., Li, F., Saagi, R., Wade, M.J., Zhu, J.J., Torfs, E., 2022. Hybrid modelling of water resource recovery facilities: status and opportunities. IWA Publishing, Japan, Water Sci. Technol. 85 (9), 2503–2524. 19969732. Available from: https://watermark.silverchair.com/wst085092503.pdf. https://doi.org/10.2166/wst.2022.115.

Shahbeik, H., Rafiee, S., Shafizadeh, A., Jeddi, D., Jafary, T., Lam, S.S., Pan, J., Tabatabaei, M., Aghbashlo, M., 2022. Characterizing sludge pyrolysis by machine learning: towards sustainable bioenergy production from wastes. Elsevier Ltd, China, Renew. Energy 199, 1078–1092. 18790682. Available from: http://www.journals.elsevier.com/renewable-and-sustainable-energy-reviews/. https://doi.org/10.1016/j.renene.2022.09.022.

Sharma, D., Misba, L., Khan, A.U., 2019. Antibiotics versus biofilm: an emerging battleground in microbial communities. BioMed Central Ltd., India, Antimicrob. Resist. Infect. Control 8 (1). 20472994. Available from: http://www.aricjournal.com/. https://doi.org/10.1186/s13756-019-0533-3.

Sin, G., Al, R., 2021. Activated sludge models at the crossroad of artificial intelligence—a perspective on advancing process modeling. Nature Research, Denmark, NPJ Clean Water 4 (1). 20597037. Available from: https://www.nature.com/npjcleanwater/. https://doi.org/10.1038/s41545-021-00106-5.

Singh, N.K., Yadav, M., Singh, V., Padhiyar, H., Kumar, V., Bhatia, S.K., Show, P.L., Artificial intelligence and machine learning-based monitoring and design of biological wastewater treatment systems. Elsevier Ltd, India Bioresour. Technol. 18732976. 369 (2023), Available from: www.elsevier.com/locate/biortech. https://doi.org/10.1016/j.biortech.2022.128486.

Smolders, G.J.F., Klop, J.M., van Loosdrecht, M.C.M., Heijnen, J.J., 1995. A metabolic model of the biological phosphorus removal process: I. Effect of the sludge retention time. Biotechnol. Bioeng. 48 (3), 222–233. https://doi.org/10.1002/bit.260480309. 10970290.

Stanescu, D., Chen-Charpentier, B., 2009. Random Coefficient Differential Equation Models for Monod Kinetics. American Institute of Mathematical Sciences.

Stavrakakis, I., Remmas, N., Melidis, P., Ntougias, S., 2024. Influence of the oxidative phosphorylation uncoupler m-chlorophenol on activated sludge production, nutrients removal efficiency and bacterial community structure of a membrane reactor system. Biochem. Eng. J. 209, 109386. https://doi.org/10.1016/j.bej.2024.109386. 1369703X.

Streeter, H.W., Phelps, E.B., 1925. A Study of the Pollution and Natural Purification of the Ohio River. United States Public Health Service.

Tang, X., Fan, C., Zeng, G., Zhong, L., Li, C., Ren, X., Song, B., Liu, X., 2022. Phage-host interactions: the neglected part of biological wastewater treatment. Water Res. 226, 119183. https://doi.org/10.1016/j.watres.2022.119183. 00431354.

Tang, B., Wang, J., Gao, X., Li, Z., 2025. Strategies for efficient enrichment of anaerobic ammonia oxidizing bacteria in activated sludge. J. Environ. Sci. 151, 703–713. https://doi.org/10.1016/j.jes.2024.04.046. 10010742.

Tejaswini, E.S.S., Maheswari, P., Ambati, S.R., 2024. Integrated supervisory fuzzy control framework for biological wastewater treatment plants operation and their effect on GHG emissions. Chem. Eng. Sci. 292, 119915. https://doi.org/10.1016/j.ces.2024.119915. 00092509.

Torregrossa, D., Leopold, U., Hernández-Sancho, F., Hansen, J., 2018. Machine learning for energy cost modelling in wastewater treatment plants. Academic Press, Luxembourg, J. Environ. Manage. 223, 1061–1067. 10958630. Available from: https://www.sciencedirect.com/science/journal/03014797. https://doi.org/10.1016/j.jenvman.2018.06.092.

Vafaie, H., De Jong, K., 1992. Genetic algorithms as a tool for feature selection in machine learning 1992. In: Proceedings—International Conference on Tools With Artificial Intelligence, ICTAI. IEEE Computer Society, United States.

Van Loosdrecht, M.C.M., Lopez-Vazquez, C.M., Meijer, S.C.F., Hooijmans, C.M., Brdjanovic, D., 2015. Twenty-five years of ASM1: past, present and future of wastewater treatment modelling. IWA Publishing, Netherlands, J. Hydroinf. 17 (5), 697–718. 14647141. Available from: http://www.iwaponline.com/jh/017/0697/0170697.pdf. https://doi.org/10.2166/hydro.2015.006.

Van Veldhuizen, H.M., Van Loosdrecht, M.C.M., Heijnen, J.J., Modelling biological phosphorus and nitrogen removal in a full scale activated sludge process. Elsevier Science Ltd, Netherlands Water Res. 00431354. 33 (16) (1999), 3459–3468, Available from: www.elsevier.com/locate/watres. https://doi.org/10.1016/S0043-1354(99)00064-0.

Vanhooren, H., Meirlaen, J., Amerlinck, Y., Claeys, F., Vangheluwe, H., Vanrolleghem, P.A., 2003. WEST: modelling biological wastewater treatment. IWA Publishing, Belgium, J. Hydroinf. 5 (1), 27–50. 14651734. Available from: https://iwaponline.com/jh. https://doi.org/10.2166/hydro.2003.0003.

Vermande, S., Simpson, K., Essemiani, K., Fonade, C., Meinhold, J., 2007. Impact of agitation and aeration on hydraulics and oxygen transfer in an aeration ditch: local and global measurements. Chem. Eng. Sci. 62 (9), 2545–2555. https://doi.org/10.1016/j.ces.2005.09.016. 00092509.

Waqas, S., Bilad, M.R., Man, Z., Wibisono, Y., Jaafar, J., Indra Mahlia, T.M., Khan, A.-L., Aslam, M., 2020. Recent progress in integrated fixed-film activated sludge process for wastewater treatment: a review. Academic Press, Malaysia, J. Environ. Manage. 268. 10958630. Available from: https://www.sciencedirect.com/journal/journal-of-environmental-management. https://doi.org/10.1016/j.jenvman.2020.110718.

Wei, Y., Van Houten, R.T., Borger, A.R., Eikelboom, D.H., Fan, Y., Minimization of excess sludge production for biological wastewater treatment. Elsevier Ltd, China Water Res. 00431354. 37 (18) (2003), 4453–4467, Available from: www.elsevier.com/locate/watres. https://doi.org/10.1016/S0043-1354(03)00441-X.

Wei, M., Yang, T., An, L., Meng, X., Tan, J., Zhang, X., 2023. Kinetic, artificial neural network, and statistical modeling to optimize the parameters of the air flotation process to remove latex suspended solids in ABS wastewater. J. Water Process Eng. 56, 104417. https://doi.org/10.1016/j.jwpe.2023.104417. 22147144.

Weijers, S.R., Vanrolleghem, P.A., 1997. A procedure for selecting best identifiable parameters in calibrating Activated Sludge Model No. 1 to full-scale plant data. In: Water Science and Technology, p. 36, https://doi.org/10.1016/S0273-1223(97)00463-0. 02731223 5 69 79 Netherlands.

Wicklein, E., Batstone, D.J., Ducoste, J., Laurent, J., Griborio, A., Wicks, J., Saunders, S., Samstag, R., Potier, O., Nopens, I., 2016. Good modelling practice in applying computational fluid dynamics for WWTP modelling. IWA Publishing, United States, Water Sci. Technol. 73 (5), 969–982. 02731223. Available from: http://wst.iwaponline.com/content/ppiwawst/73/5/969.full.pdf. https://doi.org/10.2166/wst.2015.565.

Yao, Y., Li, Y., Shi, Y., Ke Shi, Y., Gao, B.,.Y., Jiang, Q., Xue, J., Cheng, D., 2024. Prediction of artificial neural network for sulfate removal from wastewater and application analysis of key factors in anaerobic biological system. J. Environ. Chem. Eng. 12 (2). https://doi.org/10.1016/j.jece.2024.112085. 22133437. 112085.

Zadeh, L.A., 1983. The role of fuzzy logic in the management of uncertainty in expert systems. Fuzzy Set. Syst. 11 (1–3), 199–227. https://doi.org/10.1016/S0165-0114(83)80081-5. 01650114.

Zaghloul, M.S., Achari, G., 2022. Application of machine learning techniques to model a full-scale wastewater treatment plant with biological nutrient removal. J. Environ. Chem. Eng. 10 (3), 107430. https://doi.org/10.1016/j.jece.2022.107430. 22133437.

Zahorodny, S.M., Webb, D., Silva, M.R., 2022. Evaluation of a water system model for educational research. Creat. Educ. 13 (02), 637–657. https://doi.org/10.4236/ce.2022.132040. 2151–4755.

Zhang, L., Chao, B., Zhang, X., 2020. Modeling and optimization of microbial lipid fermentation from cellulosic ethanol wastewater by Rhodotorula glutinis based on the support vector machine. Bioresour. Technol. 301, 122781. https://doi.org/10.1016/j.biortech.2020.122781. 09608524.

Zhao, L., Dai, T., Qiao, Z., Sun, P., Hao, J., Yang, Y., 2020. Application of artificial intelligence to wastewater treatment: a bibliometric analysis and systematic review of technology, economy, management, and wastewater reuse. Institution of Chemical Engineers, China, Process Saf. Environ. Prot. 133, 169–182. 09575820. Available from: http://www.elsevier.com/wps/find/journaldescription.cws_home/713889/description#description. https://doi.org/10.1016/j.psep.2019.11.014.

Zhao, Z., Alzubaidi, L., Zhang, J., Duan, Y., Yuantong, G., 2024. A comparison review of transfer learning and self-supervised learning: definitions, applications, advantages and limitations. Expert Syst. Appl. 242, 122807. https://doi.org/10.1016/j.eswa.2023.122807. 09574174.

Zhou, B., Wang, D., Zhao, G., Zhang, M., Liu, X., Zhang, D., Liang, J., Zhou, L., 2024. Effects of food waste hydrolysate as an external carbon source on defoaming in wastewater treatment with activated sludge process. Bioresour. Technol. 404, 130900. https://doi.org/10.1016/j.biortech.2024.130900. 09608524.

15

Advancements on the technologies for municipal wastewater excess sludge

Petteri Peltola

VTT Technical Research Centre of Finland Ltd., Ruukinmestarintie 2, Espoo, Finland

Abbreviations

AD	anaerobic digestion
BFB	bubbling fluidized bed
CFB	circulating fluidized bed
CE	circular economy
DFB	dual fluidized bed
DME	dimethyl ether
EPS	extracellular polymeric substance
ER	equivalence ratio
GHG	greenhouse gas
GWP	global warming potential
HHV	higher heating value
HTC	hydrothermal carbonization
HTG	hydrothermal gasification
HTL	hydrothermal liquefaction
HTP	hydrothermal processes
ISSA	incinerated sewage sludge ash
PCDD/F	polychlorinated dibenzodioxins/dibenzofurans
POP	persistent organic pollutants
SCWG	supercritical water gasification
SCWO	supercritical water oxidation
SCWPO	supercritical water partial oxidation
SS	sewage sludge
TRL	Technology Readiness Level
VOC	volatile organic carbon
WWTP	wastewater treatment plant

Municipal Wastewater Treatment. https://doi.org/10.1016/B978-0-443-24826-9.00015-2

15.1 Introduction

A wastewater treatment plant (WWTP) serves as a central component in urban wastewater management systems. Its primary function is to eliminate both organic and inorganic substances from municipal wastewater, yielding water clean enough to be safely reintroduced into the water cycle. The remaining contaminants are concentrated in a semi-solid byproduct known as sewage sludge (SS), traditionally categorized as biomass waste due to its high organic content and the potential risks associated with various chemical and biological pollutants. SS generation rates vary globally, with an average of 20–25 kg of dry solids (DS) per person annually (LeBlanc et al., 2008). In Europe, the estimated production of SS exceeds 13 million t/y DS, similar to the amount produced in the USA. China produces over 30 million t/y DS, while India generates more than 43 million t/y DS (Capodaglio and Callegari, 2023). With population growth, industrialization, and urban expansion driving an inevitable increase in SS production worldwide, efficiently managing this sludge has emerged as one of today's most urgent environmental issue.

Once produced at WWTPs, SS needs mechanical, chemical, and biological processing to ensure safe and cost-effective transport and disposal. The treatment methods vary by plant type, with primary and secondary sludges potentially undergoing separate or combined treatment processes. These processes may include thickening, stabilizing, dewatering, and drying. Common stabilization techniques, such as lime treatment, anaerobic digestion (AD), and composting with organic matter, are designed to remove pathogens and odors while reducing the fermentability and volume of the remaining biosolids (Anderson et al., 2021). The final step involves disposal, which may entail distributing the sludge over soil surfaces, incineration, or other disposal approaches. The processing and disposal of SS significantly impact the overall performance of WWTPs, potentially accounting for over 50% of total costs and contributing nearly 40% to energy consumption and greenhouse gas (GHG) emissions (Capodaglio and Olsson, 2020). In EU countries, the costs for sludge treatment and disposal range between 160 €/t DS and 310 €/t DS, depending on local conditions and the methods used (Kacprzak et al., 2017).

The selection of a sludge treatment and management approach is influenced by prevailing trends and regulations, which dictate the permissible uses of treated SS. Historically, agricultural reuse has been the predominant disposal pathway due to its low investment costs and nutrient recycling benefits. However, stricter regulations, limited land availability, and mounting environmental and health concerns related to harmful substances in sludge have made traditional practices such as landfilling and direct land application increasingly problematic (Sugurbekova et al., 2023). Consequently, there is a pressing demand for economical, energy-efficient, and sustainable solutions for the treatment, management, and future utilization of SS.

Amidst the increasing emphasis on emission reduction, nutrient recovery, and clean energy production, there is a noticeable shift toward adopting sustainability and circular economy (CE) principles in sewage management (Tsybina and Wuensch, 2018). In this context, it is necessary to reevaluate some of the conventional methods of treating SS and explore their largely untapped potential in terms of environmental and economic gains. Extensive efforts have been dedicated to developing innovative solutions that could mitigate the adverse impacts of current sludge treatment practices and fully exploit the potential of sludge as a renewable resource for bioenergy, nutrients, and biomaterial recovery. This endeavor is guided and supported by new policies and regulatory frameworks that address a multitude of factors including environmental protection, public health, climate change mitigation, and socio-economic benefits (Gherghel et al., 2019).

A wide range of methods and technologies have been proposed for managing SS in accordance with sustainability and circular economy principles. These approaches vary in their level of technological maturity, with some nearing large-scale demonstration while others are still in the developmental phase, not yet suitable for industrial application. This chapter delves into recent research and explores potential pathways for the sustainable treatment and utilization of SS, with a particular emphasis on novel sludge-to-energy and sludge-to-resource recovery methods. It discusses key factors and trends shaping future developments and assesses the benefits, limitations, and prospects of applicable technologies, paying specific attention to thermochemical conversion, energy and nutrient recovery, and sludge biorefinery.

15.2 Circular economy perspective

The challenges stemming from the ever-increasing population, excessive extraction and consumption of raw materials, and dwindling resources have motivated academia and industry to pursue solutions that benefit both human welfare and environmental preservation. Achieving sustainable development in waste management has become a paramount global objective. The EU, for instance, has set a target to halve final waste disposal by 2050 compared to levels in 2000. To achieve this, the EU has outlined a strategy with the following priorities (Smol et al., 2015):

(1) preventing waste,
(2) minimizing waste generation,
(3) recovering waste through reuse, recycling, and energy recovery,
(4) enhancing waste treatment conditions, and
(5) regulating waste transport.

The EU emphasized the need for more efficient waste utilization through its Zero Waste Programme for Europe. This initiative highlights the transition to a circular economy as a crucial pathway to achieving the resource efficiency objectives set out in the Europe 2020 Strategy, which aims for smart, sustainable, and inclusive growth (Gherghel et al., 2019). As a fundamental pillar of the European Green Deal, the European Commission introduced the new circular economy action plan (CEAP) in 2020 (Directorate-General for Communication European Commission, 2020). Additionally, directives on sewage sludge (1986) and urban wastewater treatment (1991/2022) serve as cornerstones of the EU's legal framework for managing water resources (Cieślik et al., 2018). With primary objectives focused on environmental and human health protection, alongside fostering economic growth and job creation, these directives laid the groundwork for ongoing endeavors to integrate circular economy principles into the water-sewage-sludge sector.

The fundamental principle of the CE is to preserve materials within the economy at their highest value for as long as possible. Biological materials are directed back into natural metabolic cycles following necessary pre-treatment, such as composting or digestion. The concept of circular economy is visualized in Fig. 15.1. Unavoidable waste is directed toward energy recovery rather than landfilling. Furthermore, incorporating secondary raw materials into the economy is viewed as beneficial as it enhances supply security, reducing risks associated with fluctuating raw material prices and unstable supply resulting from abrupt natural disasters or geopolitical shifts (Tsybina and Wuensch, 2018).

The necessity for adequate SS treatment is no longer discretionary or solely an ethical concern. A properly executed sludge treatment approach has the potential to alleviate the negative impacts of the conventional linear economy, also known as the "take-make-dispose" economy, and pave the way toward a CE characterized by "reduce-reuse-recycle" principles. Given its rich nutrient and organic content, as well as energy equivalence to

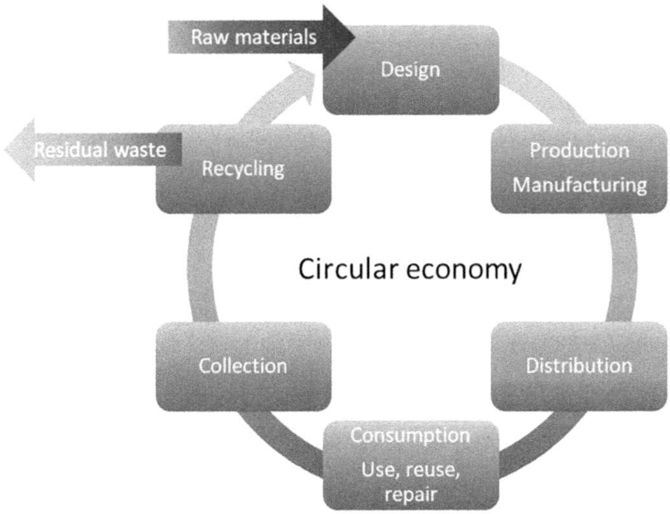

Fig. 15.1 The concept of circular economy.

woody biomass, SS emerges as a valuable secondary resource (Tsybina and Wuensch, 2018; Gherghel et al., 2019). However, the presence of hazardous substances in sludge, coupled with the evolving perspective on sludge as a resource, complicates its current treatment, utilization, and final disposal. Conventional methods such as landfill deposition, agricultural reuse, and land reclamation not only contribute to environmental pollution but also fail to fully capitalize on the opportunity to recover energy and nutrients from the sludge.

In 2017, the European Commission published a communication highlighting the importance of energy recovery from waste, particularly in the context of the circular economy (European Commission, 2017). This document emphasized technologies and procedures with the greatest potential for optimizing energy and material outputs. Thermochemical treatment methods align well with several criteria for managing SS within the circular economy framework. Thermochemical processing allows for energy recovery from sludge, elimination of harmful pollutants, conservation of natural resources such as land, and the acquisition of valuable by-products that can be used as raw materials or marketed in various industries (Jiang et al., 2021). However, it is important to note that waste incineration with limited energy recovery is classified as "disposal" rather than "recovery," unlike waste incineration with a high level of energy recovery. Therefore, as technologies evolve and new ones emerge, it is essential to evaluate not only their potential benefits and limitations, but also their compliance with these specific criteria.

15.3 Sewage sludge characteristics

Sustainability and circular economy principles emphasize the importance of efficiently harnessing the inherent resources in SS beyond conventional approaches. However, the typically high moisture content in sludge often results in increased energy consumption and recovery costs. The presence of diverse organic and inorganic pollutants, including pathogens, pharmaceuticals, microplastics, and heavy metals (e.g., Zn, Pb, Cu, Cr, Ni, Cd, Hg, and As), as well as persistent organic pollutants (POPs), poses obstacles to onward recycling (Corradini et al., 2019; Fijalkowski et al., 2017; Ivanová et al., 2018). Research into emerging contaminants has unveiled potential impacts on agricultural ecosystems, affecting everything from soil quality to human health. Managing these risks associated with pollutants demands specific treatments and restrictions on the use of untreated sludge in soil applications.

The composition and energy content of SS depend on wastewater characteristics (municipal or industrial) and the type of sludge stream (raw, digested, or waste-activated). Dried SS generally contains 21–48% volatile organics, resulting in a substantial energy content ranging from 11 to 20 MJ/kg (Shizas and Bagley, 2004). As indicated in Table 15.1, this energy equivalence places it on par with lignite and certain renewable biomass materials commonly used as fuels. The volatile matter can be converted into heat, gaseous, and/or liquid fuel products, with the carbon concentration and distribution determining the energy content of these products. A high ash content in SS is generally unfavorable, as it reduces energy conversion efficiency and may cause equipment issues.

Apart from energy, other recoverable resources include solid, liquid, and gaseous end-products such as biochar, biodiesel, and synthesis gas (syngas). Additionally, SS contains significant quantities of micro- and macronutrients, such as phosphorus (0.8–4.0% DS), nitrogen (1.5–6.0% DS), and potassium (0.1–3.0% DS), which are essential for the high crop yields characteristic of modern agriculture (Kacprzak et al., 2017). Furthermore, recent studies have explored the novel biorefinery concept, with an aim to transform SS into high-value products like polyhydroxyalkanoates (PHAs) and extracellular polymeric substances (EPS), presenting promising opportunities across multiple industries (Crutchik et al., 2020; Li et al., 2021).

Table 15.1 Calorific values of sewage sludge, coal, and biomass fuels (Syed-Hassan et al., 2017).

Fuel	HHV, dry basis (kJ/kg)
Bituminous coal	25.40–33.15
Sub-bituminous coal	20.10–29.00
Lignite	11.80–21.90
Sawdust	18.13–20.40
Wood pellet	18.30–19.60
Rice straw	14.95–15.86
Olive stone	17.54
Walnut shell	17.18
Hazelnut shell	18.40
Hazelnut husk	17.14
Chinese fir	18.38
Camphor	18.40
Switchgrass	19.60
Soybean stalk	17.50
Wheat straw	16.00
Sugarcane bagasse	17.05
Miscanthus	18.73
Sewage sludge	11.10–22.10

15.4 Resource recovery from sewage sludge

The most commonly used method in sludge treatment, AD, provides a route for partial energy recovery. AD involves the conversion of organic materials into biogas, which is primarily composed of CH_4 (60–70%) and CO_2 (30–40%), with a calorific value typically falling between 13 and 20 MJ/m^3 (Raheem et al., 2018). This conversion occurs through a batch-mode biochemical process, conducted in an oxygen-free environment and under specific temperature conditions, either mesophilic (30–40 °C) or thermophilic (49–57 °C). The biogas produced from sludge can be used for various purposes, including energy generation (both heat and electricity), injection into gas networks, or as fuel for vehicles.

The digestion process requires a long residence time (10–30 days), and only a portion (30–50%) of the organic matter undergoes conversion into biogas, with the remainder entering the residual digestate along with nutrients and potentially harmful substances that remain unaltered at the bioreactor temperature (Syed-Hassan et al., 2017). While numerous pre-, intermediate, and post-treatment technologies have been suggested to boost this conversion and elevate methane yield in biogas (Zhou et al., 2024), AD alone does not provide a comprehensive solution for sludge management. Given the need for additional processing to address digestate disposal, AD can be seen as an initial step in integrated sludge processing schemes aimed at maximizing resource recovery (Cecconet and Capodaglio, 2022).

15.4.1 Thermochemical treatment

Increasing apprehensions and more stringent regulations on conventional disposal practices have redirected considerable attention toward the energy potential of SS, giving rise to the sludge-to-energy concept (Oladejo et al., 2019). Thermochemical methods have proven effective in reducing biological materials and hazardous waste, with SS identified as a valuable feedstock for material and energy recovery. Thermochemical processing aims to minimize sludge volume and toxicity while simultaneously recovering its energy content and other resources. Advanced energy recovery technologies have shown promise in achieving lower global warming potential (GWP) compared to traditional sludge treatment methods, thus potentially contributing to global climate change mitigation efforts (Havukainen et al., 2022; Teoh and Li, 2020).

Thermochemical methods are utilized either to substantially reduce the solids content of sludge or as a pre-treatment before AD to enhance the biodegradability of its organic carbon. Similar to stabilization and drying processes, thermochemical treatment at lower temperatures yields a stable sludge product. When operated at higher temperatures for solids destruction, rather than as a pre-treatment for AD, these methods facilitate energy recovery in the form of heat or combustible gas/liquid. As depicted in Fig. 15.2, thermochemical processes are classified into two types: (i) non-oxidative processes conducted without oxygen or with substoichiometric amounts, and (ii) oxidative processes involving excessive amounts of oxygen. These methods vary in operational conditions, required feed sludge water content, and the characteristics of the resulting products (Oladejo et al., 2019).

The three main thermochemical conversion methods are pyrolysis, gasification, and incineration. Fig. 15.3 summarizes their primary products and potential applications. Incineration aims to convert the chemical energy stored in SS directly into heat, serving as a primary source of thermal energy, such as for

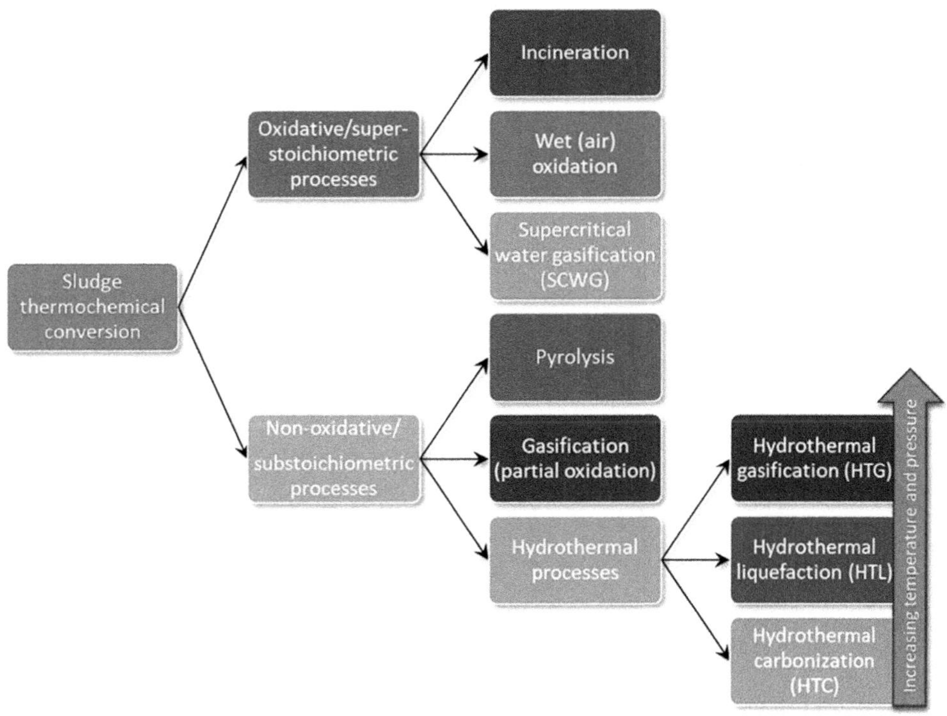

Fig. 15.2 Classification of thermochemical conversion processes.

generating electricity using steam turbines. Gasification and pyrolysis yield energy-rich end products suitable for use as fuels, providing increased adaptability in energy systems compared to incineration. These processes boast superior conversion efficiencies and operate significantly faster than AD, with processing times measured in minutes or seconds rather than days (Capodaglio and Callegari, 2023). Incineration serves currently as the predominant thermal treatment method for SS, applied at full scale in densely populated urban areas, particularly in countries like Switzerland, Germany, and the Netherlands. It is also witnessing rapid adoption across the EU, with utilization rates climbing from 19% in 2005 to 26.9% in 2010, and projected to reach 32% by 2030 (Đurđević et al., 2019). Gasification and pyrolysis are still in early developmental stages, necessitating further research to ensure both economic and environmental viability throughout the entire process, from the initial drying of mechanically dewatered sludge to the final treatment of the fluid stream.

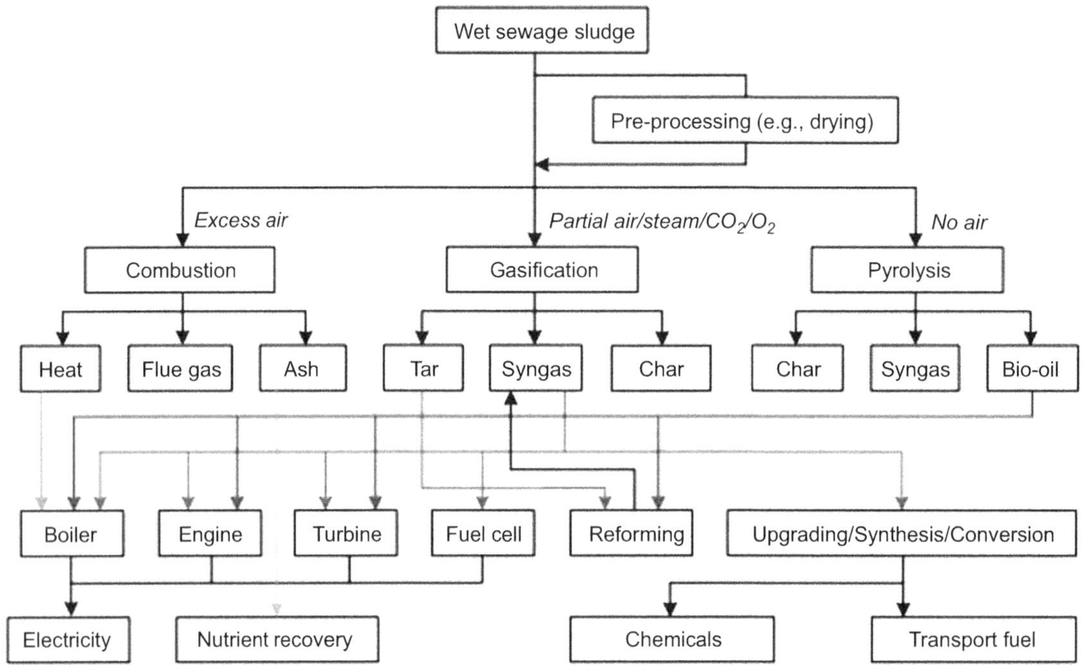

Fig. 15.3 Primary routes for thermochemical conversion of SS, their products and potential applications. Reproduced with permission from Syed-Hassan, S.S.A., Wang, Y., Hu, S., Su, S., Xiang, J., 2017. Thermochemical processing of sewage sludge to energy and fuel: Fundamentals, challenges and considerations. Renew. Sust. Energ. Rev. 80, 888–913. https://doi.org/10.1016/j.rser.2017.05.262.

Thermochemical treatment demands advanced equipment and operations but it promises to deliver superior economic performance, efficiency, and volume reduction compared to competing methods. One of the main challenges is the necessity for preliminary sludge drying to enhance the energy balance, whereas processes such as AD only require thickening as a primary pretreatment. The energy needed for sludge water removal, at 2260 kJ/kg H_2O, profoundly influences overall process performance. There is presently a growing demand for scientific research aimed at addressing efficiency issues and determining optimal technological parameters for thermal processes. Currently, the industry is witnessing the emergence of innovative solutions to tackle the many challenges and constraints associated with existing technologies. By employing advanced techniques and integrated process schemes, these novel approaches aim to maximize the value derived from SS and showcase the potential of thermal treatment on a broader scale, extending beyond individual economic entities or waste recycling plants (Peltola et al., 2023).

15.4.1.1 Drying

Raw and digested sludge are unsuitable as fuels due to their excessive water content, typically ranging between 94 and 97 wt % and 90–95 wt%, respectively. Mechanical dewatering methods, such as chamber filter presses, decanters, or centrifuges, are essential to decrease the water content in SS to around 70–80 wt%, which is a prerequisite for post-treatment in thermal processes (Schnell et al., 2020). However, to achieve autothermal operation in the incineration process (i.e., sustained by own combustion heat), an additional thermal drying stage involving simultaneous heat and mass transfer is required. This stage further reduces the sludge's water content to 10–30 wt%, thus enhancing its calorific value (Ma et al., 2016). Thermal drying processes are categorized into conduction drying, convection drying, radiation drying, and mixed forms, based on the mechanisms of heat transfer. Heat can be conveyed to the sludge via a heated surface in indirect contact with a heat transfer medium (such as steam, exhaust gases, hot water, or thermal oils), through direct contact with a heat transfer medium (such as steam, flue gas, or other hot gases), or through microwave or infrared radiation (Rao et al., 2022).

Disc and thin film dryers are frequently utilized for both complete and partial drying of SS, employing conductive heat transfer from hot liquid/gas through a wall. In belt dryers, SS is conveyed on a perforated belt and dried by hot gas flowing around and through the material. Drum dryers, which resemble rotary kilns, feature a cylindrical vessel where drying primarily occurs through convection with hot gas and sometimes through conduction via contact with the heated casing. Fluidized bed dryers combine convection and conduction, introducing heat both by hot gas used for fluidization and by steam- or water-heated heat exchangers. Direct solar drying, a method where sludge is spread over large areas in glasshouses and dried by sunlight (solar radiation), is well-suited for modest throughputs, making it particularly suitable for small WWTPs (Schnell et al., 2020).

Target water contents and techniques applied vary greatly and are specific to the technology employed. Nevertheless, due to its energy-intensive nature, thermal drying requires a significant amount of energy, typically sourced from burning natural gas, digester gas (biogas), fuel oil, or coal. Improving the efficiency of SS drying can be achieved by recovering waste heat from the drying process itself. Recent studies have introduced advanced process configurations that integrate drying with incineration and incorporate both direct and indirect drying mechanisms

(Peltola et al., 2023; Li et al., 2014). These methods leverage comprehensive thermal integration and efficient waste heat utilization, thereby reducing the dependence on auxiliary fuels and, in some cases, eliminating the need for external energy sources in the drying process. Alternatively, utilizing flue gas from a nearby thermal power plant, typically discharged at temperatures between 160 and 200 °C, can serve as a cost-efficient heating solution (Ma et al., 2012).

15.4.1.2 Incineration

Incineration, performed at high temperatures (850–1200 °C) with excess oxygen, converts carbonaceous materials into CO_2 and H_2O, producing heat, flue gas, and solid residue (ash) (Quan et al., 2022). The process is illustrated in Fig. 15.4. The organic content in SS, including organic contaminants, decompose during the process, resulting in a reduction of sludge volume by over 90%. Inorganic heavy metals remain largely unaffected and accumulate in the ash. Consequently, the resulting ash is typically disposed of in landfills or utilized in the production of concrete, cement, bricks, and asphalt (Smol et al., 2015). Moreover,

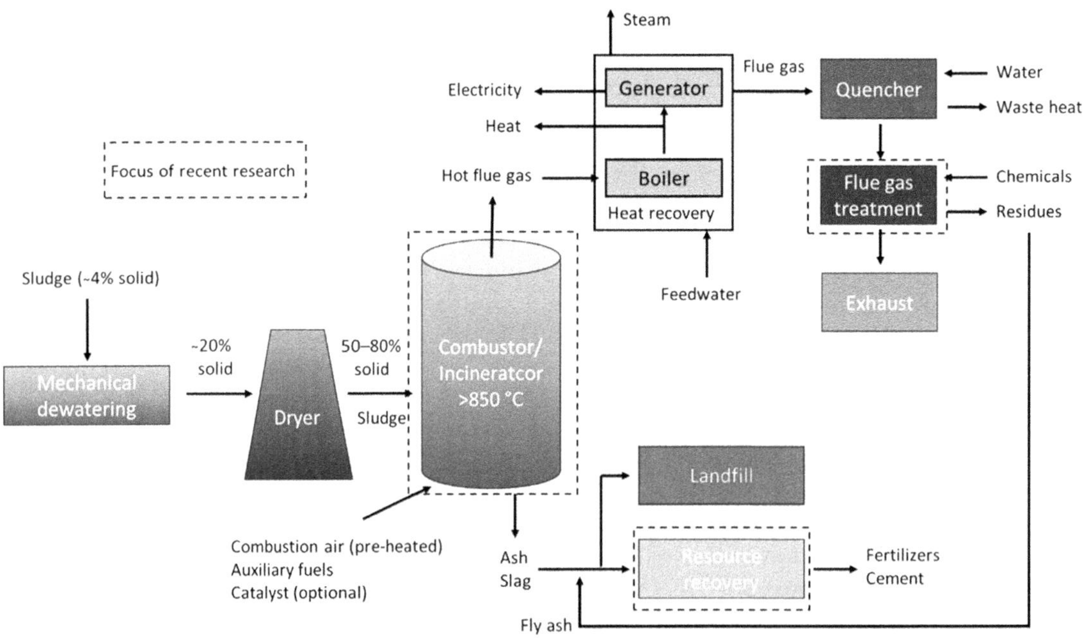

Fig. 15.4 Schematics of sludge incineration. Modified with permission from Oladejo, J., Shi, K., Luo, X., Yang, G., Wu, T., 2019. A review of sludge-to-energy recovery methods. Energies 12 (1), 60, https://doi.org/10.3390/en12010060.

incinerated SS ash has been recognized for its potential as a nutrient source, particularly phosphorus, prompting exploration into diverse technologies aimed at reclaiming these valuable nutrients for sustainable fertilizer production (Gorazda et al., 2017).

SS can be subjected to incineration in various facilities such as mono-incineration plants, thermal power plants, cogeneration (combined heat and power) plants, waste-to-energy plants, and cement kilns (Jiang et al., 2021). This method is increasingly favored to harness energy in the form of steam, which is then converted into power via steam turbines. The surplus electricity can be conveniently supplied to national grids, while heat recovery relies on the availability of local heat sinks, such as district heating systems or industrial processes. Co-incineration of sludge with other fuels like coal, biomass, or municipal solid waste leverages existing power plants and waste-treatment facilities, thereby avoiding the expense of new construction. While co-incineration may outperform mono-incineration under certain conditions, it does not facilitate the recycling of incinerated SS ash for fertilizer production due to the dilution effect caused by the co-incinerated fuels (Peccia and Westerhoff, 2015).

The two primary combustors utilized for SS incineration are multiple hearth and fluidized-bed furnaces, although rotary kilns, grate furnaces, and several types of smelting furnaces are also employed. In industrial settings, the circulating fluidized bed (CFB) combustion technology stands out for its widespread adoption due to numerous advantages over alternative configurations. These advantages include extensive fuel flexibility, robust turbulence facilitating intimate mixing of gaseous and solid reactants, temperature uniformity within the furnace, and comparatively low pollutant emissions (Werther and Ogada, 1999; Van Caneghem et al., 2012). Moreover, with the potential for autothermal operation, combustion efficiencies in CFB incinerators approach 100%.

While incineration is less demanding regarding the dry matter content of SS compared to gasification and pyrolysis, achieving autothermal combustion requires a minimum heating value of approx. 4–5 MJ/kg in wet substance. If the heating value is too low, supplementary fuel can be used, with dried sludge requiring less additional fuel. In larger waste-to-energy facilities, over 70% of the energy can typically be reclaimed, with the remainder lost through off-gas and ash discharge, as well as heat losses via radiation, convection, and conduction (Brunner and Rechberger, 2015). However, economic constraints often limit the feasibility of steam generation and electricity production. When external heat sinks are unavailable, heat recovery from flue gases can be utilized for sludge drying and pre-heating combustion air, thereby reducing the consumption of auxiliary fuel during SS incineration.

Incineration effectively minimizes biohazard risks by eliminating organic and biological contaminants within SS. However, SS incinerators can release significant quantities of air pollutants in flue gases, including NO_X, SO_X, HCl, HF, unburned hydrocarbons, and particulate matter (dust), as well as trace levels of polycyclic aromatic hydrocarbons (PAHs), polychlorinated dioxins and furans (PCDD/Fs), and volatile heavy metals such as Hg, Cd, and Pb, all of which pose serious environmental concerns (Liang et al., 2021). The formation of these pollutants is influenced by operational conditions and sludge properties. Generally, sludge incineration plants must be equipped with advanced emission control and flue gas purification systems, such as bag filters, electrostatic precipitators (ESPs), and denitrification ($DeNO_X$) and desulfurization ($DeSO_X$) units (Liew et al., 2021). Heavy metals, primarily concentrated in residual ash, mainly originate from industrial wastewater and stormwater (precipitation water), with a lesser contribution from household wastewater. Nevertheless, the heavy metal content in SS has decreased in recent decades due to the implementation of separate sewage systems for stormwater and stricter environmental regulations (Fijalkowski et al., 2017).

15.4.1.3 Pyrolysis

Pyrolysis is an endothermic decomposition process that takes place in the absence of oxygen at moderate to high temperatures (400–1200 °C), transforming various types of SS into valuable products (Fig. 15.5). This process reduces the sludge volume by over 50% but requires a moisture content of less than 10 wt%. During pyrolysis, the organic components in the sludge decompose into three main products: liquid pyrolytic oil (bio-oil), solid biochar, and non-condensable gas (syngas) (Barry et al., 2019). Commonly, the non-condensable gas stream, composed primarily of CO, CO_2, CH_4, and H_2, is combusted to supply heat for the endothermic pyrolysis process. Bio-oil is considered a promising energy source, suitable for powering boilers, combustion engines, and turbines, with an energy content of approximately 33 MJ/kg, notably surpassing the energy density of biogas obtained from AD (Syed-Hassan et al., 2017). Alternatively, bio-oil can be upgraded and refined into specialty chemicals. Biochar has demonstrated potential benefits as a phosphorus-rich soil amendment, a carbon-neutral fuel, a cost-effective adsorbent, and a substitute

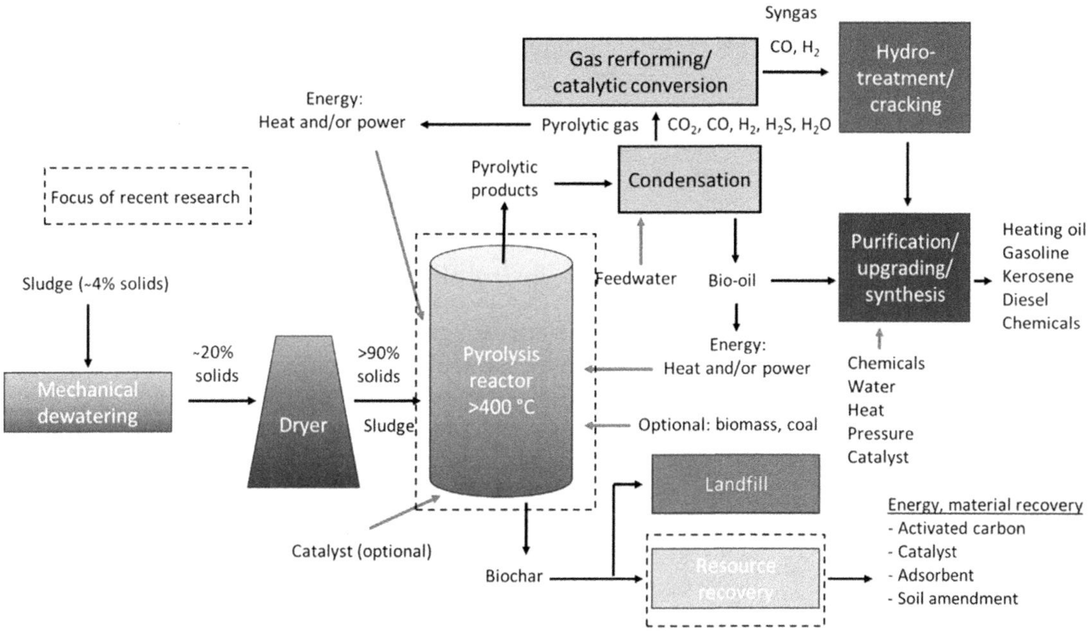

Fig. 15.5 Schematics of sludge pyrolysis. Modified with permission from Oladejo, J., Shi, K., Luo, X., Yang, G., Wu, T., 2019. A review of sludge-to-energy recovery methods. Energies 12 (1), 60, https://doi.org/10.3390/en12010060.

for carbon black, among other applications aimed at promoting environmental remediation (Haghighat et al., 2020).

The distribution and physicochemical properties of pyrolysis products depend on several factors, including reaction temperature, heating rate, residence time, and feedstock particle size. By controlling these parameters, it is possible to prioritize the production of one product over the others (Barry et al., 2019). Accordingly, pyrolysis can be categorized into three subclasses—slow pyrolysis, fast pyrolysis, and flash pyrolysis—each with unique benefits and limitations. Slow pyrolysis typically occurs at low heating rates (0.1–1.0 °C/s), long residence times (300–550 s), and low temperatures (400–500 °C), favoring the production of biochar but negatively affecting the yield and quality of bio-oil. As the pyrolysis temperature increases, biochar yield decreases, along with the energy retained in the biochar, while the carbon content, nutrient availability, and stability of heavy metals in the biochar increase. Biochar generated at lower temperatures is well-suited for agricultural purposes, although its usage may be limited by the presence of heavy metals in accordance with local regulations. On the other hand, higher temperatures can enhance the porosity of biochar, thereby improving its capacity to adsorb soil contaminants (Agrafioti et al., 2013).

Fast pyrolysis takes place at high temperatures (850–1250 °C) with rapid heating rates (10–200 °C/s), short residence times (1–10 s), and fine particles. In this process, the feedstock undergoes rapid heating and decomposition into short-chain molecules and gases, which are swiftly cooled and converted into pyrolytic oil. Fast pyrolysis is primarily utilized for bio-oil production, as the yield of oil surpasses that of char and gaseous products. Flash pyrolysis, an enhanced and modified version of fast pyrolysis, is carried out at similarly high temperatures (900–1200 °C) but with extremely rapid heating rates (>1000 °C/s), very short vapor residence times (<1 s), and very fine particles. Flash pyrolysis achieves a high biomass-to-liquid ratio and maximizes the production of gases and oils at the expense of char production (Haghighat et al., 2020).

Numerous reactor configurations have been proposed for SS pyrolysis, including fluidized bed reactors, spouted bed reactors, fixed bed reactors, batch and semi-batch reactors, microwave-assisted reactors, rotary kilns, and new innovative concepts like plasma and solar reactors. Each reactor type offers distinct advantages and limitations in terms of efficiency, scalability, and controllability. Fluidized bed reactors, for instance, are widely favored for fast and flash pyrolysis due to their superior heat transfer rates, precise temperature control, and operational flexibility. Rotary kilns, with their straightforward operation and robust design capable of handling diverse feedstocks, are considered particularly suitable for slow pyrolysis. However, practical experience with these applications is largely limited to laboratory and pilot-scale units. Pyrolysis treatment of SS currently stands at TRL (Technical Readiness Level) 5, with technical and operational challenges such as low energy conversion efficiency, limited capacity, inadequate heat transfer, uneven heat distribution, and inconsistencies in product quality. These challenges present significant obstacles to achieving economically viable pyrolysis at an industrial scale (Gao et al., 2020).

While operationally more complex than incineration, pyrolysis offers a promising route for transforming SS into transportable fuel, namely bio-oil. Unlike incineration, pyrolysis offers flexibility in energy systems by enabling the conversion of bio-oil into electrical power without on-site constraints. Bio-oil is recognized as a renewable source with lower emissions of nitrogen oxides (NO_X) and sulfur oxides (SO_X) compared to fossil fuels. However, bio-oil possesses some undesirable properties such as high water content, high viscosity, poor ignition characteristics, and corrosiveness, necessitating upgrading before effective use as fuel (Haghighat et al., 2020). Alternatively, the end-products may find utility in the chemical industry, expanding the versatility of pyrolysis-derived materials. Nonetheless, the reduced pollutant load in the outlet gas stream and the overall positive social, economic, and environmental outlook of pyrolysis as a potential zero-waste technology have sparked significant research activity in this area.

15.4.1.4 Gasification

Gasification transforms SS in an oxygen-restricted environment at elevated temperatures (700–1000 °C), producing a high yield of energy-rich product gas (syngas) through a series of chemical reactions. As illustrated in Fig. 15.6, this process occurs in the presence of a gasifying medium, which may include steam, air, oxygen, carbon dioxide, or their combinations. The resulting syngas primarily contains H_2, CO, CO_2, and CH_4, and is utilized as fuel for gas burners, cogeneration systems, gas turbines, and internal combustion engines. Additionally, it can be employed in the synthesis of Fischer–Tropsch fuels and other chemicals such as methanol and dimethyl ether (DME). Although the process shares similarities with combustion, it differs in terms of reduced tolerance for

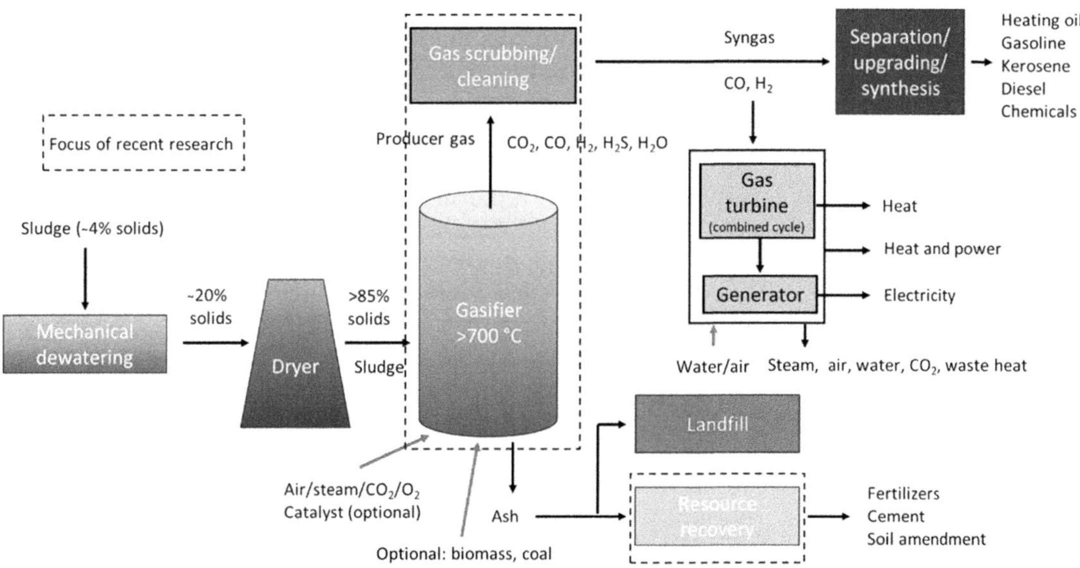

Fig. 15.6 Schematics of sludge gasification. Modified with permission from Oladejo, J., Shi, K., Luo, X., Yang, G., Wu, T., 2019. A review of sludge-to-energy recovery methods. Energies 12 (1), 60, https://doi.org/10.3390/en12010060.

feedstock moisture (<15 wt%) and the requirement for oxygen deficiency (Quan et al., 2022; Gao et al., 2020).

The chemical composition of the producer gas and its energy content are influenced by the gasification agent, process operating conditions, and feed characteristics (type, dry solids, volatile solids). Given the variability in SS properties, the distribution of products (syngas and char) and the quantities of tar and other pollutants can vary significantly. Compared to other fuels like biogas or natural gas, the calorific value of syngas is typically three to five times lower, ranging between 4 and 12 MJ/kg (Syed-Hassan et al., 2017). Besides reactor temperature, pressure, residence time, and catalytic constituents, a critical operating parameter in gasification is the equivalence ratio (ER). The ER, representing the air-to-solids ratio, affects gasifier temperature as the partial oxidation reaction is exothermic. ER values between 0.2 and 0.35 have been identified as optimal for SS gasification performance. Lower ER values result in incomplete gasification, yielding high tar and char fractions, while higher values tend to promote oxidation to CO_2 and H_2O at the expense of the syngas product (Gao et al., 2020).

Gasification utilizes reactor technologies similar to those used in pyrolysis, including fixed bed reactors, fluidized bed reactors, auger reactors, and rotary kilns among others. Fixed bed reactors are classified as either downdraft or updraft gasifiers depending on flow direction. Downdraft gasifiers typically produce high-quality syngas with lower tar contamination compared to updraft gasifiers. Certain operational challenges such as slag accumulation and increased pressure drop have been observed within fixed bed reactors (Quan et al., 2022). Bubbling fluidized bed (BFB), CFB, and dual fluidized bed (DFB) gasifiers are renowned for their high efficiency, tar removal capability, and syngas quality. However, they may face issues related to operational complexity, controllability, and high maintenance costs. Thermal plasma reactors have emerged as particularly promising for gasifying wet SS and mitigating pollutant gases at elevated temperatures. On the other hand, their industrial-scale implementation may be impeded by significant input energy requirements.

The char produced from SS gasification has potential as a resource, but the presence and stability of toxic metals within the char are major concerns. Liquid tar formation is problematic, causing clogging, fouling, and other issues in downstream processing. Removing tar from syngas or preventing its formation through catalysis or effective pre-drying of the sludge feed increases overall costs. Additionally, inorganic compounds in SS can cause sintering, leading to agglomeration and clinker formation, which necessitates more frequent reactor shutdowns and maintenance (Oladejo et al., 2019). These challenges may be addressed through the application of optimal operating conditions, appropriate catalysts, suitable reactor systems, and co-gasification with other fuels. Although gasification is expected to yield significant energy recovery from SS, current systems are at early stages of implementation, typically at TRL 5–6, and will require continued development and validation over time (Gao et al., 2020).

15.4.1.5 Alternative thermochemical processes

Torrefaction, also referred to as mild pyrolysis, is a thermal pretreatment method conducted within the temperature range of 200–300 °C under an inert atmosphere and at atmospheric pressure. During this process, water, carbon dioxide, carbon monoxide, and lightweight organic compounds are released, resulting in a solid carbon-rich product with lower moisture content, higher energy density, lower oxygen-to-carbon (O/C) ratio, and higher calorific value. When combined with drying and pelletization processes, torrefaction serves as an effective preparatory step before subsequent treatments (Syed-Hassan et al., 2017).

Hydrothermal processes (HTPs) for treating SS utilize elevated temperatures and pressures in wet conditions and without oxygen. Because HTPs can directly use water as a reaction medium, there is no need for moisture removal before processing, thus significantly reducing the energy usage of HTP compared with other thermal techniques. These processes for converting organic carbon can be categorized into three main regions of the water pressure–temperature phase diagram, based on the primary product targeted:

- hydrothermal carbonization (HTC), which predominantly yields a solid phase product,
- hydrothermal liquefaction (HTL), which predominantly yields a liquid phase product, and
- hydrothermal gasification (HTG), which primarily produces a gas phase product.

In HTC, sludge undergoes treatment under high pressure (>10–20 bar), typically at temperatures ranging between 225 and 250 °C, and for several hours in a liquid aqueous phase. This process initiates the degradation of organic constituents, followed by their subsequent polymerization through a series of reactions, resulting in the formation of a carbon-rich residue (Cebi et al., 2022). While various approaches exist for the technical implementation of HTC, the current state of development primarily involves experimental and pilot plants. The product is contaminated with heavy metals and is unsuitable for use as fertilizer or for landfilling. Consequently, HTC is not classified as a disposal method. Instead, it represents an alternative to hygienization and, depending on the successful resolution of any serious operational issues, could potentially replace thermal drying before incineration and other thermochemical processes (Schnell et al., 2020). HTL generally occurs at temperatures between 250 and 400 °C and pressures exceeding 40 bar. HTG is the highest-temperature hydrothermal process, reaching temperatures up to about 550 °C. As these temperatures surpass the critical point of water, HTG is sometimes referred to as supercritical water gasification.

Supercritical water treatment methods are categorized into supercritical water gasification (SCWG), supercritical water partial oxidation (SCWPO), and supercritical water oxidation (SCWO), depending on the oxidant quantity used. Treatment of SS in supercritical water, where both temperature and pressure surpass critical values ($T > 374$ °C, $P > 22.1$ MPa), offers advantages such as eliminating the pre-drying step required in conventional treatments and stabilizing heavy metals present in the sludge. SCWG or SCWPO processes yield H_2, CO, CO_2, CH_4, and a minor amount of C_2–C_4 gas, whereas SCWO is not able to produce any forms of biofuels. These treatments utilize both batch reactors and continuous reactors, but continuous reactors such as tubular reactors, fluidized bed reactors, and tank reactors are better suited for practical applications and have undergone pilot testing in small-scale operations (Qian et al., 2016).

15.4.2 Nutrient recovery

SS is abundant in vital plant nutrients, but it also contains heavy metal contaminants, poorly biodegradable trace organic compounds, and potentially harmful pathogens. Therefore, directly applying SS to agricultural soil is not a practical solution for nutrient recycling (Hušek et al., 2022). To safely and efficiently utilize these nutrients, alternative approaches are needed. These include advanced treatment methods to remove or neutralize contaminants and innovative techniques to extract and concentrate beneficial nutrients. Among other advantages, nutrient recovery from SS reduces dependence on conventional fossil-based fertilizers, conserves energy and raw materials, and enhances food security. Research on nutrient recycling from SS has primarily focused on phosphorus recovery, with less emphasis on other nutrients.

15.4.2.1 Phosphorus recovery

Phosphorus is an essential element for all living organisms and has no substitute in food production. As a natural resource that is unevenly distributed and continually depleting, phosphorus is classified as a critical raw material in the EU. Consequently, recovering phosphorus from waste streams, including SS, is considered a strategic priority (European Commission, 2020). Processes for phosphorus recovery from SS can be categorized into three main groups: (i) recovery from wastewater or process water, (ii) recovery directly from SS, and (iii) recovery from ash obtained after SS incineration. The phosphorus content varies across these sources, ranging from less than 0.001% in influent wastewater to around 0.8–4% in SS, and up to 5–11% in incinerated sewage sludge ash (ISSA) (Schnell et al., 2020).

During the thermochemical treatment of SS, phosphorus undergoes transformations into various forms within the final products. Currently, phosphorus recovery is primarily focused on ISSA produced from mono-incineration, as this ash typically has high phosphorus content and minimal contaminants, allowing for recovery rates of up to 90% or more. ISSA could be used directly as fertilizer on agricultural lands if heavy metal concentrations are within acceptable limits. However, if the heavy metal levels are too high, additional processing is required. While ash-based recovery is technologically more challenging, it offers greater rewards compared to other sources. For instance, wet-chemical methods recover only 5–30% of the phosphorus from untreated wastewater (Đurđević et al., 2022).

Recently, various technologies have been developed for recovering phosphorus from ISSA, including wet-chemical and thermochemical methods (Hušek et al., 2022). Wet-chemical processes involve adding acids or bases to ISSA to dissolve the biologically and chemically bound phosphorus into the liquid phase. The phosphorus is then recovered through precipitation or adsorption, resulting in the formation of phosphate compounds such as ammonium, calcium, sodium, iron, or aluminum phosphate, depending on the solvent used. These compounds are also present in mineral phosphate fertilizers (Đurđević et al., 2019). Another recovery method is to remove heavy metals from ashes using a thermochemical process. By adding salts, such as magnesium chlorine, heavy metals are transformed into chlorides at temperatures of approximately 1000 °C. Concurrently, organic pollutants are destroyed, and chemically bound phosphorus is reclaimed through the formation of solid mineral phosphate phases (Schnell et al., 2020).

Commercial adoption of phosphorus recovery has gained momentum in countries such as Germany, Japan, Canada, Sweden, and the Netherlands. Yet, despite its considerable potential, widespread adoption of phosphorus recovery from ISSA encounters a significant hurdle in the form of high production costs. Presently, the cost of phosphorus fertilizers derived from sewage sludge is up to 1.6–3 times higher than that of conventional mineral fertilizers (Liew et al., 2021). To foster broader implementation and commercialization, advancements in technological readiness, coupled with the provision of financial incentives, are essential.

15.4.2.2 Nitrogen recovery

WWTPs are mandated to remove nitrogen from wastewater to meet effluent concentration limits, but this process does not offer additional benefits beyond regulatory compliance. Unlike phosphorus, global nitrogen resources are abundant, virtually inexhaustible. However, the large-scale production of nitrogen fertilizers through the energy-intensive Haber–Bosch process, which relies on fossil gas, accounts for up to 2% of global energy consumption and results in significant greenhouse gas (GHG) emissions, raising major environmental concerns (Koskue et al., 2021). Additionally, disruptions in raw material supply have caused the price of nitrogen fertilizers to increase sixfold over the last few years, from 250 €/t of ammonium nitrate in 2021 to 1500 €/t in 2023 (Barry et al., 2019).

Wastewater streams entering WWTPs typically have high volumes but low concentrations, making them unsuitable for profitable resource recovery without a concentration step (Beckinghausen et al., 2020). However, nitrogen tends to accumulate in the activated sludge produced during the treatment process. During AD, this nitrogen is released back into the aqueous phase, and when the anaerobically digested sludge is mechanically dewatered, a nitrogen-rich liquid fraction (reject water) is produced, with NH_4-N concentrations reaching up to 1.5 g/L. Reject water is the most nitrogen-rich stream in a WWTP, containing 15–25% of the total nitrogen content but representing less than 5% of the total influent wastewater volume. The remaining solid fraction (SS) contains up to 6% (DS) nitrogen (Koskue et al., 2021).

Existing and emerging methods for recovering ammonia from liquid streams, such as reject water, include struvite precipitation, ammonia stripping, membrane separation, ion exchange- and adsorption-based processes, and bioelectrochemical systems.

However, many of these approaches lack the necessary energetic and cost analyses for upscaling and implementation in WWTPs (Beckinghausen et al., 2020). When SS is incinerated, the nitrogen in the sludge reacts to form either nitrogen oxides or elemental nitrogen, which are then released into the atmosphere, eliminating any possibility of recovery. Therefore, nitrogen recovery should be implemented before sludge incineration, which is often preceded by thermal drying to reduce the water content. During thermal drying, a significant portion of the nitrogen in SS is released into the drying fumes. Condensing these fumes produces a nitrogen-rich condensate that can be combined with reject water from mechanical dewatering to optimize nitrogen recovery (Saud et al., 2023).

15.4.3 Biorefinery

A biorefinery is a versatile approach that combines various established and innovative processes and technologies. Its purpose is to produce a wide range of value-added products, such as biofuels and biochemicals, from different biomass feedstocks, including plant-derived materials and organic waste. By replacing fossil energy sources and chemicals, biorefineries play a key role in closing the biomass production and consumption loop. Since biomass is renewable, biorefineries are also regarded as a viable solution for mitigating GHG emissions (Cecconet and Capodaglio, 2022).

Traditional methods for treating and disposing of sludge often entail high costs, energy demands, and adverse environmental impacts. The biorefinery concept seeks to revamp sludge treatment facilities to capitalize on potential value-generating streams and mitigate sludge disposal costs. Rather than solely focusing on waste volume reduction, these facilities can be meticulously designed to optimize resource recovery throughout the treatment process. Drawing inspiration from oil and petrochemical refineries, where processes are engineered for maximum efficiency and profitability, biorefineries can use SS as feedstock to produce a variety of valuable products. By paving the way for non-conventional resource recovery, the biorefinery concept has the potential to revolutionize sludge management, turning it from a necessary expense into a lucrative opportunity and fostering the implementation of an SS-based circular economy.

A biorefinery encompasses any technology used to process SS for biofuel and/or resource recovery, yielding a range of final products (energy and/or materials) through integrated processes. The design of an SS biorefinery can incorporate both thermochemical and biochemical processes. AD, which combines sludge stabilization with on-site biogas production, represents one of the earliest instances of resource recovery and remain an important step within many biorefinery schemes. Similarly, thermal treatment of SS for energy recovery results in diverse energy products, aligning with the principles of biorefinery. Additionally, innovative technologies have been proposed for extracting valuable substances from sludge, including biopolymers such as polyhydroxyalkanoates (PHAs) and extracellular polymeric substances (EPS), along with proteins, enzymes, biopesticides, and other compounds. Despite undergoing testing and validation at laboratory or pilot scales, these novel technologies are still distant from industrial application, necessitating further optimization in terms of operation and cost (Capodaglio, 2023; Battista et al., 2020).

An integrated biorefinery system focuses on recovery and reuse options instead of disposal pathways. Effective process integration can optimize the recovery potential of SS by generating diverse end products. This approach ensures adaptability and operational efficiency, accommodating potential variations in SS quantities, composition, and characteristics over time. In an ideal scenario, treatment processes and operational adjustments could be incorporated within WWTPs to enhance the properties of SS thus expanding the range of recovery options available. Nevertheless, any approach should emphasize simplifying the process rather than introducing multiple units that add unnecessary complexity (Capodaglio, 2023).

15.5 Conclusions and future trends

The increasing focus on sustainability and circular economy principles underscores the importance of efficiently harnessing the diverse resources in SS beyond conventional methods, unlocking significant economic and environmental benefits. Presently, the two components in SS deemed technically and economically viable for recycling are nutrients (primarily phosphorus and nitrogen) and energy (carbon). Both biological and thermochemical pathways exist for resource recovery from SS. Among these, AD stands out as a well-established biological method with relatively simple and low-cost operation, unaffected by the high water content in sludge. However, its application is constrained by challenges such as prolonged sludge retention times, poor conversion efficiencies, and adverse environmental effects resulting from pollutant emissions. Due to its limited energy recovery capacity and the need for additional treatment of the remaining digestate, AD is best viewed as the initial step in comprehensive sludge processing schemes aimed at fully exploiting the recovery potential of SS.

Thermochemical processes enable the rapid conversion of SS into a variety of energy-rich products for diverse applications. Sludge mono-incineration is a proven technology that allows for phosphorus recovery from the resulting ash, while alternative methods such as pyrolysis and gasification are still in their infancy. These methods offer numerous advantages over traditional approaches, including significant volume reduction, improved energy recovery, and the elimination of biohazards. However, they also face operational challenges and economic constraints, particularly with high-moisture feedstocks like SS. Besides the high costs associated with pre-treating SS to reduce moisture content, common issues include the need for conditioning and purifying flue gases in incineration, syngas in gasification, and the requirement to upgrade pyrolysis-derived bio-oil before its use as fuel. Nevertheless, ongoing efforts are focused on developing innovative solutions to address these limitations. Integrating AD with thermochemical processes stands as a potential strategy for beneficial sludge valorization, provided that process economics are enhanced through improved reactor design, optimized operating conditions, and reduced energy requirements.

Amidst a severe raw material crisis caused by global supply chain disruptions, the bioeconomy is becoming increasingly vital. An essential aspect of the bioeconomy lies in optimizing the utilization of scarce nutrient resources by recovering them from various waste streams. Given the substantial volume of SS generated in WWTPs, it holds promise as a significant raw material for producing organic-mineral phosphorus fertilizers sustainably, utilizing locally available renewable resources and minimizing carbon footprint. Direct phosphorus recovery from untreated wastewater or SS is less efficient compared to recovery from ISSA, primarily due to the need for processing larger material volumes with lower phosphorus concentrations. However, despite the existence of technologies capable of high-rate phosphorus extraction from ISSA, their broader implementation is hindered by excessive production costs. This underscores the imperative for further R&D efforts to enable widespread phosphorus recovery accompanied by financial benefits. Furthermore, recent challenges in nitrogen fertilizer production, triggered by increased raw material prices and limited access to natural gas, have redirected attention toward WWTPs as renewable biological nitrogen mines. This trend is anticipated to catalyze the development of new approaches and technologies for nitrogen recovery from different streams produced during wastewater treatment, including excess SS.

Apart from energy and nutrients, SS harbors a substantial amount of renewable organic matter with significant potential for recovery and economic use. Among the various methods for utilizing these resources, biorefinery is notable for its ability to combine both established and novel processes to enhance material recovery and minimize the environmental impact of sludge disposal. The diverse and sustainable products derived from biorefinery present new opportunities to reduce dependence on fossil fuels. However, while many of the proposed technologies have been validated in laboratory or pilot settings, they remain in the initial stages of industrial deployment. Future research efforts should prioritize achieving technical maturity and cost-effectiveness in the application of these technologies. Although progress has been gradual, the use of state-of-the-art technologies to valorize SS is starting to gain momentum. As a result, WWTPs are increasingly viewed not just as "pollution removal systems" but as "energy, nutrient, and material recovery plants".

References

Agrafioti, E., Bouras, G., Kalderis, D., Diamadopoulos, E., 2013. Biochar production by sewage sludge pyrolysis. J. Anal. Appl. Pyrolysis 101, 72–78. https://doi.org/10.1016/j.jaap.2013.02.010.

Anderson, N., Snaith, R., Madzharova, G., Bonfait, J., Doyle, L., Godley, A., 2021. Sewage Sludge and the Circular Economy.

Barry, D., Barbiero, C., Briens, C., Berruti, F., 2019. Pyrolysis as an economical and ecological treatment option for municipal sewage sludge. Biomass Bioenergy 122, 472–480. https://doi.org/10.1016/j.biombioe.2019.01.041.

Battista, F., Frison, N., Pavan, P., Cavinato, C., Gottardo, M., Fatone, F., Eusebi, A.L., Majone, M., Zeppilli, M., Valentino, F., Fino, D., Tommasi, T., Bolzonella, D., 2020. Food wastes and sewage sludge as feedstock for an urban biorefinery producing biofuels and added-value bioproducts. J. Chem. Technol. Biotechnol. 95 (2), 328–338. https://doi.org/10.1002/jctb.6096.

Beckinghausen, A., Odlare, M., Thorin, E., Schwede, S., 2020. From removal to recovery: an evaluation of nitrogen recovery techniques from wastewater. Appl. Energy 263, 114616. https://doi.org/10.1016/j.apenergy.2020.114616.

Brunner, P.H., Rechberger, H., 2015. Waste to energy—key element for sustainable waste management. Waste Manag. 37, 3–12. https://doi.org/10.1016/j.wasman.2014.02.003.

Capodaglio, A.G., 2023. Biorefinery of sewage sludge: overview of possible value-added products and applicable process technologies. Water 15 (6), 1195. https://doi.org/10.3390/w15061195.

Capodaglio, A.G., Callegari, A., 2023. Energy and resources recovery from excess sewage sludge: a holistic analysis of opportunities and strategies. Resour. Conserv. Recycl. Advances 19, 200184. https://doi.org/10.1016/j.rcradv.2023.200184.

Capodaglio, A., Olsson, G., 2020. Energy issues in sustainable urban wastewater management: use, demand reduction and recovery in the urban water cycle. Sustainability **12** (1), 266. https://doi.org/10.3390/su12010266.

Cebi, D., Celiktas, M.S., Sarptas, H., 2022. A review on sewage sludge valorization via hydrothermal carbonization and applications for circular economy. Circ. Econ. Sustain. 2 (4), 1345–1367. https://doi.org/10.1007/s43615-022-00148-y.

Cecconet, D., Capodaglio, A.G., 2022. Sewage sludge biorefinery for circular economy. Sustainability 14 (22), 14841. https://doi.org/10.3390/su142214841.

Cieślik, B.M., Świerczek, L., Konieczka, P., 2018. Analytical and legislative challenges of sewage sludge processing and management. Monatsh. Chem. 149 (9), 1635–1645. https://doi.org/10.1007/s00706-018-2255-2.

Corradini, F., Meza, P., Eguiluz, R., Casado, F., Huerta-Lwanga, E., Geissen, V., 2019. Evidence of microplastic accumulation in agricultural soils from sewage sludge disposal. Sci. Total Environ. 671, 411–420. https://doi.org/10.1016/j.scitotenv.2019.03.368.

Crutchik, D., Franchi, O., Caminos, L., Jeison, D., Belmonte, M., Pedrouso, A., del Rio, A.V., Mosquera-Corral, A., Campos, J.L., 2020. Polyhydroxyalkanoates (PHAs) production: a feasible economic option for the treatment of sewage sludge in municipal wastewater treatment plants? Water 12 (4), 1118. https://doi.org/10.3390/w12041118.

Directorate-General for Communication (European Commission), 2020. Circular Economy Action Plan—For a Cleaner and More Competitive Europe. https://doi.org/10.2779/05068.

Đurđević, D., Blecich, P., Jurić, Ž., 2019. Energy recovery from sewage sludge: the case study of croatia. Energies 12 (10), 1927. https://doi.org/10.3390/en12101927.

Đurđević, D., Žiković, S., Blecich, P., 2022. Sustainable sewage sludge management technologies selection based on techno-economic-environmental criteria: case study of croatia. Energies 15 (11), 3941. https://doi.org/10.3390/en15113941.

European Commission, 2017. The Role of Waste-to-Energy in the Circular Economy (Communication COM/2017/034). European Commission.

European Commission, 2020. Critical Raw Materials Resilience: Charting a Path Towards Greater Security and Sustainability (Communication COM/2020/474). European Commission.

Fijalkowski, K., Rorat, A., Grobelak, A., Kacprzak, M.J., 2017. The presence of contaminations in sewage sludge—the current situation. J. Environ. Manag. 203, 1126–1136. https://doi.org/10.1016/j.jenvman.2017.05.068.

Gao, N., Kamran, K., Quan, C., Williams, P.T., 2020. Thermochemical conversion of sewage sludge: a critical review. Prog. Energy Combust. Sci. 79. https://doi.org/10.1016/j.pecs.2020.100843.

Gherghel, A., Teodosiu, C., De Gisi, S., 2019. A review on wastewater sludge valorisation and its challenges in the context of circular economy. J. Clean. Prod. 228, 244–263. https://doi.org/10.1016/j.jclepro.2019.04.240.

Gorazda, K., Tarko, B., Wzorek, Z., Kominko, H., Nowak, A.K., Kulczycka, J., Henclik, A., Smol, M., 2017. Fertilisers production from ashes after sewage sludge combustion—a strategy towards sustainable development. Environ. Res. 154, 171–180. https://doi.org/10.1016/j.envres.2017.01.002.

Haghighat, M., Majidian, N., Hallajisani, A., Samipourgiri, M., 2020. Production of bio-oil from sewage sludge: a review on the thermal and catalytic conversion by pyrolysis. Sustain. Energy Technol. Assessments 42, 100870. https://doi.org/10.1016/j.seta.2020.100870.

Havukainen, J., Saud, A., Astrup, T.F., Peltola, P., Horttanainen, M., 2022. Environmental performance of dewatered sewage sludge digestate utilization based on life cycle assessment. Waste Manag. 137, 210–221. https://doi.org/10.1016/j.wasman.2021.11.005.

Hušek, M., Moško, J., Pohořelý, M., 2022. Sewage sludge treatment methods and P-recovery possibilities: current state-of-the-art. J. Environ. Manag. 315, 115090. https://doi.org/10.1016/j.jenvman.2022.115090.

Ivanová, L., Mackuľak, T., Grabic, R., Golovko, O., Koba, O., Staňová, A.V., Szabová, P., Grenčíková, A., Bodík, I., 2018. Pharmaceuticals and illicit drugs—a new threat to the application of sewage sludge in agriculture. Sci. Total Environ. 634, 606–615. https://doi.org/10.1016/j.scitotenv.2018.04.001.

Jiang, G., Donghai, X., Botian Hao, L., Liu, S.W., Zhiqiang, W., 2021. Thermochemical methods for the treatment of municipal sludge. J. Clean. Prod. 311, 127811. https://doi.org/10.1016/j.jclepro.2021.127811.

Kacprzak, M., Neczaj, E., Fijałkowski, K., Grobelak, A., Grosser, A., Worwag, M., Rorat, A., Brattebo, H., Almås, Å., Singh, B.R., 2017. Sewage sludge disposal strategies for sustainable development. Environ. Res. 156, 39–46. https://doi.org/10.1016/j.envres.2017.03.010.

Koskue, V., Rinta-Kanto, J.M., Freguia, S., Ledezma, P., Kokko, M., 2021. Optimising nitrogen recovery from reject water in a 3-chamber bioelectroconcentration cell. Sep. Purif. Technol. 264, 118428. https://doi.org/10.1016/j.seppur.2021.118428.

LeBlanc, J., Matthews, P., Richard, R., 2008. Global Atlas of Excreta, Wastewater Sludge, and Biosolids Management: Moving Forward the Sustainable and Welcome Uses of a Global Resource. UN-Habitat.

Li, S., Li, Y., Lu, Q., Zhu, J., Yao, Y., Bao, S., 2014. Integrated drying and incineration of wet sewage sludge in combined bubbling and circulating fluidized bed units. Waste Manag. 34 (12), 2561–2566. https://doi.org/10.1016/j.wasman.2014.08.018.

Li, J., Hao, X., Gan, W., van Loosdrecht, M.C.M., Yuanyuan, W., 2021. Recovery of extracellular biopolymers from conventional activated sludge: potential, characteristics and limitation. Water Res. 205, 117706. https://doi.org/10.1016/j.watres.2021.117706.

Liang, Y., Xu, D., Feng, P., Hao, B., Guo, Y., Wang, S., 2021. Municipal sewage sludge incineration and its air pollution control. J. Clean. Prod. 295, 126456. https://doi.org/10.1016/j.jclepro.2021.126456.

Liew, C.S., Kiatkittipong, W., Lim, J.W., Lam, M.K., Ho, Y.C., Ho, C.D., Ntwampe, S.K.O., Mohamad, M., Usman, A., 2021. Stabilization of heavy metals loaded sewage sludge: reviewing conventional to state-of-the-art thermal treatments in achieving energy sustainability. Chemosphere, 277, 130310. https://doi.org/10.1016/j.chemosphere.2021.130310.

Ma, X.W., Weng, H.X., Su, M.H., Pan, L., 2012. Drying sewage sludge using flue gas from power plants in China. Environ. Earth Sci. 65 (6), 1841–1846. https://doi.org/10.1007/s12665-011-1166-x.

Ma, J., Zhang, L., Li, A., 2016. Energy-efficient co-biodrying of dewatered sludge and food waste: synergistic enhancement and variables investigation. Waste Manag. 56, 411–422. https://doi.org/10.1016/j.wasman.2016.06.007.

Oladejo, J., Shi, K., Luo, X., Yang, G., Wu, T., 2019. A review of sludge-to-energy recovery methods. Energies 12 (1), 60. https://doi.org/10.3390/en12010060. 1996–1073.

Peccia, J., Westerhoff, P., 2015. We should expect more out of our sewage sludge. Environ. Sci. Technol. 49 (14), 8271–8276. https://doi.org/10.1021/acs.est.5b01931.

Peltola, P., Ruottu, L., Larkimo, M., Laasonen, A., Myöhänen, K., 2023. A novel dual circulating fluidized bed technology for thermal treatment of municipal sewage sludge with recovery of nutrients and energy. Waste Manag. 155, 329–337. https://doi.org/10.1016/j.wasman.2022.11.017.

Qian, L., Wang, S., Donghai, X., Guo, Y., Tang, X., Wang, L., 2016. Treatment of municipal sewage sludge in supercritical water: a review. Water Res. 89, 118–131. https://doi.org/10.1016/j.watres.2015.11.047.

Quan, L.M., Kamyab, H., Yuzir, A., Ashokkumar, V., Hosseini, S.E., Balasubramanian, B., Kirpichnikova, I., 2022. Review of the application of gasification and combustion technology and waste-to-energy technologies in sewage sludge treatment. Fuel 316, 123199. https://doi.org/10.1016/j.fuel.2022.123199.

Raheem, A., Sikarwar, V.S., He, J., Dastyar, W., Dionysiou, D.D., Wang, W., Zhao, M., 2018. Opportunities and challenges in sustainable treatment and resource reuse of sewage sludge: a review. Chem. Eng. J. 337, 616–641. https://doi.org/10.1016/j.cej.2017.12.149.

Rao, B., Wang, G., Peng, X., 2022. Recent advances in sludge dewatering and drying technology. Dry. Technol. 40 (15), 3049–3063. https://doi.org/10.1080/07373937.2022.2043355.

Saud, A., Havukainen, J., Peltola, P., Horttanainen, M., 2023. Environmental performance of nitrogen recovery from reject water of sewage sludge treatment based on life cycle assessment. Recycling 8 (2), 43. https://doi.org/10.3390/recycling8020043.

Schnell, M., Horst, T., Quicker, P., 2020. Thermal treatment of sewage sludge in Germany: a review. J. Environ. Manag. 263, 110367. https://doi.org/10.1016/j.jenvman.2020.110367.

Shizas, I., Bagley, D.M., 2004. Experimental determination of energy content of unknown organics in municipal wastewater streams. J. Energy Eng. 130 (2), 45–53. https://doi.org/10.1061/(ASCE)0733-9402(2004)130:2(45).

Smol, M., Kulczycka, J., Henclik, A., Gorazda, K., Wzorek, Z., 2015. The possible use of sewage sludge ash (SSA) in the construction industry as a way towards a circular economy. J. Clean. Prod. 95, 45–54. https://doi.org/10.1016/j.jclepro.2015.02.051.

Sugurbekova, G., Nagyzbekkyzy, E., Sarsenova, A., Danlybayeva, G., Anuarbekova, S., Kudaibergenova, R., Frochot, C., Acherar, S., Zhatkanbayev, Y., Moldagulova, N., 2023. Sewage sludge management and application in the form of sustainable fertilizer. Sustainability 15 (7), 6112. https://doi.org/10.3390/su15076112.

Syed-Hassan, S.S., Wang, Y., Hu, S., Su, S., Xiang, J., 2017. Thermochemical processing of sewage sludge to energy and fuel: Fundamentals, challenges and considerations. Renew. Sust. Energ. Rev. 80, 888–913. https://doi.org/10.1016/j.rser.2017.05.262.

Teoh, S.K., Li, L.Y., 2020. Feasibility of alternative sewage sludge treatment methods from a lifecycle assessment (LCA) perspective. J. Clean. Prod. 247, 119495. https://doi.org/10.1016/j.jclepro.2019.119495.

Tsybina, A., Wuensch, C., 2018. Analysis of sewage sludge thermal treatment methods in the context of circular economy. Detritus 2, 3–15. https://doi.org/10.31025/2611-4135/2018.13668.

Van Caneghem, J., Brems, A., Lievens, P., Block, C., Billen, P., Vermeulen, I., Dewil, R., Baeyens, J., Vandecasteele, C., 2012. Fluidized bed waste incinerators: design, operational and environmental issues. Prog. Energy Combust. Sci. 38 (4), 551–582. https://doi.org/10.1016/j.pecs.2012.03.001.

Werther, J., Ogada, T., 1999. Sewage sludge combustion. Prog. Energy Combust. Sci. 25 (1), 55–116. https://doi.org/10.1016/s0360-1285(98)00020-3.

Zhou, P., Li, D., Zhang, C., Ping, Q., Wang, L., Li, Y., 2024. Comparison of different sewage sludge pretreatment technologies for improving sludge solubilization and anaerobic digestion efficiency: a comprehensive review. Sci. Total Environ. 921, 171175. https://doi.org/10.1016/j.scitotenv.2024.171175.

Index

Note: Page numbers followed by *f* indicate figures, *t* indicate tables, and *b* indicate boxes.

N

T

9780443248269